T0328292

Ices in the Solar System

Ices in the Solar System
A Volatile-Driven Journey from the Inner Solar System to its Far Reaches

Edited by

Richard J. Soare
Department of Geography, Dawson College, Montreal, QC, Canada

Jean-Pierre Williams
Department of Earth, Planetary, and Space Sciences, University of California, Los Angeles, CA, United States

Caitlin J. Ahrens
NASA Goddard Space Flight Center, Greenbelt, MD, United States

Frances E.G. Butcher
Department of Geography, University of Sheffield, Sheffield, United Kingdom

Mohamed Ramy El-Maarry
Space and Planetary Science Center and Department of Earth Sciences, Khalifa University, Abu Dhabi, United Arab Emirates

ELSEVIER

Elsevier
Radarweg 29, PO Box 211, 1000 AE Amsterdam, Netherlands
The Boulevard, Langford Lane, Kidlington, Oxford OX5 1GB, United Kingdom
50 Hampshire Street, 5th Floor, Cambridge, MA 02139, United States

Notices

Knowledge and best practice in this field are constantly changing. As new research and experience broaden our understanding, changes in research methods, professional practices, or medical treatment may become necessary.

Practitioners and researchers must always rely on their own experience and knowledge in evaluating and using any information, methods, compounds, or experiments described herein. In using such information or methods they should be mindful of their own safety and the safety of others, including parties for whom they have a professional responsibility.

To the fullest extent of the law, neither the Publisher nor the authors, contributors, or editors, assume any liability for any injury and/or damage to persons or property as a matter of products liability, negligence or otherwise, or from any use or operation of any methods, products, instructions, or ideas contained in the material herein.

ISBN: 978-0-323-99324-1

For information on all Elsevier publications
visit our website at https://www.elsevier.com/books-and-journals

Publisher: Candice Janco
Acquisitions Editor: Peter Llewellyn
Editorial Project Manager: Sara Valentino
Production Project Manager: Paul Prasad Chandramohan
Cover Designer: Mark Rogers

Typeset by STRAIVE, India

To my mother Martha,
who brought me to the stars
but passed before I came to see them.
RJS

Contents

CHAPTER 6   Ice Exploration on Mars: Whereto and when? **193**

*James B. Garvin, Richard J. Soare, Adam J. Hepburn, Michelle Koutnik,
and E. Godin*

Contributors

Caitlin J. Ahrens
NASA Goddard Space Flight Center, Greenbelt, MD, United States

Carver J. Bierson
Arizona State University, Tempe, AZ, United States

Stephen Brough
School of Environmental Sciences, University of Liverpool, Liverpool, United Kingdom

Frances E.G. Butcher
Department of Geography, University of Sheffield, Sheffield, United Kingdom

Julie Castillo-Rogez
Jet Propulsion Laboratory, California Institute of Technology, Pasadena, CA, United States

S.J. Conway
CNRS, UMR 6112 Planetology and Geodynamics Laboratory, University of Nantes, Nantes, France

F. Costard
Geosciences Paris Saclay, University of Paris-Saclay, Orsay, France

R.D. Dhingra
NASA Jet Propulsion Laboratory, Pasadena, CA, United States

Mohamed Ramy El-Maarry
Space and Planetary Science Center and Department of Earth Sciences, Khalifa University, Abu Dhabi, United Arab Emirates

K.K. Farnsworth
NASA Goddard Space Flight Center, Greenbelt, MD, United States

Colman Gallagher
UCD School of Geography; UCD Earth Institute, University College Dublin, Dublin, Ireland

Anna Grau Galofre
Laboratoire de Planétologie et Géosciences, CNRS UMR 6112, Nantes Université, Université d'Angers, Le Mans Université, Nantes, France

James B. Garvin
NASA Goddard Space Flight Center, Greenbelt, MD, United States

E. Godin
Northern Studies Centre, Laval University, Quebec City, QC, Canada

Adam J. Hepburn
European Space Astronomy Centre, European Space Agency, Madrid, Spain

Samuel M. Howell
NASA Jet Propulsion Laboratory, California Institute of Technology, Pasadena, CA, United States

Bryn Hubbard
Department of Geography and Earth Sciences, Aberystwyth University, Aberystwyth, United Kingdom

Klára Kalousová
Charles University, Prague, Czechia

Lauren E. Mc Keown
NASA Jet Propulsion Laboratory, California Institute of Technology, Pasadena, CA, United States

Michelle Koutnik
Department of Earth and Space Sciences, University of Washington, Seattle, WA, United States

Margaret E. Landis
Laboratory for Atmospheric and Space Physics, University of Colorado, Boulder, CO, United States

Jeremie Lasue
Institut de Recherche en Astrophysique et Planétologie, Université de Toulouse, CNRS, CNES, Toulouse, France

Erin Leonard
NASA Jet Propulsion Laboratory, California Institute of Technology, Pasadena, CA, United States

Carey M. Lisse
Applied Physics Laboratory, Johns Hopkins University, Laurel, MD, United States

L.O. Magaña
John Hopkins Applied Physics Laboratory, Laurel, MD, United States

E.M. Nathan
Department of Earth, Environmental, and Planetary Sciences, Brown University, Providence, RI, United States

Asmin Pathare
Planetary Science Institute, Tucson, AZ, United States

M. Philippe
CNRS, UMR 6112 Planetology and Geodynamics Laboratory, University of Nantes, Nantes, France

Lior Rubanenko
Department of Geological Sciences, Stanford University, Stanford, CA, United States; Department of Geospatial Engineering, Technion, Israel Institute of Technology, Haifa, Israel

Kat Scanlon
Planetary Geology, Geophysics, and Geochemistry Laboratory, NASA Goddard Space Flight Center, Greenbelt, MD, United States

Richard J. Soare
Department of Geography, Dawson College, Montreal, QC, Canada

Gregor Steinbrügge
NASA Jet Propulsion Laboratory, California Institute of Technology, Pasadena, CA, United States

D. Stillman
Department of Space Studies, Southwest Research Institute, Boulder, CO, United States

Jean-Pierre Williams
Department of Earth, Planetary, and Space Sciences, University of California, Los Angeles, CA, United States

Natalie Wolfenbarger
Jackson School of Geosciences, University of Texas at Austin, Austin, TX, United States

Ice song

This evening, I took the path that wound
through woods. The birds—once closer
than skin—kept to their veil of leaves,
odd voices counterpointing daylight stars.
I half-recalled a poem about birches bent
by ice storms, and ice turned deftly
into an eager child swinging branches up
and down. The whoosh, the weighted
dip to ground yourself on Earth, in love again.
So, those solitary calls became a boy, a girl
—an octave of poised notes rhymed like arms
cast high in air, hands shading eyes—
a girl, a boy listening for the sister
music of far stars, their colours, ices,
gasses. Or even for children, perched
in the topmost branches of some unearthly tree,
singing back to them like birds. Songs
sang down to me. A boy looked up.

David Kinloch

Note: the 'half-recalled poem' is Robert Frost's 'Birches.'

The Solar System's ices and their origin

Sean N. Raymond

Bordeaux Astrophysical Laboratory (LAB; UMR 5804), University of Bordeaux, CNRS, Pessac, France

Why are ices important to understanding the Solar System's formation and evolution? Volatiles are found across the Solar System, mostly in the form of ices. While water is the most prominent, there is a diversity of ices found in the planets, moons, and small bodies. The rocky planets only have a sprinkling of water, each with no more than 1 part in 1000 water by mass (Earth being the most water-rich). Earth and Mars have polar ice caps and glaciers. Both planets may have large additional reservoirs of water buried in their interiors (e.g., Hirschmann, 2006; Scheller et al., 2021). Of course, Venus' meager (present-day) water budget is mainly found as atmospheric vapor (although it may potentially have hydrated silicates; Zolotov et al., 1997). Water ice has been detected in perpetually shadowed areas of the Moon's polar regions (Li et al., 2018). NASA's MESSENGER mission has confirmed previous reports of localized water ice deposits on Mercury's surface (Deutsch et al., 2017, 2021). The asteroids are heterogeneous. Inner-belt asteroids are generally quite dry, with any water locked up in silicate minerals (Alexander et al., 2018). In contrast, outer-belt asteroids (like C- and D-type) contain up to 10% water by mass (in the form of ice and hydrated minerals), and comet-like activity has been detected in some asteroids (Hsieh & Jewitt, 2006). Ceres, the target of NASA's Dawn mission, contains up to 20–25% water by mass (McCord & Sotin, 2005). This constitutes roughly a third of the mass of the entire asteroid belt, making it large enough to be classified as a dwarf planet.

Generally speaking, the Solar System beyond the asteroid belt is colder and icier than the inner Solar System. The gas giants' water is trapped as vapor, but their moons are predominantly ice-rich. Despite being more than an order of magnitude less massive than Earth, several massive Jovian/Saturnian moons (Ganymede, Callisto, Europa, and Titan) contain as much or more total water (Titan also has methane ice in clouds and perhaps in its subsurface). Given their title as the "ice giants," Uranus and Neptune are thought to contain a large fraction of water ice, although the exact breakdown is uncertain (Helled et al., 2011). Comets and trans-Neptunian objects inhabit the outermost reaches of the Solar System and are thought to be largely made of ices.

The cosmic origins of ice date back to the *Big Bang* roughly 13.8 billion years ago. Hydrogen, the most abundant element, was formed during recombination, a few hundred thousand years after the Big Bang. Oxygen continues to be produced by nuclear fusion within stars significantly more massive than the Sun and expelled into the Galaxy via supernova explosions. Water ice is observed in the interstellar medium and plays a central role in the gas phase and dust-grain chemistry as clumps of molecular gas collapse to form stars and circumstellar disks (van Dishoeck et al., 2014).

Gas-dominated dusty disks are observed around virtually all young stars, lasting up to 10 million years before they dissipate (Williams & Cieza, 2011). As the cradles of planet formation, the structure and evolution of these disks are central in determining the compositions of small bodies and planets (Armitage, 2011). Disks are heated by irradiation from the central star as well as from viscous heating in some regions (Turner et al., 2014). The *snow line* is the radial location from the central star where the temperature drops below the condensation temperature for water, roughly 170 K in such low-pressure environments. Beyond the snow line, water ice is a solid and likely a central ingredient in dust grains, as the total mass ratio of water ice to silicates is roughly 1:1 based on the Solar composition (Lodders, 2003). This means that ice is intermixed with dust of different compositions past the snow line, but closer-in dust is desiccated. Like water, other ices also only exist as solids beyond their respective condensation fronts (or ice lines). The composition of solids at a given radial distance within the disk is therefore determined by which species are solid at that local temperature (Dodson-Robinson et al., 2009). Therefore, the relative concentrations of different ices within planetesimals vary with radial distance within the disk. Only a few species are likely to have been abundant enough to have an effect on the structure of the gas disk, by creating pressure bumps that may serve as sites for planetesimal formation (see below); from the inside out, these include silicates, water, and carbon monoxide (see Izidoro et al., 2022).

Planets form in a series of steps (see Fig. 1). Dust particles grow by mutual collisions until one of several growth barriers is reached, typically at millimeter to centimeter sizes (Birnstiel et al., 2016). These large dust grains are referred to as *pebbles*. Pebbles have enough inertia not to be dragged along with the gas within the disk but instead feel a headwind that extracts orbital energy and causes them to spiral inward (Johansen & Lambrechts, 2017). Drifting pebbles become concentrated at localized zones, or *bumps*, of increased pressure. Corresponding ring-shaped accumulations are commonly observed in telescope-based images of planet-forming disks (ALMA Partnership et al., 2015). When sufficiently concentrated, pebbles directly clump via the streaming instability to form hundred km-scale planetesimals (Johansen et al., 2014). These are the macroscopic building blocks of the planets and the precursors of asteroids and comets.

Planetesimals grow by colliding with other planetesimals and also by accreting drifting pebbles. The cores of the giant planets grew to 10–20 Earth masses within a few million years of the start of planet formation by accreting ice-rich pebbles efficiently (Johansen & Lambrechts, 2017). The cores of Jupiter and Saturn accreted gas gravitationally from the disk (Pollack et al., 1996). At first, this occurred slowly and then in a runaway mode until they carved annular gaps in the gaseous disk (Lin & Papaloizou, 1986). Uranus and Neptune are often considered 'failed' gas giants that reached the core phase but accreted gas too slowly. In contrast, Hafnium-Tungsten isotopes indicate that Earth required 50–100 million years to complete its growth (Kleine et al., 2009), perhaps because the pebbles were smaller and pebble accretion was less efficient (Morbidelli et al., 2015). After Moon- to Mars-sized "planetary embryos" were formed, the late phases of Earth's growth were characterized by giant impacts. The last of these impacts spun out a disk that coalesced into our Moon (Canup & Asphaug, 2001). The giant planets' large moons accreted within gas-rich disks around the growing planets in a fashion analogous to the planets' formation around the young Sun (Canup & Ward, 2006; there are a few exceptions, such as Triton, which is thought to have been captured—see Agnor & Hamilton, 2006).

It remains a challenge to assemble a sequence of events that explains the Solar System's orbital architecture (Raymond et al., 2020). The terrestrial planets comprise a total of ~2 Earth masses, and the cores of the giant planets contain a total of ~70 Earth masses. The asteroid belt is almost empty—the total mass of all asteroids combined is only ~0.05% of Earth's mass (DeMeo & Carry, 2013). It is unclear whether the asteroid belt originally contained a large amount of mass that was almost entirely removed,

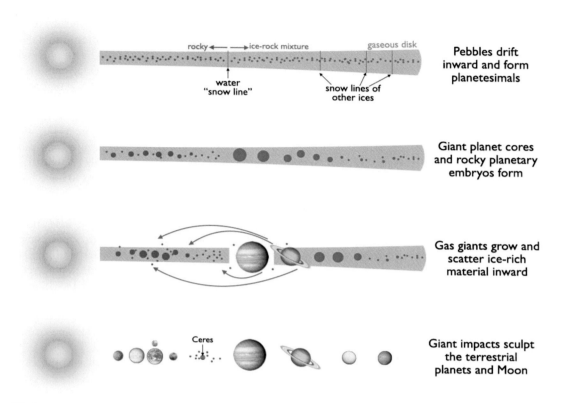

rocky ←——→ ice-rock mixture   gaseous disk

water "snow line"   snow lines of other ices

Pebbles drift inward and form planetesimals

Giant planet cores and rocky planetary embryos form

Gas giants grow and scatter ice-rich material inward

Ceres

Giant impacts sculpt the terrestrial planets and Moon

FIG. 1

Cartoon view of the current paradigm of planet formation (see Raymond et al., 2020 for details).

Images of the planets are from NASA.

or whether it was born nearly devoid of planetesimals (Raymond & Nesvorny, 2020). Regardless, there is strong circumstantial evidence from isotopic studies of meteorites that the volatile-rich asteroids formed beyond Jupiter and were implanted into the belt (Kruijer et al., 2020); moreover, some volatile-poor asteroids could have been implanted from the terrestrial planet-forming region (Raymond & Izidoro, 2017a). It is quite possible that the volatile-rich C-type asteroids that dominate the outer main belt were implanted as a by-product of the gas giants' growth and orbital migration, and a fraction of planetesimals that were gravitationally scattered by Jupiter had their orbits re-circularized due to aerodynamic gas drag (Raymond & Izidoro, 2017b). The detection of ammoniated phyllosilicates on Ceres (de Sanctis et al., 2015) supports the idea that it originated in the outer Solar System (Raymond & Izidoro, 2017b; Ribeiro de Sousa et al., 2022; Takir et al., 2023), beyond the condensation front for ammonia, which would explain its large inventory of ice.

The origin of Earth's water is a long-standing question (see, e.g., Morbidelli et al., 2000; Alexander et al., 2012; Meech & Raymond, 2020). It has been known for decades that there is a similarity in hydrogen and nitrogen isotopes between Earth's water and carbonaceous chondrite meteorites—sourced from C-type asteroids (Marty, 2012). Newer, more accurate measurements show that enstatite chondrite meteorites—thought to be close in composition to the planetary building blocks at Earth's orbital distance—are an even closer (but still imperfect) isotopic match to Earth's hydrogen and nitrogen (Piani et al., 2020). The best match is obtained if Earth was built of ~95% enstatite chondrite-like material, with a ~5% contribution from carbonaceous chondrites (Burkhardt et al., 2021). Taking several isotopes into account, this implies that ~70% of Earth's water was locally sourced from noncarbonaceous material, and the remaining ~30% was delivered by carbonaceous impactors (Savage et al., 2022; Steller et al., 2022). This supports new models of Solar System formation, in which Earth was built mostly from local material, with a contribution of C-type (carbonaceous) planetesimals that originated beyond Jupiter and scattered inward during its growth and perhaps migration (Raymond & Izidoro, 2017b; Walsh et al., 2011). This contrasts with the 'classical model,' in which Earth's water was thought to have been sourced from the outer asteroid belt, at the tail of our planet's feeding zone (Morbidelli et al., 2000).

It is worth keeping in mind that a large amount of water was lost. The most prolific heat source in the early Solar System was aluminum-26 (Al-26), which was implanted into the Sun's planet-forming disk from a nearby massive star from winds or a supernova explosion (Hester et al., 2004). With a half-life of 700,000 years, Al-26 is thought to have desiccated any planetesimals that accreted within the first two million years after the start of planet formation (Monteux et al., 2018). In addition, collisions between growing planets do not always result in perfect mergers. Rather, the outer layers, which contain lower-density materials such as ice, are more likely to be removed in giant impacts (Marcus et al., 2010). Late impacts can also act to strip planets' atmospheres and oceans (Genda & Abe, 2005; Lock & Stewart, 2023).

One might wonder about the prospects for ices on exoplanets. We expect any distant planet to contain ices based simply on its formation temperature, and many planets with orbital distances comparable to the giant planets have been detected with radial velocity, direct imaging, and microlensing techniques (Bowler, 2016; Mayor et al., 2011; Suzuki et al., 2016). Unfortunately, observation of the presence of ices is not feasible with current instruments except in rare cases. For instance, the existence of water-rich bodies was inferred in disrupted planetesimals around some white dwarfs (Farihi et al., 2013); these represent the leftovers of planets around dead stars rather than the precursors of ocean- or ice-rich planets. Water vapor has been detected in the atmosphere of a *sub-Neptune*-sized planet in the habitable zone of its host star (Tsiaras et al., 2019). The presence of water or other volatiles may be observable in the atmospheres of selected rocky exoplanets using instruments such as NASA James Webb Space Telescope (see Gillon et al., 2020).

The existence of a rocky planet like ours with a small but non-zero water content may depend on the specific evolution of the dominant gas giant in the system. Had Jupiter's core grown more slowly or not at all, ice-rich pebbles could have drifted past its orbit and rained down on the growing Earth (Morbidelli et al., 2015). In this scenario, Earth would have grown much faster (and much wetter; Sato et al., 2016); yet it likely would have become massive enough to undergo large-scale migration to become a *super-Earth* on a short-period orbit, similar to those found around a large fraction of stars (Lambrechts et al., 2019). Had Jupiter and Saturn grown faster and interacted more strongly with one another, they almost certainly would have undergone a very strong dynamical instability like the ones invoked for ~90% or more of giant exoplanet systems (Jurić & Tremaine, 2008). This instability would have ejected one of the gas giants, and in the process would have driven Earth's building blocks into the Sun (Raymond et al., 2011). Among the population of known giant exoplanets, there do exist numerous gas giants with relatively wide and circular orbits, although they only exist around ~1% of Sun-like stars (Morbidelli & Raymond, 2016). These may be the best candidates for systems like our own, inclusive of similar ice-rich worlds.

References

Agnor, C. B., & Hamilton, D. P. (2006). Neptune's capture of its moon Triton in a binary-planet gravitational encounter. *Nature*, *441*(7090), 192–194.

Alexander, C. M. O., Bowden, R., Fogel, M. L., Howard, K. T., Herd, C. D. K., & Nittler, L. R. (2012). The provenances of asteroids, and their contributions to the volatile inventories of the terrestrial planets. *Science*, *337*, 721.

Alexander, C. M. O., McKeegan, K. D., & Altwegg, K. (2018). Water reservoirs in small planetary bodies: Meteorites, asteroids, and comets. *Space Science Reviews*, *214*, 36.

ALMA Partnership, Brogan, C. L., Pérez, L. M., Hunter, T. R., Dent, W. R. F., Hales, A. S.,... Tatematsu, K. (2015). The 2014 ALMA Long Baseline Campaign: First results from high angular resolution observations toward the HL Tau region. *The Astrophysical Journal Letters*, *808*(1), L3.

Armitage, P. J. (2011). Dynamics of protoplanetary disks. *Annual Review of Astronomy and Astrophysics*, *49*, 195–236.

Birnstiel, T., Fang, M., & Johansen, A. (2016). Dust evolution and the formation of Planetesimals. *Space Science Reviews*, *205*, 41–75.

Bowler, B. P. (2016). Imaging Extrasolar Giant Planets. *Publications of the Astronomical Society of the Pacific*, *128*, 102001.

Burkhardt, C., Spitzer, F., Morbidelli, A., Budde, G., Render, J. H., Kruijer, T. S., & Kleine, T. (2021). Terrestrial planet formation from lost inner solar system material. *Science Advances*, *7*(52), eabj7601.

Canup, R. M., & Asphaug, E. (2001). Origin of the moon in a giant impact near the end of the Earth's formation. *Nature*, *412*, 708–712.

Canup, R. M., & Ward, W. R. (2006). A common mass scaling for satellite systems of gaseous planets. *Nature*, *441*, 834–839.

de Sanctis, M. C., et al. (2015). Ammoniated phyllosilicates with a likely outer solar system origin on (1) Ceres. *Nature*, *528*, 241–244.

DeMeo, F. E., & Carry, B. (2013). The taxonomic distribution of asteroids from multi-filter all-sky photometric surveys. *Icarus*, *226*, 723–741.

Deutsch, A. N., Head, J. W., Parman, S. W., Wilson, L., Neumann, G. A., & Lowden, F. (2021). Degassing of volcanic extrusives on Mercury: Potential contributions to transient atmospheres and buried polar deposits. *Earth and Planetary Science Letters*, *564*, 116907.

Deutsch, A. N., Neumann, G. A., & Head, J. W. (2017). New evidence for surface water ice in small-scale cold traps and in three large craters at the north polar region of mercury from the mercury laser altimeter. *Geophysical Research Letters*, *44*, 9233–9241.

Dodson-Robinson, S. E., Willacy, K., Bodenheimer, P., Turner, N. J., & Beichman, C. A. (2009). Ice lines, planetesimal composition and solid surface density in the solar nebula. *Icarus*, *200*, 672–693.

Farihi, J., Gänsicke, B. T., & Koester, D. (2013). Evidence for water in the rocky debris of a disrupted extrasolar minor planet. *Science*, *342*, 218–220.

Genda, H., & Abe, Y. (2005). Enhanced atmospheric loss on protoplanets at the giant impact phase in the presence of oceans. *Nature*, *433*, 842–844.

Gillon, M., et al. (2020). *The TRAPPIST-1 JWST community initiative*. arXiv e-prints arXiv:2002.04798.

Helled, R., Anderson, J. D., Podolak, M., & Schubert, G. (2011). Interior models of Uranus and Neptune. *The Astrophysical Journal, 726*, 15.

Hester, J. J., Desch, S. J., Healy, K. R., & Leshin, L. A. (2004). The cradle of the solar system. *Science, 304*, 1116–1117.

Hirschmann, M. M. (2006). Water, melting, and the deep earth H2O cycle. *Annual Review of Earth and Planetary Sciences, 34*, 629–653.

Hsieh, H. H., & Jewitt, D. (2006). A population of comets in the Main Asteroid Belt. *Science, 312*, 561–563.

Izidoro, A., Dasgupta, R., Raymond, S. N., Deienno, R., Bitsch, B., & Isella, A. (2022). Planetesimal rings as the cause of the solar System's planetary architecture. *Nature Astronomy, 6*, 357–366.

Johansen, A., Blum, J., Tanaka, H., Ormel, C., Bizzarro, M., & Rickman, H. (2014). The multifaceted planetesimal formation process. *Protostars and Planets, VI*, 547.

Johansen, A., & Lambrechts, M. (2017). Forming planets via pebble accretion. *Annual Review of Earth and Planetary Sciences, 45*, 359–387.

Jurić, M., & Tremaine, S. (2008). Dynamical origin of extrasolar planet eccentricity distribution. *The Astrophysical Journal, 686*, 603–620.

Kleine, T., Touboul, M., Bourdon, B., Nimmo, F., Mezger, K., Palme, H., Jacobsen, S. B., Yin, Q.-Z., & Halliday, A. N. (2009). Hf-W chronology of the accretion and early evolution of asteroids and terrestrial planets. *Geochimica et Cosmochimica Acta, 73*(17), 5150–5188.

Kruijer, T. S., Kleine, T., & Borg, L. E. (2020). The great isotopic dichotomy of the early solar system. *Nature Astronomy, 4*, 32–40.

Lambrechts, M., Morbidelli, A., Jacobson, S. A., Johansen, A., Bitsch, B., Izidoro, A., & Raymond, S. N. (2019). Formation of planetary systems by pebble accretion and migration. How the radial pebble flux determines a terrestrial-planet or super-earth growth mode. *Astronomy and Astrophysics, 627*, A83.

Li, S., Lucey, P. G., Milliken, R. E., Hayne, P. O., Fisher, E., Williams, J.-P., Hurley, D. M., & Elphic, R. C. (2018). Direct evidence of surface exposed water ice in the lunar polar regions. *Proceedings of the National Academy of Science, 115*, 8907–8912.

Lin, D. N. C., & Papaloizou, J. (1986). On the tidal interaction between protoplanets and the protoplanetary disk. III. Orbital migration of protoplanets. *The Astrophysical Journal, 309*, 846.

Lock, S. J., & Stewart, S. T. (2023). Atmospheric loss in giant impacts depends on pre-impact surface conditions. eprint arXiv:2309.16399. doi:10.48550/arXiv.2309.16399.

Lodders, K. (2003). Solar system abundances and condensation temperatures of the elements. *The Astrophysical Journal, 591*, 1220–1247.

Marcus, R. A., Sasselov, D., Stewart, S. T., & Hernquist, L. (2010). Water/icy super-earths: giant impacts and maximum water content. *The Astrophysical Journal, 719*, L45–L49.

Marty, B. (2012). The origins and concentrations of water, carbon, nitrogen and noble gases on Earth. *Earth and Planetary Science Letters, 313*, 56–66.

Mayor, M., et al. (2011). *The HARPS search for southern extra-solar planets XXXIV. Occurrence, mass distribution and orbital properties of super-Earths and Neptune-mass planets*. arXiv e-prints arXiv:1109.2497.

McCord, T. B., & Sotin, C. (2005). Ceres: Evolution and current state. *Journal of Geophysical Research (Planets), 110*, E05009.

Meech, K., & Raymond, S. N. (2020). Origin of Earth's water: Sources and constraints. In V. S. Meadows, G. N. Arney, B. E. Schmidt, & D. J. Des Marais (Eds.), *Planetary astrobiology. Space science series*. University of Arizona Press.

Monteux, J., Golabek, G. J., Rubie, D. C., Tobie, G., & Young, E. D. (2018). Water and the interior structure of terrestrial planets and icy bodies. *Space Science Reviews, 214*, 39.

Morbidelli, A., Chambers, J., Lunine, J. I., Petit, J. M., Robert, F., Valsecchi, G. B., & Cyr, K. E. (2000). Source regions and time scales for the delivery of water to earth. *Meteoritics and Planetary Science, 35*, 1309–1320.

Morbidelli, A., Lambrechts, M., Jacobson, S., & Bitsch, B. (2015). The great dichotomy of the solar system: Small terrestrial embryos and massive giant planet cores. *Icarus, 258*, 418–429.

Morbidelli, A., & Raymond, S. N. (2016). Challenges in planet formation. *Journal of Geophysical Research (Planets), 121*, 1962–1980.

Piani, L., Marrocchi, Y., Rigaudier, T., Vacher, L., Thomassin, D., & Marty, B. (2020). Earth's water may have been inherited from material similar to enstatite chondrite meteorites. *Science, 369*, 1110–1113.

Pollack, J. B., Hubickyj, O., Bodenheimer, P., Lissauer, J. J., Podolak, M., & Greenzweig, Y. (1996). Formation of the giant planets by concurrent accretion of solids and gas. *Icarus, 124*(1), 62–85.

Raymond, S. N., Armitage, P. J., Moro-Martín, A., Booth, M., Wyatt, M. C., Armstrong, J. C., Mandell, A. M., Selsis, F., & West, A. A. (2011). Debris disks as signposts of terrestrial planet formation. *Astronomy and Astrophysics, 530*, A62.

Raymond, S. N., & Izidoro, A. (2017a). The empty primordial asteroid belt. *Science Advances, 3*, e1701138.

Raymond, S. N., & Izidoro, A. (2017b). Origin of water in the inner solar system: Planetesimals scattered inward during Jupiter and Saturn's rapid gas accretion. *Icarus, 297*, 134–148.

Raymond, S. N., Izidoro, A., & Morbidelli, A. (2020). Solar system formation in the context of extrasolar planets. *Planetary Astrobiology, 287*.

Raymond, S. N., & Nesvorny, D. (2020). Origin and dynamical evolution of the asteroid belt. In S. Marchi, C. A. Raymond, & T. Christopher (Eds.), *Vesta and Ceres. Insights from the Dawn Mission for the Origin of the Solar System* (p. 227). Cambridge, UK: Cambridge University Press, ISBN:978-1-108-47973-8. 2022.

Ribeiro de Sousa, R., Morbidelli, A., Gomes, R., Vieira Neto, E., Izidoro, A., & Atanasio Alves, A. (2022). Dynamical origin of the dwarf planet Ceres. *Icarus, 379*, 114933.

Sato, T., Okuzumi, S., & Ida, S. (2016). On the water delivery to terrestrial embryos by ice pebble accretion. *Astronomy & Astrophysics, 589*, A15.

Savage, P. S., Moynier, F., & Boyet, M. (2022). Zinc isotope anomalies in primitive meteorites identify the outer solar system as an important source of Earth's volatile inventory. *Icarus, 386*, 115172.

Scheller, E. L., Ehlmann, B. L., Hu, R., Adams, D. J., & Yung, Y. L. (2021). Long-term drying of Mars by sequestration of ocean-scale volumes of water in the crust. *Science, 372*, 56–62.

Steller, T., Burkhardt, C., Yang, C., & Kleine, T. (2022). Nucleosynthetic zinc isotope anomalies reveal a dual origin of terrestrial volatiles. *Icarus, 386*, 115171.

Suzuki, D., et al. (2016). The exoplanet mass-ratio function from the MOA-II survey: Discovery of a break and likely peak at a Neptune mass. *The Astrophysical Journal, 833*, 145.

Takir, D., Neumann, W., Raymond, S. N., Emery, J. P., & Trieloff, M. (2023). Late accretion of Ceres-like asteroids and their implantation into the outer main belt. *Nature Astronomy, 7*, 524–533.

Tsiaras, A., Waldmann, I. P., Tinetti, G., Tennyson, J., & Yurchenko, S. N. (2019). Water vapour in the atmosphere of the habitable-zone eight-earth-mass planet K2-18 b. *Nature Astronomy, 3*, 1086–1091.

Turner, N. J., Fromang, S., Gammie, C., Klahr, H., Lesur, G., Wardle, M., & Bai, X.-N. (2014). Transport and accretion in planet-forming disks. *Protostars and Planets, VI*, 411.

van Dishoeck, E. F., Bergin, E. A., Lis, D. C., & Lunine, J. I. (2014). Water: From clouds to planets. *Protostars and Planets, VI 835*.

Walsh, K. J., Morbidelli, A., Raymond, S. N., O'Brien, D. P., & Mandell, A. M. (2011). A low mass for Mars from Jupiter's early gas-driven migration. *Nature, 475*(7355), 206–209.

Williams, J. P., & Cieza, L. A. (2011). Protoplanetary disks and their evolution. *Annual Review of Astronomy and Astrophysics, 49*, 67–117.

Zolotov, M. Y., Fegley, B., & Lodders, K. (1997). Hydrous silicates and water on Venus. *Icarus, 130*, 475–494.

The ice frontier for science in the upcoming decades: A strategy for Solar System exploration?

James B. Garvin

NASA Goddard Space Flight Center, Greenbelt, MD, United States

For more than two decades a catchphrase associated with one of the major crosscutting strategies in the exploration of our solar system has been "follow the water" (*FTW*), perhaps derived from the "follow the money" approach in criminal investigations. This theme has been employed in developing scientific mission concepts for exploring Mars and other worlds where habitability and the search for signs of life as well as atmosphere-climate evolution are established priorities. Embedded within the *FTW* pathway is the multi-disciplinary investigation of ices within the solar system as a new frontier in planetology. This "ice frontier" may appear to be a sub-set of *FTW*; however, it extends beyond water ice to embrace all frozen volatiles as systems for understanding and predicting key behaviors and processes that are widespread within our solar system, including a so-called "ice cycle." Thus, a key question is "Where are we going and why"?

Over the past ~25 years the critical role ices play as clues to environments and chemical processes have catalyzed missions to the "ice frontiers" of Mercury and the Moon, the polar regions of Mars, Europa, Titan, and other bodies relatively small or large, distal or close. Now is the time to take stock of where we have come since the late 1990s when missions such as *Cassini/Huygens*, *Mars Polar Lander*, *Lunar Prospector*, and *ICESat-1* were launched as icy pathfinders.

A particularly compelling treatment of ices in the solar system was showcased in a 1999 issue of *The Planetary Society*'s monthly magazine, as the permanent icecaps of Mars, the icy moons of Jupiter and Saturn, and the role of ice at the poles of the Moon and Mercury were being explored for the first time. By 2000, a new awareness of planetary ices and the "ice cycle" was emergent, as was the idea of volatile sources and sinks as a connecting theme in the exploration of the solar system. Consequently, more and more missions of discovery were proposed to target locations far and near, i.e., from the Pluto-Charon binary system to the poles of Mercury.

This was underlined formally by the first "Decadal Survey" of planetary science in 2002. The Decadal Survey process is organized periodically by the US National Academy of Sciences, Engineering, and Medicine. Its aim was/is to identify scientific priorities for planetary sciences and does so by soliciting inputs from individuals within the community, town halls, and panels of experts. The end-product was/is a framework of recommendations for science priorities and potential missions through a 10-year time horizon.

At the turn of the century, increasingly focused attention was paid to the reconnaissance and inventorying of remotely observed icy materials at a suite of targets throughout the solar system. No less important was the idea of collecting, retrieving, and, possibly, returning icy samples to Earth. From 2000 to 2014 numerous mission plans were drawn up or implemented with these disparate but related goals in mind: *Cassini/Huygens* (Titan), the *Phoenix* polar lander and related orbiters, i.e., the *ESA*'s *Mars Express*, *NASA*'s *Odyssey* and *Mars Reconnaissance Orbiter* (Mars), *NASA*'s *Lunar Reconnaissance Orbiter* (the Moon), the *ESA*'s *Rosetta* mission (comet 67P/*Churyumov-Gerasimenko*), *NASA*'s *MESSENGER* mission (Mercury), and *NASA*'s *New Horizons* mission (Pluto-Charon).

The ice frontier for fundamental science as well as for the applied science of human exploration has become a key element in solar system exploration because it integrates the interplay between habitability, volatile sources and sinks, and dynamic planetary environments and processes (i.e., the ice cycle and climate connections). The in situ analytical chemistry of planetary ices as exemplified by the *Phoenix* lander on Mars has catalyzed consideration of new technologies and instruments to support the cryogenic sample return of such materials to Earth and their laboratory analysis by techniques not yet ready for space. For example, several *Apollo* lunar samples (e.g., from Apollo 17) that were archived to preserve them for ~50 years are now being opened via the *ANGSA* program at *NASA*. The volatile samples entrapped and even frozen therein are being studied by techniques not available when lunar sampling occurred in the early 1970s.

In addition, advanced studies of terrestrial ice sheets and reservoirs are underway as part of critical climate-change research using integrated remote sensing (*ICESat-2*, *EOS*, *GRACE Follow-On*, etc.) with field- and laboratory-based investigations that can be applied to ice-bearing worlds as close as Mars. Some terrestrial analogues offer fertile field-to-laboratory experiences that benefit from well-characterized environments relevant to planetary sciences.

The next 25 years of robotic and human-based planetary exploration will embrace the ice frontier in ways not imagined even 25 years ago, as techniques for the detection, access, measurement, and return of such materials (to Earth) are implemented. To some, a heyday for the informed scientific exploration of the ever-expanding ice frontier is at hand thanks to a suite of robotic missions that are approved for flight before 2030, as well as new laboratory-based ice-related simulation facilities. These transdisciplinary, international missions will catalyze paradigm shifts in the state of knowledge of ice reservoirs and transport processes across our solar system. This is invaluable, in and of itself, and no less so with regard to the current and imminent discovery of exoplanets with possible atmosphere-climate systems (i.e., via the *James Webb Space Telescope* and others to come in the 2030s).

Closer to home, *NASA*'s *VIPER* mobile exploration of a lunar polar region will build on more than a decade of measurements from missions such as the *Lunar Reconnaissance Orbiter* to ground truth the character of ices on the Moon. It will be followed by human-based missions as part of *NASA*'s *Artemis* program which will explore potentially ice-bearing localities off the planet for the first time in human history in the mid-2020s.

The *ESA*'s *JUICE* mission to the Galilean satellites of Jupiter will expand the remote sensing and reconnaissance of icy objects in our solar system, with special attention being focused on the outer ice-silicate shell of Ganymede and its hypothesized sub-surface "ocean." This interaction between ice-rich crusts and sub-crustal oceans is a critical area of research as part of the Ocean Worlds strategy and applies to Ganymede, Europa, Enceladus, Titan, and others as potentially habitable ice-ocean worlds. The *NASA*'s *Europa Clipper* multi-flyby mission to Europa is equipped to make major strides in the measurement and understanding of the interplay between a sub-surface ocean with an ice-rich crust and the possibility of preserved organics from dynamic exchange processes including plumes.

In addition, the capabilities of the recently launched *James Webb Space Telescope* (*JWST*), with its high-spectral resolution infrared spectroscopy, may detect chemical signatures of molecules within ice-bearing plumes from Europa, for example, that connect to ice-ocean exchanges and perhaps to currently habitable environments. In this way, ices are the "tablet" for preserving and recording the cycles of interaction between liquid volatiles and their solid (ice) phases across time. This "ice cycle" varies across the solar system but is a new cosmic imperative within science with broad applicability.

With missions such as ESA's *JUICE*, NASA's *Europa Clipper*, and NASA's *Dragonfly* mission (to Titan's surface with an octo-copter chemistry laboratory) slated to launch in the 2020s and provide ground-breaking new observations in the 2030s, the ice frontier for science will expand our knowledge base and animate further inquiries on the search for signs of life.

One key target for comprehensive ice reconnaissance is Mars. Here, potentially vast ice reservoirs could be lurking, perhaps only a few meters under the active surface layer of dust and sand. Advanced ground-penetrating radars such as the *RIMFAX* on *NASA*'s *Perseverance* rover are sensing the vertical structure of the uppermost tens of meters of Mars today. This provides boundary conditions for how frozen volatiles could be sequestered across the planet at new scales in space and time.

The observation by *NASA's Mars Reconnaissance Orbiter* of ice layers in cliff-side exposures as well as exposed ice within fresh impact-craters points highlight the possibility that the buried icescapes of Mars could be used as a benchmark for climate and habitability research. For example, there are plans afoot for a multi-national *Mars Ice Mapper* orbiter mission to inventory the accessible ice reservoirs of Mars by the early 2030s. This mission would employ a highly sensitive polarimetric synthetic-aperture radar (Canadian), complemented perhaps by other sensors to map the buried ice deposits and to identify possible "old ice" regions. Such a mission could engender the planning of landed missions associated with astrobiological priorities, i.e., looking for the preservation of extant life biomarkers or the preservation of molecular biosignatures of past life. Learning about the origin and geological history of the preserved record of ice on Mars will improve our understanding of temperature and atmospheric temperature variances, as well as the hydrological cycling of water when Mars' surface and/or near surface could have been more favorable to habitability than today.

To many scientists, the return to Earth and the laboratory-based study of planetary, lunar, or small bodies is the "holy grail" and an end-member of what could be called the "ice frontier." Science advisory bodies have long called for cryogenically preserved samples of comet nuclei as one potentially achievable goal in the upcoming decades, especially after the return of samples from Japan's *Hyabusa-2* and NASA's OSIRIS-REx (samples returned to Earth on September 24, 2023).

It is more than likely that the ice-sample return either from comets or from buried but near-surface ice deposits on Mars will be possible by the 2030s, supplemented by in situ analytical instruments that are nearing readiness for robotic deployment today. Human-tended in situ analysis on the lunar surface as part of *Artemis'* "basecamp" is another possibility in which astronauts could interrogate lunar ice samples without returning them to Earth. By moving state-of-the-art laboratory instrumentation into space, i.e., with the *International Space Station* and, hopefully, the *Lunar Gateway*, some of the breakthrough measurements associated with planetary ices could be undertaken before robotic sample returns can be funded and implemented.

Overall, the next 25 years will expand awareness of ice reservoirs (as sources and sinks), and transport processes (the "ice cycle"). Additionally, the next generation of measurements required to use these priceless scientific resources as forensic clues associated with environments (and climate states), habitability, and even as part of the search for signs of life will be developed. Agnostic biosignatures require attention to what can be discovered within such ices at multiple scales, so the return of relevant ice samples to Earth will become an imperative especially after the first samples from the lunar poles and the surface of Mars are available for testing, here on Earth. While "science on the rocks" (and of the rocks) will remain a vital part of the scientific exploration of our solar system, the ice frontier is a key "next step" as we prepare for the study of exoplanets, increasingly resolvable thanks to the breakthroughs in astrophysics beginning with the *JWST* and others to come in its footsteps.

To some, the distribution and significance of ices within our solar system may remain just one of many questions to be pursued as we discover our place in space; but if the role of ice on Earth and throughout Earth's history is any indication, then these frozen materials will serve as time capsules connecting us with many places and bodies small and large within our solar system. Planetary ices link climates, oceans, and possible preservation of biosignatures to the next wave of missions now under development. Perhaps cryo-ocean worlds abound in our galaxy and understanding those at hand here within our solar system will guide us in interpreting foreseen and unforeseen discoveries through the next 50 years. Let us *never wait to wonder* as we pursue new science on the ice!

Introduction

Ices in the Solar System: A Volatile-Driven Journey from the Inner Solar System to its Far Reaches investigates the origins, development, and distribution of various ice species throughout the solar system. The book also identifies and constrains fundamental science questions about these *ices* and filters these questions through the lens of human exploration, especially as it concerns the pursuit of H_2O ice (Garvin, 2024, pp. xxvii–xxx; Garvin et al., 2024, pp. 193–220).

Spatially, *Ices in the Solar System: A Volatile-Driven Journey from the Inner Solar System to its Far* migrates outwardly from Mercury, the Earth and the Moon, Mars, then Ceres and other icy, volatile-bearing small bodies in the inner solar system; this perambulation continues outwardly from the sun by exploring some of the icy moons of Jupiter, Saturn, Uranus, and Neptune, as well as the Pluto/Charon system and other Trans-Neptunian/Kuiper Belt Objects.

Conceptually, the book describes the often-enigmatic geological histories of ice species such as H_2O, CO_2, and CH_4 located throughout the solar system on a diverse set of planets, moons, and small bodies of lesser mass (e.g., Raymond, 2024, pp. xix–xxv). This is underlined, for example, by chronicling and evaluating the interaction of these ice species with a wide range of surface, near-subsurface, interior, and atmospheric processes (e.g., Ahrens et al., 2024, pp. 357–376; Farnsworth et al., 2024, pp. 315–356; Howell et al., 2024, pp. 283–314).

Each chapter addresses a broad and engaging sweep of questions. For example, why is there H_2O ice on Mercury and the Moon, where absent atmospheres and hostile surface/near-surface thermal environments seemingly are inconsistent with stable water ice (Williams & Rubanenko, 2024, pp. 1–30)? Is the origin of the water ice on these bodies primordial, or might it be rooted in the dynamic evolution and displacement of ices in the solar system that postdates the formation of these bodies (Williams & Rubanenko, 2024, pp. 1–30)?

Why and by what means does Mars currently host the concurrent presence of H_2O and CO_2? And, no less importantly, does the waxing and waning of glaciation on Mars resemble glacial cycles on Earth (Gallagher, 2024, pp. 31–72; Grau Galofre et al., 2024, pp. 73–100; Koutnik et al., 2024, pp. 101–142)? What are the principal landforms and landscape features thought to be the result of Martian glaciation? What is their consonance or dissonance with glacial landscapes and landforms on Earth (Gallagher, 2024, pp. 31–72; Grau Galofre et al., 2024, pp. 73–100; Koutnik et al., 2024, pp. 101–142)? Is periglaciation, i.e., the freeze-thaw cycling of water at or near the Martian surface, and associated landforms such as clastically sorted and non-sorted patterned ground, thermokarst-like depressions, and perennial ice-cored mounds or pingos, as ubiquitous on Mars as on Earth (Soare et al., 2024, pp. 143–192)?

What enables dwarf planets such as Ceres and Pluto to host possible cryovolcanism and ice-associated endogenic processes thought to be energetically inconsistent with the relatively small mass and old age of these bodies (Ahrens et al., 2024, pp. 357–376; Landis et al., 2024, pp. 221–260)?

Comets, some asteroids, and even some dwarf planets such as Ceres within the inner solar system are icy (El-Maarry, 2024, pp. 261–282; Landis et al., 2024, pp. 221–260). Wherefrom are they? Did they originate at or proximal to their present-day orbital distance from the sun, or does their genesis lie further afield in the Trans-Neptunian region of the solar system, if not beyond? Moreover, does the recently identified iciness of many of these bodies blur one of the most commonplace distinctions between them, i.e., "rocky" vs "icy?" If so, then to what extent should these bodies be, or not be, categorized as co-members of an icy continuum (El-Maarry, 2024, pp. 261–282)?

What of the moons in the outer solar system?

In the Jovian system, Europa's icy surface and its plate tectonics, Ganymede's dichotomous terrain and icy shell, and Callisto's densely cratered icy blanket of dark material comprise a geologically unique set of moons, each of which possibly hosts a subsurface ocean that could be a safe haven for simple life (Howell et al., 2024, pp. 283–314).

Ice is omni-present in the Saturnian system of moons as well. Water ice occurs on the surface of Titan, Enceladus, Iapetus, Hyperion, Dione, Rhea, Tethys, and Mimas; carbon dioxide, methane, and cyanide ice along with methane clathrates have also been identified or are hypothesized to exist on the surface or within the interior of a few of these icy moons. Titan, in particular, exhibits a plethora of ice compositions due to the complex chemistry arising from its surface-atmosphere interaction. As ocean worlds, both Titan and Enceladus are considered high-priority targets for astrobiology (Farnsworth et al., 2024, pp. 315–356).

The moons of Uranus and Neptune as well as the Pluto-Charon system contain, in addition to water, CO_2, CO, N_2, NH_3, and CH_4 ices. On Pluto's surface, volatile species such as nitrogen, methane, and carbon monoxide are mobilized seasonally, and many ice bodies in the outer reaches of the solar system display a diversity of surface processes and modifications. Both Pluto and Triton possess features that have been interpreted as cryovolcanic in origin and may harbor subsurface oceans, joining the growing list of hypothesized ocean worlds (Ahrens et al., 2024, pp. 357–376).

It is clear from this holistic review of ices throughout the solar system that volatiles have played an integral role in the thermal, atmospheric, and geological evolution of the planets, moons, and other objects in the solar system.

Richard J. Soare
Department of Geography, Dawson College, Montreal, QC, Canada

Jean-Pierre Williams
Department of Earth, Planetary, and Space Sciences, University of California, Los Angeles, CA, United States

Frances E.G. Butcher
Department of Geography, University of Sheffield, Sheffield, United Kingdom

Mohamed Ramy El-Maarry
Space and Planetary Science Center and Department of Earth Sciences, Khalifa University, Abu Dhabi, United Arab Emirates

References

Ahrens, C. J., Lisse, C. M., Williams, J.-P., & Soare, R. J. (2024). *Geocryology of Pluto and the icy moons of Uranus and Neptune*. In R. J. Soare, J.-P. Williams, C. Ahrens, F. E. G. Butcher, & M. R. El-Maarry (Eds.), *Ices in the solar system, a volatile-driven journey from the inner solar system to its far reaches*. Elsevier Books.

El-Maarry, M. R. (2024). *Small icy bodies in the inner Solar System.* In R. J. Soare, J.-P. Williams, C. Ahrens, F. E. G. Butcher, & M. R. El-Maarry (Eds.), *Ices in the solar system, a volatile-driven journey from the inner solar system to its far reaches.* Elsevier Books.

Farnsworth, K. K., Dhingra, R. D., Ahrens, C. J., Nathan, E. M., & Magaña, L. O. (2024). *Titan, Enceladus, and other icy moons of Saturn.* In R. J. Soare, J.-P. Williams, C. Ahrens, F. E. G. Butcher, & M. R. El-Maarry (Eds.), *Ices in the solar system, a volatile-driven journey from the inner solar system to its far reaches.* Elsevier Books.

Gallagher, C. (2024). *Glaciation and glacigenic geomorphology on Earth in the Quaternary Period.* In R. J. Soare, J.-P. Williams, C. Ahrens, F. E. G. Butcher, & M. R. El-Maarry (Eds.), *Ices in the solar system, a volatile-driven journey from the inner solar system to its far reaches.* Elsevier Books.

Garvin, J. B. (2024). *Prologue II: The ice frontier for science in the upcoming decades: A strategy for Solar System exploration?* In R. J. Soare, J.-P. Williams, C. Ahrens, F. E. G. Butcher, & M. R. El-Maarry (Eds.), *Ices in the solar system, a volatile-driven journey from the inner solar system to its far reaches.* Elsevier Books.

Garvin, J. B., Soare, R. J., Hepburn, A. J., Koutnik, M., & Godin, E. (2024). *Ice Exploration on Mars: Where to and when?* In R. J. Soare, J.-P. Williams, C. Ahrens, F. E. G. Butcher, & M. R. El-Maarry (Eds.), *Ices in the solar system, a volatile-driven journey from the inner solar system to its far reaches.* Elsevier Books.

Grau Galofre, A., Lasue, J., & Scanlon, K. (2024). *Ice on Noachian and Hesperian Mars: Atmospheric, surface, and subsurface processes.* In R. J. Soare, J.-P. Williams, C. Ahrens, F. E. G. Butcher, & M. R. El-Maarry (Eds.), *Ices in the solar system, a volatile-driven journey from the inner solar system to its far reaches.* Elsevier Books.

Howell, S. M., Bierson, C. J., Kalousová, K., Leonard, E., Steinbrügge, G., & Wolfenbarger, N. (2024). *Jupiter's ocean worlds: Dynamic ices and the search for life.* In R. J. Soare, J.-P. Williams, C. Ahrens, F. E. G. Butcher, & M. R. El-Maarry (Eds.), *Ices in the solar system, a volatile-driven journey from the inner solar system to its far reaches.* Elsevier Books.

Koutnik, M., Butcher, F., Soare, R., Hepburn, A., Hubbard, B., Brough, S., … Pathare, A. (2024). *Glacial deposits, remnants, and landscapes on Amazonian Mars: Using setting, structure, and stratigraphy to understand ice evolution and climate history.* In R. J. Soare, J.-P. Williams, C. Ahrens, F. E. G. Butcher, & M. R. El-Maarry (Eds.), *Ices in the solar system, a volatile-driven journey from the inner solar system to its far reaches.* Elsevier Books.

Landis, M. E., Castillo-Rogez, J., & Ahrens, C. (2024). *Ceres—A volatile-rich dwarf planet in the asteroid belt.* In R. J. Soare, J.-P. Williams, C. Ahrens, F. E. G. Butcher, & M. R. El-Maarry (Eds.), *Ices in the solar system, a volatile-driven journey from the inner solar system to its far reaches.* Elsevier Books.

Raymond, S. N. (2024). *Prologue I: The Solar System's Ices and their Origin.* In R. J. Soare, J.-P. Williams, C. Ahrens, F. E. G. Butcher, & M. R. El-Maarry (Eds.), *Ices in the solar system, a volatile-driven journey from the inner solar system to its far reaches.* Elsevier Books.

Soare, R. J., Costard, F., Williams, J.-P., Gallagher, C., Hepburn, A. J., Stillman, D., … Godi, E. (2024). *Evidence, arguments, and cold-climate geomorphology that favour periglacial cycling at the Martian mid-to-high latitudes in the Late Amazonian Epoch.* In R. J. Soare, J.-P. Williams, C. Ahrens, F. E. G. Butcher, & M. R. El-Maarry (Eds.), *Ices in the solar system, a volatile-driven journey from the inner solar system to its far reaches.* Elsevier Books.

Williams, J.-P., & Rubanenko, L. (2024). *Cold-trapped ices at the poles of Mercury and the Moon.* In R. J. Soare, J.-P. Williams, C. Ahrens, F. E. G. Butcher, & M. R. El-Maarry (Eds.), *Ices in the solar system, a volatile-driven journey from the inner solar system to its far reaches.* Elsevier Books.

Cold-trapped ices at the poles of Mercury and the Moon

1

Jean-Pierre Williams[a] and Lior Rubanenko[b,c]

[a]Department of Earth, Planetary, and Space Sciences, University of California, Los Angeles, CA, United States,
[b]Department of Geological Sciences, Stanford University, Stanford, CA, United States, [c]Department of Geospatial Engineering, Technion, Israel Institute of Technology, Haifa, Israel

Abstract

Ice deposits have been detected within permanently shadowed regions (PSRs) near the poles of both Mercury and the Moon. Observations of Mercury's poles by Earth-based radar, and subsequent observations by the MESSENGER (MErcury Surface, Space ENvironment, GEochemistry, and Ranging) mission, reveal what appear to be thick ice deposits in permanently shadowed craters near the poles. Surface reflectance observations show surface materials that are approximately collocated with radar-bright areas within permanently shadowed north polar craters. Thermal models predict these areas are associated with mean temperatures less than ~100 K suggesting the distribution of regions with high backscatter is due to the presence of water ice which is not thermally stable against sublimation in a vacuum at temperatures higher than ~100 K. Permanently shadowed craters near the poles of the Moon represent similar thermal environments, however, evidence for ice in these regions has proven to be more elusive and the abundance and distribution more challenging to characterize. Evidence for surficial water frost has been inferred from reflected ultraviolet spectral characteristics, infrared absorption features, and an increase in reflectance, which correspond with cryogenic temperatures within permanently shadowed areas. Neutron spectrometers on the Lunar Prospector and Lunar Reconnaissance Orbiter show enhanced hydrogen abundance at both poles, and the Lunar Crater Observation and Sensing Spacecraft (LCROSS) impact experiment confirmed the presence of water and other volatile species within the permanently shadowed crater of Cabeus. However, the thick (>10 m) near-pure ice deposits that are observed on Mercury have not been identified on the Moon where the nature of the ice distribution remains ambiguous. Why the deposits of polar ices on these two solar system bodies differ, and what the source of these ices are, remain a fundamental outstanding question in planetary science.

1 Introduction

The low obliquity of the spin axis relative to the ecliptic plane of Mercury (0.034°) and the Moon (1.54°) results in topographic depressions near their poles to form permanently shadowed regions (PSRs) (e.g., Bussey et al., 1999, 2003; Margot et al., 1999; Mazarico et al., 2011; Paige et al., 1992; Thomas, 1974; Urey, 1952; Watson et al., 1961a). These regions can become very cold as they radiate directly to space with the loss of heat primarily balanced by scattered, indirect illumination and thermal emission from nearby sunlit surfaces, residual heat from the planetary interior, and for the Moon, Earthshine (Ingersoll et al., 1992; Paige et al., 1992, 2010b; Salvail & Fanale, 1994; Vasavada et al., 1999). The modest 1.54°

obliquity of the lunar axis also results in seasonal variations in PSR temperatures and in persistently shadowed areas that expand and contract with the seasonal migration of the subsolar latitude (Kloos et al., 2019; Schorghofer & Williams, 2020; Williams et al., 2019).

Molecules of water and other volatile substances are thought to travel in ballistic hops in the faint, gravitationally bound exosphere of the Moon (Watson et al., 1961a). The residence time of volatile molecules is an exponential function of the surface temperature and the hop length is proportional to temperature; in illuminated, equatorial regions, the residence time is significantly smaller than the molecule ballistic flight time (Schorghofer, 2015). However, molecules landing in shadowed, polar regions may become cold trapped inside PSRs for extended periods of time, and accumulate into deposits over geologic timescales where surface temperatures are cold enough for the sublimation rate to become exceedingly low ($E \lesssim 1$ mm per billion years) (Fig. 1) (Watson et al., 1961a; Zhang & Paige, 2009).

The net accumulation rate of cold-trapped volatiles inside PSRs depends not only on the influx of molecules, but also on the temperature-dependent sublimation rate as well as temperature-independent processes such as photodissociation and micrometeorite bombardment that can erode surface deposits at rates comparable to their deposition rate (Hurley et al., 2012b; Lawrence, 2017; Morgan & Shemansky, 1991; Moses et al., 1999). Compared to exposed surface ice, subsurface ice is significantly more durable; models (Schorghofer & Taylor, 2007) have demonstrated the loss rate of ice buried beneath a layer of 1 cm of regolith decreases by nearly three orders of magnitude (Fig. 1). In addition, buried ice is far less susceptible to temperature-independent erosive processes which typically do not penetrate the physical barrier.

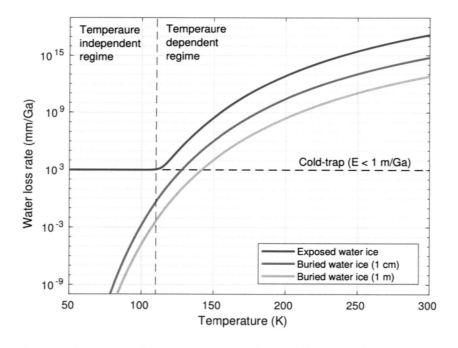

FIG. 1

Sublimation rates of cold-trapped surface and subsurface water ice deposits, based on the model of Schorghofer and Taylor (2007).

The lateral size of cold traps is limited at the largest scale by the size of the largest PSRs that harbor them and at the smallest scales by the thermal properties of the regolith, which is thermally insulating on scales >1 cm (Bandfield et al., 2015). Recent studies show that despite their smaller size, these "micro" cold traps contain approximately 10–20% of the total cold trap area fraction on the Moon and Mercury (Hayne et al., 2021a; Rubanenko et al., 2018).

In this chapter, we discuss the observational evidence for ices in cold traps, the possible sources and delivery of ices to the cold traps, the accumulation and destruction of ice within cold traps, and outstanding questions and future missions.

2 Observations of ices in cold traps

2.1 Initial observations of water ice

Early work concluded that volatiles would not survive for long periods of time on the lunar surface, particularly water due to its low molecular weight and low ionization threshold (Herring & Licht, 1959; Öpik & Singer, 1960). However, based on infrared observations demonstrating nighttime temperatures on the Moon as low as 120 K (Pettit & Nicholson, 1930), Watson et al. (1961a, 1961b) showed that the sublimation rate of water ice in permanently shadowed regions near the lunar poles would be slow enough that water ice could be stable over the lifetime of the Moon. Thomas (1974) further suggested this could also occur in PSRs on Mercury. Arnold (1979), considering possible source, transport, and destruction mechanisms for water, concluded that an average concentration of 1–10 wt% mixed within the top 2 m of regolith in PSRs was plausible. Lanzerotti et al. (1981) countered that the loss rate of water ice by solar wind and magnetospheric ions was comparable in magnitude to the accumulation rate estimated by Arnold (1979) and, thus, concluded that no significant deposits of ice could exist in the PSRs.

The first observational evidence for water ice in PSRs came from Earth-based radar observations of Mercury in the early 1990s (Harmon & Slade, 1992; Slade et al., 1992). Rader mapping of Mercury using the Goldstone 70 m antenna as a transmitter and the Very Large Array (VLA) as a receiver at 3.5 cm wavelengths identified a highly reflective region at the north pole with a high same sense-to-opposite sense circular polarization ratio (CPR) (Butler et al., 1993; Slade et al., 1992). This feature was confirmed with the Arecibo 12.6 cm (S-band) radar along with the identification of a smaller radar-bright feature at the south pole primarily confined to the floor of crater Chao Meng-Fu (Harmon & Slade, 1992).

Similarly, high reflectivity and circular polarization ratio (CPR) had previously been observed for the icy Galilean satellites (Campbell et al., 1978; Goldstein & Green, 1980; Goldstein & Morris, 1975; Ostro & Shoemaker, 1990), Titan (Muhleman et al., 1990), and the south polar cap of Mars (Muhleman et al., 1991) and more recently have been observed for terrestrial ice sheet (Rignot, 1995). Such characteristics can result from coherent backscatter of water ice where the radar signal propagates through the weakly absorbing ice and is volume scattered by heterogeneities (Hapke, 1990; Hapke & Blewitt, 1991). The interpretation that the radar observations resulted from water ice deposits was strengthened by thermal model calculations that showed maximum temperatures within permanently shadowed craters on Mercury are expected to be significantly colder than required for water ice to be stable for billions of years (Ingersoll et al., 1992; Paige et al., 1992).

Subsequent radar observations and data processing resolved the polar anomalies into numerous features that, after making small corrections to the positions of the poles, aligned with observed craters in the hemisphere imaged by the Mariner 10 spacecraft (Harmon et al., 1994). The correlation of

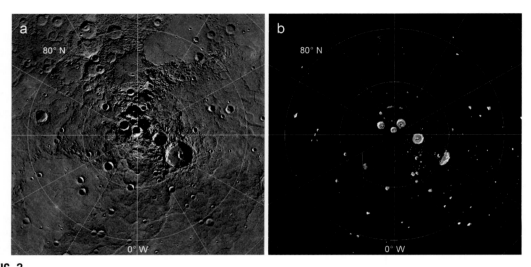

FIG. 2

Mercury's north polar regions. (a) MESSENGER Mercury Dual Imaging System (MDIS) moderate incidence angle monochrome mosaic (BDR) (NASA/Johns Hopkins University Applied Physics Laboratory/Carnegie Institution of Washington). (b) Arecibo S-band radar observations from Harmon et al. (2011) of the same region as (a) showing radar-bright regions interpreted to be ice deposits.

radar-bright materials with craters continued to be refined over time with improvements to analysis techniques and additional observations, including upgrades to Arecibo, further supporting the conclusion that the PSRs on Mercury contained ice (Harmon, 2007; Harmon et al., 2001, 2011) (Fig. 2).

Radar observations of the Moon, by comparison, have been ambiguous and inconclusive (Harmon, 1997). Bistatic measurements using the Clementine spacecraft and the Deep Space Network showed a backscatter enhancement consistent with the presence of ice (Nozette et al., 1994, 1996). However, observations using Arecibo showed many of the high CPR areas occurred in regions that were at least occasionally sunlit leading Stacy et al. (1997) to conclude that the radar signal was a result of increased surface roughness. A detailed reanalysis of the Clementine bistatic data was unable to reproduce the results of Nozette et al. (1996) and found that the enhancements in backscattering could be attributed to surface roughness and topographic variations (Simpson & Tyler, 1999). However, a mixture of ice and regolith with a mixing ratio less than 1% would not be inconsistent with the observations. However, ice deposits of similar quantities and scattering properties as observed for Mercury could be ruled out.

Neutron spectroscopy provided unambiguous evidence that water ice could be present in the lunar polar regions. Neutron data from the Lunar Prospector (LP) (Binder, 1998), which orbited the Moon from January 1998 to July 1999, showed a decrease in epithermal neutrons at both lunar poles consistent with buried water ice (Feldman et al., 1998). Epithermal neutrons, with energies between 0.5 and 0.5 MeV are sensitive to the presence of hydrogen, resulting in a depression of the observed neutron counts (Feldman et al., 1991; Lingenfelter et al., 1961; Metzger & Drake, 1990) and Feldman et al. (1998) broadly estimated a total equivalent water at each pole of 3×10^9 metric tons assuming pure water ice deposits beneath a 40 cm layer of dry regolith extending to a depth of 2 m. With improved data resolution during the later months of the mission, achieved by lowering the spacecraft from a

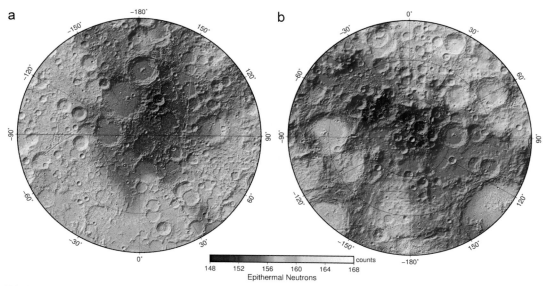

FIG. 3

Map of epithermal neutron count rate measured in 8s for the low orbit portion of the Lunar Prospector mission (Feldman et al., 2001) for the lunar (a) north and (b) south polar regions poleward of 75° latitude superposed on shaded relief derived from Lunar Orbiter Laser Altimeter (LOLA; Smith et al., 2010) digital elevation model (DEM).

100 ± 20 km altitude to a 30 ± 15 km altitude orbit, Feldman et al. (2000, 2001) showed that the areas where the largest neutron suppressions were observed overlay PSRs (Fig. 3). Combined with area estimates of the large south polar PSRs, hydrogen concentrations of 1670 ± 890 ppm, or 1.5 ± 0.8 wt% water equivalent hydrogen (WEH) assuming the hydrogen is in the form of H_2O, were estimated. While the interpretation of the neutron data as a unique detection of hydrogen, and by extension, water has been questioned by Hodges (2002) who suggested variations in abundances of other elements could explain the results, subsequently refined modeling confirmed that the presence of hydrogen provided the best explanation for the observed decrease in epithermal neutrons over the lunar poles (Lawrence et al., 2006). Analysis of the LP neutron data using deconvolution algorithms provided improved resolution of hydrogen distribution around the poles, further refining the locations to be concentrated within PSRs (Eke et al., 2009; Elphic et al., 2007; Teodoro et al., 2010).

2.2 Detailed observations of PSRs

A number of spacecraft over the last decade and a half have provided a much more detailed look at PSRs on both Mercury and the Moon. This includes several NASA missions as well as several international missions by India, Japan, and China.

2.2.1 MESSENGER at Mercury

After several flybys of Mercury, the MESSENGER mission (Solomon et al., 2001, 2007) entered orbit around the planet in 2011, providing the first global view of the surface. The only prior spacecraft to visit Mercury was the Mariner 10 spacecraft which conducted three flybys during 1974–1975 imaging less than half of the planet (e.g., Spudis & Guest, 1988; Strom, 1979). MESSENGER was inserted into a highly elliptical, near-polar orbit with a periapsis altitude of ~200 km and a subspacecraft periapsis latitude of 60°N, and an apoapsis ~15,200 km in the southern hemisphere. Periapsis altitudes were maintained within 200–500 km before being lowered during the second extended mission (Solomon & Byrne, 2019). As a result of the orbital eccentricity, with a periapsis in the northern hemisphere, observations from the Neutron Spectrometer (NS) (Goldsten et al., 2007) and Mercury Laser Altimeter (MLA) (Cavanaugh et al., 2007) were only obtained in the northern hemisphere.

In spite of the challenging orbital configuration for neutron spectrometer and altimetry observations, essential measurements for determining the nature of polar volatile deposits in the north polar region were obtained. Lawrence et al. (2013) found the epithermal neutron count rate was consistent with the expected rate if the north polar PSRs contained high concentrations of water ice (50–100 wt% WEH). This supported the interpretation of the Earth-based radar observation that radar-bright enhancements in PSRs were likely due to water ice deposits. MLA also provided a geodetically controlled topographic model of the northern hemisphere (Zuber et al., 2012b) and measured surface reflectance at 1064 nm (Neumann et al., 2013). The reflectance measurements of the PSRs revealed anomalously dark and bright materials deposited within the PSRs that spatially correlated with the radar-bright areas (Deutsch et al., 2017; Neumann et al., 2013). Using the topographic model, Paige et al. (2013) modeled the polar temperatures and found that the distribution of radar-bright features strongly correlated with areas that experienced average temperatures less than ~100 K consistent with the volatility of water ice (Fig. 4). The anomalously dark reflectance areas were observed to be more spatially extensive than the radar-bright features and the thermal modeling predicted these surfaces to experience a wider range of temperatures. These surfaces have been suggested to be a sublimation lag of carbonaceous organic materials remnant from a previous greater extent of surface water ice deposits (Paige et al., 2013). Additionally, the topography data has been used by Rivera-Valentín et al. (2022) to simulate incidence angles of new Arecibo S-band observations, and applying radar-scattering models showed that some of the ice deposits display a gradation in radar properties that likely reflect transitions in ice purity.

Visible images of the PSR interiors were obtained by MESSENGER's Mercury Dual Imaging System (MDIS) (Hawkins III et al., 2007). The wide-angle camera, equipped with a broadband clear filter, successfully imaged the floors of many of the PSRs using long exposures and showed that they contained extensive regions with distinctive reflectance properties suggesting the presence of multiple volatile species with differing reflectances and cold-trapping temperatures (Chabot et al., 2014, 2016) (Fig. 4). The deposits appear to overlay preexisting small impact craters and the lack of regolith cover and distinct, well-defined boundaries suggest the deposits have experienced minimal disruption and mixing by impact gardening. This is consistent with the geologically recent emplaced of the materials or an ongoing process that replenishes the deposits.

2.2.2 Lunar Reconnaissance Orbiter and other recent spacecraft at the Moon

A sustained, international campaign of lunar exploration began in the middle of the last decade with successful lunar missions from the United States, China, India, and Japan with numerous missions planned over the next decade, including commercial delivery services as part of the Commercial Lunar Payload Services (CLPS) initiative (Bussey et al., 2019; Voosen, 2018; Williams et al., 2022), and the campaign to return humans to the lunar surface through the Artemis program (Smith et al., 2020).

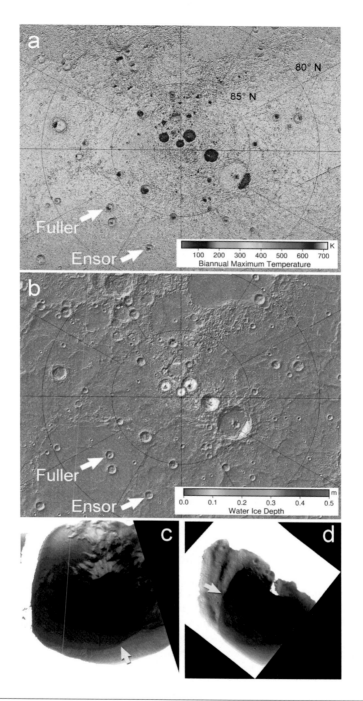

FIG. 4

Thermal model results from Paige et al. (2013) using an updated MLA DEM (Chabot et al., 2018a) showing (a) the biannual maximum temperatures and (b) the water ice stability depth accounting for insulating regolith cover of the north polar region of Mercury. White regions in (b) are where water ice is stable at the surface. MDIS WAC broadband images of PSRs in (c) Fuller crater (27 km diameter, 82.63°N, 317.35°E, image: EW1062695957B) and (d) Ensor crater (25 km diameter, 82.32°N, 342.47°E, image: EW1051458815B) showing low-reflectance regions, inferred to be thin deposits of dark, organic-rich volatile materials, which can be seen to extend up crater walls (*yellow arrows*). *White areas* in images are saturated pixels due to directly illuminated surfaces.

The NASA Lunar Reconnaissance Orbiter (LRO) (Chin et al., 2007; Tooley et al., 2010; Vondrak et al., 2010) and Lunar Crater Observation Sensing Satellite (LCROSS) (Colaprete et al., 2012) have provided a multitude of new observation that has refined our understanding of ices within the lunar polar regions. LRO was inserted into a near-polar orbit in 2009 and continues to operate, providing over a decade of observations from its payload of seven instruments.

Early in the LRO mission, the LCROSS mission, an accompanying payload on the LRO launch vehicle, provided ground truth as a kinetic impacting experiment whereby a Shepherding Spacecraft (SSC) guided the spent Atlas V Centaur upper stage into a PSR in October 2009 (Colaprete et al., 2012). The Cabeus PSR was targeted and the impact and ejecta plume were observed by the SSC, LRO, and Earth-based instruments. The impact is estimated to have formed a ~20 m diameter crater with a ~160 m wide ejecta region at −84.6796°S, −48.7093°E (Marshall et al., 2012). The impact experiment confirmed the presence of water in Cabeus along with other volatiles including CO_2, CO, light hydrocarbons, and sulfur compounds (Colaprete et al., 2010; Gladstone et al., 2010; Hayne et al., 2010; Hurley et al., 2012a; Luchsinger et al., 2021; Schultz et al., 2010).

Polar topography measurements were made by the LRO Lunar Orbital Laser Altimeter (LOLA) (Smith et al., 2010, 2017) instrument along with laser altimeters aboard the Japanese Kaguya (Araki et al., 2009) and the Chinese Chang'E-1 (Ping et al., 2009) orbiters. Illumination models constructed from the elevation data provided improved estimates of PSR areas. Mazarico et al. (2011) and McGovern et al. (2013) estimated the total PSR area to be 1.2–1.3 × 10^4 km^2 in the north and 1.6–1.7 × 10^4 km^2 in the south poleward of 80° latitude. Additional refinement to our understanding of the polar illumination conditions has been provided by repeat imaging of the polar regions by the Lunar Reconnaissance Orbiter Camera (LROC) Wide Angle Camera (WAC) at pixel scales of 100 m (Robinson et al., 2010; Speyerer & Robinson, 2013).

The LRO payload included a neutron spectrometer the Lunar Exploration Neutron Detector (LEND) (Mitrofanov et al., 2010). The LEND instrument included a collimator composed of neutron-absorbing materials to narrow the field of view of the neutron sensor in order to increase the spatial resolution. While the performance of the collimated data has been debated (Eke et al., 2012; Lawrence et al., 2011; Miller et al., 2012; Teodoro et al., 2014), the uncollimated epithermal data is consistent with the Lunar Prospector neutron data and several studies concluded that the collimated epithermal neutron data could be used successfully to map neutron suppression at a higher resolution than the Lunar Prospector data (Boynton et al., 2012; Litvak et al., 2012; Livengood et al., 2018; Mitrofanov et al., 2011, 2012; Sanin et al., 2012).

Orbital radar observations were made by the Miniature Synthetic Aperture Radar (Mini-SAR) on Chandraayan-1 (Spudis et al., 2009) and the Miniature Radio-Frequency (Mini-RF) instrument on LRO (Nozette et al., 2010). Results from these experiments remained comparatively ambiguous with regards to ice detections relative to the radar observation of Mercury's polar regions where interpretations of radar data provided more definitive conclusions. Many young craters on the Moon are found to be associated with high CPR signatures both within the craters and exterior to their rims consistent with rough surfaces derived from impact melts and clastic, brecciated materials. An anomalous class of craters was observed with high CPR only within the crater interiors (Spudis et al., 2010, 2013). As many of these craters are PSRs and correlated with hydrogen-rich regions observed in the neutron data, these features were interpreted to be deposits of water ice within craters (Spudis et al., 2010, 2013). However, a subsequent systematic survey of craters showed that anomalous CPR craters were distributed across the Moon. Fa and Eke (2018) suggested that these craters represented an intermediate stage in crater evolution and were not evidence for water ice deposits. A survey of crater depth-to-diameter ratios does show, however, that craters become distinctly shallower at higher polar southern latitudes suggesting that ice deposits may be present within craters, a trend that is also observed at Mercury where evidence for thick water ice deposits is more conclusive (Rubanenko et al., 2019), and bistatic observations between Mini-RF and the Arecibo Observatory have shown an opposition effect within a portion of the floor of Cabeus crater that is consistent with near-surface water ice (Patterson et al., 2017).

The Diviner Lunar Radiometer Experiment (Diviner) (Paige et al., 2010a) aboard LRO has provided global brightness temperature measurements across multiple infrared spectral bands (Williams et al., 2017). This has enabled the temperatures of the polar regions to be mapped along with regions of ice stability (Landis et al., 2022; Paige et al., 2010b; Schorghofer et al., 2021b; Schorghofer & Williams, 2020; Williams et al., 2019) (Figs. 5 and 6). These observations have been complimented with ice diffusion models to map the stability of surface and subsurface ices (Paige et al., 2010b; Schorghofer & Aharonson, 2014; Schorghofer & Williams, 2020). Additionally, regolith thermal modeling that accounts for polar illumination along with predicted subsurface ice stability shows large areas of stability and has been useful in showing how past obliquity and true polar wander may have resulted in changes to the ice distribution, and can explain some aspects of the hydrogen distribution observed by the neutron spectrometers better than present-day maximum surface temperatures alone (Gläser et al., 2021; Siegler et al., 2011, 2015, 2016).

The Diviner temperatures have also been utilized to map the thermal stability of other volatile species in addition to H_2O. Using sublimation rates averaged over a draconic year to account for diurnal and seasonal temperature variation, Schorghofer et al. (2021b) identified a cumulative area of ~200 km^2 of spatially contiguous regions where CO_2 ice would be thermally stable within several south polar PSRs including the Amundsen, Haworth, de Gerlache and Cabeus craters. Landis et al. (2022), using the Diviner maximum temperatures, a more conservative stability criterion than the average sublimation rate, mapped out the areas of surface thermal stability for a variety of volatile species for both the north and the south polar region using stability temperatures calculated by Zhang and Paige (2009, 2010) (Table 1). In addition to water and carbon dioxide ice, they identified a variety of other species that may be thermally stable given their volatility, such as carbon monoxide, ammonia, and other compounds. Many of these species were observed in the LCROSS ejecta plume, which included, in decreasing abundance after water: H_2S, NH_3, SO_2, C_2H_4, and CO_2 (Colaprete et al., 2010).

Several lines of evidence suggest that surficial exposures of water ice or frost may be present within PSRs, albeit in apparent heterogeneous, scattered distributions. Reflectance measurement at far-ultraviolet wavelengths using data from the Lyman Alpha Mapping Project (LAMP) instrument on LRO showed a higher reflectance at longer wavelengths compared to non-PSR regions consistent with the presence of ~1–2 wt% water frost at the surface (Gladstone et al., 2012). Hayne et al. (2015) additionally found UV spectral features consistent with water ice within south polar PSRs where annual maximum temperatures were below 110 K, as observed by Diviner, and possibly a colder population of ice deposits at temperatures below ~65 K which could be CO_2 ice. Water ice concentrations of ~0.1–2 wt% are inferred assuming the water ice is intimately mixed with dry regolith.

Active reflectance measurements at 1064 nm using LOLA show that PSRs are systematically brighter than non-PSRs (Lucey et al., 2014; Qiao et al., 2019), particularly within Shackleton crater (Lucey et al., 2014; Zuber et al., 2012a). Fisher et al. (2017) found that the reflectance changes mirrored the results of Hayne et al. (2015) in the south polar region with reflectance rapidly increasing for surfaces with maximum temperatures below ~110 K. Most of these areas also have been found to

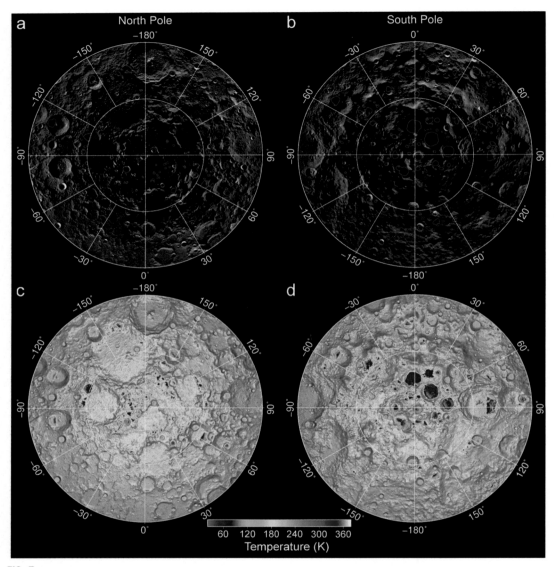

FIG. 5

LROC Wide Angle Camera (WAC) image mosaic of (a) the north pole and (b) the south pole with PSR boundaries outlined in *red* for PSRs $\geq 10\,km^2$ and maximum bolometric temperatures observed by LRO Diviner for (c) the north polar region and (d) the south polar region poleward of 80° latitude. WAC mosaic uses 643 nm band images acquired near the summer solstice, i.e., subsolar latitudes near the maximum extent of 1.5° north for (a) or 1.5° south for (b) (Speyerer et al., 2020). PSR boundaries are derived from illumination simulations using LOLA DEMs (Mazarico et al., 2011) and maximum temperatures are taken from summer season polar stereographic gridded Diviner data (Williams et al., 2019).

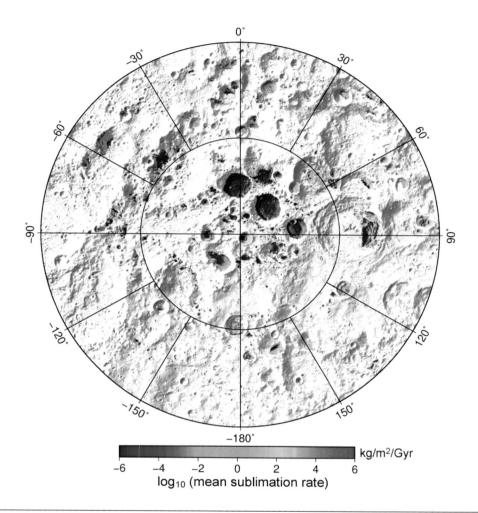

FIG. 6

Long-term average of the potential sublimation rate of water ice from Schorghofer and Williams (2020) for the south polar region (poleward of 80°S) with a logarithmic color scale superposed on shaded topographic relief.

exhibit near-infrared spectral features diagnostic of water ice in data from the Chandrayaan-1 Moon Mineralogy Mapper (M^3) instrument (Li et al., 2018), though the distribution is patchy with only ~3.5% of the cold traps exhibiting surface-exposed ice at relatively low abundances (~30 wt% if intimately mixed with dry regolith). Differential emissivities between PSR and non-PSR areas within the Amundsen crater in the far-infrared Diviner spectral bands could also be related to the presence of ice (Sefton-Nash et al., 2019).

Visible images of PSRs, however, have not shown anomalous features attributed to the presence of volatiles as has been observed within Mercury's PSRs (Chabot et al., 2014). High reflectance surfaces

Table 1 Stability temperature and predicted thermal stability area of cold-trapped species on the Moon.

Chemical formula	Name	T_v (K)[a]	Area, North (km²)[b]	Area, South (km²)[b]
S	Sulfur	201.5	48,782	64,837
NH_4HCO_3	Ammonium bicarbonate	113.3	7,008	14,818
$NH_4CO_2NH_2$	Ammonium carbonate	107.4	5,700	12,238
H_2O	Water	106.6	5,513	11,888
$C_5H_{10}O_2$	Ethyl propionate	103.6	4,804	10,629
NH_4SH	Ammonium hydrosulfide	96.1	3,102	7,887
NH_4CN	Ammonium cyanide	93.8	2,629	7,104
$C_5H1_{10}O$	3-Pentanone	92.8	2,432	6,753
C_7H_5	Toluene	87.6	1,589	5,134
HCN	Hydrogen cyanide	80.5	896	3,328
CS_2	Carbon disulfide	74.4	516	1,991
C_5H_{12}	Pentane	73.6	472	1,810
NH_3	Ammonia	65.5	99	554
SO_2	Sulfur dioxide	62.3	38	273
CO_2	Carbon dioxide	54.3	0.18	20
H_2S	Hydrogen sulfide	50.6	–	5

[a] Stability temperature, T_v, is defined as the temperature at which $\leq 1\,mm\,Gyr^{-1}$ of sublimation will occur. Values are from Zhang and Paige (2009, 2010) and listed in increasing volatility.
[b] Stability areas are calculated for latitudes $\geq 60°$ by Landis et al. (2022) using maximum observed bolometric temperatures from Diviner.

were observed by LOLA within the interior of Shackleton crater (Zuber et al., 2012a); however, images from Kaguya's Terrain Camera of the interior of Shackleton crater showed no evidence of exposed water ice (Haruyama et al., 2008) and data from the Kaguya Multiband Imager suggest the high reflectance surfaces are the result of an anorthosite-rich composition (Haruyama et al., 2013). LROC conducted a series of long exposure imaging campaigns of PSRs to image their interiors using its Narrow Angle Camera (NAC). Reflectance anomalies that could be attributed to surface frost or permafrost-like land-forms have not been identified (Brown et al., 2020; Cisneros et al., 2017; Koeber et al., 2014), though it has been noted that surfaces within several PSRs where water ice is thermally stable appear smoother than adjacent warmer surfaces which have been suggesting to be the result of terrain softening from subsurface ice (Deutsch et al., 2021; Moon et al., 2021).

In spite of the vast improvement in characterizing the lunar PSRs with these more recent missions, the quantity, distribution, and nature of ices in the lunar polar regions remains uncertain. Unlike Mercury, where several lines of evidence strongly support relatively thick deposits of volatiles within PSRs, abundances on the Moon appear to be much smaller with distributions that are heterogeneous and noncontiguous.

3 Sources and delivery of polar volatiles

Understanding the nature of polar ices and why they differ between the Moon and Mercury requires an understanding of how volatiles are delivered, transported, lost, and sequestered at the poles. Detection of a near-IR hydration spectral absorption feature around 3 μm at the Moon by spectrometers on the

Chandrayaan-1, Cassini, and Deep Impact missions (Clark, 2009; Pieters et al., 2009; Sunshine et al., 2009) suggested global surface hydration as predicted by Zeller et al. (1966) who had previously proposed that hydroxyl could be produced by solar wind protons reacting with oxygen at the lunar surface. The 3 μm absorption feature can result from the presence of both hydroxyl and molecular water and translating the strength of the absorption feature to abundances is complicated by the mixture of both reflected visible and emitted thermal emissions at these wavelengths. How these components are separated is dependent on modeling with studies coming to differing conclusions regarding the distribution and temporal variability of hydration (Bandfield et al., 2018; Grumpe et al., 2019; Honniball et al., 2020; Li & Milliken, 2017; Lucey et al., 2022; McCord et al., 2011; Wöhler et al., 2017). However, recently, a spectral feature at 6 μm produced by a fundamental vibration of molecular water has been observed using the NASA/DLR Stratospheric Observatory for Infrared Astronomy (SOFIA), which has confirmed the presence of molecular water at a localized area on the lunar surface at high latitudes with an estimated abundance of ~100–400 $\mu g\,g^{-1}$ H_2O (Honniball et al., 2021).

Water is transported through the exosphere where it can become cold trapped in the PSRs near the poles. This is hypothesized to occur through ballistic migration whereby water molecules with enough energy to overcome the activation energy of desorption are released from the surface to ballistically propagate through the exosphere and either escape the Moon (Jean's escape) or remain gravitationally bound and reencounter the surface where the residence time will depend on the temperature (see recent reviews by Lucey et al., 2022; Schorghofer et al., 2021a, and references therein). Observations suggest an active exchange of water with the regolith and migration through the exosphere (Hendrix et al., 2012, 2019) and model calculations predict that the rapid rise in surface temperature with the sunrise result in the release of absorbed water with some of the mobilized water falling back onto the night side of the advancing dawn terminator resulting in concentrations of water molecules on the surface and exosphere near the dawn terminator (Prem et al., 2018; Schorghofer et al., 2017). Surface roughness is predicted to provide colder sloped surfaces at higher latitudes that can provide temporary cold trap reservoirs (Prem et al., 2018) as well as seasonally shadowed regions (Kloos et al., 2019; Williams et al., 2019) and micro cold traps (potentially centimeter scale at the smallest sizes) (Hayne et al., 2021a). These cold traps can stall the migration of molecules and if the residence times are long relative to the photodestruction timescales, water may be prevented from reaching PSRs.

The primary contributor to volatiles in the surface environments of the Moon and Mercury are impactors, solar wind, and outgassing from the interior through volcanism (see Prologue I by Raymond (2024, pp. xix–xxv) for further discussion of the origin of ices in the Solar System). Extensive volcanism occurred on the Moon predominately before ~3 Ga and declined to significantly lower levels between 3 and 1 Ga (Head & Wilson, 2017; Hiesinger et al., 2011). During periods of mare emplacement, transient collisional atmospheres may have formed with small fractions of the volatiles released migrating to PSRs (Needham & Kring, 2017; Stern, 1999). Head et al. (2020) concluded that the frequency and duration of eruptions were likely too low to sustain such atmospheres long enough for them to be efficient sources of polar volatiles, however, recent modeling by Wilcoski et al. (2022) found that erupted H_2O, assuming the eruption timeline of Head et al. (2020), could condense ~10^{16} kg of water ice at the poles, though the total accumulation of erupted H_2O will depend on the water abundance of the preerupted melt and that a size-frequency distribution of eruptions dominated by smaller eruptions would result in smaller fractions of erupted H_2O being deposited as ice. Lucey et al. (2022) suggested the volcanic contribution to be roughly 10^{16}–10^{17} g, predominately during the Imbrian. They estimated impactors to be the largest contributor to lunar volatiles delivering an estimated ~10^{17}–10^{18} g during the Copernican, and larger amounts in earlier epochs when impact rates were higher. This is orders of magnitude more than the amount of water estimated to be derived from the solar wind ($\sim 10^{14}$–10^{15} g). Studies suggest transient collisional atmospheres formed by hydrated impactors may efficiently transport cometary and asteroidal volatiles to the PSRs (Prem et al., 2015, 2019; Stewart et al., 2011). This is supported by the species observed by the LCROSS impact experiment which are generally consistent with asteroidal and cometary sources (Berezhnoy et al., 2012). Mandt et al. (2022) found that elemental abundances were consistent with a cometary origin for the volatiles while a high nitrogen abundance ruled out a volcanic atmosphere as the source.

The delivery history of the lunar ice is not clear. Absolute model ages of south polar craters that host PSRs with water ice detections show that the majority of the surface ice is contained within craters older than 3 Gyr (Deutsch et al., 2020a). A population of smaller, morphologically fresh craters has also been identified which host water ice requiring more recent water delivery (Deutsch et al., 2020a), and calculations by Farrell et al. (2019) of the erosional loss of ice in the space environment due to plasma sputtering and meteoritic impact vaporization and ejection suggest the observed surface frosts to be <2000 kyr old. However, PSRs within craters estimated to be Copernican in age, based on roughness properties, do not appear to contain surface ice (Deutsch et al., 2020b). PSR lacking water ice detections also exist within ancient craters suggesting the delivery and retention of ice has been complicated.

Periods of high obliquity up to $\sim 77°$ early in the lunar history during a transition in Cassini states (Ward, 1975; Wisdom, 2006; Siegler et al., 2011, 2015) or reorientation of the lunar spin axis (true polar wander) (Siegler et al., 2016) complicate the picture for the earliest delivery of ices as such orbital changes would have substantially altered the thermal conditions in the polar regions. It is unlikely ice, even interstitial pore ice at depth, could survive the high obliquities predicted to have occurred during the Cassini state transition which likely occurred ~ 3–4 billion years ago. Therefore, the sequestration and retention of volatiles at the poles early in lunar history would depend on the timing of the transition relative to their delivery.

The delivery of the observed ice to the poles of Mercury may be recent. Age estimates of the surfaces of polar ices suggest the ice was delivered within the last ~ 300 Myr (Deutsch et al., 2019) consistent with regolith gardening models that estimate ice deposits must have been emplaced within the last 200 Myr as ices and lag deposits would be overturned and well mixed with regolith over this length of time (Costello et al., 2020; Crider & Killen, 2005). The spatially coherent deposits with relatively sharp reflectance boundaries are consistent with young age and minimal impact gardening. Given the age constraints, relative purity of the ice, total volume requiring a rapid delivery rate, and uneven distribution indicating a discrete event rather than a steady, sustained process, it has been suggested that the emplacement of ice most likely occurred via a recent single impact (Chabot et al., 2018b; Costello et al., 2020; Deutsch et al., 2019; Moses et al., 1999) and a study by Ernst et al. (2018) suggests the relatively young 97 km diameter Hoskusai crater (16.8°E, 57.7°N) as a candidate. This is further supported by the presence of micro cold traps, small-scale cold traps on scales of 1–10 m, inferred from the darkening of terrain at high latitudes observed in MLA reflectance data (Rubanenko et al., 2018). The preservation of such spatially limited deposits in the presence of impact gardening supports a recent ($\lesssim 100$ Ma) depositional event.

4 The accumulation and loss of ice in cold traps

Molecules of volatile substances delivered to the surfaces of the Moon and Mercury are hypothesized to travel in ballistic hops in their putative exospheres. The typical hop length d of molecules can be obtained by comparing their thermal energy with their gravitation potential energy

$$kT = mgh,$$

where T is the surface temperature, m is the mass of the molecule, h is hop height, g is the acceleration due to gravity, and k is the Boltzmann constant. For a ballistic hop at a launching angle of $45°$, $h = d/2$, and $d = 2kT/mg$ (Schorghofer, 2015; Watson et al., 1961b). For water molecules in illuminated regions on the Moon and Mercury (i.e., at daytime temperatures), $d \sim 100$–200 km, much greater than the typical size of cold traps. Consequently, water molecules traveling into cold traps can be thought of as raindrops accumulating in buckets; the probability of a molecule landing inside a polar cold trap depends on its surface area.

Hopping molecules may escape the surface if their thermal velocity exceeds the escape velocity, or be destroyed by in-flight photodissociation. Butler et al. (1993) and Butler (1997) determined that 5–15% of molecules in an exosphere will survive to reach polar cold traps. Hydrodynamic models that simulate the transient collisional vapor plume produced during a comet impact found that most of the volatile mass is lost during the impact itself; however, ~0.1% of the original mass of the comet is retained inside cold traps, though this is strongly dependent on the impact velocity (Ong et al., 2010; Stewart et al., 2011). For example, a 5 km comet will deposit a layer that is ~1 cm thick (0.1% volatile mass retention) to ~10 cm thick (5% volatile mass retention). Ignoring erosion of cold-trapped volatiles and using modeled impact rates for comets and asteroids (10^{-7} yr^{-1}, e.g., Shoemaker et al., 1990), 1–10 m of water ice should accumulate in cold traps over 1 Ga.

The sublimation rate of cold-trapped ice exponentially depends on the surface temperature (Fig. 6). Historically (e.g., in Ingersoll et al., 1992; Landis et al., 2022; Paige et al., 1992; Rubanenko & Aharonson, 2017; Schorghofer & Taylor, 2007; Watson et al., 1961b), cold traps are defined as regions in which the loss by sublimation does not exceed 1 mm/Ga or 1 m/Ga, which, for water ice, occurs at a temperature of 106 or 120 K, respectively (Fig. 1). However, cold-trapped molecules are still susceptible to impact gardening and dissociation by UV photons and cosmic rays, a process which may be enhanced on Mercury due to focusing induced by its magnetic cusps (Delitsky et al., 2017).

Models show that temperature-independent loss rates of volatiles from cold traps range between 10 cm/Ga and 1 m/Ga (Farrell et al., 2019; Lanzerotti et al., 1981; Morgan & Shemansky, 1991). As a result, over long timescales, the surface loss rate from regions colder than ~110–120 K is most likely temperature-independent (Schorghofer & Aharonson, 2014; Schorghofer & Taylor, 2007). The sublimation rate of buried ice is significantly lower than that of surface ice due to the dissipation of the amplitude of the thermal wave, but also since the overlaying regolith acts as a diffusion barrier (Schorghofer & Taylor, 2007). As a result, the loss rate of ice buried beneath 1 cm of regolith is three orders of magnitude lower than surface ice found at the same temperature (Fig. 1). Despite the higher durability of subsurface ice, the mechanisms by which surface ice is buried are poorly understood. In regions where the surface temperature temporarily exceeds the subsurface temperature, molecules may migrate into the subsurface (Schorghofer & Aharonson, 2014). However, the burial fluxes produced by this "thermal ice pump" are relatively low. Impacts may also bury ice by covering it with regolith or gardening it into the subsurface (e.g., Hurley et al., 2012b). However, it is currently unknown whether impact gardening processes protect ice by burying it, eroding it directly via heating, or indirectly, by transporting it to the surface, where it may be destroyed through one of the mechanisms mentioned above.

The comparable accumulation and erosion rates of surface ice over geologic timescales, along with the presence of surface water on Mercury and the Moon indicates that episodic deposition, by, e.g., a comet impact, is more likely than slow, continuous accumulation by, e.g., micrometeorites. Additionally, while the solar wind may add to the balance of cold-trapped volatiles, it cannot be the sole source of volatiles on Mercury, where darkened carbonaceous surface ices are widespread (Chabot et al., 2014, 2016) and subsurface ice deposits are radar bright, and thus highly pure (Harmon & Slade, 1992; Slade et al., 1992).

5 Summary and future missions

The Moon and Mercury both possess PSRs with temperatures low enough to cold trap a range of volatile species (Landis et al., 2022; Zhang & Paige, 2009). The presence of ice within PSRs has been confirmed on both planetary objects (Colaprete et al., 2012; Feldman et al., 1998; Lawrence, 2017; Slade et al., 1992). Mercury's PSRs contain large amounts of ice with surface exposures while the nature of the lunar ices appears to be more challenging to quantify due to a nonuniform distribution at lower abundances. Why the quality and quantity of the ices are so markedly different between the two objects in spite of being similar planetary environments has become a fundamental question that has emerged as a more refined picture of their polar regions has come into focus. Resolving this question requires a deeper understanding of the processes that deliver, sequester, and remove volatiles over time and has implications for the source and evolution of water and other volatile species on other airless bodies within the solar system.

Evidence suggests a possible explanation, that Mercury experienced a recent, volatile-rich impact that filled much of its thermally stable, ice "reservoirs" (Costello et al., 2020; Deutsch et al., 2019), while the lunar ice deposits are ancient by comparison with a distribution possibly shaped by changes in the lunar spin characteristics (Siegler et al., 2016; Ward, 1975; Wisdom, 2006). A candidate crater on Mercury has even been identified as the remnant of such an event (Ernst et al., 2018) and such an impact can provide the water and dark organic materials observed in the PSRs (Chabot et al., 2018b). This would then suggest the fate of the ices at Mercury will follow those of the Moon where impact gardening increasingly mixes ices with regolith materials to create a heterogeneous patchwork of icy regolith deposits.

However, this question remains far from being definitively answered, and much about the source, distribution, and composition of ices on the two objects remain unclear. The relationship between surface and subsurface ice remains unknown, making estimates of the inventory and distribution of ice highly uncertain (Brown et al., 2022; Cannon et al., 2020). This is further confounded by discrepancies among measurements of the ice distribution at the Moon. The apparent spatial variability of the ice requires understanding processes such as impact gardening and delivery history. Impact gardening can redistribute and bury ice as well as remove volatiles through vaporization. Differences in environmental factors may be influencing the ices in ways that are not fully comprehended. The PSRs of Mercury are generally warmer, with some surfaces within lunar PSRs being among the coldest temperatures measured in the solar system (Paige et al., 2010b, 2013; Williams et al., 2019). Therefore, temperature-dependent processes that influence volatile stability and mobility at the surface and within the regolith may operate differently (Schorghofer & Aharonson, 2014). The space environments and the interactions of energetic particles with the ices may be influenced by the presence of a present-day intrinsic magnetic field at Mercury (Delitsky et al., 2017) or a paleo-magnetic field in the past at the Moon (Garrick-Bethell et al., 2019). Mercury's closer proximity to the Sun will influence solar wind intensity and impact rates and energies. The resulting elevated surface temperatures and solar wind proton flux make water derived from solar wind (recombinative desorption) more favorable on Mercury as a possible steady source of ice (Jones et al., 2020). Differences in the abundance of endogenic volatiles may also be playing a role. MESSENGER showed that Mercury is surprisingly enriched in volatile elements (Nittler et al., 2011; Peplowski et al., 2011) and numerous, relatively young volcanic features, designated hollows, may have formed from the rapid loss of volatiles providing a means of transporting volatile species from the planetary interior into the surface environment where they can migrate to the poles (Blewett et al., 2011).

Future orbital and landed missions in the next decade will provide a wealth of data from the polar regions of the Moon and Mercury (see also Prologue II by Garvin, 2024, pp. xxvii–xxx). The international Bepi Colombo mission, jointly built and operated by the European Space Agency (ESA) and the Japanese Space Agency (JAXA), will orbit two spacecraft around Mercury increasing the global coverage compared to the MESSENGER mission by providing higher resolution images and altimetry in the southern hemisphere and will include thermal and neutron measurements (Benkhoff et al., 2021).

The international effort to explore the Moon will continue with a growing interest and participation of the private sector (Hayne et al., 2021b; Lemelin et al., 2021). A number of orbiters will possess instruments designed to further characterize volatiles and PSRs. In particular, the joint NASA and Korea Aerospace Research Institute (KARI) mission, the Korean Pathfinder Lunar Orbiter (KPLO), formally named Danuri in 2022, has recently begun imaging and mapping the interiors of PSRs at a high resolution ($1.7\,\mathrm{m}\ \mathrm{pix}^{-1}$) with the ShadowCam instrument (Robinson et al., 2022). Lunar Trailblazer will map thermal and spectral characteristics of the lunar surface from a ~100 km polar orbit with a spectral range and resolution designed to differentiate between different forms of hydration (OH, H_2O, and H_2O ice) and determine its abundance, distribution, and time variability (Ehlmann et al., 2021).

In addition to orbital missions, landed missions are planned that include the polar regions, such as NASA's Volatiles Investigating Polar Exploration Rover (VIPER), capable of roving into PSRs and analyzing subsurface sampled materials (Colaprete et al., 2020). The exploration of the lunar poles will include the first crewed mission to the surface since Apollo 17 with the Artemis III mission in the south polar region (Smith et al., 2020). Such in situ exploration will provide the opportunity to directly sample beneath the surface and quantify ice deposits with depth. The stratigraphy of regolith and ice will provide a history of ice delivery (Cannon & Britt, 2020; Cannon et al., 2020) and the inventory and distribution of water along with other volatile species will be indicative of their sources (Landis et al., 2022). In addition to the anticipated future missions on the horizon, laboratory experiments, returned sample analysis, and theoretical modeling will provide a more comprehensive understanding of ices on airless bodies in the inner solar system.

Acknowledgments

This work was funded in part by the NASA Lunar Reconnaissance Orbiter project and the Diviner science investigation, under contract with NASA, and the Korea Pathfinder Lunar Orbiter (KPLO) Participating Scientist Program (PSP) under award number 80NSSC21K0711.

References

Araki, H., Tazawa, S., Noda, H., Ishihara, Y., Goossens, S., Sasaki, S., et al. (2009). Lunar global shape and polar topography derived from Kaguya-LALT laser altimetry. *Science, 323*(5916), 897–900. https://doi.org/10.1126/science.1164146.

Arnold, J. R. (1979). Ice in the lunar polar regions. *Journal of Geophysical Research: Solid Earth, 84*, 5659–5668. https://doi.org/10.1029/JB084iB10p05659.

Bandfield, J. L., Hayne, P. O., Williams, J.-P., Greenhagen, B. T., & Paige, D. A. (2015). Lunar surface roughness derived from LRO diviner radiometer observations. *Icarus, 248*, 357–372. https://doi.org/10.1016/j.icarus.2014.11.009.

Bandfield, J. L., Poston, M. J., Klima, R. L., & Edwards, C. S. (2018). Widespread distribution of OH/H_2O on the lunar surface inferred from spectral data. *Nature Geoscience, 11*(3), 173–177. https://doi.org/10.1038/s41561-018-0065-0.

Benkhoff, J., Murakami, G., Baumjohann, W., Besse, S., Bunce, E., Casal, E. M., et al. (2021). BepiColombo—Mission overview and science goals. *Space Science Reviews, 217*, 90. https://doi.org/10.1007/s11214-021-00861-4.

Berezhnoy, A. A., Kozlova, E. A., Sinitsyn, M. P., Shangaraev, A. A., & Shevchenko, V. V. (2012). Origin and stability of lunar polar volatiles. *Advances in Space Research, 50*(12), 1638–1646. https://doi.org/10.1016/j.asr.2012.03.019.

Binder, A. B. (1998). Lunar prospector: Overview. *Science, 281*(5382), 1475–1476. https://doi.org/10.1126/science.281.5382.1475.

Blewett, D. T., Chabot, N. L., Denevi, B. W., Ernst, C. M., Head, J. W., Izenberg, N. R., et al. (2011). Hollows on Mercury: MESSENGER evidence for geologically recent volatile-related activity. *Science, 333*(6051), 1856–1859. https://doi.org/10.1126/science.1211681.

Boynton, W. V., Droege, G. F., Mitrofanov, I. G., McClanahan, T. P., Sanin, A. B., Litvak, M. L., et al. (2012). High spatial resolution studies of epithermal neutron emission from the lunar poles: Constraints on hydrogen mobility. *Journal of Geophysical Research: Planets, 117*, E00H33. https://doi.org/10.1029/2011JE003979.

Brown, H. M., Boyd, A. K., Denevi, B. W., Henriksen, M. R., Manheim, M. R., Robinson, M. S., Speyerer, E. J., & Wagner, R. V. (2022). Resource potential of lunar permanently shadowed regions. *Icarus, 377*, 114874. https://doi.org/10.1016/j.icarus.2021.114874.

Brown, H. M., Robinson, M. S., & Boyd, A. K. (2020). Morphologic landforms in Shoemaker and Faustini lunar permanently shadowed craters. In *51st Lunar and planetary science conference, abstract 1765*.

Bussey, B., Clarke, S. W., Jenkins, J., & Bailey, B. E. (2019). NASA's lunar discovery and exploration program. *AGUFM, 2019*, PA54B-11.

Bussey, D. B. J., Lucey, P. G., Steutel, D., Robinson, M. S., Spudis, P. D., & Edwards, K. D. (2003). Permanent shadow in simple craters near the lunar poles. *Geophysical Research Letters, 30*(6), 1278. https://doi.org/10.1029/2002GL016180.

Bussey, D. B. J., Spudis, P. D., & Robinson, M. S. (1999). Illumination conditions at the lunar south pole. *Geophysical Research Letters, 26*(9), 1187–1190. https://doi.org/10.1029/1999GL900213.

Butler, B. J. (1997). The migration of volatiles on the surfaces of Mercury and the Moon. *Journal of Geophysical Research: Planets, 98*, 19283–19291. https://doi.org/10.1029/97JE01347.

Butler, B. J., Muhleman, D. O., & Slade, M. A. (1993). Mercury—Full-disk radar images and the detection and stability of ice at the North-Pole. *Journal of Geophysical Research: Planets, 98*, 15003–15023. https://doi.org/10.1029/93JE01581.

Campbell, D. B., Chandler, J. F., Ostro, S. J., Pettengill, G. H., & Shapiro, I. I. (1978). Galilean satellites—1976 radar results. *Icarus, 34*, 254–267. https://doi.org/10.1016/0019-1035(78)90166-5.

Cannon, K. M., & Britt, D. T. (2020). A geologic model for lunar ice deposits at mining scales. *Icarus, 347*(113), 778. https://doi.org/10.1016/j.icarus.2020.113778.

Cannon, K. M., Deutsch, A. N., Head, J. W., & Britt, D. T. (2020). Stratigraphy of ice and ejecta deposits at the lunar poles. *Geophysical Research Letters*, *46*. https://doi.org/10.1029/2020GL088920. e2020GL088920.

Cavanaugh, J. F., Smith, J. C., Sun, X., Bartels, A. E., Ramos-Izquierdo, L., Krebs, D. J., et al. (2007). The Mercury Laser Altimeter instrument for the MESSENGER mission. *Space Science Reviews*, *131*, 451–479. https://doi.org/10.1007/s11214-007-9273-4.

Chabot, N. L., Ernst, C. M., Denevi, B. W., Nair, H., Deutsch, A. N., Blewett, D. T., et al. (2014). Images of surface volatiles in Mercury's polar craters acquired by the MESSENGER spacecraft. *Geology*, *42*, 1051–1054. https://doi.org/10.1130/G35916.1.

Chabot, N. L., Ernst, C. M., Paige, D. A., Nair, H., Denevi, B. W., Blewett, D. T., et al. (2016). Imaging Mercury's polar deposits during MESSENGER's low-altitude campaign. *Geophysical Research Letters*, *43*, 9461–9468. https://doi.org/10.1002/2016GL070403.

Chabot, N. L., Lawrence, D. J., Neumann, G. A., Feldman, W. C., & Paige, D. A. (2018a). Mercury's polar deposits. In S. C. Solomon, et al. (Eds.), *Mercury: The view after MESSENGER* (pp. 346–370). Cambridge University Press.

Chabot, N. L., Shread, E. E., & Harmon, J. K. (2018b). Investigating Mercury's south polar deposits: Arecibo radar observations and high-resolution determination of illumination conditions. *Journal of Geophysical Research: Planets*, *123*, 666–681. https://doi.org/10.1002/2017JE005500.

Chin, G., Brylow, S., Foote, M., Garvin, J., Kasper, J., Keller, J., et al. (2007). Lunar Reconnaissance Orbiter overview: The instrument suite and mission. *Space Science Reviews*, *129*(4), 391–419. https://doi.org/10.1007/s11214-007-9153-y.

Cisneros, E., Awumah, A., Brown, H. M., Martin, A. C., Paris, K. N., Povilaitis, R. Z., et al. (2017). Lunar Reconnaissance Orbiter Camera permanently shadowed region imaging—Atlas and controlled mosaics. In *48th Lunar and planetary science conference, abstract 2469*.

Clark, R. N. (2009). Detection of adsorbed water and hydroxyl on the moon. *Science*, *326*(5952), 562–564. https://doi.org/10.1126/science.1178105.

Colaprete, A., Elphic, R. C., Heldmann, J., & Ennica, K. (2012). An overview of the lunar crater observation and sensing satellite (LCROSS). *Space Science Reviews*, *167*, 3–22. https://doi.org/10.1007/s11214-012-9880-6.

Colaprete, A., Elphic, R. C., Shirley, M., Ennico-Smith, K., Heldmann, J., & Lim, D. S. S. (2020). The volatiles investigating polar exploration rover (VIPER) mission: Measurement goals and traverse planning. In *51st Lunar and planetary science conference, abstract 2326*.

Colaprete, A., Schultz, P., Heldmann, J., Wooden, D., Shirley, M., Ennico, K., et al. (2010). Detection of water in the LCROSS ejecta plume. *Science*, *330*(6003), 463–468. https://doi.org/10.1126/science.1186986.

Costello, E. S., Ghent, R. R., Hirabayashi, M., & Lucey, P. G. (2020). Impact gardening as a constraint of the age, source, and evolution of ice on mercury and the moon. *Journal of Geophysical Research: Planets*, *125*. https://doi.org/10.1029/2019JE006172. e2019JE006172.

Crider, D., & Killen, R. M. (2005). Burial rate of Mercury's polar volatile deposits. *Geophysical Research Letters*, *32*, L12201. https://doi.org/10.1029/2005GL022689.

Delitsky, M. L., Paige, D. A., Siegler, M. A., Harju, E. R., Schriver, D., Johnson, R. E., & Travnicek, P. (2017). Ices on Mercury: Chemistry of volatiles in permanently cold areas of Mercury's north polar region. *Icarus*, *281*, 19–31. https://doi.org/10.1016/j.icarus.2016.08.006.

Deutsch, A. N., Head, J. W., & Neumann, G. A. (2019). Age constraints of Mercury's polar deposits suggest recent delivery of ice. *Earth and Planetary Science Letters*, *520*, 26–33. https://doi.org/10.1016/j.epsl.2019.05.027.

Deutsch, A. N., Head, J. W., & Neumann, G. A. (2020a). Analyzing the ages of south polar craters on the Moon: Implications for the sources and evolution of surface water ice. *Icarus*, *336*, 113455. https://doi.org/10.1016/j.icarus.2019.113455.

Deutsch, A. N., Head, J. W., Neumann, G. A., Kreslavsky, M. A., & Barker, M. K. (2020b). Assessing the roughness properties of circumpolar Lunar craters: Implications for the timing of water-ice delivery to the Moon. *Geophysical Research Letters*, *47*. https://doi.org/10.1029/2020GL087782. e2020GL087782.

Deutsch, A. N., Heldmann, J. L., Colaprete, A., Cannon, K. M., & Elphic, R. C. (2021). Analyzing surface ruggedness inside and outside of ice stability zones at the lunar poles. *The Planetary Science Journal, 2*, 213. https://doi.org/10.3847/PSJ/ac24ff.

Deutsch, A. N., Neumann, G. A., & Head, J. W. (2017). New evidence for surface water ice in small-scale cold traps and in three large craters at the north polar region of Mercury from the Mercury Laser Altimeter. *Geophysical Research Letters, 44*, 9233–9241. https://doi.org/10.1002/2017GL074723.

Ehlmann, B. L., Klima, R. L., Bennett, C. L., Blaney, D., Bowles, N., Calcutt, S., et al. (2021). Lunar Trailblazer: A pioneering smallsat for lunar water and lunar geology. In *52nd Lunar and planetary science conference, abstract #1740*.

Eke, V. R., Teodoro, L. F. A., & Elphic, R. C. (2009). The spatial distribution of polar hydrogen deposits on the Moon. *Icarus, 200*, 12–18. https://doi.org/10.1016/j.icarus.2008.10.013.

Eke, V. R., Teodoro, L. F. A., Lawrence, D. J., Elphic, R. C., & Feldman, W. C. (2012). A quantitative comparison of lunar orbital neutron data. *Astrophysical Journal, 747*, 6. https://doi.org/10.1088/0004-637x/747/1/6.

Elphic, R. C., Eke, V. R., Teodoro, L. F. A., Lawrence, D. J., & Bussey, D. B. J. (2007). Models of the distribution and abundance of hydrogen at the lunar south pole. *Geophysical Research Letters, 34*, L13204. https://doi.org/10.1029/2007GL029954.

Ernst, C. M., Chabot, N. L., & Barnouin, O. S. (2018). Examining the potential contribution of the Hokusai impact to water ice on Mercury. *Journal of Geophysical Research: Planets, 123*, 2628–2646. https://doi.org/10.1029/2018JE005552.

Fa, W., & Eke, V. R. (2018). Unravelling the mystery of lunar anomalous craters using radar and infrared observations. *Journal of Geophysical Research: Planets, 123*, 2119–2137. https://doi.org/10.1029/2018JE005668.

Farrell, W. M., Hurley, D. M., Poston, M. J., Hayne, P. O., Szalay, J. R., & McLain, J. L. (2019). The young age of the LAMP-observed frost in lunar polar cold traps. *Geophysical Research Letters, 46*, 8680–8688. https://doi.org/10.1029/2019GL083158.

Feldman, W. C., Lawrence, D. J., Elphic, R. C., Barraclough, B. L., Maurice, S., Genetay, I., & Binder, A. B. (2000). Polar hydrogen deposits on the Moon. *Journal of Geophysical Research: Planets, 105*, 4175–4195. https://doi.org/10.1029/1999JE001129.

Feldman, W. C., Maurice, S., Binder, A. B., Barraclough, B. L., Elphic, R. C., & Lawrence, D. J. (1998). Fluxes of fast and epithermal neutrons from Lunar Prospector: Evidence for water ice at the lunar poles. *Science, 281*(5382), 1496–1500. https://doi.org/10.1126/science.281.5382.1496.

Feldman, W. C., Maurice, S., Lawrence, D. J., Little, R. C., Lawson, S. L., Gasnault, O., et al. (2001). Evidence for water ice near the lunar poles. *Journal of Geophysical Research: Planets, 106*, 23231–23251. https://doi.org/10.1029/2000JE001444.

Feldman, W. C., Reedy, R. C., & McKay, D. S. (1991). Lunar neutron leakage fluxes as a function of composition and hydrogen content. *Geophysical Research Letters, 18*, 2157–2160. https://doi.org/10.1029/91GL02618.

Fisher, E. A., Lucey, P. G., Lemelin, M., Greenhagen, B. T., Siegler, M. A., Mazarico, E., et al. (2017). Evidence for surface water ice in the lunar polar regions using reflectance measurements from the Lunar Orbiter Laser Altimeter and temperature measurements from the Diviner Lunar Radiometer Experiment. *Icarus, 292*, 74–85. https://doi.org/10.1016/j.icarus.2017.03.023.

Garrick-Bethell, I., Poppe, A. R., & Fatemi, S. (2019). The lunar paleo-magnetosphere: Implications for the accumulation of polar volatile deposits. *Geophysical Research Letters, 46*, 5778–5787. https://doi.org/10.1029/2019GL082548.

Garvin, J. B. (2024). *The ice frontier for science in the upcoming decades: A strategy for solar system exploration?* In R. J. Soare, J.-P. Williams, C. Ahrens, F. E. G. Butcher, & M. R. El-Maarry (Eds.), *Ices in the solar system, a volatile-driven journey from the inner solar system to its far reaches*. Elsevier Books.

Gladstone, G. R., Hurley, D. M., Retherford, K. D., Feldman, P. D., Pryor, W. R., Chaufray, J.-Y., et al. (2010). LRO-LAMP observations of the LCROSS impact plume. *Science, 330*(6003), 472–476. https://doi.org/10.1126/science.1186474.

Gladstone, G. R., Retherford, K. D., Egan, A. F., Kaufmann, D. E., Miles, P. F., Parker, J. W., et al. (2012). Far-ultraviolet reflectance properties of the Moon's permanently shadowed regions: Albedo of Moon's permanently shadowed regions. *Journal of Geophysical Research: Planets, 117*, E00H04. https://doi.org/10.1029/2011JE003913.

Gläser, P., Sanin, A., Williams, J.-P., Mitrofanov, I., & Oberst, J. (2021). Temperatures near the lunar poles and their correlation with hydrogen predicted by LEND. *Journal of Geophysical Research: Planets, 126*. https://doi.org/10.1029/2020JE006598. e2020JE006598.

Goldstein, R. M., & Green, R. R. (1980). Ganymede—Radar surface characteristics. *Science, 207*(4427), 179–180. https://doi.org/10.1126/science.207.4427.179.

Goldstein, R. M., & Morris, G. A. (1975). Ganymede: Observations by radar. *Science, 188*, 1211–1212. https://doi.org/10.1126/science.188.4194.1211.

Goldsten, J. O., Rhodes, E. A., Boynton, W. V., Feldman, W. C., Lawrence, D. J., Trombka, J. I., et al. (2007). The MESSENGER gamma-ray and neutron spectrometer. *Space Science Reviews, 131*, 339–391. https://doi.org/10.1007/s11214-007-9262-7.

Grumpe, A., Wöhler, C., Berezhnoy, A. A., & Shevchenko, V. V. (2019). Time-of-day-dependent behavior of surficial lunar hydroxyl/water: Observations and modeling. *Icarus, 321*, 486–507. https://doi.org/10.1016/j.icarus.2018.11.025.

Hapke, B. (1990). Coherent backscatter and the radar characteristics of outer planet satellites. *Icarus, 88*, 407–417. https://doi.org/10.1016/0019-1035(90)90091-M.

Hapke, B., & Blewitt, D. (1991). Coherent backscatter model for the unusual radar reflectivity of icy satellites. *Nature, 352*, 46–47. https://doi.org/10.1038/352046a0.

Harmon, J. K. (1997). Mercury radar studies and lunar comparisons. *Advances in Space Research, 19*, 1487–1496. https://doi.org/10.1016/S0273-1177(97)00347-5.

Harmon, J. K. (2007). Radar imaging of Mercury. *Space Science Reviews, 132*, 307–349. https://doi.org/10.1007/s11214-007-9234-y.

Harmon, J. K., Perillat, P. J., & Slade, M. A. (2001). High-resolution radar imaging of Mercury's north pole. *Icarus, 194*, 1–15. https://doi.org/10.1006/icar.2000.6544.

Harmon, J. K., & Slade, M. A. (1992). Radar mapping of Mercury—Full-disk images and polar anomalies. *Science, 258*(5082), 640–643. https://doi.org/10.1126/science.258.5082.640.

Harmon, J. K., Slade, M. A., & Rice, M. S. (2011). Radar imagery of Mercury's putative polar ice: 1999–2005 Arecibo results. *Icarus, 211*(1), 37–50. https://doi.org/10.1016/j.icarus.2010.08.007.

Harmon, J. K., Slade, M. A., Velez, R. A., Crespo, A., Dryer, M. J., & Johnson, J. M. (1994). Radar mapping of Mercury's polar anomalies. *Nature, 369*(6477), 213–215. https://doi.org/10.1038/369213a0.

Haruyama, J., Ohtake, M., Matsunaga, T., Morota, T., Honda, C., Yokota, Y., et al. (2008). Lack of exposed ice inside lunar south pole Shackleton crater. *Science, 322*(5903), 938–939. https://doi.org/10.1126/science.1164020.

Haruyama, J., Yamamoto, S., Yokota, Y., Ohtake, M., & Matsunaga, T. (2013). An explanation of bright areas inside Shackleton crater at the lunar south pole other than water-ice deposits. *Geophysical Research Letters, 40*, 3814–3818. https://doi.org/10.1002/grl.50753.

Hawkins, S. E., III, Boldt, J. D., Darlington, E. H., Espiritu, R., Gold, R. E., Gotwols, B., et al. (2007). The Mercury dual imaging system on the MESSENGER spacecraft. *Space Science Reviews, 131*, 247–338. https://doi.org/10.1007/s11214-007-9266-3.

Hayne, P. O., Aharonson, O., & Schörghofer, N. (2021a). Micro cold traps on the Moon. *Nature Astronomy, 5*(2), 169–175. https://doi.org/10.1038/s41550-020-1198-9.

Hayne, P. O., Greenhagen, B. T., Foote, M. C., Siegler, M. A., Vasavada, A. R., & Paige, D. A. (2010). Diviner lunar radiometer observations of the LCROSS impact. *Science, 330*(6003), 477–479. https://doi.org/10.1126/science.1197135.

Hayne, P. O., Hendrix, A., Sefton-Nash, E., Siegler, M. A., Lucey, P. G., Retherford, K. D., et al. (2015). Evidence for exposed water ice in the Moon's south polar regions from Lunar Reconnaissance Orbiter ultraviolet albedo and temperature measurements. *Icarus, 255*, 58–69. https://doi.org/10.1016/j.icarus.2015.03.032.

Hayne, P. O., Paige, D. A., Ingersoll, A. P., Aharonson, O., Byrne, S., & Cohen, B. (2021b). New approaches to lunar ice detection and mapping. Planetary science and astrobiology decadal survey 2023-2032 white paper. *Bulletin of the American Astronomical Society*, *53*(4), 251. https://doi.org/10.3847/25c2cfeb.bacb0f3b.

Head, J. W., & Wilson, L. (2017). Generation, ascent and eruption of magma on the Moon: New insights into source depths, magma supply, intrusions and effusive/explosive eruptions (Part 2: Predicted emplacement processes and observations). *Icarus*, *283*, 176–223. https://doi.org/10.1016/j.icarus.2016.05.031.

Head, J. W., Wilson, L., Deutsch, A. N., Rutherford, M. J., & Saal, A. E. (2020). Volcanically induced transient atmospheres on the moon: Assessment of duration, significance, and contributions to polar volatile traps. *Geophysical Research Letters*, *47*. https://doi.org/10.1029/2020GL089509. e2020GL089509.

Hendrix, A. R., Hurley, D. M., Farrell, W. M., Greenhagen, B. T., Hayne, P. O., Retherford, K. D., et al. (2019). Diurnally migrating lunar water: Evidence from ultraviolet data. *Geophysical Research Letters*, *46*, 2417–2424. https://doi.org/10.1029/2018GL081821.

Hendrix, A. R., Retherford, K. D., Randall Gladstone, G., Hurley, D. M., Feldman, P. D., Egan, A. F., et al. (2012). The lunar far-UV albedo: Indicator of hydration and weathering. *Journal of Geophysical Research: Planets*, *117*(E12). https://doi.org/10.1029/2012JE004252.

Herring, J. R., & Licht, A. L. (1959). Effect of the solar wind on the lunar atmosphere. *Science*, *130*(3370), 266. https://doi.org/10.1126/science.130.3370.266.

Hiesinger, H., Head, J. W., III, Wolf, U., Jaumann, R., & Neukum, G. (2011). Ages and stratigraphy of lunar mare basalts: A synthesis. *GSA Special Papers*, *477*, 1–51. https://doi.org/10.1130/2011.2477(01).

Hodges, R. R. (2002). Reanalysis of Lunar prospector neutron spectrometer observations over the lunar poles. *Journal of Geophysical Research: Planets*, *107*(E12), 5125. https://doi.org/10.1029/2000JE001483.

Honniball, C. I., Lucey, P. G., Ferrari-Wong, C. M., Flom, A., Li, S., Kaluna, H. M., & Takir, D. (2020). Telescopic observations of lunar hydration: Variations and abundance. *Journal of Geophysical Research: Planets*, *125*(9). https://doi.org/10.1029/2020JE006484. e2020JE006484.

Honniball, C. I., Lucey, P. G., Li, S., Shenoy, S., Orlando, T. M., Hibbitts, C. A., et al. (2021). Molecular water detected on the sunlit Moon by SOFIA. *Nature Astronomy*, *5*(2), 121–127. https://doi.org/10.1038/s41550-020-01222-x.

Hurley, D. M., Gladstone, G. R., Stern, S. A., Retherford, K. D., Feldman, P. D., Pryor, W., et al. (2012a). Modeling of the vapor release from the LCROSS impact: 2. Observations from LAMP. *Journal of Geophysical Research: Planets*, *117*(E12), E00H07. https://doi.org/10.1029/2011JE003841.

Hurley, D. M., Lawrence, D. J., Bussey, D. B. J., Vondrak, R. R., Elphic, R. C., & Gladstone, G. R. (2012b). Two-dimensional distribution of volatiles in the lunar regolith from space weathering simulations. *Geophysical Research Letters*, *39*(9), L09203. https://doi.org/10.1029/2012GL051105.

Ingersoll, A. P., Svitek, T., & Murray, B. C. (1992). Stability of polar frosts in spherical bowl-shaped craters on the Moon, Mercury, and Mars. *Icarus*, *100*, 40–47. https://doi.org/10.1016/0019-1035(92)90016-Z.

Jones, B. M., Sarantos, M., & Orlando, T. M. (2020). A new in situ quasi-continuous solar-wind source of molecular water on Mercury. *The Astrophysical Journal Letters*, *891*, L43. https://doi.org/10.3847/2041-8213/ab6bda.

Kloos, J. L., Moores, J. E., Sangha, J., Nguyen, T. G., & Schorghofer, N. (2019). The temporal and geographic extent of seasonal cold trapping on the moon. *Journal of Geophysical Research: Planets*, *124*(7), 1935–1944. https://doi.org/10.1029/2019JE006003.

Koeber, S. D., Robinson, M. S., & Speyerer, E. J. (2014). LROC observations of permanently shadowed regions on the Moon. In *45th Lunar and planetary science conference, abstract 2811*.

Landis, M. E., Hayne, P. O., Williams, J.-P., Greenhagen, B. T., & Paige, D. A. (2022). Spatial distribution and thermal diversity of surface volatile cold traps at the lunar poles. *The Planetary Science Journal*, *3*, 39. https://doi.org/10.3847/PSJ/ac4585.

Lanzerotti, L. J., Brown, W. L., & Johnson, R. E. (1981). Ice in the polar-regions of the Moon. *Journal of Geophysical Research: Planets*, *86*, 3949–3950. https://doi.org/10.1029/JB086iB05p03949.

Lawrence, D. J. (2017). A tale of two poles: Toward understanding the presence, distribution, and origin of volatiles at the polar regions of the Moon and Mercury. *Journal of Geophysical Research: Planets, 122*, 21–52. https://doi.org/10.1002/2016JE005167.

Lawrence, D. J., Eke, V. R., Elphic, R. C., Feldman, W. C., Funsten, H. O., Prettyman, T. H., & Teodoro, L. F. A. (2011). Technical comment on "hydrogen mapping of the lunar south pole using the LRO neutron detector experiment LEND". *Science, 334*(6059), 1058. https://doi.org/10.1126/science.1203341.

Lawrence, D. J., Feldman, W. C., Elphic, R. C., Hagerty, J. J., Maurice, S., McKinney, G. W., & Prettyman, T. H. (2006). Improved modeling of Lunar prospector neutron spectrometer data: Implications for hydrogen deposits at the lunar poles. *Journal of Geophysical Research: Planets, 111*, E08001. https://doi.org/10.1029/2005JE002637.

Lawrence, D. J., Feldman, W. C., Goldsten, J. O., Maurice, S., Peplowski, P. N., Anderson, B. J., et al. (2013). Evidence for water ice near Mercury's north pole from MESSENGER Neutron Spectrometer measurements. *Science, 339*(6117), 292–296. https://doi.org/10.1126/science.1229953.

Lemelin, M., Li, S., Mazarico, E., Siegler, M. A., Kring, D. A., & Paige, D. A. (2021). Framework for coordinated efforts in the exploration of volatiles in the south polar region of the Moon. *The Planetary Science Journal, 2*, 103. https://doi.org/10.3847/PSJ/abf3c5.

Li, S., Lucey, P. G., Milliken, R. E., Hayne, P. O., Fisher, E., Williams, J.-P., et al. (2018). Direct evidence of surface exposed water ice in the lunar polar regions. *Proceedings of the National Academy of Sciences, 115*(36), 8907–8912. https://doi.org/10.1073/pnas.1802345115.

Li, S., & Milliken, R. E. (2017). Water on the surface of the moon as seen by the moon mineralogy mapper: Distribution, abundance, and origins. *Science Advances, 3*(9), e1701471. https://doi.org/10.1126/sciadv.1701471.

Lingenfelter, R. E., Hess, W. N., & Canfield, E. H. (1961). The lunar neutron flux. *Journal of Geophysical Research: Planets, 66*, 2665–2671. https://doi.org/10.1029/JZ066i009p02665.

Litvak, M. L., Mitrofanov, I. G., Sanin, A., Malakhov, A., Boynton, W. V., Chin, G., & Zuber, M. T. (2012). Global maps of lunar neutron fluxes from the lend instrument. *Journal of Geophysical Research: Planets, 117*. https://doi.org/10.1029/2011je003949. E00H22.

Livengood, T. A., Mitrofanov, I. G., Chin, G., Boynton, W. V., Bodnarik, J. G., Evans, L. G., et al. (2018). Background and lunar neutron populations detected by LEND and average concentration of near-surface hydrogen near the Moon's poles. *Planetary and Space Science, 162*, 89–104. https://doi.org/10.1016/j.pss.2017.12.004.

Lucey, P. G., Neumann, G. A., Riner, M. A., Mazarico, E., Smith, D. E., Zuber, M. T., et al. (2014). The global albedo of the Moon at 1064 nm from LOLA: The global albedo of the Moon from LOLA. *Journal of Geophysical Research: Planets, 119*, 1665–1679. https://doi.org/10.1002/2013JE004592.

Lucey, P. G., Petro, N., Hurley, D. M., Farrell, W. M., Prem, P., Costello, E. S., et al. (2022). Volatile interations with the lunar surface. *Geochemistry*. https://doi.org/10.1016/j.chemer.2021.125858.

Luchsinger, K. M., Chanover, N. J., & Strycker, P. D. (2021). Water within a permanently shadowed lunar crater: Further LCROSS modeling and analysis. *Icarus, 354*, 114089. https://doi.org/10.1016/j.icarus.2020.114089.

Mandt, K. E., Mousis, O., Hurley, D., Bouquet, A., Retherford, K. D., Magaña, L. O., & Luspay-Kuti, A. (2022). Exogenic origin for the volatiles sampled by the Lunar CRater Observation and Sensing Satellite impact. *Nature Communications, 13*, 642. https://doi.org/10.1038/s41467-022-28289-6.

Margot, J. L., Campbell, D. B., Jurgens, R. F., & Slade, M. A. (1999). Topography of the lunar poles from radar interferometry: A survey of cold trap locations. *Science, 284*(5420), 1658–1660. https://doi.org/10.1126/science.284.5420.1658.

Marshall, W., Shirley, M., Moratto, Z., Colaprete, A., Neumann, G., Smith, D., et al. (2012). Locating the LCROSS impact craters. *Space Science Reviews, 167*, 71–92. https://doi.org/10.1007/s11214-011-9765-0.

Mazarico, E., Neumann, G. A., Smith, D. E., Zuber, M. T., & Torrence, M. H. (2011). Illumination conditions of the lunar polar regions using LOLA topography. *Icarus, 211*(2), 1066–1081. https://doi.org/10.1016/j.icarus.2010.10.030.

McCord, T. B., Taylor, L. A., Combe, J. P., Kramer, G., Pieters, C. M., Sunshine, J. M., & Clark, R. N. (2011). Sources and physical processes responsible for OH/H_2O in the lunar soil as revealed by the Moon Mineralogy Mapper (M^3). *Journal of Geophysical Research: Planets*, *116*(E6), E00G05. https://doi.org/10.1029/2010JE003711.

McGovern, J. A., Bussey, D. B., Greenhagen, B. T., Paige, D. A., Cahill, J. T. S., & Spudis, P. D. (2013). Mapping and characterization of non-polar permanent shadows on the lunar surface. *Icarus*, *223*(1), 566–581. https://doi.org/10.1016/j.icarus.2012.10.018.

Metzger, A. E., & Drake, D. M. (1990). Identification of lunar rock types and search for polar ice by gamma ray spectroscopy. *Journal of Geophysical Research: Planets*, *95*, 449–460. https://doi.org/10.1029/JB095iB01p00449.

Miller, R. S., Nerurkar, G., & Lawrence, D. J. (2012). Enhanced hydrogen at the lunar poles: New insights from the detection of epithermal and fast neutron signatures. *Journal of Geophysical Research: Planets*, *117*, E11007. https://doi.org/10.1029/2012JE004112.

Mitrofanov, I. G., Boynton, W. V., Litvak, M. L., Sanin, A. B., & Starr, R. D. (2011). Response to comment on "hydrogen mapping of the lunar south pole using the LRO neutron detector experiment lend". *Science*, *334*(6059), 1058. https://doi.org/10.1126/science.1203483.

Mitrofanov, I., Litvak, M., Sanin, A., Malakhov, A., Golovin, D., Boynton, W., et al. (2012). Testing polar spots of water-rich permafrost on the Moon: LEND observations onboard LRO. *Journal of Geophysical Research: Planets*, *117*, E00H27. https://doi.org/10.1029/2011JE003956.

Mitrofanov, I., Sanin, A. B., Boynton, W. V., Chin, G., Garvin, J. B., Golovin, D., et al. (2010). Hydrogen mapping of the Lunar south pole using the LRO neutron detector experiment LEND. *Science*, *330*(6003), 483–486. https://doi.org/10.1126/science.1185696.

Moon, S., Paige, D. A., Siegler, M. A., & Russell, P. S. (2021). Geomorphic evidence for the presence of ice deposits in the permanently shadowed regions of Scott-E crater on the Moon. *Geophysical Research Letters*, *48*. https://doi.org/10.1029/2020GL090780. e2020GL090780.

Morgan, T. H., & Shemansky, D. E. (1991). Limits to the lunar atmosphere. *Journal of Geophysical Research: Space Physics*, *96*(A2), 1351–1367. https://doi.org/10.1029/90JA02127.

Moses, J. I., Rawlins, K., Zahnle, K., & Dones, L. (1999). External sources of water for Mercury's putative ice deposits. *Icarus*, *137*(2), 197–221. https://doi.org/10.1006/icar.1998.6036.

Muhleman, D. O., Butler, B. J., Grossman, A. W., & Slade, M. A. (1991). Radar images of Mars. *Science*, *253*, 1508–1513. https://doi.org/10.1126/science.253.5027.1508.

Muhleman, D. O., Grossman, A. W., Butler, B. J., & Slade, M. A. (1990). Radar reflectivity of titan. *Science*, *248*, 975–980. https://doi.org/10.1126/science.248.4958.975.

Needham, D. H., & Kring, D. A. (2017). Lunar volcanism produced a transient atmosphere around the ancient moon. *Earth and Planetary Science Letters*, *15*(478), 175–178. https://doi.org/10.1016/j.epsl.2017.09.002.

Neumann, G. A., Cavanaugh, J. F., Sun, X., Mazarico, E. M., Smith, D. E., Zuber, M. T., et al. (2013). Bright and dark polar deposits on Mercury: Evidence for surface volatiles. *Science*, *339*(6117), 296–300. https://doi.org/10.1126/science.1229764.

Nittler, L. R., Starr, R. D., Weider, S. Z., McCoy, T. J., Boynton, W. V., Ebel, D. S., et al. (2011). The major-element composition of Mercury's surface from MESSENGER X-ray spectrometry. *Science*, *333*(6051), 1847–1850. https://doi.org/10.1126/science.1211567.

Nozette, S., Lichtenberg, C. L., Spudis, P., Bonner, R., Ort, W., Malaret, E., Robinson, M., & Shoemaker, E. M. (1996). The Clementine bistatic radar experiment. *Science*, *274*(5292), 1495–1498. https://doi.org/10.1126/science.274.5292.1495.

Nozette, S., Rustan, P., Pleasance, L. P., Kordas, J. F., Lewis, I. T., Park, H. S., et al. (1994). The Clementine Mission to the Moon—Scientific overview. *Science*, *266*(5192), 1835–1839. https://doi.org/10.1126/science.266.5192.1835.

Nozette, S., Spudis, P., Bussey, B., Jensen, R., Raney, K., Winters, H., et al. (2010). The lunar reconnaissance orbiter miniature radio frequency (Mini-RF) technology demonstration. *Space Science Reviews, 150*, 285–302. https://doi.org/10.1007/s11214-009-9607-5.

Ong, L., Asphaug, E. I., Korycansky, D., & Coker, R. F. (2010). Volatile retention from cometary impacts on the Moon. *Icarus, 207*(2), 578–589. https://doi.org/10.1016/j.icarus.2009.12.012.

Öpik, E. J., & Singer, S. F. (1960). Escape of gases from the moon. *Journal of Geophysical Research, 65*, 3065–3070. https://doi.org/10.1029/JZ065i010p03065.

Ostro, S. J., & Shoemaker, E. M. (1990). The extraordinary radar echoes from Europa, Ganymede, and Callisto: A geological perspective. *Icarus, 85*, 335–345. https://doi.org/10.1016/0019-1035(90)90121-o.

Paige, D. A., Foote, M. C., Greenhagen, B. T., Schofield, J. T., Calcutt, S., Vasavada, A. R., et al. (2010a). The lunar reconnaissance orbiter diviner lunar radiometer experiment. *Space Science Reviews, 150*(1–4), 125–160. https://doi.org/10.1007/s11214-009-9529-2.

Paige, D. A., Siegler, M. A., Harmon, J. K., Neumann, G. A., Mazarico, E. M., Smith, D. E., et al. (2013). Thermal stability of volatiles in the North Polar Region of Mercury. *Science, 339*(6171), 300–303. https://doi.org/10.1126/science.1231106.

Paige, D. A., Siegler, M. A., Zhang, J. A., Hayne, P. O., Foote, E. J., Bennett, K. A., et al. (2010b). Diviner Lunar radiometer observations of cold traps in the Moon's south polar region. *Science, 330*, 479–482. https://doi.org/10.1126/science.1187726.

Paige, D. A., Wood, S. E., & Vasavada, A. R. (1992). The thermal-stability of water ice at the poles of mercury. *Science, 258*, 643–646. https://doi.org/10.1126/science.258.5082.643.

Patterson, G. W., Stickle, A. M., Turner, F. S., Jensen, J. R., Bussey, D. B. J., Spudis, P., et al. (2017). Bistatic radar observations of the Moon using Mini-RF on LRO and the Arecibo observatory. *Icarus, 283*, 2–19. https://doi.org/10.1016/j.icarus.2016.05.017.

Peplowski, P. N., Evans, L. G., Hauck, S. A., McCoy, T. J., Boynton, W. V., Gillis-Davis, J. J., et al. (2011). Radioactive elements on Mercury's surface from MESSENGER: Implications for the planet's formation and evolution. *Science, 333*(6051), 1850–1852. https://doi.org/10.1126/science.1211576.

Pettit, E., & Nicholson, S. B. (1930). Lunar radiation and temperatures. *The Astrophysical Journal, 71*(71), 102–135. https://doi.org/10.1086/143236.

Pieters, C. M., Goswami, J. N., Clark, R. N., Annadurai, M., Boardman, J., Buratti, B., et al. (2009). Character and spatial distribution of OH/H_2O on the surface of the Moon seen by M3 on Chandrayaan-1. *Science, 326*(5952), 568–572. https://doi.org/10.1126/science.1178658.

Ping, J. S., Huang, Q., Yan, J. G., Cao, J. F., Tang, G. S., & Shu, R. (2009). Lunar topographic model CLTM-s01 from Chang'E-1 laser altimeter. *Science in China Series G: Physics Mechanics and Astronomy, 52*(7), 1105–1114. https://doi.org/10.1007/s11433-009-0144-8.

Prem, P., Artemieva, N. A., Goldstein, D. B., Varghese, P. L., & Trafton, L. M. (2015). Transport of water in a transient impact-generated lunar atmosphere. *Icarus, 255*, 148–158. https://doi.org/10.1016/j.icarus.2014.10.017.

Prem, P., Goldstein, D. B., Varghese, P. L., & Trafton, L. M. (2018). The influence of surface roughness on volatile transport on the moon. *Icarus, 299*, 31–45. https://doi.org/10.1016/j.icarus.2017.07.010.

Prem, P., Goldstein, D. B., Varghese, P. L., & Trafton, L. M. (2019). Coupled DSMC-Monte Carlo radiative transfer modeling of gas dynamics in a transient impact-generated lunar atmosphere. *Icarus, 326*, 88–104. https://doi.org/10.1016/j.icarus.2019.02.036.

Qiao, L., Ling, Z., Head, J. W., Ivanov, M. A., & Liu, B. (2019). Analyses of Lunar Orbiter Laser Altimeter 1,064-nm albedo in permanently shadowed regions of polar crater flat floors: Implications for surface water ice occurrence and future in situ exploration. *Earth and Space Science, 6*, 467–488. https://doi.org/10.1029/2019EA000567.

Raymond, S. N. (2024). *The solar system's ices and their origin*. In R. J. Soare, J.-P. Williams, C. Ahrens, F. E. G. Butcher, & M. R. El-Maarry (Eds.), *Ices in the solar system, a volatile-driven journey from the inner solar system to its far reaches*. Elsevier Books.

Rignot, E. (1995). Backscatter model for the unusual radar properties of the Greenland Ice Sheet. *Journal of Geophysical Research: Atmospheres, 100*, 9389–9400. https://doi.org/10.1029/95JE00485.

Rivera-Valentín, E. G., Meyer, H. M., Taylor, P. A., Mazarico, E., Bhiravarasu, S. S., Virkki, A. K., et al. (2022). Arecibo S-band radar characterization of local-scale heterogeneities within Mercury's north polar deposits. *The Planetary Science Journal, 3*, 62. https://doi.org/10.3847/PSJ/ac54a0.

Robinson, M. S., Brylow, S. M., Tschimmel, M., Humm, D., Lawrence, S. J., Thomas, P. C., et al. (2010). Lunar reconnaissance orbiter camera (LROC) instrument overview. *Space Science Reviews, 150*, 81–124. https://doi.org/10.1007/s11214-010-9634-2.

Robinson, M. S., Mahanti, P., Brylow, S. M., Bussey, D. B. J., Carter, L. M., Clark, M. J., et al. (2022). ShadowCam: Seeing in the Moon's shadows. In *53rd Lunar and planetary science conference, abstract #1659*.

Rubanenko, L., & Aharonson, O. (2017). Stability of ice on the Moon with rough topography. *Icarus, 296*, 99–109. https://doi.org/10.1016/j.icarus.2017.05.028.

Rubanenko, L., Mazarico, E., Neumann, G. A., & Paige, D. A. (2018). Ice in micro cold traps on Mercury: Implications for age and origin. *Journal of Geophysical Research: Planets, 123*(8), 2178–2191. https://doi.org/10.1029/2018JE005644.

Rubanenko, L., Venkatraman, J., & Paige, D. A. (2019). Thick ice deposits in shallow simple craters on the Moon and Mercury. *Nature Geoscience, 12*, 597–601. https://doi.org/10.1038/s41561-019-0405-8.

Salvail, J. R., & Fanale, F. P. (1994). Near-surface ice on Mercury and the Moon: A topographic thermal model. *Icarus, 111*, 441–455. https://doi.org/10.1006/icar.1994.1155.

Sanin, A. B., Mitrofanov, I. G., Litvak, M. L., Malakhov, A., Boynton, W. V., Chin, G., et al. (2012). Testing lunar permanently shadowed regions for water ice: LEND results from LRO. *Journal of Geophysical Research: Planets, 117*, E00H26. https://doi.org/10.1029/2011JE003971.

Schorghofer, N. (2015). Two-dimensional description of surface-bounded exospheres with application to the migration of water molecules on the Moon. *Physical Review E, 91*(5), 052154. https://doi.org/10.1103/PhysRevE.91.052154.

Schorghofer, N., & Aharonson, O. (2014). The lunar thermal ice pump. *The Astrophysical Journal, 788*(169), 7. https://doi.org/10.1088/0004-637X/788/2/.

Schorghofer, N., Benna, M., Berezhnoy, A. A., Greenhagen, B., Jones, B. M., Li, S., et al. (2021a). Water group exospheres and surface interactions on the Moon, Mercury, and Ceres. *Space Science Reviews, 217*, 74. https://doi.org/10.1007/s11214-021-00846-3.

Schorghofer, N., Lucey, P., & Williams, J.-P. (2017). Theoretical time variability of mobile water on the Moon and its geographic pattern. *Icarus, 298*, 111–116. https://doi.org/10.1016/j.icarus.2017.01.029.

Schorghofer, N., & Taylor, G. J. (2007). Subsurface migration of H2O at lunar cold traps. *Journal of Geophysical Research: Planets, 112*, E02010. https://doi.org/10.1029/2006JE002779.

Schorghofer, N., & Williams, J.-P. (2020). Mapping of ice storage processes on the Moon with time-dependent temperatures. *The Planetary Science Journal, 1*, 54. https://doi.org/10.3847/PSJ/abb6ff.

Schorghofer, N., Williams, J.-P., Martinez-Camacho, J., Paige, D. A., & Siegler, M. A. (2021b). Carbon dioxide cold traps on the Moon. *Geophysical Research Letters, 48*. https://doi.org/10.1029/2021GL095533. e2021GL095533.

Schultz, P. H., Hermalyn, B., Colaprete, A., Ennico, K., Shirley, M., & Marshall, W. S. (2010). The LCROSS cratering experiment. *Science, 330*(6003), 468–472. https://doi.org/10.1126/science.1187454.

Sefton-Nash, E., Williams, J.-P., Greenhagen, B., Warren, T., Bandfield, J., Aye, K.-M., et al. (2019). Evidence for ultra-cold traps and surface water ice in the lunar south polar crater Amundsen. *Icarus, 332*, 1–13. https://doi.org/10.1016/j.icarus.2019.06.002.

Shoemaker, E. M., Shoemaker, C. S., & Wolfe, R. F. (1990). Asteroid and comet flux in the neighborhood of the Earth. In *Global catastrophes in Earth history; An interdisciplinary conference on impacts, volcanism, and mass mortality, GSA special papers, vol. 247* (pp. 155–170). https://doi.org/10.1130/SPE247-p155.

Siegler, M. A., Bills, B. G., & Paige, D. A. (2011). Effects of orbital evolution on lunar ice stability. *Journal of Geophysical Research: Planets, 116*(E3), E03010. https://doi.org/10.1029/2010JE003652.

Siegler, M. A., Miller, R. S., Keane, J. T., Laneuville, M., Paige, D. A., Matsuyama, I., et al. (2016). Lunar true polar wander inferred from polar hydrogen. *Nature, 531*(7595), 480–484. https://doi.org/10.1038/nature17166.

Siegler, M. A., Paige, D., Williams, J.-P., & Bills, B. (2015). Evolution of lunar polar ice stability. *Icarus, 255*, 78–87. https://doi.org/10.1016/j.icarus.2014.09.037.

Simpson, R. A., & Tyler, G. L. (1999). Reanalysis of Clementine bistatic radar data from the lunar south pole. *Journal of Geophysical Research: Planets, 104*, 3845–3862. https://doi.org/10.1029/1998JE900038.

Slade, M. A., Butler, B. J., & Muhleman, D. O. (1992). Mercury radar imaging—Evidence for polar ice. *Science, 258*(5082), 635–640. https://doi.org/10.1126/science.258.5082.635.

Smith, M., Craig, D., Herrmann, N., Mahoney, E., Krezel, J., McIntyre, N., & Goodliff, K. (2020). The Artemis program: An overview of NASA's activities to return humans to the Moon. *IEEE Aerospace Conference, 2020*, 1–10. https://doi.org/10.1109/AERO47225.2020.9172323.

Smith, D. E., Zuber, M. T., Jackson, G. B., Cavanaugh, J. F., Neumann, G. A., Riris, H., et al. (2010). The Lunar Orbiter Laser Altimeter investigation on the lunar reconnaissance orbiter mission. *Space Science Reviews, 150*, 209–241. https://doi.org/10.1007/s11214-0 09-9512-y.

Smith, D. E., Zuber, M. T., Neumann, G. A., Mazarico, E., Lemoine, F. G., Head, J. W., et al. (2017). Summary of the results from the Lunar Orbiter Laser Altimeter after seven years in Lunar orbit. *Icarus, 283*, 70–91. https://doi.org/10.1016/j.icarus.2016.06.006.

Solomon, S. C., & Byrne, P. K. (2019). The exploration of Mercury by spacecraft. *Elements, 15*, 15–20. https://doi.org/10.2138/gselements.15.1.15.

Solomon, S. C., McNutt, R. L., Gold, R. E., Acuña, M. H., Baker, D. N., Boynton, W. V., et al. (2001). The MESSENGER mission to Mercury: Scientific objectives and implementation. *Planetary and Space Science, 49*, 1445–1465. https://doi.org/10.1016/S0032-0633(01)00085-X.

Solomon, S. C., McNutt, R. L., Gold, R. E., & Domingue, D. L. (2007). MESSENGER—Mission overview. *Space Science Reviews, 131*(1–4), 3–39. https://doi.org/10.1007/s11214-007-9247-6.

Speyerer, E. J., & Robinson, M. S. (2013). Persistently illuminated regions at the lunar poles: Ideal sites for future exploration. *Icarus, 222*, 122–136. https://doi.org/10.1016/j.icarus.2012.10.010.

Speyerer, E. J., Robinson, M. S., Boyd, A., Wagner, R. V., & Henriksen, M. R. (2020). Exploration of the lunar south pole with LROC data products. In *Lunar surface science workshop, abstract #5132*.

Spudis, P. D., Bussey, D. B. J., Baloga, S. M., Butler, B. J., Carl, D., Carter, L. M., et al. (2010). Initial results for the north pole of the Moon from mini-SAR, Chandrayaan-1 mission. *Geophysical Research Letters, 37*, L06204. https://doi.org/10.1029/2009GL042259.

Spudis, P. D., Bussey, D. B. J., Baloga, S. M., Cahill, J. T. S., Glaze, L. S., Patterson, G. W., et al. (2013). Evidence for water ice on the Moon: Results for anomalous polar craters from the LRO mini-RF imaging radar. *Journal of Geophysical Research: Planets, 118*, 2016–2029. https://doi.org/10.1002/jgre.20156.

Spudis, P. D., & Guest, J. E. (1988). Stratigraphy and geologic history of Mercury. In F. Vilas, C. R. Chapman, & M. S. Matthews (Eds.), *Mercury*. Tucson: University of Arizona Press.

Spudis, P. D., Nozette, S., Bussey, B., Raney, K., Winters, H., Lichtenberg, C., Marinelli, W., Crusan, J., & Gates, M. M. (2009). Mini-SAR: An imaging radar experiment for the Chandrayaan-1 mission to the Moon. *Current Science, 96*(4), 533–539. https://www.jstor.org/stable/24105465.

Stacy, N. J. S., Campbell, D. B., & Ford, P. G. (1997). Arecibo radar mapping of the lunar poles: A search for ice deposits. *Science, 276*(5318), 1527–1530. https://doi.org/10.1126/science.276.5318.1527.

Stern, S. A. (1999). The lunar atmosphere: History, status, current problems, and context. *Reviews of Geophysics*, *37*(4), 453–491. https://doi.org/10.1029/1999RG900005.

Stewart, B. D., Pierazzo, E., Goldstein, D. B., Varghese, P. L., & Trafton, L. M. (2011). Simulations of a comet impact on the moon and associated ice deposition in polar cold traps. *Icarus*, *215*, 1–6. https://doi.org/10.1016/j.icarus.2011.03.014.

Strom, R. G. (1979). Mercury: A post-mariner 10 assessment. *Space Science Reviews*, *24*, 3–70. https://doi.org/10.1007/BF00221842.

Sunshine, J. M., Farnham, T. L., Feaga, L. M., Groussin, O., Merlin, F., & Milliken, R. E. (2009). Lunar hydration as observed by the deep impact spacecraft. *Science*, *326*(5952), 565–568. https://doi.org/10.1126/science.1179788.

Teodoro, L. F. A., Eke, V. R., & Elphic, R. C. (2010). Spatial distribution of lunar polar hydrogen deposits after KAGUYA (SELENE). *Geophysical Research Letters*, *37*, L12201. https://doi.org/10.1029/2010GL042889.

Teodoro, L. F. A., Eke, V. R., Elphic, R. C., Feldman, W. C., & Lawrence, D. J. (2014). How well do we know the polar hydrogen distribution on the Moon? *Journal of Geophysical Research: Planets*, *119*, 574–593. https://doi.org/10.1002/2013JE004421.

Thomas, G. E. (1974). Mercury—Does its atmosphere contain water. *Science*, *183*(4130), 1197–1198. https://doi.org/10.1126/science.183.4130.1197.

Tooley, C. R., Houghton, M. B., Saylor, R. S., Peddie, C., Everett, D. F., Baker, C. L., & Safdie, K. N. (2010). Lunar Reconnaissance Orbiter mission and spacecraft design. *Space Science Reviews*, *150*(1–4), 23–62. https://doi.org/10.1007/978-1-4419-6391-8_4.

Urey, H. C. (1952). *The planets: Their origin and development*. New Haven, CT: Yale University Press (245 pp).

Vasavada, A. R., Paige, D. A., & Wood, S. E. (1999). Near-surface temperatures on Mercury and the Moon and the stability of polar ice deposits. *Icarus*, *141*(2), 179–193. https://doi.org/10.1006/icar.1999.6175.

Vondrak, R., Keller, J., Chin, G., & Garvin, J. (2010). Lunar Reconnaissance Orbiter (LRO): Observations for lunar exploration and science. *Space Science Reviews*, *150*(1–4), 7–22. https://doi.org/10.1007/978-1-4419-6391-8_3.

Voosen, P. (2018). NASA to pay private space companies for moon rides. *Science*, *362*(6417), 875–876. https://doi.org/10.1126/science.362.6417.875.

Ward, W. R. (1975). Past orientation of the lunar spin axis. *Science*, *189*(4200), 377–379. https://doi.org/10.1126/science.189.4200.377.

Watson, K., Brown, H., & Murray, B. (1961a). On possible presence of ice on Moon. *Journal of Geophysical Research*, *66*, 1598–1600. https://doi.org/10.1029/JZ066i005p01598.

Watson, K., Brown, H., & Murray, B. C. (1961b). Behavior of volatiles on lunar surface. *Journal of Geophysical Research*, *66*, 3033–3045. https://doi.org/10.1029/JZ066i009p03033.

Wilcoski, A. X., Hayne, P. O., & Landis, M. E. (2022). Polar ice accumulation from volcanically induced transient atmospheres on the Moon. *The Planetary Science Journal*, *3*, 99. https://doi.org/10.3847/PSJ/ac649c.

Williams, J.-P., Greenhagen, B. T., Bennett, K. A., Paige, D. A., Kumari, N., Ahrens, C. J., et al. (2022). Temperatures of the lacus mortis region of the Moon. *Earth and Space Science*, *9*. https://doi.org/10.1029/2021EA001966. e2021EA001966.

Williams, J.-P., Greenhagen, B. T., Paige, D. A., Schorghofer, N., Sefton-Nash, E., Hayne, P. O., et al. (2019). Seasonal polar temperatures on the Moon. *Journal of Geophysical Research: Planets*, *124*, 2505–2521. https://doi.org/10.1029/2019JE006028.

Williams, J.-P., Paige, D. A., Greenhagen, B. T., & Sefton-Nash, E. (2017). The global surface temperatures of the Moon as measured by the diviner lunar radiometer experiment. *Icarus*, *283*, 300–325. https://doi.org/10.1016/j.icarus.2016.08.012.

Wisdom, J. (2006). Dynamics of the lunar spin axis. *The Astronomical Journal*, *131*, 1864–1871. https://doi.org/10.1086/499581.

Wöhler, C., Grumpe, A., Berezhnoy, A. A., & Shevchenko, V. V. (2017). Time-of-day–dependent global distribution of lunar surficial water/hydroxyl. *Science Advances*, *3*(9), e170128. https://doi.org/10.1126/sciadv.1701286.

Zeller, E. J., Ronca, L. B., & Levy, P. W. (1966). Proton-induced hydroxyl formation on the lunar surface. *Journal of Geophysical Research, 71*, 4855–4860. https://doi.org/10.1029/JZ071i020p04855.

Zhang, J. A., & Paige, D. A. (2009). Cold-trapped organic compounds at the poles of the Moon and Mercury: Implications for origins. *Geophysical Research Letters, 36*, L16203. https://doi.org/10.1029/2009GL038614.

Zhang, J. A., & Paige, D. A. (2010). Correction to "Cold-trapped organic compounds at the poles of the Moon and Mercury: Implications for origins". *Geophysical Research Letters, 37*, L03203. https://doi.org/10.1029/2009GL041806.

Zuber, M. T., Head, J. W., Smith, D. E., Neumann, G. A., Mazarico, E., Torrence, M. H., et al. (2012a). Constraints on the volatile distribution within Shackleton crater at the lunar south pole. *Nature, 486*, 378–381. https://doi.org/10.1038/nature11216.

Zuber, M. T., Smith, D. E., Phillips, R. J., Solomon, S. C., Neumann, G. A., Hauck, S. A., et al. (2012b). Topography of the northern hemisphere of Mercury from MESSENGER laser altimetry. *Science, 336*(6078), 217–220. https://doi.org/10.1126/science.1218805.

Glaciation and glacigenic geomorphology on Earth in the Quaternary Period

2

Colman Gallagher

UCD School of Geography, University College Dublin, Dublin, Ireland
UCD Earth Institute, University College Dublin, Dublin, Ireland

Abstract

During the Quaternary Period, ~30% of Earth's northern hemisphere land area, and all of Antarctica, became covered by glaciers and ice sheets during intervals of planetary cooling (glacials). Shorter intervals of glacial contraction (interglacials) alternated with glacials. Likewise, permafrost areas oscillated between ~35 and 24 million km^2. Glacial-interglacial alternation occurred in lockstep with rhythmic insolation variability. During glacials, mean sea surface temperatures fell by ~2°C but polar or glacially affected seas experienced 10°C reductions. Permanent sea ice extended into the mid-latitudes and vast iceberg discharges periodically calved from marine-terminating glacial systems of continental size. Integrated with glacial-interglacial oscillations, freshwater discharges changed in timing, duration, and quantity. Hence, glacigenic factors modulated the volume, temperature, and salinity of the oceans and, through glacio-isostasy, caused complex variations in relative sea level. These endogenic changes impacted ocean circulation, tuning upwelling, and ocean-atmosphere CO_2 exchange to exogenic insolation variation. Reduced atmospheric temperatures, cryospheric evolution, and global responses in sea level and ocean-atmosphere interaction produced an interlocked, oscillating environmental system grounded in insolation variability. These are the characteristics of *glaciation*, the total condition of the Quaternary Earth—especially the potential for glacier and ice sheet formation but encompassing the entire Earth system. Focussing on the magnitude of change in Earth's glacial system raises two key questions. (1) How does the glacial system integrate with the broader planetary environment? (2) How does that integration lead to a systematic geomorphology that can be read to reconstruct and understand the Quaternary environmental change? This chapter provides insights into these questions and a benchmark against which glacial processes on other solar system bodies can be understood through geomorphology. The figures in the chapter are integral and, therefore, intended to represent a genetically ordered sequence of process environments, and their landscape products, as explanatory analogues for candidate glacigenic landforms on other planets.

1 Introduction

The Quaternary Period, starting ~2.6 million years ago (Ma), is an ongoing global environmental interval in which Earth has been dominated by extreme oscillations in temperature, the volume of glacial ice on land, and of water in the oceans. Glaciers (landborne ice capable of motion due, at least, to plastic deformation) are the hallmark of the Quaternary Earth (but not its only significant characteristic), ranging in size from continental ice sheets (>50,000 km^2, almost completely covering and filling all the underlying topography, often terminating in floating ice shelves), ice caps/ice fields (<50,000 km^2, filling most valleys and basins, with some peaks visible), outlet glaciers (flowing mainly through the pre-existing valleys from ice sheets/ice caps onto *piedmont* plains), *corrie/cirque* glaciers (accumulations of glacial ice occupying and overdeepening small topographic basins), valley glaciers (longer outlet glaciers, and independent glaciers, that overspill and extend from *corries/cirques*), hanging glaciers (tributaries of larger valley glaciers, altitudinally discordant with the larger glacier—unlike tributary rivers), *piedmont* (mountain foot) glaciers (emerging from ice sheet outlets and spreading radially outwards on a peripheral plain), tidewater glaciers (outlets of ice sheets/ice caps and valley glaciers terminating in the ocean, producing icebergs by "calving" due to tidal cracking and wave attack) and ice shelves (thick, extensive masses—up to 10^3 m thick, 10^5 km^2—of floating ice, in places grounded on the seabed or islands, supplied by marine terminating glacial systems, especially ice sheets/outlets). The environmental alternations defining the Quaternary have been characterised by times when the global area of land-ice expanded by 200%, compared with the present. In that context, however, while Antarctica has been very stable for the past ~11 Ma, changing areally by only ~10% through these oscillations, North America has experienced repeated alternations in glacial area, from ~16×10^6 km^2 (larger than the present Antarctic ice sheets) to ~0.23×10^6 km^2 at present (~99% loss of area) (Fig. 1). In lockstep with Antarctica and North America, Northern Europe (Fennoscandia, Britain, and Ireland) experienced oscillations in glacial area of even greater relative intensity, from ~6.7×10^6 km^2 to only ~0.004×10^6 km^2 (>99.9% loss of area).

During periods of maximum glacial area, ice sheets widely reached km-scale thickness, with the giant systems of Antarctica and North America accumulating to maximum thicknesses of ~4 km. Up to 10^7 km^3 of ocean water, *via* the hydrological cycle, are required to build Earth's glacial system at times of maximum area and volume. Consequently, coupled changes in ocean-water volume lead to repeated, cyclical changes in the glacio-eustatic sea level (sea level due to the volume of water either abstracted from, or returned to, the oceans as a function of the expansion or contraction of Earth's ice sheets) of ~120 m. Geomorphology has been key to revealing and explaining these changes in the glacial and marine components of the environment—and in uncovering the complexity of the controls, interactions, and feedbacks among these and other key environmental systems (still in operation and at the centre of concerns about anthropogenic global warming). The magnitude and global reach of these changes in the glacial components of Earth's environment raise the questions at the focus of this chapter: how does the global glacial system (GGS) integrate with the broader planetary environment; how does that integration of process systems lead to a systematic geomorphology; how can that geomorphology be read to reconstruct past environments and infer the controls on environmental change? The aim of this chapter is to provide foundational answers, or at least insights, to these questions, and, in the wider context of the book, provide a benchmark against which glacial processes on other solar system bodies, especially Mars (see Grau Galofre et al., 2024, pp. 73–100), can be understood through geomorphology. Integral to these aims, the figures are intended to lead the reader through a genetically ordered sequence of process environments, and their landscape products, as explanatory analogues for candidate glacigenic landforms and landscapes on other planets.

2 The global glacial system (GGS) and the cryosphere

The GGS has played a central role in the functioning of the Quaternary cryosphere and in shaping the Quaternary Period. To understand the significance of the GGS in characterising the Quaternary Earth, it is useful to consider the environmental changes required to put Earth into a state of glaciation. It is important to emphasise that the term *glaciation* does not mean that the entire surface of Earth is covered by glaciers. Rather, Earth is affected by a very significant growth in the coverage by, or action of, glaciers (*glacierisation*).

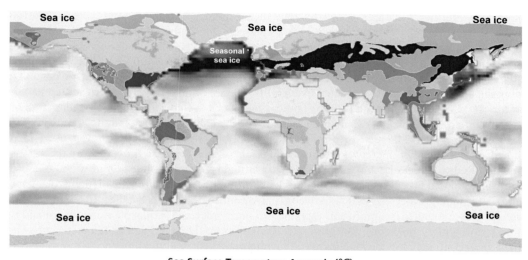

Sea Surface Temperature Anomaly (°C)

-10 -8 -6 -4 -2 0 2 4 6

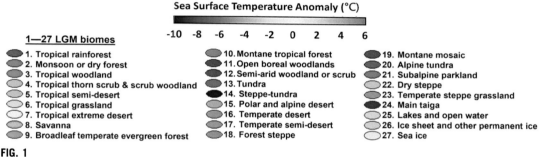

1—27 LGM biomes

1. Tropical rainforest
2. Monsoon or dry forest
3. Tropical woodland
4. Tropical thorn scrub & scrub woodland
5. Tropical semi-desert
6. Tropical grassland
7. Tropical extreme desert
8. Savanna
9. Broadleaf temperate evergreen forest

10. Montane tropical forest
11. Open boreal woodlands
12. Semi-arid woodland or scrub
13. Tundra
14. Steppe-tundra
15. Polar and alpine desert
16. Temperate desert
17. Temperate semi-desert
18. Forest steppe

19. Montane mosaic
20. Alpine tundra
21. Subalpine parkland
22. Dry steppe
23. Temperate steppe grassland
24. Main taiga
25. Lakes and open water
26. Ice sheet and other permanent ice
27. Sea ice

FIG. 1

Last glacial maximum (LGM; ~24,000 to 18,000 years ago) ice sheet systems, summer sea surface temperature (SST) anomaly (°C difference *vs* present), and biomes. Within the extensively glacierised regions, the present-day land mass outlines are in *pale grey*. Continental shorelines reflect the glacio-eustatic sea level at −120 m *vs* the present mean sea level.

Data credit; CLIMAP 18K (credits NOAA, National Geophysical Data Center, NASA), biome spatial rendering after Ray and Adams (2001).

This in turn effects changes in the rest of Earth's environmental system, co-involving planetary albedo, sea level, atmospheric and ocean circulation, and the concentration of key greenhouse gases (CO_2, CH_4). Hence, *glaciation* is a complex term, meaning both the widespread existence of glaciers and the total set of environmental consequences arising from the development of an Earth system permitting the presence of significant volumes of glacial ice on the continents. The environmental changes required to put Earth into a state of widespread *glacierisation* first demand that the planet cools enough for the likelihood of precipitation in the form of snow at low altitudes and latitudes to increase significantly. Secondly, for glaciers to form, more snow must survive than be lost through ablation, from winter through summer over many cycles. Thirdly, the accumulating mass and thickness of snow (density ~0.4 g cm^{-3}) must increase and become sufficiently compressed (through expulsion of air) to become glacial ice (density ~0.9 g cm^{-3}). Change in glacial ice volume depends on the mass balance between accumulated and ablated ice.

Generally, during climatic cooling, the zone of net ice mass accumulation expands into lower altitudes. Under warming climatic conditions, it contracts into higher altitudes. Hence, the ablation zone (with a net mass deficit) descends during climatic cooling but ascends during warming. The altitude of the zone in which accumulation is balanced by ablation, the equilibrium line, is significant. Firstly, the equilibrium line altitude (ELA) is determined by the glacier-specific climatic context (including the adiabatic lapse rate, moisture availability, precipitation delivery, patterns of temperature, solar aspect, wind, and continentality). Neighbouring glacial systems have a significant variation in ELA, owing to the interplay between these factors. The availability of snowfall, and its survivability into successive winters, largely governs glacial expansion *vs* contraction. However, other factors are involved. As elevation increases, atmospheric temperatures decrease (governed by the adiabatic lapse rate). Reduced pressure and temperature with altitude reduce absolute humidity and, therefore, cloud formation and precipitation. Regional to local differences in effective insolation (solar aspect, albedo) determining water flux and radiative flux also contribute to mass balance. In addition, glacial oceanicity/continentality is significant. Oceanic proximity increases atmospheric moisture and heat (*via* emission and evaporation); maritime glaciers are more dynamic, and accumulate mass more quickly, than continental glaciers. However, they are more sensitive to climatic warming; maritime glaciers terminating in their bounding continental shelf (i.e. tidewater glaciers, ice shelves) can experience catastrophic ice mass loss. Glaciers also have complex relationships with sea level and atmospheric temperature due to glacio-isostatic adjustment. Glaciers cause crustal depression (δ_{elev}) of ~1/4 ice thickness, due to changed crustal buoyancy reflecting the relative specific gravities of the underlying asthenosphere and overlying glacial ice (δ_{elev} ~ $3.7\,\mathrm{g\,cm^{-3}}$/~$0.9\,\mathrm{g\,cm^{-3}}$). Glacio-isostatic depression increases RSL (modifying subglacial hydrology and ice marginal stability) but also reduces glacier surface altitude; the surface of a 4 km-thick ice sheet will enter an atmospheric region ~10°C warmer than expected without glacio-isostatic depression (assuming an adiabatic lapse rate ~$0.01°\mathrm{C\,m^{-1}}$). ELA also represents the location at which ice velocity, discharge, and erosion are maximised—delineating where glacial erosion becomes most effective in landscape production (producing overdeepened longitudinal valley profiles). On Earth, the combination of erosive glaciers occupying valleys experiencing tectonic uplift results in the glacial buzzsaw effect, in which extremely overdeepened glacial valleys are produced.

Although ELA partitions glaciological responses to climate, Furbish and Andrews (1984) showed that the hypsometric characteristics of glaciers (glacier area as a function of altitude) modulate their sensitivity to climate. Hypsometry is crucial in the response of glacier termini to changes in ELA. In situations where these factors cannot all be measured (e.g. in planetary geomorphology), median glacial elevation (MGE) is a reliable proxy for ELA (Braithwaite & Raper, 2009). Glacier area *vs* MGE is quantified by the hypsometric index (HI), an integral describing the weighting of glacier area with respect to MGE. Furbish and Andrews (1984) showed that terminus altitude is highly sensitive to hypsometry. Bottom-heavy glaciers (with the lower 50% of their area concentrated into a narrower altitude range than the upper 50%) tend to be very sensitive to warming and rising ELA. However, top-heavy glaciers can also become sensitive if their ELA rises enough to significantly reduce their accumulation areas and overall distribution of mass with altitude (Davies et al., 2012; McGrath et al., 2017). Overall, glaciers with different hypsometries respond differently to the same climatic stimulus. Consequently, a combined understanding of ELA, MGE, and HI is key to understanding glaciology and palaeo-glacierisation. In studies of relict glacial landscapes, where ice wastage has exposed former subglacial beds, determining these hypsometric parameters is important in understanding glacigenic landscapes as proxies for palaeoclimate.

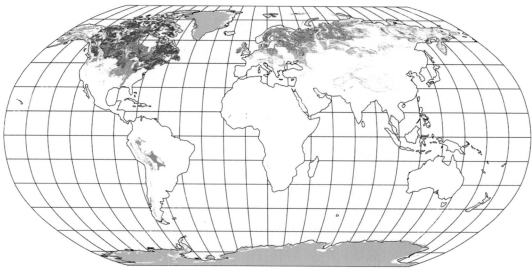

FIG. 2

Distribution of *glacigenic terrain* produced directly and indirectly by ice sheets and glaciers. *Dark blue fills*; tills. *Green fills*; glacio-aqueous deposits. *Orange fills*; loess. *Blue fills*; ice sheets/glaciers. The map compares the maximal *vs* present scale of the glacigenic process environments.

Open access data from Börker et al. (2018; Creative Commons Attribution 3.0) and from the National Snow and Ice Data Center.

The presence of a significant cryosphere (glaciers, sea ice, snow, permafrost) differentiates the Quaternary from the preceding 10^7–10^8 Ma. However, of all cryosphere components, glaciers have had the most significant effect on the landscape (including the rock cycle), atmospheric energy balance, sea level, and both oceanic temperature and salinity. Glaciers are therefore the defining environmental process system of the last 2.6 Ma, having produced the single most widespread land surface assemblage on Earth—*glacigenic* terrain (Fig. 2). Consequently, they have great significance in the functioning of the Quaternary Earth system, with which they interact through six process environments (times and/or locations in which distinct Earth surface processes are likely to be produced, resulting in distinct landform assemblages). The defining processes are driven by the environmental context of their location and/or time of operation, now globally contextualised by the oxygen isotope stratigraphies of pelagic marine sediments, and ice core (isotopic) chemistry from Antarctica and Greenland (Fig. 3).

3 The process environments of the GGS that integrate it with climate and the broader global environment

3.1 The subglacial process environment

The subglacial process environment involves interactions between basal ice and the rock or sediment substrate (Fig. 4a, b). Through this interaction, the GGS communicates and interacts with the broader environment *via* processes that modify the behaviour or physical properties of the subglacial

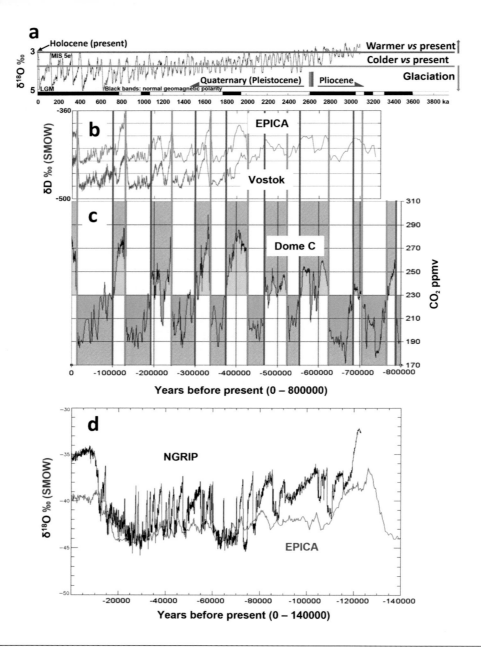

FIG. 3

(a) Rhythmic changes in the ice storage/glacio-eustatic sea level (and temperature) as represented by the changing $^{18}O:^{16}O$ in benthic sediments. (b) Changes in deuterium (δD ‰), EPICA, and Vostok ice cores (Antarctica). Reductions in δD equate to reductions in temperature. (c) Variations in atmospheric CO_2, Dome C ice core, Antarctica. *Blue lines* (a and b); transitions into glacial climate (<230 ppmv CO_2). *Orange vertical lines* (b and c); glacial terminations. (d) $\delta^{18}O$ curves, Antarctica (EPICA) *vs* Greenland (NGRIP). Note very short-term temperature oscillations (Dansgaard-Oeschger Events) in NGRIP.

Panel (a) After Lisiecki and Raymo (2005). Panel (b) Creative Commons licence https://commons.wikimedia.org/w/index. php?curid=10683795. Data credit: [EPICA] Lüthi et al. (2008); [Vostok] Laboratoire de Glaciologie et Geophysique de l'Environnement, Grenoble, France and Petit et al. (1999). Panel (c) Creative Common licence https://commons.wikimedia.org/w/in-dex.php?curid=16147504. Data credit: EPICA community members (2004). Panel (d) Data credit: Creative Common licence https:// commons.wikimedia.org/w/index.php?curid=6710392. Data credit: NGRIP, Eicher et al. (2016).

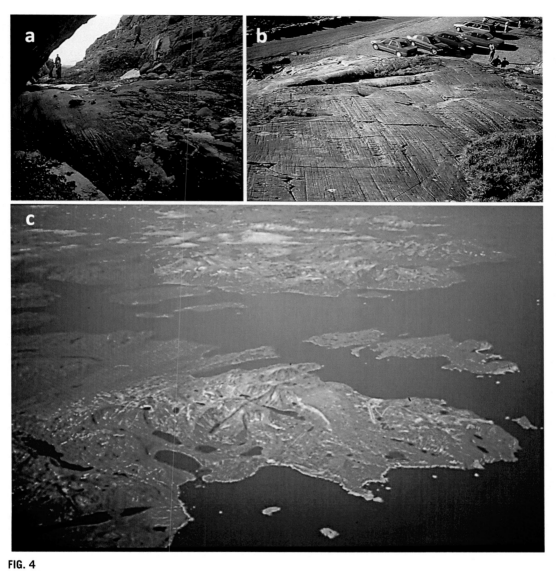

FIG. 4

(a) Subglacial substrate; scoured, striated bedrock (Iceland). (b) Glacially scoured and striated bedrock (Ireland). (c) Subglacially scoured landscape (linear troughs, corries, and basins), Labrador.

Photos: C. Gallagher.

lithosphere. The most obvious product of the subglacial process environment is distinctive terrain. A subglacial terrain is produced by scour and deposition, but deposits largely originate in erosion (often preceded by weathering and/or mass wasting). During glacial erosion, bedrock or pre-existing sediments, are entrained, transported, and comminuted (reduced in size, from larger clasts to minute matrix particles) by particle-bedrock abrasion or particle-particle attrition. The geographical distribution of these terrain types, basally scoured and deposited (Fig. 2), is indicative of the very large proportion of Earth that is, has been, and probably will again be, glacierised in the Quaternary (Fig. 3). It is also an indication of the thermal characteristics of the glaciers and ice sheets that formerly operated across this enormous area of the mid-to-high latitudes, for glacial erosion and deposition are most intense beneath warm-based glacial systems in which liquid water can persist at the ice base. Cold-based glacial systems, largely frozen to their substrate, move almost entirely by creep of the ice column. Although cold-based glaciers are not as erosive as warm-based, and therefore transport and deposit relatively little rock debris, they can achieve some erosion (Cuffey et al., 2000; Hambrey & Fitzsimmons, 2010). Entrainment at subfreezing temperatures is enabled by interfacial water films, which permit sliding and, therefore, abrasion—albeit at rates significantly less than under warm-based glaciers. While basal sliding, owing to the presence of basal meltwater, is required for glaciers to be regionally erosive, if sufficient pressure is achieved at a cold base, pressure melting can occur, giving rise to polythermal basal regimes. Polythermal systems are typically frozen and dry-based around thin margins but wet-based in their interiors or where they impinge on large bedrock obstacles. From a Clapeyron-type derivation, Radd and Oertle (1968) showed that the melting point of ice is a linear function of pressure when the ice is under pressure but water at the basal contact is unconfined. Christoffersen and Tulaczyk (2003) showed that, in the Clapeyron equation, ice-water phase changes depend further on solute concentration and capillary forces. However, the main control on basal thermal regime is ice thickness coupled to geothermal heat supply.

Cold-based glaciers produce meltwater, but their drainage systems are mainly supraglacial and englacial, more rarely subglacial (Hambrey & Glasser, 2012). Hence, polythermal glaciers can move partly by basal sliding or, where subglacial porewater pressure is high, by limited subglacial deformation in which a substrate of liquefied rock debris is entrained. Sugden (1973) interpreted landscapes of selective linear abrasion as indicative of basal sliding in limited contexts where ice thickness allowed basal pressure melting.

In the Dry Valleys of Victoria Land, Antarctica, where, with basal temperatures as low as $-16°C$, glaciers entrain and transport older deposits and regolith by folding and thrusting, producing both erosional and depositional features, and distinctive landscapes. Broad scrapes and grooves (10^0 cm deep, 10^1 cm wide, up to 10^2 cm long) have been incised into bedrock by rock tools, without the involvement of water, around the margins of cold-based glaciers (Atkins et al., 2002). Also, boulders embedded in the compressed desert pavement, but projecting into the base of overriding cold-based glaciers, can be abraded by entrained basal debris (Atkins et al., 2002; Bockheim, 2010). Meltwater channels can be incised by small flows of summertime supraglacial meltwater at the margins of cold-based glaciers. However, channels incised by summer meltwater alternate with the unaltered ground, previously protected by the cold-based glaciers—exemplifying the ability of cold-based glaciers both to protect surfaces and cause their erosion (Atkins et al., 2002). For example, Antarctic cold-based glaciers preserved Pliocene (3 Ma) eskers and valleys (Näslund, 1997). Similarly, Antarctic drifts containing recycled ventifacts and buried soils dating from the late Quaternary to the Miocene (~15 Ma) survived cold-based glacierisation, with reworked clasts from several glaciations within individual sedimentary units (Bockheim, 2010). Landscapes of Antarctic cold-based glacierisation therefore reflect significant, very long-term inheritance and recycling, combined with very low rates of denudation (by largely ineffectual glacial erosion and limited incision by meltwater flows). By contrast, warm-based systems, move by both creep and sliding on a basal film of water, or water and finely comminuted rock debris (and larger clasts). This means that warm-based glacial systems are responsible for having produced most of the erosional and depositional landscapes of Quaternary glacierisation. In the context of reconstructing glacial thermal regime from landscape evidence, particularly from remote sensing imaging data, the suite of six erosional landforms in Figs 4–9 is considered diagnostic of warm-based glaciation (after Glasser & Bennett, 2004).

FIG. 5

(a) Glacial trough with degrading glacier and proglacial meltwater river. (Alaska). (b) Large glacial trough (Kaskawulsh Glacier ~4 km width). Note confluent systems, short troughs, and corries (Yukon). (c) Linear U-shaped troughs and fiords (Greenland). (d) Steep, smoothed wall of a glacially eroded trough (Ireland).

Photos: C. Gallagher.

FIG. 6

(a) High altitude corrie inundated by the St. Elias Ice Cap (Alaska). (b) Low elevation corries devoid of glaciers, Vatnajökull (Iceland). (c) Remnant Holocene glacier in a Pleistocene corrie (Alaska). (d) Deglaciated corrie with moraine ridges (Younger Dryas stadial, ~11 ka) (Ireland).

Photos: C. Gallagher.

At the sole of warm-based glaciers, a layer of melting ice, eroded bedrock clasts, and fine-grained comminuted rock debris is mobilised by drag, and moved upwards (Fig. 10a). In warm-based glaciers, basal melting is due to mainly geothermal heat, strain heating (Lüthi et al., 2015), shearing friction (Yuen & Schubert, 1979) and both cryostatic (overburden) pressure and cryo-dynamic force (exerted by ice movement against bedrock protuberances), all of which reduce the melting point of ice. The geothermal heat flux averages $60\,mW\,m^{-2}$ but can rise to $>100\,W\,m^{-2}$ in glacierised volcanic areas. Cryostatic pressure melting occurs as a function of glacier thickness, with the freezing temperature falling by 1°C for every 13 MPa (130 atm). Hence, under ice sheets ~1 km thick, basal cryostatic pressure is ~9 MPa and water will remain liquid to a temperature of approximately −0.7°C. The pressure melting point decreases to a minimum of −21.9°C at 209.9 MPa but would require a glacier ~23 km thick—an impossibility, given that the glacier thickness presently does not exceed ~5 km. However, cryo-dynamic pressure (force) can induce pressure melting in relatively thin ice on the upflow side of obstacles; pressure-meltwater refreezes (*regelates*) once the ice passes around the obstacle and the pressure diminishes.

Deposits largely unaffected by post-depositional, non-direct glacial processes (but originating from warm-based glaciers) are *primary tills*. These can be further classified genetically into the five main

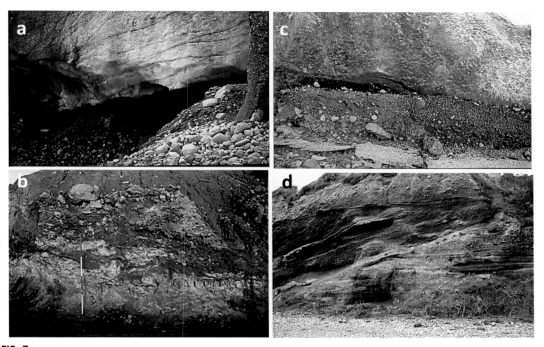

FIG. 7

(a) Subglacial cavity and basal meltout debris, mass wasting deposits, and varied clast sorting (Alaska). (b) Cavity fill by flows into a subglacial accommodation space (cavity or short tunnel, post-depositional glacio-tectonism). (Iceland). (c) "Floating" bedrock masses supported by injected fine-grained cavity slurry (Ireland; Meehan et al., 1996). (d) Large subaqueous fan. Meltwater-deposited foresets truncate horizontal, older bottomsets. Overlying lodgement till truncates foresets (Ireland).

Photos: C. Gallagher.

basal tills (Dreimanis, 1988) described next. All are associated with warm-based or polythermal glaciers. Post-depositionally, primary tills can be modified, becoming secondary tills created by mass wasting processes.

    Lodgement tills originate as an abrasive mix of meltwater and comminuted fine-grained rock debris (Fig. 10a, b). Below a critical threshold of viscosity, this fluid maintains motion and is erosive. However, the fluid becomes lodged by being plastered onto a static glacial substrate once a critical threshold of friction is exceeded (Fig. 10c). This can occur because of loss of water from, or an increase in the concentration of, basal rock debris. The process of till lodgement is associated with dynamic glaciers and is closely related to basal entrainment and transport of bedrock debris, which provide tools that cause abrasion on other mobile clasts and/or bedrock (Fig. 10d). **Basal meltout tills**, by contrast, form below stagnant glaciers due to the loss, through melting, of ice that surrounded and supported basal clasts and matrix comminuted during ice motion (Fig. 11a). However, both lodgement tills and basal meltout tills originate in basal abrasion, reflecting the erosion of the substrate by sliding glaciers and ice sheets. Hence, extensive deposits of melted-out englacial layers occur (Fig. 11b). **Supraglacial meltout tills**

FIG. 8

(a) LGM Roche moutonnée (Ireland): ice flow from right to left; stoss side (right side), abraded, smoothed; lee side (left side) rougher, plucked. (b) Streamlined, smoothed, convex-up (whaleback) bedrock morphology (Ireland). (c) Transverse fractures cross-cutting linear abrasions reflect local side-slipping glacial motion, common in warm-based systems (Ireland). (d) Large glacial grooves at Kelleys Island, Ohio, USA (people for scale).

Panels (a and c) Photo: C. Gallagher. Panel (d) Creative Commons Licence: https://en.wikipedia.org/wiki/File:Glacial_grooves.jpg.

are typically coarse and angular, due to freeze-thaw fracturing and the lack of inter-clast contact at the glacial surface and, in the near surface (Fig. 12a). These tills can accumulate to a significant extent and thickness, where debris supply is high and transport rates are low—common characteristics of the ablation zones of many mountain glaciers. These debris covered glaciers are often characterised by dark, low albedo surfaces (an important glacial impact on the atmospheric energy balance) (Fig. 12b). Supraglacial debris is generally transported further than subglacial material. Owing to fewer inter-clast impacts in the supraglacial environment, far-travelled supraglacial clasts retain angular joint-block morphology (Fig. 12c). However, supraglacial debris accumulations can result also from thrusting and emergence of basal and englacial debris layers, in zones of glacial deceleration and compression (Fig. 12d). In contrast to supraglacially transported debris, these emerged basal deposits are matrix enriched (although susceptible to post-depositional winnowing and wind deflation) and contain edge-rounded clasts. **Subglacial deformation tills** (Fig. 13a) are produced when warm-based glaciers slide over a mobile substrate, liquefied due to high porewater pressures. Dewatered substrates become rigid but retain deformation structures emplaced during the liquefied phase, resulting in a till composed of geotechnically weak materials but with persistent soft-sediment characteristics e.g. folds and *boudins*.

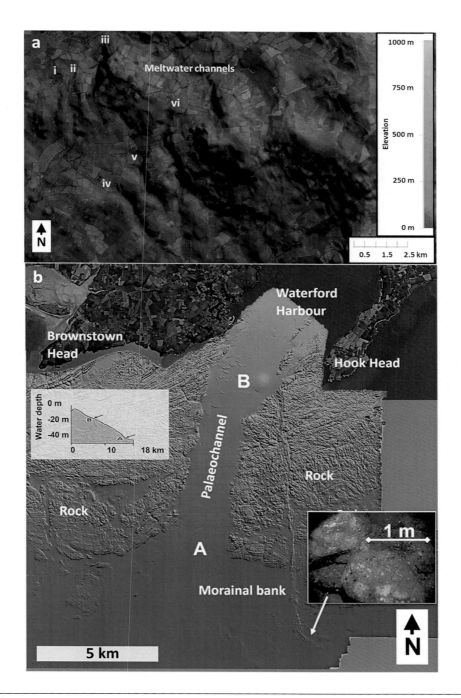

FIG. 9

(a) The Ballyfoyle meltwater channels (Ireland). (b) Tunnel valley (Ireland), submerged due to the postglacial sea level rise (Gallagher et al., 2004, 2018). Inserted panel; video frame of glacially abraded boulders comprising the morainal bank (credit G. Sutton, in Gallagher et al., 2004).

Panel (a) Image: Digital Globe. DEM: EUDEM 25 m/px (tile eudem_dem_5deg_n50w010).Panel (b) Data credit: DEM, EUDEM 25 m/ px (tile eudem_dem_5deg_n50w010); multibeam sonar: Geological Survey of Ireland data. Processing: C. Gallagher.

FIG. 10

(a) Fine-grained basal debris layers (Iceland). Note that the direction of dip of the isolated clast is coherent in three dimensions with the fine-grained debris planes. (b) The finest-grained matrix material winnowed by inter-layer meltwater is flushed out of the basal environment as rock flour particles (10^{-1}–10^1 μm) (Alaska). (c) Lodgement till; fine-grained matrix, parallel thin layers/laminations, supporting isolated clasts (Ireland). Limestone clast, typically bullet-shaped (i.e. streamlined) (d) Glacially streamlined, bullet-shaped, faceted boulder, ~2 m a-axis length (Ireland). (e) Texturally immature till—plucking and sub-sole drag beneath a polythermal glacier (Ireland). Thin joint blocks moved <5 cm (O-O'). Clasts at the mobile layer base moved O_2-O_2'. The longitudinal (L) and vertical (V) clast displacements ($T(O$-$O')$) are equivalent to the resultant (R_1) in the plane of the photo. Integrating transverse displacement (towards the reader; $R_T \sim 1$ cm) produces a 3D clast fabric. Translation $T(O_2$-$O_2')$ equivalent to resultant (R_2) and $R_2 \approx R_1$. Below the bedrock contact, the joint blocks have zero displacements; $T(O_3) = (0,0)$.

Photos: C. Gallagher.

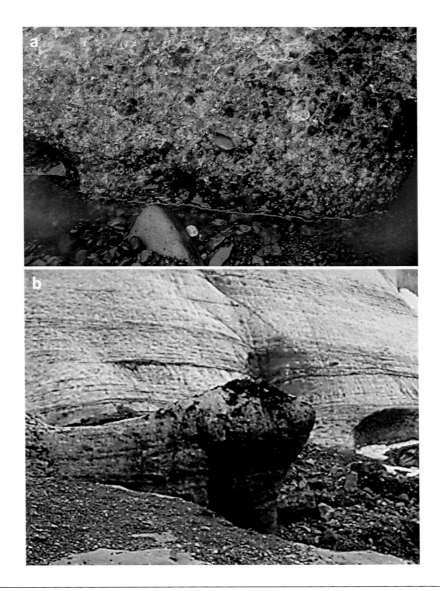

FIG. 11

(a) Basal meltout tills have a higher clast-to-matrix ratio and less consolidation than texturally mature lodgement tills. (b) A degree of directional fabric can remain in meltout tills.

Photos (Iceland): C. Gallagher.

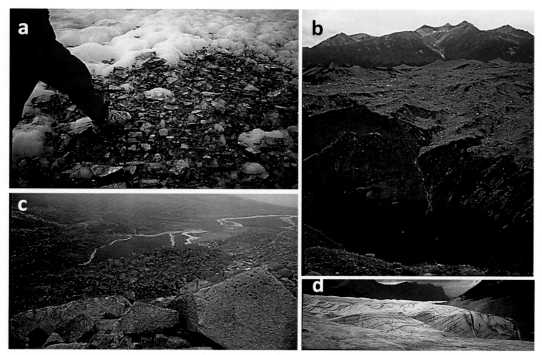

FIG. 12

(a) Supraglacial debris, distinctively coarse and angular (Alaska). (b) Supraglacial debris can become thick but is normally consolidated and clast supported. (c) Far-travelled clasts are common and can retain joint-block geometry (Alaska). (d) Supraglacial debris resulting from upward movement and the emergence of basal and englacial debris layers in compressive flow zones (Alaska).

Photos: C. Gallagher.

However, highly developed deformation tills are structureless and diamictic (i.e. thoroughly mixed). A **glacitectonic till** (Fig. 13b) develops when bedrock experiences plucking and rafting by ice with a variable basal thermal regime, including periods when, or locations where, basal ice freezes onto bedrock and experiences thrust. This process induces fracturing of bedrock along joints, and displacement down-ice. Displaced clasts may dip steeply in the up-ice direction or be rotated *in situ* if not displaced (Fig. 13b). Glacitectonism can entrain very large bedrock "rafts" and transport them with more far-travelled material, including very distantly entrained erratics (Fig. 13c). **Secondary tills** (Fig. 14) form when primary tills become post-depositionally modified by gravity and liquid flows, especially when glacial disintegration begins and: (1) deposits lose ice-support; (2) there is an abundance of meltwater and surface runoff (from snowmelt and rain), and; (3) a high density of steep slopes develop on ice-failure faces and dumped rock debris (common processes are detachment, faulting, fracturing and rotation of debris-rich ice blocks, slumping and liquid flow of melted-out debris, winnowing of fines and, consequently, the passive concentration of clasts.

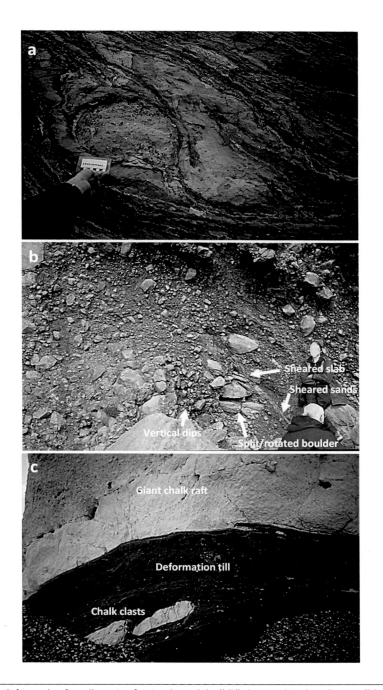

FIG. 13

(a) Sub-glacially deformed soft sediments of estuarine origin (UK); incomplete boudinage. (b) Glacitectonism affects more rigid materials, particularly jointed bedrock (Ireland)—consistent with a variable thermal regime. (c) A giant raft of Cretaceous chalk overlying a Pleistocene deformation till (UK).

Photos: C. Gallagher.

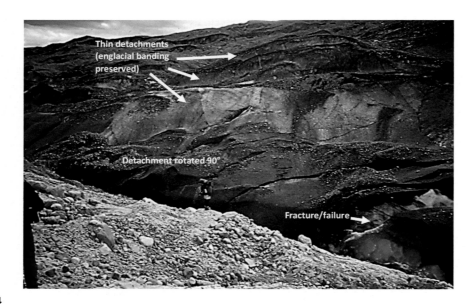

FIG. 14

Primary glacial structures and tills (lodgement and meltout), modified by gravity and liquid flows when glacial disintegration begins, become secondary tills.

Photo: C. Gallagher.

The ice substrate margin has a significance extending below the bedrock contact for the GGS affects Earth's crust and upper mantle through glacio-isostasy and volcanic outgassing (Fig. 15a). The glacio-isostatic loading imposed by glaciers and ice sheets is significant due to the magnitude of crustal depression they cause ($\sim h_g/4$, where h_g is glacier thickness) due to rheological displacement of underlying mantle material. In Antarctica, where the average thickness of its ice sheets is ~2.2 km, and can exceed 4.7 km, the crust is depressed by ~0.5–1 km. Glacio-isostatic depression therefore adjusts the relationship between the elevation of glacierised, or formerly glacierised, land masses and sea surface elevation. This adjustment is a function of the global volumetric magnitude of glacierisation, which determines glacio-eustatic sea level (global ocean volume controlled by the volume of ocean water abstracted or returned during the cycles of glacierisation and decay). The LGM glacio-eustatic sea level was ~120 m lower than at present ($>50 \times 10^6$ km^3 of water were abstracted from the global ocean to construct the GGS). However, due to local/regional-scale variation in glacier thickness (Fig. 15b), glacio-isostatic crustal depression is not uniform. The resulting RSL, therefore, is dynamic, reflecting both global and local/regional glacial evolution. A consequence of this spatially and temporally complex relationship, contingent in part on glacio-isostatic crustal depression, is that the local-scale sea level can be out of phase with the global sea level. For example, in Ireland, LGM ice sheets were up to ~1 km thick, and glacio-isostatic crustal depression was ~250 m. Consequently, RSL was greater than the present mean sea level (Fig. 15a). So, in the Quaternary, the main factor causing sea level variation around the margins of glaciated land masses is determined by the interplay between glaciers and the crust. The result is a complex, still evolving, RSL around North America N of ~40°, and around Europe N of ~50° (and around other glacially affected land masses) (Roy & Peltier, 2015; Tushingham & Peltier, 1992).

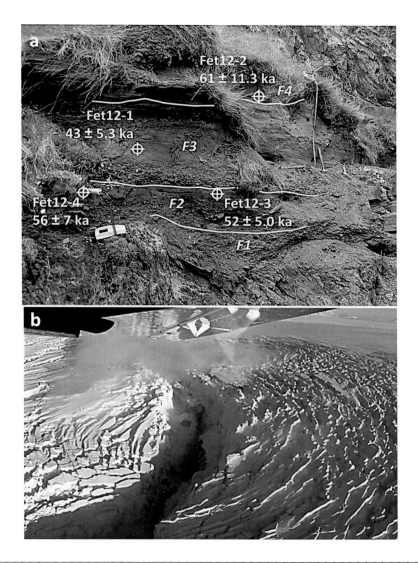

FIG. 15

(a) Raised beach deposited during interstadial, high RSL conditions ~52 ka (Ireland; Gallagher et al., 2015). (b) Inward collapsing crevasses around the subsidence bowl, Grimsvötn (Vatnajokull, Iceland) subglacial volcanic eruption (1996). Steam was released into the atmosphere, following the main ash-rich eruption, with CO_2 and H_2S.

Photos: C. Gallagher.

Another consequence of the stress imposed by glaciers and ice sheets on Earth's crust and upper mantle is an alteration in the spatial distribution and frequency of volcanic events, particularly volcanic outgassing (Fig. 15b). In addition to H_2S, two of the most important gases produced in volcanic eruptions or effusions are CO_2 and water vapour, both of which are significant greenhouse gases. Consequently, glaciers and ice sheets, through their effects on the crust and upper mantle, play a role in the amount of greenhouse gases transferred from Earth's interior to its atmosphere. Along with changes in continental palaeogeography and ocean circulation, it is possible that reductions in volcanic CO_2 outgassing in the Paleogene (65–23 Ma) were associated with the beginnings of global cooling that led to the first continental glaciers in Antarctica, ~46–35 Ma. By contrast, volcanic CO_2 outgassing owing to glacial unloading of active magmatic provinces (e.g. Praetorius et al., 2016; Sternai et al., 2020) contributes to increased atmospheric CO_2 concentrations characterising Quaternary interglacials. In this way, the GGS instigates a significant positive feedback with deglaciation through its correlation with increased, in-phase mid-latitude and high-latitude insolation. This is regarded as the primary determinant of the rhythmicity of stadial-to-interstadial transitions (alternations between intervals of glacial expansion and contraction) of 23 ka and 41 ka periodicity, and of full-blown 100 ka glacial terminations (after the mid-Pleistocene transition, ~1.2–0.6 Ma).

3.2 The supraglacial process environment

The supraglacial process environment involves the surface of glaciers and ice sheets in contact with the atmosphere. Hence, the supraglacial interface of the GGS affects the energy balance of the atmosphere through its influence on albedo (Fig. 16) and on the Atmospheric General Circulation (AGC; the pattern of dominant wind systems driven by the distribution of heat from the equator to higher latitudes, and by the rotation of Earth). When the global climate system is conducive to widespread glacial growth, increased glacierisation and the consequent increase in albedo act as a positive feedback; more solar radiation is reflected back to space before it can be absorbed by surface materials and emitted to the atmosphere as infrared (heat) radiation. Increased ice cover/albedo therefore reduces atmospheric temperature and either stabilises or accelerates glacierisation (given sufficient moisture supply). However, if global climate is trending towards general ice mass loss, in the context of global warming, the consequent reduction in albedo amplifies warming by reducing the reflection of solar radiation, while increasing the absorption and subsequent infrared emission from exposed, formerly glacierised surfaces (e.g. Zeitz et al., 2021).

These albedo feedbacks, forced by the total GGS surface area, permit glaciers and ice sheets to communicate important thermal influences on the atmosphere, affecting the AGC. However, the supraglacial glacial process environment imparts significant influences on the thermal characteristics of the atmosphere also through changes in albedo brought about by the meltout of englacial debris onto glacial surfaces. This process reduces the GGS-scale albedo during intervals of significant ice mass loss associated with increased melt season (summer half-year) intensity and duration. The emergence of formerly englacial rock debris, due to melting and consequent ice surface lowering, leads to the development of an extensive, often complete, debris cover that forms a supraglacial till (Figs 12 and 14) (Zeitz et al., 2021). During deglacial episodes, the increased supraglacial debris significantly reduces albedo, leading to a potentially positive feedback with increasing insolation. This would amplify ablation due to greater absorption of solar radiation, leading to downwasting, increased supraglacial debris emergence, and greater infrared emission of heat into the colder glacial interior from the atmospherically warmed supraglacial debris cover. In turn, this could reduce

FIG. 16

(a) High albedo surface of the St. Elias Ice Cap (Alaska/Yukon); high-altitude corrie with crevassed glacial ice discharging (L to R) to the main ice cap. (b) Low albedo deglaciated corrie complex (Alaska).

Photos: C. Gallagher.

the emission of geothermal heat, usually conducted towards the ice surface before emission to the atmosphere. Hence, the increasing supraglacial debris can initiate a positive feedback cycle that accelerates glacial ablation. However, this process and the resulting ablation rates are significantly affected by the thickness of supraglacial debris (Collier et al., 2015; Juen et al., 2014; Östrem, 1959). Thick debris layers buffer the ice within from climate warming—thin debris layers amplify it, causing enhanced surface melting and lowering, manifested by the production of supraglacial meltwater lakes (Figs 17 and 18).

FIG. 17

(a) Ice-contact lake (mid-ground) fed by direct inputs of meltwater (Vatnajökull, Iceland). Foreground lake basins formed at previous ice termini. (b) Ice-contact lake, Vatnajökull. Roofs of subglacial meltwater portals visible (right mid-ground).

Photos: C. Gallagher.

FIG. 18

(a) Glaciofluvial delta (Ireland); tripartite sedimentary sequence (topsets, foresets, bottomsets) represents inputs of meltwater to a lake *via* subaerial flows. (b) Subaqueous fan (horizontally bedded sands (uppermost). Deformed lowermost sands and gravels are esker sediments (Ireland).

Panel (a) Photo: UCD School of Geography. Panel (b) Photo: C. Gallagher (Ireland).

3.3 The glacial Lake process environment

3.3.1 Supraglacial lakes

Supraglacial lakes (SGL; Fig. 19a,b,c) are ephemeral bodies of standing water on glacial surfaces, formed by summertime insolation or melting through interception of seasonal/monsoonal rain. They can be associated with organised supraglacial drainage channels that follow lateral troughs between ridges formed by crevasse closure in compressive flow zones (Fig. 19a). Even in these zones, however, some crevasses remain open, forming topographic lows through which some supraglacial meltwater enters the englacial domain (Fig. 19d). Owing to the meltout of near-surface englacial debris during SGL formation, local albedo can be reduced in comparison with surrounding ice (Fig. 19a), driving a positive feedback between thermal absorption and lake expansion (contrast Fig. 19a,b). Although temporally inconstant, SGL can form repeatedly in the same location. Earth's largest SGL ($\sim 72\,\mathrm{km}^2$) forms during consecutive Antarctic summers on the Amery Glacier. However, most SGL diameters are 10^2–$10^3\,\mathrm{m}$, with areas of 10^{-1}–$10^0\,\mathrm{km}^2$ (Fitzpatrick et al., 2014). SGL drainage happens by overspill or drainage into the glacial interior *via moulins*, the mouths of shafts that supply supraglacial meltwater to the englacial and subglacial routing system (Fig. 20a–e). *Moulins* form in melt-enlarged crevasses (Fig. 19d), which can become extended to the glacier base by hydrofracturing (hydrostatic pressure

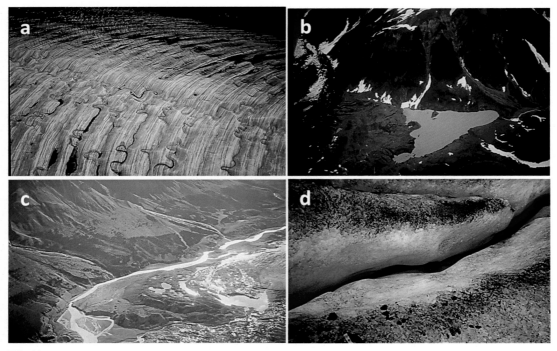

FIG. 19

(a) Supraglacial streams, melt ponds, and *moulins* (Yukon). (b) Supraglacial lake in dead-ice (Alaska).
(c) Supraglacial lake with dead-ice, inter-moraine lake complex, and proglacial river system (Yukon).
(d) Melt-enlarged crevasse (Alaska).

Photos: C. Gallagher.

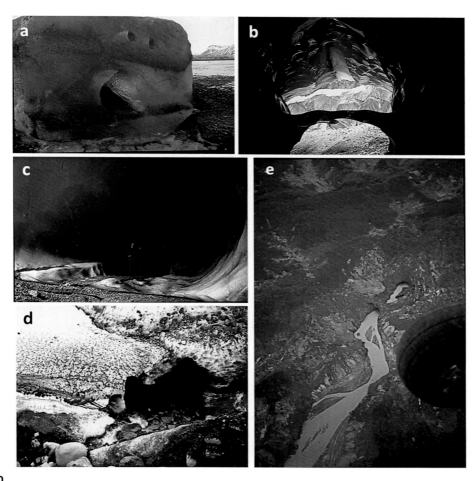

FIG. 20

(a) Ice margin fragment, the southern terminus of the Vatnojökull (Iceland) with three small englacial conduits (~15–45 cm diameter). (b) Subglacial tunnel system interior; vertical debris-curtain indicative of crevasse cut-and-closure processes (Alaska). Deposits inside tunnels become eskers when exposed to glacial wastage. (c) Portal of the tunnel system in (b). Tunnel-full flow velocity ~5–10 m s^{-1}. (d) Small glaciofluvial portal ~1 m diameter (Iceland). Clasts (~8 cm) imply transporting flows ~0.8 m s^{-1}. (e) Subglacial river emerging from tunnel portal becomes an ice-walled channel (glacier surface vegetated by trees). The ice bridge is a tunnel roof remnant.

Photos: C. Gallagher.

exerted against crevasse walls). On Antarctic ice shelves, hydrofractured crevasses can extend vertically to the sub-shelf cavity occupied by seawater. They can then elongate into extended surface fractures, contributing to an ice shelf break-up. Moreover, supraglacial freshwater drained into the sub-shelf cavity changes local salinity and circulation, leading to the freezing-on of some freshwater to the ice shelf base (part of the shelf mass balance not available to terrestrial glaciers) and contributing locally to the circum-Antarctic overturning circulation. SGL drainage to the glacier base can be extremely rapid ($<10^1$ d) and, with terrestrial substrates, can significantly reduce basal friction through substrate liquefaction. This can trigger a glacier surge in which ice flow velocity may increase to $10^1\,\mathrm{m\,d^{-1}}$. Sediments accumulated in SGL can be stratified but, owing to mass wasting of the surrounding ice by glacial thermokarst processes, also contain in-fallen supraglacial clasts and remobilised supraglacial till—and can be deformed by both "cut and closure" of crevasses and by impacts of supraglacial debris. The resulting deposits survive as kames—formed in dead-ice (glacial thermokarst) and characterised by a complex, often chaotic morphology and sedimentology (Fig. 21a,b,c).

3.3.2 Subglacial lakes

Subglacial lakes occur beneath ice sheets, ice caps or ice streams, at the ice-bedrock contact. Most subglacial lakes (~400) have been discovered in Antarctica, where subglacial drainage basins are overlain by ice sheets or occur near ice streams. Lake Vostok is the largest, occurring beneath ~3700 m of ice and with a water volume of $5400\,\mathrm{km^3}$ (with depths of ~500–900 m). Occurring also in Greenland, Iceland, and northern Canada, subglacial lakes are estimated to contain ~15% of all liquid freshwater on Earth (Dowdeswell & Siegert, 2003). The water in subglacial lakes is produced by basal pressure melting of overlying ice, geothermal heating of basal ice layers, or by the introduction of meltwater to subglacial basins through subglacial meltwater routes. Meltwater in subglacial lakes remains liquid because geothermal heating balances heat loss at the ice surface or because of the presence of hypersaline water. The ice roof of a subglacial lake occurs where the pressure melting point exceeds the temperature of the basal ice. The water in a subglacial lake is constrained when the margins of the basal ice roof (which can have a vault-like, convex-up form) intersect the equipotential surface. However, the floating level of the lake can significantly exceed the subaerial ice surface. Consequently, if the hydrostatic seal is breached (e.g. due to rapid rates of meltwater production or introduction) (Fig. 15b), extreme outburst floods (*jökulhlaups*) can be triggered, producing proglacial discharges of 10^4–$10^6\,\mathrm{m^3\,s^{-1}}$. Subglacial lakes are revealed by radio echo-sounding (RES) and radar altimetry (Ridley et al., 1993). Large examples can be inferred by their characteristically flat subaerial surface, which is due to reduced basal friction, increased sliding velocity, and surface flattening due to extensional flow (although compressional flow at the downstream margin imparts a small topographic rise there; Arnold et al., 2022; Mantripp et al., 1992; Remy et al., 1989). By contrast, small subglacial lakes are manifest by an undulation along the ice flowline. Livingstone et al. (2022) found that 80% of subglacial lakes are stable, closed systems or approximately hydrologically balanced. However, active subglacial lakes in Greenland and Iceland have higher discharge rates for a given lake volume than Antarctic lakes. An underlying control on the lake refilling rate is implied by the faster recharge rate of larger compared with smaller subglacial lakes. Enhanced supraglacial melting and rainfall inputs to the subglacial environment modulate the fill-drain cycles, increase the potential for *jökulhlaups*, and introduce oxygen, sediment, microorganisms, and nutrients to the subglacial environment (Livingstone et al., 2022). Some subglacial lakes contain active communities of extremophiles (cold-adapted microbes viable in nutrient-deficient conditions). Bio-geochemical cycles in subglacial lakes fully isolated from the external environment are driven by interactions between ice, water, sediments (including reservoirs of ancient carbon and methane clathrates liberated from basal ice), and organisms, together supplying energy not dependent on insolation (Christner et al., 2008). Subglacial lakes and their biology are therefore potential analogues of subcrustal oceans on icy satellites (e.g. Europa and Enceladus), subglacial lakes under the polar caps of Mars (Orosei et al., 2018), and the bases of former wet-based glaciers on Mars.

FIG. 21

(a) Dead-ice topography—kames and kettle holes (Alaska). (b) Large subglacial tunnel, ~10 m diameter; maximum tunnel-full flow velocities ~40 m s^{-1} (Alaska) (c) Kame mound (Ireland) reflecting flowing water, standing water, and mass wasting. (d) Tunnel-fill esker cross section (Ireland). Adverse longitudinal gradient indicative of pumped subglacial hydrology. (e) Longitudinal esker section; rhythmic, coarsening upward sands-to-boulders (Ireland).

Photos: C. Gallagher.

3.3.3 Proglacial Lakes

Proglacial lakes develop where a glacier or ice sheet terminates in meltwater impounded by a combination of rock and glacial margins or solely by converging glacial margins (Fig. 17a,b) (Ashley, 1988). Many proglacial lakes existed in the last glacial, with areas ranging from giant continental interior systems of 10^5 km^2 to small seasonal ponds. In Ireland, the landscape evidence for the distribution and size of post-LGM proglacial lakes, mainly in the form of proglacial glaciofluvial deltas and ice-contact sub-aqueous fans (Fig. 18a), indicates that, as ice sheets started to separate many meltwater lakes up to 10^3 km^2 formed at the retreating ice margins. The entry points of sub-aerial meltwater rivers into non-ice-contact lakes are represented by deltas composed of a tripartite sedimentary sequence of (a) topsets, (b) foresets, and (c) bottomsets (reflecting [a] the course of the subaerial meltwater river, [b] its entry to the lake basin at the water surface, the local base level, and [c] the quiescent lakebed away from the direct influence of the inflowing meltwater). By contrast, subglacial meltwater discharges into ice-contact lakes enter the water below its surface, producing subaqueous fans. These landforms consist of a bipartite sedimentary sequence of foresets and bottomsets, lacking the topsets that would represent subaerial, overland flow. However, subaqueous fans can be the terminal forms of eskers (Warren & Ashley, 1994) (Fig. 18b)—sinuous sedimentary accumulations deposited in subglacial tunnels, melted and carved upwards into overlying glacial ice. Judging from the sedimentary geomorphology, nearly every location in Ireland that had been glacierised experienced a temporary phase of ice-contact to proglacial lacustrine activity. Globally, most such glacigenic lakes were small but some became so large as to be of planetary significance e.g. Glacial Lake Missoula (GLM, Montana, USA, 17.5–14.5 ka, maximum volume ~2000 km^3), Glacial Lake Agassiz (GLA, central to eastern Canada, 13–8.5 ka, maximum volume ~19,000 km^3), and the Baltic Ice Lake (BIL, Baltic/North Sea, 12.6–8 ka, maximum volume ~29,000 km^3).

3.4 The glaciofluvial process environment

The glaciofluvial process environment is characterised by the development of supraglacial channels (Fig. 20a), englacial conduits, and both subglacial and proglacial rivers. Proglacial discharges are routed through englacial, including subglacial, pathways. These are supplied with liquid water by atmospheric precipitation and both supraglacial and englacial (including subglacial) melt. The glaciofluvial process environment is globally significant, for it is the route through which glacial mass is returned to the hydrological cycle following up to 10^4 years in ice storage. Although significant meltwater volumes are produced in the subglacial glaciofluvial process environment (by geothermal heat, pressure melting, and strain heating), most are produced in the supraglacial process environment (Fig. 20a).

3.4.1 The englacial (including subglacial) glaciofluvial process environment

Supraglacial meltwaters and precipitation may be discharged directly off a glacial surface into the proglacial subaerial environment or can enter the glacial system *via* moulins and crevasses. These waters flow towards the glacial base through inclined englacial conduits (Fig. 20a), ultimately becoming subglacial tunnels (Rothlisberger channels melted into the overlying ice; Figs 20b,c,d,e and 21b), or Nye channels incised into the substrate. Sediments deposited in Rothlisberger channels can be preserved as eskers (Figs 18b and 21d,e)—sinuous ridges of sands, gravels, cobbles, and boulders. Eskers are significant in that they unequivocally reflect the operation of warm-based subglacial thermal regimes, their morphologies, and sedimentologies representing variations in melt and subglacial discharge. Owing to the connection between subglacial meltwater discharge and supraglacial meltwater generation, meltwaters discharged through the englacial-subglacial route to the proglacial environment (Figs 12c and 19b) have rhythmicities that broadly reflect insolation variation.

Glaciofluvial sediments are dominantly derived from the erosion of bedrock, till, and englacial debris, mainly by channelised flows of meltwater. However, non-channelised, subglacial sheet floods also can move subglacial debris towards zones of lower hydrostatic pressure in subglacial channels or tunnels. Meltwater erosion rates are characteristically high, so deposition occurs mainly in diminishing or recessional flow phases. Consequently, particle-size variation is the most important lithological variable in glaciofluvial deposits, reflecting relationships between flow velocity, flow regime, and particle fall velocity. Textural variation is also important. During glaciofluvial transport, reworking of particles begins immediately, due to both shear stresses in traction currents and impacts during saltation and reptation. These processes cause cleavage and comminution of mineral grains and larger clasts. Owing to the constant lowering of the glacier base in warm-based thermal regimes, and the unloading of basal sediments, subglacial glaciofluvial sediments tend to be transported over short distances. Consequently, glaciofluvial particles (grains and clasts) are less edge-rounded than fluvial sediments. Similarly, local bedrock petrographies dominate. By contrast, englacial glaciofluvial sediments, which moved higher into the glacier during transport, are less affected by basal lowering and can be transported further.

3.4.2 The proglacial glaciofluvial process environment

In many formerly glacierised regions, most land surfaces consist of proglacial glaciofluvial deposits, not direct glacial sediments (Fig. 2) or rock. Hence, sedimentary processes in the proglacial environment are extremely important geomorphological agents. Ice contact landforms occur mainly as extensive mounds or plains of sands and gravels. These are first produced in direct ice contact, as melted-out accumulations of glacigenic debris (kames; Fig. 20a) with various degrees of hydraulic transport and sorting. Beyond the ice margin, meltwaters can flow considerable distances in the form of medial subaerial braided rivers before becoming single-channel distal rivers. At any point in this sequence, temporary storage of meltwater in distal lakes is common.

In proglacial environments, variations in particle size and sedimentary structure are controlled by relationships between hydraulic energy and flow regime (both diminishing with increasing distance from the ice margin), and by particle fall velocity. Compared with subglacial glaciofluvial sediments, proglacial sediments generally attain longer transport distances. Hence, they can be characterised by significant downflow sedimentary fining and sorting. Proximal glaciofluvial rivers, operating close to the ice margin, are typically coarse-grained, poorly sorted bar sediments deposited between braided channels. These characteristics reflect seasonally variable flood discharges and depth, and the abundance of poorly sorted glacigenic sediments in the ice-contact and proximal proglacial depositional environments. Beyond the ice margin, hydraulic sorting improves, and there is a general fining in sediment calibre with increasing ice marginal distality (Fig. 22a–h). However, glaciofluvial discharges are highly seasonal. Increasing meltwater supply as the melt season progresses also causes sediment grade and flux to increase, generally resulting in coarsening-upwards sedimentary sequences. Moreover, if glacial re-advance occurs, post-erosional flow phases will be followed by the deposition of coarser sedimentary units superposing finer, better-sorted forerunners (Figs 21e and 22h). Lateral migration of glaciofluvial flow courses also leads to erosion of older sequences, with replacement by younger deposits. In this situation, younger sediments can be bounded by terraces sinuously alternating from

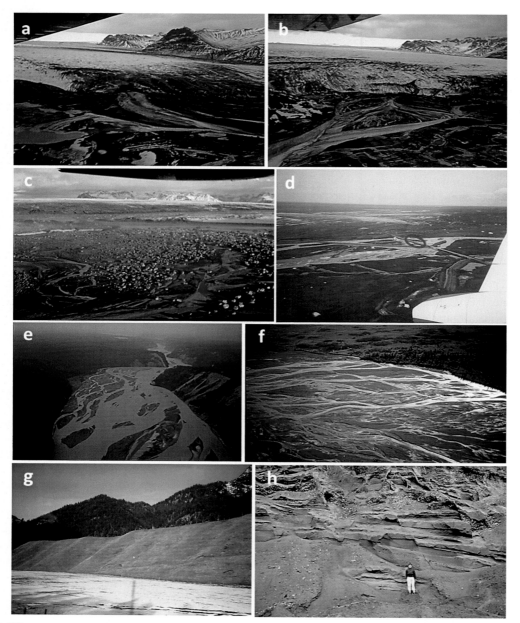

FIG. 22

(a) Elementary sandur (Iceland); deposition from meltwaters emerging from a glacial portal (subglacial, ice-contact, and proglacial processes). (b) Composite sandur; coalescence of ice-proximal outwash rivers from glacial portals (Iceland). (c) Broad expanse of the Vatnajökull composite outwash plain (Iceland), two days after the Grimsvötn subglacial volcanic eruption and flood (*jökullhlaup*). (d) Six months after the jökulhlaup; most icebergs melted, meltwaters contained within outwash channels (*red ellipse* highlights a passenger coach for scale). (e) Inter-ridge medial outwash channel/sandur (Alaska), shortly after midsummer; braid-bars are submerged in meltwater. (f) Late summer outwash channel/sandur (Yukon); braid bar surfaces are exposed (trees ~5 m). (g) Outwash terrace edge (Austria)—outwash channel beds, abandoned after vertical incision. (h) Outwash sediments (Ireland); stacked, rhythmic braid channel beds and bars overlain by suspended load units (pervasive post-depositional faulting).

Photos: C. Gallagher.

one channel side to the other. By contrast, when sediment supply diminishes but discharge remains unchanged, vertical channel incision results in terrace pairs on both channel sides. Hence, glaciofluvial sequences are extremely variable in form, texture, and environment of deposition, including being time transgressive (where the basal contact of a younger unit, representing a single time, cross-cuts the upper surface of underlying units representing several times; Fig. 7d).

3.5 The tidewater and Fjord glacimarine process environment

3.5.1 Controls and feedbacks

The glacimarine process environment is characterised by glacial ice that terminates in the ocean. Glacimarine ice can be supplied by ice sheets and valley glaciers but some is supplied by ice streams—rapidly flowing ice corridors (10^1 km in width, up to 10^2 km in length, flowing up to ~3 km year^{-1}) within an ice sheet or constrained by a bedrock trough. Glaciers that transgress their continental margins develop floating ice tongues or ice shelves (Figs 23a–d and 24a), which interact directly with the relatively warm and perpetually dynamic ocean (tides, currents, and waves). Consequently, marine-terminating glacial systems produce large amounts of icebergs. The last point at which the glacial base contacts the terrestrial substrate, beyond which there is either open water, a floating ice tongue, or a larger ice shelf, is the glacial grounding line (Fig. 23a). Grounding line stability is significant to the entire glacial system up-ice, for it restricts glacial extension by advection of floating ice. These considerations apply also to glacial systems terminating in large proglacial lakes. Changes in the intensity of physical wave attack on marine-terminating ice change the sensitivity to other factors, notably glacial thickness over the grounding line, sea ice cover, sea surface temperature (SST), and atmospheric temperature (maritime-induced excess/deficit compared with direct insolation). In turn, glacio-eustatic sea level varies due to the variation in iceberg melting and associated changes in glacial meltwater production. Increased glacio-eustatic sea level constitutes a positive feedback to the initial climate impulse, imposing cycles of greater uplift on the floating glacial margin, and more pervasive lateral cracking, resulting in accelerated iceberg calving.

Grounding line instability is controlled by ice thickness and ice-proximal water depth. Ice thickness reflects surface melting as a function mainly of insolation. However, grounding line instability is also directly proportional to water temperature (generally insolation-controlled). Therefore, grounding line instability depends on positive feedbacks involving insolation, surface melting, and glacio-eustatic sea level. Ice marginal instability is affected also by the likelihood of seawater incursions beneath glacimarine ice margins (Konrad et al., 2018). Incursions of external water increase the local base level, reducing basal friction and the efficiency of subglacial meltwater routing. Together, these factors contribute to rapid ice marginal slip associated with liquefaction of subglacial sediments near the ice margin and, therefore, to reduced basal friction. Although ice streams achieve higher velocity due to reduced basal friction owing to the presence of subglacial water and/or the wet subglacial till (Fig. 24a), but their activity/velocity increases also in direct proportion to the increasing sea level.

Beyond the ice shelf margins, icebergs export freshwater and ice-rafted debris (IRD) well beyond the glacimarine margins (Fig. 24b). Increased (decreased) meltwater and iceberg fluxes force increases (decreases) in glacio-eustatic sea level, initiating an amplifying feedback loop that accelerates (slows) glacio-eustatic sea level rise, ice mass loss, and ice marginal retreat. Beneath grounded glacimarine systems, incursions of seawater wedges of between 10^2 and 10^3 m thickness can increase ice volume loss

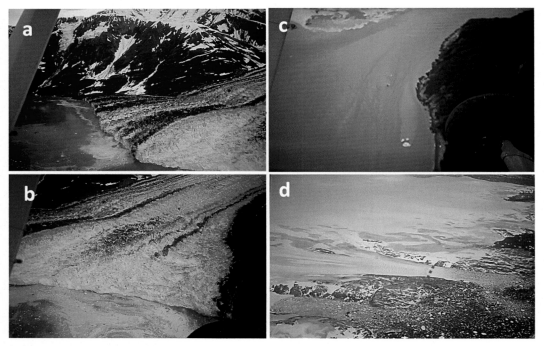

FIG. 23

(a) Floating glacimarine ice margin (Alaska). Grounding line (*dotted*) manifest by reduced surface slope and upward convexity of the glacial surface. Extremely high crevasse concentration beyond the grounding line. (b) Iceberg calving (ice mass detachment) and iceberg rainout are common. (c) Icebergs (foreground), dense brash ice and bergy bits (extreme background); turbid sediment plumes (midground). (d) Outlet fiord glaciers, Greenland, calving icebergs. Margin of the glacier floating beyond the red dotted line. Dense crevasse field up-ice of the red line.

Photos: C. Gallagher.

by between 10% and 100%, depending on the glacial substrate type (with greater incursion distances and effects on hard substrate systems compared with soft) (Robel et al., 2021). This potentially forces a transition from efficient, channelised subglacial discharge (Rothlisberger, 1972) to inefficient, linked cavity subglacial discharge (Creyts & Schoof, 2009). These transitions in turn can increase substrate liquefaction and reduce basal friction, inducing deforming bed conditions and glacial acceleration.

Although these positive feedback loops can be restricted to individual valley glaciers, they can apply to entire ice sheets *via* interactions between sea level and ice stream behaviour. Although ice streams exist today only in Antarctica and Greenland, in the LGM they were more common, given the greater extent of glacierisation (Bennett, 2003; Krabbendam et al., 2016) (e.g. Fig. 24c). Their significance to the stability of the GGS and global climate is that ice streams discharge the largest proportion of ice drained from their land mass interiors to the ocean. For example, the Antarctic ice sheets contain at least 48 ice streams (Rignot et al., 2011; Rignot & Thomas, 2002), through which 90% of Antarctic ice that reaches the Southern Ocean is drained (Fig. 24a). As ice streams are responsible for such a large

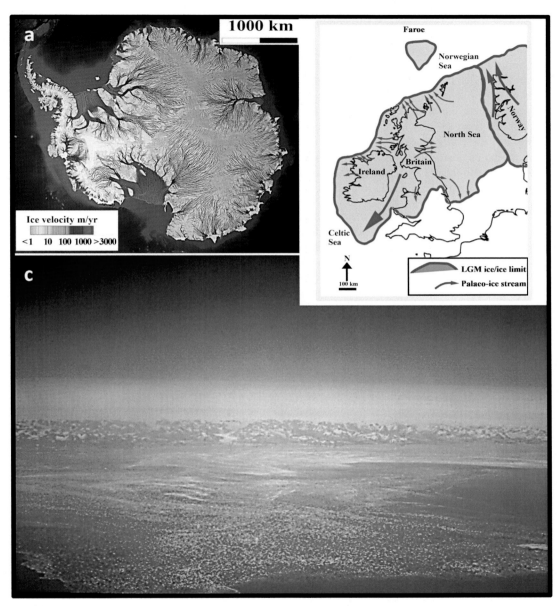

FIG. 24

(a) Flowlines and velocities of Antarctic ice streams; 1 m year^{-1} (*brown* to *green*) to 3000 m year^{-1} (*green* to *blue* and *pink*). (b) Palaeo-ice streams of Irish and Scottish glacial systems (after Krabbendam et al., 2016; Elsevier, Creative Commons Licence). (c) Icebergs calved from the Greenland Ice Sheet; the edge of the iceberg zone is ~100 km offshore in this photo.

Panel (a) Image credit: NASA/Goddard Space Flight Center Scientific Visualization Studio. Data credit: CSA, JAXA, and ESA. Data processing credit: UC Irivine/NASA Research Team. NASA open access granted. Panel (c) Photo: C. Gallagher.

proportion of ice mass discharged from ice sheet interiors to the ocean, as ice stream velocity increases, ice sheet thickness/elevation is/are reduced by the process of interior downdraw. As glacio-eustatic sea level rises in direct proportion to the mass of ice drained to the ocean, with ice streams playing a disproportionately large role, these relationships are the basis of a powerful positive feedback loop that can accelerate ice mass loss through increased glacio-eustatic sea level. When less melt occurs during times of glacial accumulation and growth, glacio-eustatic sea level falls, meaning that glacial bases become more strongly fixed to their substrate, with slower sliding velocities and less deforming bed flow. In this context, ice shelves form buttresses to the interior ice upstream and, therefore, slow the drainage of the entire glacial system to the ocean. Hence, an opposite positive feedback is generated that contributes positively to mass balance and ice sheet interior elevations. With reduced rates of ice mass drainage to the ocean, glacio-eustatic sea level falls and the feedback loop is intensified in favour of accelerated ice mass accumulation.

3.5.2 The glacimarine sedimentary system

The cycles of GGS expansion and contraction inherent in Quaternary glaciation have controlled the activity of the glacimarine sedimentary system *via* changes in mass balance and sea level (glacio-eustatic and RSL). During glacial maximum conditions, marine-terminating ice sheets and ice streams conveyed glacigenic sediments, meltwater, and icebergs to the continental shelf edge. By contrast, most glacimarine sedimentation now occurs at fjord heads, including meltout from floating glacier bases, ice front collapse, iceberg rainout, glaciofluvial sediment plumes (from englacial, supra- and subglacial meltwater fluxes), gravity flow resedimentation, iceberg reworking, bioturbation, and fjord-wall processes (including non-glacial mass wasting and snowmelt processes). In general, ice streams and glacimarine margins of valley-guided outlet glaciers produce the highest sediment flux, especially in contrast to less dynamic inter-ice stream areas (Dowdeswell et al., 2016). Tidewater glacimarine depositional process environments are characterised by a continuum from: (1) at the ice-front, sub-glacial depositional processes, mass wasting, and subaerial/englacial glaciofluvial discharges that enter the marine water column at its surface; (2) proximal-to-distal subaqueous outwash (where glacial meltwater enters the water column below its surface), and (3) distal (quiet water) marine processes. Corresponding sediments vary from: (1) tills, poorly sorted/stratified gravels, and resedimented sub-aerial meltwater sediments; (2) sorted/stratified sands and gravels fining to laminated silts and sands, to; (3) massive/laminated clays and silts. In addition, tidewater glaciers can be bounded by regressive sedimentary sequences of organic muds and sands deposited in tidal flats and beaches. Reworking (ploughing, deformation, thrusting, fluidised injection) by grounded glacial bases and iceberg keels is very common but diminishes with increasing water depth/distality.

3.5.3 The glacimarine land system

Corresponding to variations in water depth, the potential for basal grounding, and the relative distality of sedimentation processes, distinctive glacimarine land systems are produced (which vary as the scale of glacierisation and both glacio-eustatic and RSL change). In fjords, these include (proximal-to-distal) submerged moraines, submarine slide scars, transverse sediment ridges, and wider arcuate ridges at fjord mouths (Dowdeswell et al., 2016). Between fjord mouths and the shelf break, large retreat moraine complexes can occur, along with streamlined glacial lineations. Hummocky terrain occurs at the shelf break and, on the shelf slope, mass wasting deposits and iceberg plough marks are produced. In the glacimarine environment beyond ice stream termini, small transverse ridges are replaced (with increasing distality) by grounding-zone wedges (rapidly accumulated glacigenic debris deposited at ice stream grounding lines during retreat-phase still-stands) and mega-scale glacial lineations. The shelf break in this context is characterised by proglacial submarine debris flows, troughs incised by dense underflows and turbidity currents and trough-mouth fans. In both contexts, the occurrence of submarine glacial lineations as far as the shelf break evidences the significant extent of direct glacigenic processes in the nearshore shelf environment. However, while direct processes cannot occur far beyond the shelf break, due to the increasing water depth, iceberg rafted detritus/deposits (IRD) are common in this deeper-water process environment.

3.6 The pelagic glacimarine process environment

Ice streams, marine-terminating glacial systems and ice shelves have a disproportionate influence on the stability of the GGS due to positive feedbacks between the glacial and marine environments. These relationships extend into the pelagic zone—the open ocean. Presently, $36–80 \times 10^6 \, \text{km}^2$ of the world ocean experiences glacimarine sedimentation; at the LGM, this area was close to $150 \times 10^6 \, \text{km}^2$. The discovery in the North Atlantic Ocean of repeated IRD layers, originating in Canada and south Greenland (Heinrich, 1988), testified to the reach of the GGS beyond the terrestrial environment. The presence of pelagic IRD layers is a consequence of the enormous scale of both iceberg discharges and their parent ice sheets. The petrographic mix in the IRD maps the transport history of the parent ice sheets before reaching the intertidal glacimarine iceberg calving zone. For example, Canadian components of the LIS transported rock debris from east-central Canada to the North Atlantic Ocean, calving huge numbers of icebergs that carried terrestrial glacial debris ~3000 km across the Atlantic, as far east as Ireland. These iceberg storms affected the open ocean by reducing water temperature and salinity, increasing turbidity, and changing the physical character of the seabed—all of which are environmentally significant with respect to climate and/or biology (e.g. in the Last Glacial, the concentration of the planktonic foraminfera *Neogloquadrina pachyderma* increased very significantly as a function of water temperature reduction induced by each Heinrich Event). On the eastern side of the North Atlantic, the Heinrich Events were matched by IRD events from Ireland, Scotland, and Scandinavia.

4 Synthesis—The global glacial system, Earth's cryosphere, and planetary analogues

Earth's cryosphere is not only a consequence of the functioning of the climate system but also feeds back into that system, becoming the driver of many of the key processes that generate geomorphologies characteristic of the Quaternary Period. In that sense, the identification of water ice in the polar caps, viscous flow forms, and near-surface regolith of Mars, together with associated morphological analogues of Quaternary cryogenic landforms on Earth, offers immense potential to understand planetary environmental evolution in the solar system more broadly. For example, a fundamental question concerns identifying and understanding the interaction between the ice sources and sinks on Mars that operate as genetic analogues of Earth's oceans and the GGS. This viewpoint is amplified, for example, by the discovery of nitrogen-methane glacial morphological analogues on Pluto (Witze, 2015; Ahrens et al., 2024, pp. 357–376). This finding brings with it a series of deep implications, not least the requirement that the Pluto system must involve an atmosphere that experiences fluxes of methane and nitrogen (probably modulated by orbitally induced variations in temperature) and oscillations between methane and nitrogen sources and sinks (Earle & Binzel, 2015), broadly analogous to the hydrological exchanges between Earth's oceans and GGS.

Taking another perspective, the stability of Earth's permafrost (ground—rock or sediment—that remains at a temperature $\leq 0°C$ for at least two consecutive years) is of central importance to the functioning of the planet's total environment due to the amount of carbon it stores in the form of frozen-in biomass. Over the past ~750 ka (about seven glacial-interglacial cycles), Earth's permafrost has stored 1.7×10^{12} t (1700 Gt) of carbon (Miner et al., 2022). If the permafrost were to thaw fully, this carbon would be released into the atmosphere by the production of methane through bacterial processing of the biomass (which does not occur in its frozen state). Consequently, the atmospheric concentration of carbon would increase from the present ~400 to ~1200 ppmv. The last time Earth had an atmosphere with such a high carbon concentration was during the Cretaceous Period (ending 66 Ma). In the Cretaceous, Earth had no cryosphere: average atmospheric temperatures were 10°C higher than at present; ocean temperatures reached 35°C at the equator and 15°C in polar latitudes, and sea levels reached 100 m higher than at present. The geomorphology produced by alternate cycles of freeze-thaw in permafrost containing excess ice (an ice volume greater than the pore space capacity of the associated permafrost sediments), particularly the spatial distribution of ice-wedge casts (French & Millar, 2014) and the patterned ground generated by their growth and degradation, indicates that the permafrost zone expands and contracts in phase with the GGS. The Last Permafrost Maximum (LPM) and the LGM were coeval and involved a ~52% permafrost expansion to $~35 \times 10^6$ km^2 (Lindgren et al., 2016), constituting a major morphogenetic agent. Patterned ground, closely analogous morphologically to landforms associated with freeze-thaw cycling in periglacial terrains on Earth (see Soare et al., 2024, pp. 143–192), has been found extensively across the mid-latitudes of Mars (e.g. Balme et al., 2013). However, the volume of ice at the surface and in the subsurface of Mars (to ~1–2 km depth) is not well constrained or even characterised—a significant proportion of it appears likely to be buried glacial ice (e.g. Karlsson et al., 2015), rather than just excess segregation ice. However, the balance between its stability and dynamics is likely to be crucial in determining atmospheric moisture and pressure on Mars, and in reorganising the pathways connecting planetary moisture sources to sinks. Hence, aiming to understand Earth's cryospheric integration with the total planetary environment in the Quaternary should not be a scientific goal that runs parallel to the quest to characterise and understand ice on other planets but, rather, an intimately related heuristic venture. For instance, by studying assemblages of landforms in the mid- and high-latitudes of Mars that include not only morphological analogues, but also important heterologues, of glacial and periglacial landform assemblages on Earth, the similarities and differences can be used to test morphogenetic hypotheses, infer the genesis of the Martian assemblage, and test models of atmospheric processes and climate variations. Already, it is clear that liquid water, generated from glacial melt and ground ice thaw, has been more abundant and of greater geomorphological importance on Mars than previously hypothesised or currently modelled (see Koutnik et al., 2024, pp. 101–142). Through geomorphology, the water sources and generation mechanisms can be hypothesised, and broader environmental implications can be considered, by reference to morphologically associated terrestrial analogues. However, although this methodology is robust, it is difficult (probably impossible) to find process environments on Earth (either active or relict) that are complete analogues of Martian settings—neither with respect to the processes represented in the landscape nor regarding the dimensions, quantities, and rates of key morphogenetic parameters. This problem is even deeper with respect to understanding cryogenic morphological analogues on Pluto. Hence, it is important to distinguish between landscapes on Earth that are good process analogues and those that are good climate analogues, as well as recognising that, while morphology may be process-specific, processes are rarely climatically specific.

On the basis of geomorphology, the recent morphogenetic action of liquid water on Mars seems probable but apparently requires special circumstances, possibly including some combination of: obliquity-driven climate change in the recent past; optically thin snow and ice deposits to create a solid-state greenhouse effect; and the presence of regolith salts that can depress the freezing point of the water (the presence of ammonia has similar importance in the melting behaviour of nitrogen ice glaciers on Pluto). However, in what spatial and temporal contexts could these processes operate (or co-operate) on Mars, and are geomorphological observations alone capable of adequately answering that question? Even (or, perhaps, especially) on Earth, landscapes may display an imprint of former climatic conditions but resolving specific imprints remotely, or from remote sensing data alone, is challenging. Also, experience from Earth shows that it is unlikely that any single landform can be used as a quantitative indicator of past and/or present climatic conditions. So, considering the wisdom granted by understanding the Earth system, what are the keys to a better understanding of cryo-morphogenesis on other planets, especially Mars? First, climate models and geomorphology must be reconciled for key periods e.g. the last ~20 Ma on Mars in the context of obliquity and other orbital parameters. Second, modelling of meso-scale and local relief-dependent microclimates must advance but must accommodate contradictory observations that point to divergent conclusions based on Earth analogue studies. Third, ice emplacement and precipitation patterns from global to regional scale must be better understood and reconciled with improved mapping of the spatial distribution of ice sinks. Fourth, in the case of Mars, the chronology of liquid generation from ice sinks must be determined, especially any periodicity in the processes of glacial melt and the freeze-thaw cycling of water from ground-ice sinks. Fifth, the applicability of global climate models (GCM) to local-scale morphogenesis must be questioned, with lessons taken from understanding the complexity and dimensionally hierarchic nature of process-to-landscape pathways on Earth.

5 Conclusions

The GGS has played a central role in the functioning of the Quaternary cryosphere and in shaping the Quaternary Period—and will continue to do so in a future dominated by anthropogenic climate change. The Earth-surface materials and landforms produced in this environmental context are sufficiently distinctive and different from those associated with the warmer, dominantly ice-free periods of the previous ~250 Ma that is clear that Earth's cryosphere, and the GGS that dominates it, are consequences of dynamic external controls and complex internal feedbacks. These are important insights with respect to understanding the Earth system, reconciling paleo-climate models with geomorphology, and in using Earth as an analogue to understand cryo-geomorphology on other planets as a proxy for their total environmental evolution.

Acknowledgements

The author expresses his sincere thanks to Michelle Koutnik for reading earlier versions of this chapter and making many helpful recommendations. He is also grateful to the formal reviewers of the chapter for their time, effort, and constructive criticisms.

References

Ahrens, C. J., Lisse, C. M., Williams, J.-P., & Soare, R. J. (2024). *Geocryology of Pluto and the icy moons of Uranus and Neptune*. In R. J. Soare, J.-P. Williams, C. Ahrens, F. E. G. Butcher, & M. R. El-Maarry (Eds.), *Ices in the solar system, a volatile-driven journey from the inner solar system to its far reaches*. Elsevier Books.

Arnold, N. S., Butcher, F. E. G., Conway, S. J., Gallagher, C., & Balme, M. R. (2022). Surface topographic impact of subglacial water beneath the south polar ice cap of Mars. *Nature astronomy* (in press, https://doi.org/10.1038/s41550-022-01782-0).

Ashley, G. M. (1988). Classification of glaciolacustrine sediments. In R. P. Goldthwaite & C. L. Matsch (Eds.), *Genetic classification of Glacigenic deposits* (pp. 243–260). Rotterdam: Balkema.

Atkins, C. B., Barrett, P. J., & Hicock, S. R. (2002). Cold glaciers erode and deposit: Evidence from Allan Hills, Antarctica. *Geology*, *30*, 659–662.

Balme, M. R., Gallagher, C. J., & Hauber, E. (2013). Morphological evidence for geologically young thaw of ice on Mars: A review of recent studies using high-resolution imaging data. *Progress in Physical Geography*, *37*(3), 289–324. doi.org/10.1177/0309133313477123.

Bennett, M. R. (2003). Ice streams as the arteries of an ice sheet: Their mechanics, stability and significance. *Earth-Science Reviews*, *61*(3–4), 309–339. https://doi.org/10.1016/S0012-8252(02)00130-7.

Bockheim, J. G. (2010). Soil preservation and ventifact recycling from dry-based glaciers in Antarctica. *Antarctic Science*, *22*, 409–417. https://doi.org/10.1017/s0954102010000167.

Börker, J., Hartmann, J., Amann, T., & Romero-Mujalli, G. (2018). Global unconsolidated Sediments Map Database v1.0 (shapefile and gridded to 0.5 spatial resolution). *PANGAEA*. https://doi.org/10.1594/PANGAEA.884822. Supplement to: Börker, J et al. (2018): Terrestrial sediments of the earth: Development of a global unconsolidated sediments map database (GUM). Geochemistry, Geophysics, Geosystems 19(4), 997–1024, https://doi.org/10.1002/2017GC007273.

Braithwaite, R., & Raper, S. C. B. (2009). Estimating equilibrium-line altitude (ELA) from glacier inventory data. *Annals of Glaciology*, *50*, 127–132. https://doi.org/10.3189/172756410790595930.

Christner, B. C., Skidmore, M. L., Priscu, J. C., Tranter, M., & Foreman, C. M. (2008). Bacteria in subglacial environments. In R. Margesin, F. Schinner, J. C. Marx, & C. Gerday (Eds.), *Psychrophiles: From biodiversity to biotechnology*. Berlin, Heidelberg: Springer. https://doi.org/10.1007/978-3-540-74335-4_4.

Christoffersen, P., & Tulaczyk, S. (2003). Response of subglacial sediments to basal freeze-on 1. Theory and comparison to observations from beneath the West Antarctic Ice Sheet. *Journal of Geophysical Research*, *108*(B4), 2222. https://doi.org/10.1029/2002JB001935.

Collier, E., Maussion, F., Nicholson, L. I., Molg, T., Immerzeel, W. W., & Bush, A. B. G. (2015). Impact of debris cover on glacier ablation and atmosphere–glacier feedbacks in the Karakoram. *The Cryosphere*, *9*, 1617–1632. https://doi.org/10.5194/tc-9-1617-2015.

Creyts, T. T., & Schoof, C. G. (2009). Drainage through subglacial water sheets. *Journal of Geophysical Research: Earth Surface*, *114*, 2009.

Cuffey, K., Conway, H., Gades, A., Hallet, B., Lorrain, R., Severinghaus, J., Steig, E., Vaughn, B., & White, J. (2000). Entrainment at cold glacier beds. *Geology*, *28*, 351–354. https://doi.org/10.1130/0091-7613(2000)28<351:EACGB>2.0.CO;2.

Davies, B., Carrivick, J., Glasser, N., Hambrey, M., & Smellie, J. (2012). Variable glacier response to atmospheric warming, northern Antarctic Peninsula, 1988-2009. *The Cryosphere*, *6*(5), 1031–1048. https://doi.org/10.5194/tc-6-1031-2012.

Dowdeswell, J. A., Canals, M., Jakobsson, M., Todd, B. J., Dowdeswell, E. K., & Hogan, K. A. (2016). Geological society. *London, Memoirs*, *46*(1), 3. https://doi.org/10.1144/M46.171.

Dowdeswell, J. A., & Siegert, M. J. (2003). The physiography of modern Antarctic subglacial lakes. *Global and Planetary Change*, *35*(34), 221236. https://doi.org/10.1016/S0921-8181(02)00128-5.

Dreimanis, A. (1988). Tills: Their genetic terminology and classification. In R. P. Goldthwaite & C. L. Matsch (Eds.), *Genetic classification of Glacigenic deposits* (pp. 17–83). Rotterdam: Balkema.

Earle, A. M., & Binzel, R. P. (2015). Pluto's insolation history: Latitudinal variations and effects on atmospheric pressure. *Icarus, 250*, 405–412. https://doi.org/10.1016/j.icarus. 2014.12.028.2015.

Eicher, O., Baumgartner, M., Schilt, A., Schmitt, J., Schwander, J., Stocker, T. F., & Fischer, H. (2016). *NOAA/ WDS paleoclimatology - NGRIP ice Core 120,000 year Total air content data. [indicate subset used]*. NOAA National Centers for Environmental Information. https://doi.org/10.25921/n0d0-ng26. Accessed 26 May 2022.

EPICA community members. (2004). Eight glacial cycles from an Antarctic ice core. *Nature, 429*, 623–628. https://doi.org/10.1038/nature02599.

Fitzpatrick, A. A. W., Hubbard, A. L., Box, J. E., Quincey, D. J., van As, D., Mikkelsen, A. P. B., Doyle, S. H., Dow, C. F., Hasholt, B., & Jones, G. A. (2014). A decade (2002−2012) of supraglacial lake volume estimates across Russell Glacier, West Greenland. *The Cryosphere, 8*, 107–121. https://doi.org/10.5194/tc-8-107-2014.

French, H. M., & Millar, S. W. S. (2014). Permafrost at the time of the last glacial maximum (LGM) in North America. *Boreas, 43*, 667–677. https://doi.org/10.1111/bor.12036. ISSN 0300-9483.

Furbish, D. J., & Andrews, J. (1984). The use of hypsometry to indicate long-term stability and response of valley glaciers to changes in mass transfer. *Journal of Glaciology, 30*, 199–211. https://doi.org/10.3189/s0022143000005931.

Gallagher, C., Balme, M., & Clifford. (2018). Discriminating between the roles of late Pleistocene palaeodischarge and geological-topographic inheritance in fluvial longitudinal profile and channel development. *Earth Surface Processes and Landforms, 43*, 444–462.

Gallagher, C., Sutton, G., & Bell, T. (2004). Submerged ice marginal forms in the Celtic Sea off Waterford harbour, Ireland: Implications for understanding regional glaciation and sea level changes following the last glacial maximum in Ireland. *Irish Geography, 37*(2), 145–165. doi.org/10.1080/00750770409555839.

Gallagher, C., Telfer, M. W., & Cofaigh, C.Ó. (2015). A marine isotope stage 4 age for Pleistocene raised beach deposits near Fethard, southern Ireland. *Journal of Quaternary Science, 30*(8), 754–763.

Gallagher, C. (2024). *Glaciation and glacigenic geomorphology on Earth in the Quaternary Period*. In R. J. Soare, J.-P. Williams, C. Ahrens, F. E. G. Butcher, & M. R. El-Maarry (Eds.), *Ices in the solar system, a volatile-driven journey from the inner solar system to its far reaches*. Elsevier Books.

Glasser, N. F., & Bennett, M. R. (2004). Glacial erosional landforms: Origins and significance for palaeoglaciology. *Progress in Physical Geography, 28*, 43–75.

Grau Galofre, A., Lasue, J., & Scanlon, K. (2024). *Ice on Noachian and Hesperian Mars: Atmospheric, surface, and subsurface processes*. In R. J. Soare, J.-P. Williams, C. Ahrens, F. E. G. Butcher, & M. R. El-Maarry (Eds.), *Ices in the solar system, a volatile-driven journey from the inner solar system to its far reaches*. Elsevier Books.

Hambrey, M. J., & Fitzsimmons, S. J. (2010). Development of sediment-landform associations at cold glacier margins, Dry Valleys, Antarctica. *Sedimentology, 57*, 857–882.

Hambrey, M. J., & Glasser, N. F. (2012). Discriminating glacier thermal regimes in the sedimentary record. *Sedimentary Geology, 251*, 1–33.

Heinrich, H. (1988). Origin and consequences of cyclic ice rafting in the Northeast Atlantic Ocean during the past 130,000 years. *Quaternary Res, 29*(2), 142–152. https://doi.org/10.1016/0033-5894(88)90057-9.

Juen, M., Mayer, C., Lambrecht, A., Han, H., & Liu, S. (2014). Impact of varying debris cover thickness on ablation: A case study for Koxkar glacier in the Tien Shan. *The Cryosphere, 8*(2), 377–386. https://doi.org/10.5194/tc-8-377-2014.

Karlsson, N. B., Schmidt, L. S., & Hvidberg, C. S. (2015). Volume of Martian mid-latitude glaciers from radar observations and ice-flow modelling. *Geophysical Research Letters, 2015*. https://doi.org/10.1002/2015GL063219.

Konrad, H., Shepherd, A., Gilbert, L., Hogg, A. E., McMillan, M., Muir, A., & Slater, T. (2018). Net retreat of Antarctic glacier grounding lines. *Nature Geoscience, 11*, 258–262. https://doi.org/10.1038/s41561-018-0082-z.

Koutnik, M., Butcher, F., Soare, R., Hepburn, A., Hubbard, B., Brough, S., … Pathare, A. (2024). *Glacial deposits, remnants, and landscapes on Amazonian Mars: Using setting, structure, and stratigraphy to understand ice evolution and climate history.* In R. J. Soare, J.-P. Williams, C. Ahrens, F. E. G. Butcher, & M. R. El-Maarry (Eds.), *Ices in the solar system, a volatile-driven journey from the inner solar system to its far reaches.* Elsevier Books.

Krabbendam, M., Eyles, N., Putkinen, N., Bradwell, T., & Arbelaez-Moreno, L. (2016). Streamlined hard beds formed by palaeo-ice streams: A review. *Sedimentary Geology, 338*, 24–50. https://doi.org/10.1016/J.SEDGEO.2015.12.007 (Elsevier, open access under CC Licence).

Lindgren, A., Hugelius, G., Kuhry, P., Christensen, T. R., & Vandenberghe, J. (2016). GIS-based maps and area estimates of northern hemisphere permafrost extent during the last glacial maximum. *Permafrost and Periglacial Processes, 27*, 6–16. https://doi.org/10.1002/ppp.1851.

Lisiecki, L. E., & Raymo, M. E. (2005). A Pliocene-Pleistocene stack of 57 globally distributed benthic δ^{18}O records. *Paleoceanography, 20*, PA1003. https://doi.org/10.1029/2004PA001071.

Livingstone, S. J., Li, Y., Rutishauser, A., et al. (2022). Subglacial lakes and their changing role in a warming climate. *Nature Reviews Earth & Environment, 3*, 106124. https://doi.org/10.1038/s43017-021-00246-9.

Lüthi, D., Le Floch, M., Bereiter, B., et al. (2008). High-resolution carbon dioxide concentration record 650,000–800,000 years before present. *Nature, 453*, 379–382. https://doi.org/10.1038/nature06949.

Lüthi, M. P., Ryser, C., Andrews, L. C., Catania, G. A., Funk, M., Hawley, R. L., Hoffman, M. J., & Neumann, T. A. (2015). Heat sources within the Greenland ice sheet: Dissipation, temperate paleo-firn and cryo-hydrologic warming. *The Cryosphere, 9*, 245–253. https://doi.org/10.5194/tc-9-245-2015, 2015.

Mantripp, D. N., Ridley, J. K., & Rapley, C. G. (1992). Antarctic map from the Geosat Radar Altimeter Geodetic Mission. *Eos Transactions American Geophysical Union, 37*(38), 610.

McGrath, D., Sass, L., O'Neel, S., Arendt, A., & Kienholz, C. (2017). Hypsometric control on glacier mass balance sensitivity in Alaska and northwest Canada. *Earth's Future, 5*, 324–336. https://doi.org/10.1002/2016EF000479.

Meehan, R. T., Warren, W. P., & Gallagher, C. J. (1996). The sedimentology of a Late Pleistocene drumlin near Kingscourt, Ireland. *Sedimentary Geology, 111*(1–4), 91–105. https://doi.org/10.1016/S0037-0738(97)00008-0.

Miner, K. R., Turetsky, M. R., Malina, E., Bartsch, A., Tamminen, J., Maguire, A. D., Fix, A., Sweeney, C., Elder, C. D., & Miller, C. E. (2022). Permafrost carbon emissions in a changing Arctic. *Nature Reviews Earth & Environment, 3*, 55–67. https://doi.org/10.1038/s43017-021-00230-3.

Näslund, J. O. (1997). Subglacial preservation of valley morphology at Amundsenisen, western Dronning Maud Land, Antarctica. *Earth Surface Processes and Landforms, 22*, 441–455. Quaternary Science Reviews, 21, 8–9, 879–887.

Orosei, R., Lauro, S. E., Pettinelli, E., Cicchetti, A., Coradini, M., Cosciotti, B., di Paolo, F., Flamini, E., Mattei, E., Pajola, M., Soldovieri, F., Cartacci, M., Cassenti, F., Frigeri, A., Giuppi, S., Martufi, R., Masdea, A., Mitri, G., Nenna, C., … Seu, R. (2018). Radar evidence of subglacial liquid water on Mars. *Science, 361*, eaar7268. https://doi.org/10.1126/science.aar7268.

Östrem, G. (1959). Ice melting under a thin layer of moraine, and the existence of ice cores in moraine ridges. *Geografiska Annaler, 41*(4), 228–230. http://www.jstor.org/stable/4626805.

Petit, J. R., Jouzel, J., Raynaud, D., Barkov, N. I., Barnola, J. M., et al. (1999). Climate and atmospheric history of the past 420,000 years from the Vostok ice Core, Antarctica. *Nature, 399*, 429–436.

Praetorius, S. K., Mix, B., Jensen, D., Froese, G., Milne, M., Wolhowe, J., Addison, F., & Prahl. (2016). Interaction between climate, volcanism, and isostatic rebound in Southeast Alaska during the last deglaciation. *Earth and Planetary Science Letters, 452*, 79–89.

Radd, F., & Oertle, D. (1968). Melting point behaviour of glacier ice. *Nature, 218*, 1242. https://doi.org/10.1038/2181242a0.

Ray, N., & Adams, J. M. (2001). A GIS-based vegetation map of the world at the last glacial maximum (25,000-15,000 BP). *Internet Archaeology, 11*. https://doi.org/10.11141/ia.11.2.

Remy, F., Mazzega, P., Houry, S., Brossier, C., & Minster, J. F. (1989). Mapping of the topography of continental ice by inversion of satellite-altimeter data. *Journal of Glaciology, 35*, 98107.

Ridley, J. K., Cudlip, W., & Laxon, S. W. (1993). Identification of subglacial lakes using ERS-1 radar altimeter. *Journal of Glaciology, 39*, 625634.

Rignot, E., Mouginot, J., & Scheuchl, B. (2011). Ice flow of the Antarctic ice sheet. *Science, 333*, 1427–1430. https://doi.org/10.1126/science.1208336.

Rignot, E., & Thomas, R. H. (2002). Mass balance of polar ice sheets. *Science, 297*, 1502–1506. https://doi.org/10.1126/science.1073888.

Robel, A., Wilson, E., & Seroussi, H. (2021). *Layered seawater intrusion and melt under grounded ice*. https://doi.org/10.5194/tc-2021-262.

Rothlisberger, H. (1972). Water pressure in intra- and subglacial channels. *Journal of Glaciology, 11*, 172–203.

Roy, K., & Peltier, W. R. (2015). Glacial isostatic adjustment, relative sea level history and mantle viscosity: Reconciling relative sea level model predictions for the U.S. East coast with geological constraints. *Geophysical Journal International, 201*, 1156–1181. https://doi.org/10.1093/gji/ggv066.

Soare, R. J., Costard, F., Williams, J.-P., Gallagher, C., Hepburn, A. J., Stillman, D., … Godi, E. (2024). *Evidence, arguments, and cold-climate geomorphology that favour periglacial cycling at the Martian mid-to-high latitudes in the Late Amazonian Epoch*. In R. J. Soare, J.-P. Williams, C. Ahrens, F. E. G. Butcher, & M. R. El-Maarry (Eds.), *Ices in the solar system, a volatile-driven journey from the inner solar system to its far reaches*. Elsevier Books.

Sternai, P., Caricchi, L., Pasquero, C., Garzanti, E., van Hinsbergen, D. J. J., & Castelltort, S. (2020). Magmatic forcing of Cenozoic climate? *Journal of Geophysical Research: Solid Earth, 125*, e2018JB016460. https://doi.org/10.1029/2018JB016460.

Sugden, D. (1973). Landscapes of glacial erosion in Greenland and their relationship to ice, topographic and bedrock conditions. *Institute of British Geographers Special Publication, 7*, 177–195.

Tushingham, A. M., & Peltier, W. R. (1992). Validation of the ICE-3G model of Wurm-Wisconsin deglaciation using a global data-base of relative sea-level histories. *Journal of Geophysical Research: Solid Earth, 97*(B3), 3285–3304.

Warren, W. P., & Ashley, G. M. (1994). Origins of the ice-contact stratified ridges (eskers) of Ireland. *Journal of Sedimentary Research, A64*, 433–444.

Witze, A. (2015). Nitrogen glaciers flow on Pluto. *Nature*. https://doi.org/10.1038/nature.2015.18062.

Yuen, D., & Schubert, G. (1979). The role of shear heating in the dynamics of large ice masses. *Journal of Glaciology, 24*(90), 195–212. https://doi.org/10.3189/S002214300001474X.

Zeitz, M., Reese, R., Beckmann, J., Krebs-Kanzow, U., & Winkelmann, R. (2021). Impact of the melt–albedo feedback on the future evolution of the Greenland ice sheet with PISM-dEBM-simple. *The Cryosphere, 15*(12), 5739–5764. https://doi.org/10.5194/tc-15-5739-2021.

Ice on Noachian and Hesperian Mars: Atmospheric, surface, and subsurface processes

Anna Grau Galofre[a],*, Jeremie Lasue[b],*, and Kat Scanlon[c],*

[a]*Laboratoire de Planétologie et Géosciences, CNRS UMR 6112, Nantes Université, Université d'Angers, Le Mans Université, Nantes, France,* [b]*Institut de Recherche en Astrophysique et Planétologie, Université de Toulouse, CNRS, CNES, Toulouse, France,* [c]*Planetary Geology, Geophysics, and Geochemistry Laboratory, NASA Goddard Space Flight Center, Greenbelt, MD, United States*

Abstract

Much of the Noachian-Hesperian (i.e., >~3.0 Ga) geology of Mars remains exposed at the surface, providing a unique opportunity for insight into the early history of an Earth-like planet. Water ice, an important component of present-day Martian geology, was even more abundant in the ancient past. This chapter presents an integrated view of ice on early Mars, discussing the spatial extent, glacio-geology, and climate implications of an early Martian cryosphere, and discusses repercussions for the water budget and its fate that are relevant to later stages of Mars' history (Koutnik et al., 2024, pp. 101–142). In this chapter, we review the predicted distribution of surface ice under various early Mars climate scenarios, compare these hypothetical ice sheets to the locations on Mars where geological evidence of ancient glaciations has been described, and consider the extent to which valley networks are consistent with an ice-related origin. Finally, we consider the sequestration of ice beneath the surface of Mars as the climate cooled following the time of the valley network incision. The abundance of new data from Mars exploration by satellites and rovers in recent decades, along with advances in climate modeling, has allowed for novel insights about the role of ice on early Mars.

1 Introduction

Mars is a cold planet relative to Earth (mean global surface temperature of approximately −65°C), located between Earth and the asteroid belt and at 1.5 AU from the Sun. The geological history of over 4.5 Ga (Fig. 1) is better preserved on Mars than on the Earth, given a lack of plate tectonics and exceedingly slow erosion rates (Carr, 2007). With a large water inventory (thought to have been even more significant in early periods of its history; e.g., as observed in Carr & Head, 2015; Jakosky, 2021; Kite & Daswani, 2019; Lasue et al., 2013), a thicker early atmosphere, and chaotic obliquity/eccentricity variations throughout history (e.g., Baker, 2001; Carr & Head, 2015; Laskar et al., 2004; Wordsworth, 2016), Mars poses a fascinating case study for the evolution of ice on a terrestrial planet and its influence in shaping surface morphology.

---

* Authors are listed in alphabetical order and contributed equally to this work.

Ices in the Solar System. https://doi.org/10.1016/B978-0-323-99324-1.00005-5

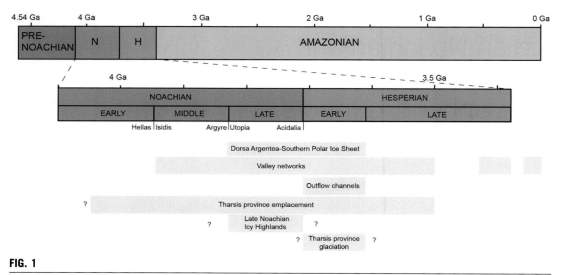

FIG. 1

Geologic time line showing the three major divisions in Mars' history (Noachian, Hesperian, Amazonian), with a focus on the 'early Mars' Noachian and Hesperian eras, described in this chapter. A summary of the most representative events, as well as the timing of major glacial episodes, is shown in the figure. Age boundaries presented follow the Neukum chronology system (Hartmann & Neukum, 2001).

The topography of Mars (Fig. 2) is primarily a product of volcanism and impact cratering (e.g., Smith et al., 2001), rather than of plate tectonics as on Earth (e.g., Black et al., 2017). In the absence of a sea level, elevations for points on Mars are given relative to a reference equipotential surface (Smith et al., 2001). The dichotomy boundary, coinciding roughly with the 0m reference elevation, divides the planet into lowland plains in most of the northern hemisphere and crater-studded highlands in the southern hemisphere. The crustal dichotomy is generally thought to be the result of a large impact or series of impacts in the early Noachian era (Marinova et al., 2008; Pan et al., 2017), but a mantle plume hypothesis is also plausible (e.g., Citron et al., 2018; Šrámek & Zhong, 2012). The equatorial Tharsis bulge (e.g., Bouley et al., 2016; Phillips et al., 2001) is a volcanic construct with an elevation of up to +7km and upon which the Tharsis Montes volcanoes stand up to +18km. Two large impact basins, Hellas and Argyre, occur in the southern hemisphere (e.g., Bernhardt et al., 2016; Hiesinger & Head, 2002). The floor of the Hellas basin, with elevations ~5–8km below the reference surface, is the lowest point on the planet.

This chapter explores the glacio-geological history of early Mars, including atmospheric, glacial, and deep cryospheric processes. We use the term "glaciation" to describe the suite of processes involving the accumulation of surface crystalline ice. We use the term "deep cryosphere" to describe ice and ice-rich substrate located below the depth of direct interaction with the atmosphere. The Noachian and Hesperian eras are thought to have been characterized by a thicker atmosphere, lower insolation, and a larger water budget than the present day (Carr & Head, 2015; Wordsworth, 2016). Characteristics such as a lower surface gravity than Earth, a north-south topographic dichotomy, and chaotic obliquity/eccentricity variations have remained throughout Mars'

FIG. 2

Geography of Mars as referenced in this chapter. THEMIS daytime IR 100 m global base map (Edwards et al., 2011) with Mars Orbiter Laser Altimeter (MOLA) elevation color overlay (Smith et al., 2003). Valley networks as mapped by Hynek et al. (2010) are shown in *blue-green*.

history. To describe the current knowledge of glacial and ground ice on early Mars, we consider three questions:

1. *How was ice distributed at the surface of Mars in the Noachian and Hesperian?* In the thin present-day Martian atmosphere, the ground surface temperature at any point on the planet is determined almost entirely by the sunlight absorbed at that point (e.g., Read & Lewis, 2004). As such, at the present-day spin-axis obliquity of ~25°, surface ice is stable only at the north and south poles. If early Mars had a higher surface pressure, as geomorphologic, geochemical, isotopic, and escape rate evidence suggest it did (e.g., Bristow et al., 2017; Jakosky, 2021; Kite, 2019; Kite et al., 2014; Manga et al., 2012; Warren et al., 2019), the greater heat capacity of the atmosphere would allow much more heat exchange between the atmosphere and the surface (or the greater greenhouse effect would result in more longwave radiation to the surface from the atmosphere; cf. Fan et al., 2022). In this case, ground surface temperatures would decrease with altitude as atmospheric temperatures do, and ice would be expected to accumulate at high altitudes (Forget et al., 2013; Wordsworth et al., 2013). We will describe the scenarios currently under study by climate modelers, and the geologic evidence for ice in Noachian and Hesperian terrains (Fig. 3).

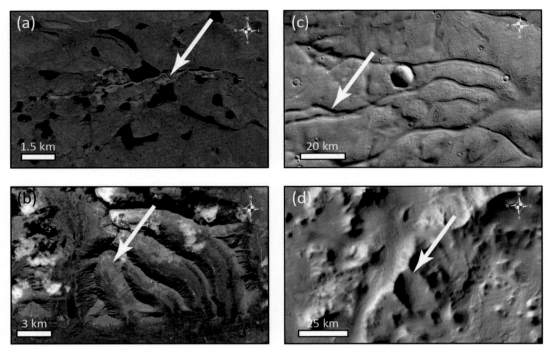

FIG. 3

Example terrestrial analogue features (a, b) and Martian equivalent landforms (c, d). (a) Terrestrial esker centered around 66.21 N, 111.78 W (Kugluktuk land, NU, Canada). (b) Glacial cirques at 62.64 N, 128.57 W (Nááts'įhch'oh National Park Reserve, AK, United States). (c) Dorsa Argentea eskers, centered around 76.81S, 317.9E. (d) Putative glacial cirques in the Charitum Montes, southern Argyre rim (center: 55.41S, 307.8E).

Panel (b): Esri/Maxar/ArcGIS; panel (c): Credit: CTX/MSSS/NASA; panel (d): THEMIS/ASU/JPL-Caltech/NASA.

2. *What fraction of observed fluvial geology on Mars is due to ice melt?* The crater lake systems and valley networks in the equatorial and southern highlands of Mars are generally agreed to have formed by the action of liquid water (e.g., Carr, 1995, 1999; Craddock & Howard, 2002; Fassett & Head, 2008b; Goudge et al., 2015; Gulick, 2001; Hynek et al., 2010), but there is less consensus as to whether that water fell to the surface as rain or resulted from snow and/or ice melt. Some researchers have argued that the Martian valley networks and crater lakes could have been carved primarily by meltwater, either as snowmelt (e.g., Carr & Head, 2003; Wordsworth et al., 2015), as top-down melting from a hypothesized ice sheet covering the highlands (e.g., Fastook & Head, 2015), or as subglacial channelized melt from a similar environment (e.g., Grau Galofre et al., 2020b). Others hold that they indicate a relatively temperate climate in the Noachian-Hesperian with rainfall and/or groundwater seepage (e.g., Carr, 1995; Craddock & Howard, 2002; Gulick, 2001; Howard et al., 2005; Hynek et al., 2010; Irwin et al., 2011, 2005;

Kite, 2019; Ramirez & Craddock, 2018). We will review the evidence for both end-members and consider intermediate scenarios.

3. *How did the subsurface water inventory and cryosphere evolve with time, and what is its effect on the global water inventory?* Finally, we consider the possibility of ice beneath the surface of Mars, modeling the thickness and depth of the subsurface cryosphere as a function of latitude, brine composition, and orbital parameters. We discuss the proportion of the total Martian water inventory that could be confined in the subsurface and the observational evidence on which these calculations are based. This informs the availability of ice for processes on recent Mars (as discussed by Koutnik et al., 2024, pp. 101–142).

To preface our in-depth discussion of early Mars glacial geology, we note that precise dating of glacial and periglacial landscapes is not without controversy. In addition to the broader uncertainties in dating the surface of Mars with impact crater size-frequency distributions (e.g., Robbins et al., 2013), three problems arise with the dating of glacial terrains on Mars. First, these techniques provide the modeled age of terrain, not of landforms within it, which can be much younger. Buffered crater counting methods, such as those developed for valley networks and eskers (e.g., Fassett & Head, 2008a; Kress & Head, 2015), provide more accurate constraints on the age of a landform than crater counting of the underlying geology but are not always feasible given the small areal extent of individual landforms. Second, glacial landscapes formed in the Noachian and Hesperian have undergone over 3 Ga of degradation, including crater gardening, mass wasting, intermittent glacial/fluvial erosion, and mantling, among other processes. Therefore, the interpretation of glacial landform assemblages as part of a group of the same age is further hindered due to the different levels of degradation of individual features, including older craters. The third issue involves the role of an ice cover in preventing smaller impacts from penetrating the surface. The decoupling of surface and atmosphere by an ice layer is indeed a major issue for dating glacial landscapes even on Earth, where isotopic dating of terrains provides ages of ice retreat and not erosion rates or dates of landform emplacement (e.g., Balco, 2011 and references within). Modeled ages derived from crater counting in regions with extensive ice cover may record the time-integrated effect of intermittent surface cover, resulting in a crater counting age much younger than the true age of the landscape due to ice cover filtering of smaller impacts. Furthermore, crater counting dates would reflect the timing of ice retreat from a subglacial landform and not of subglacial landform emplacement. This ice-crater filtering effect depends at least on ice sheet thickness and projectile momentum upon impact, but concrete numbers require modeling efforts that have not yet been performed for Mars and are thus a necessary future line of research.

Fig. 4 illustrates these issues. The age of the surface is 4 Ga (panel a), but crater dating landforms (1) and (2) yield at best ice retreat ages (3.6 Ga; panel c) and not the time of emplacement (3.8 Ga; panel b). Degradation of craters and subsequent resurfacing (represented by color changes; panel d) further complicates the dating of individual landforms. Landforms (1) and (2) could have formed simultaneously under the ice sheet, but their age dates according to the crater counting techniques differ by 200 Myr, due to the different exposure times. Similarly, any dating of glacial landforms will refer to "exposure dates" and not reflect the time of emplacement or formation. This challenge becomes more acute if a surface has been influenced by multiple glacial cycles and/or postglacial revision. Accurate dating of glacial landforms is a major unresolved issue in early Mars glacial geomorphology.

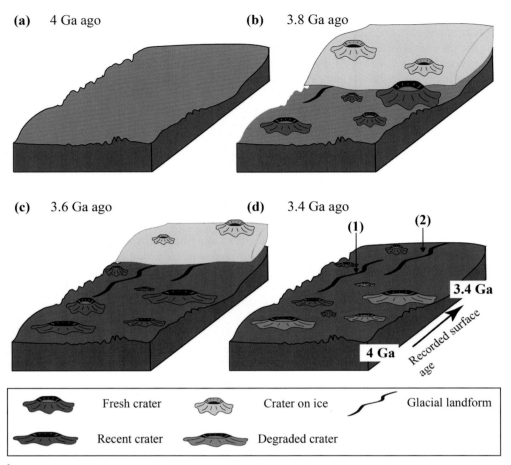

FIG. 4

Cartoon illustrating the issues of crater dating surface modification in ice-covered surfaces. Panel (a) shows an originally emplaced, pristine surface 4 Ga ago, aged 4 Ga. Panel (b) corresponds to the same surface covered by an ice sheet at 3.8 Ga. Fresh craters on land and ice are shown, and a glacial landform (e.g., esker) emerges from under the ice. Panel (c) shows ice retreat 3.6 Ga ago, exposing more of the original landform and a second one. Fresh and recent craters are shown on land and ice. Panel (d) shows the final surface after ice retreat, at $t = 3.4$ Ga. Crater size distribution and density change from the surface that was always fully exposed, to that covered by an ice layer, yielding apparently different age dates. The emplacement ages of eskers (1) and (2) are unknown, whereas exposure ages are given by crater counting techniques.

2 Question 1: How was ice distributed at the surface of Mars in the Noachian and Hesperian?

While the MAVEN mission confirmed that the inventory of water on early Mars was 5–20 times larger than on present-day Mars (Jakosky, 2021), geologic constraints on the atmospheric composition and pressure of early Mars are few and imprecise (Kite, 2019). However, climate hypotheses can also be

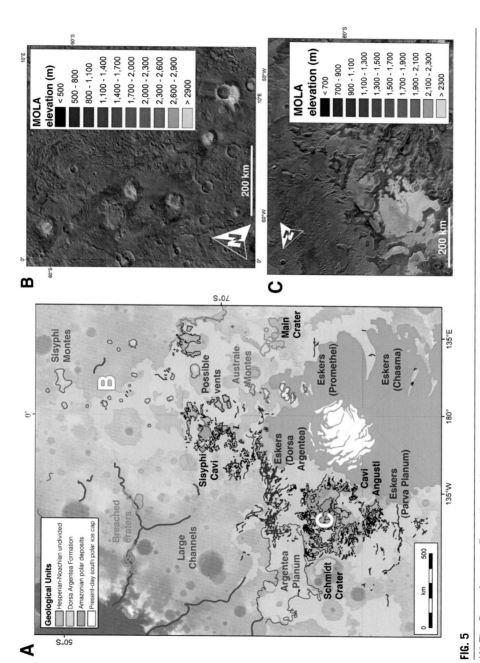

FIG. 5

(A) The Dorsa Argentea Formation, with units as mapped by Tanaka et al. (2014) and hypothesized glacial, glaciovolcanic, and glaciofluvial features as mapped by Scanlon et al. (2018). (B) The flat-topped, steep-sided Sisyphi Montes are interpreted as subglacial volcanoes, or tuyas (Ghatan & Head, 2004); MOLA altitude shading (Smith et al., 2003) on THEMIS daytime IR 100 m global base map (Edwards et al., 2011). (C) The Cavi Angusti are interpreted as glacial melt features (Ghatan et al., 2003); MOLA altitude shading on THEMIS daytime IR 100 m global base map.

after the completion of true polar wander due to low-obliquity excursions or the melting of ice that had been emplaced before true polar wander. Alternatively, true polar wander may not be necessary to explain the distribution of valley networks (cf. Guzewich et al., 2021).

For reasonable estimates of the total Noachian/Hesperian water inventory, some degree of warming above the CO_2/H_2O baseline was likely required for basal melting of the south polar ice sheet (Scanlon et al., 2018). Along with the strong likelihood of Tharsis glaciation in any warm climate, this underscores that even if the climate of early Mars was warm enough for rainfall in some regions, glacial processes would still have played an important role in others. It is increasingly clear (Guzewich et al., 2021; Kamada et al., 2020; Schmidt et al., 2022; Wordsworth et al., 2018) that both the "warm and wet" and "cold and icy" end-members are tools for understanding the complex whole of ancient Mars paleoclimate rather than complete pictures in themselves.

3 Question 2: What fraction of observed fluvial geology on Mars is due to ice melt?

Thousands of valley networks incise the southern Martian highlands, speaking of a past with surface water activity. Whereas the community has largely interpreted these drainage systems in terms of a "warm and wet" environment, the possibility that a more or less significant fraction of these valleys formed through glaciofluvial processes related to an "Icy Highlands" ice sheet has been increasingly considered (Bouquety et al., 2019, 2020; Cassanelli & Head, 2019; Fastook & Head, 2015; Grau Galofre et al., 2020b). In this scenario, the formation of valley networks involves sporadic or continued production of meltwater, encompassing different degrees of ice cover that include melting of snowpacks, ice-marginal streams sourced from supraglacial meltwater, and subglacial streams. This hypothesis is supported by statistical analyses of valley network morphometry, by erosion rates of analog snow and ice melt valleys in Antarctica, and by the presence of morphological characteristics (constant width downstream, longitudinal profile undulations, discontinuous valley segments, etc.) that are consistent with subglacial, but not fluvial, erosion (Cassanelli & Head, 2019; Fastook & Head, 2015; Grau Galofre et al., 2020b; Lee et al., 1999).

In the "warm and wet" hypothesis, the origin of the Martian valleys is generally attributed to rainfall and surface runoff, which may or may not involve contributions from groundwater seepage (Craddock & Howard, 2002; Gulick, 2001; Howard et al., 2005; Hynek et al., 2010; Irwin et al., 2011; Lamb et al., 2006; Malin & Carr, 1999). Support for this hypothesis comes from the similarity in planform morphology to possible analog rivers on Earth, as well as morphometrical quantities such as drainage density, network topology, and relationship with drainage divides (e.g., Craddock & Howard, 2002; Hynek et al., 2010). This environmental end-member often contemplates the existence of a northern ocean, which would have occupied the northern lowlands extending north from the dichotomy boundary, along two proposed shorelines: Deuteronilus and Arabia. This hypothesis is consistent with the distribution of deltas, the elevation of the equipotential curve, and possible tsunami deposits (Baker et al., 1991; Costard et al., 2017, 2019; De Blasio, 2020; Di Achille & Hynek, 2010; Perron et al., 2007; Rodriguez et al., 2016). However, rainfall and/or melt of snowpacks fails to explain aspects of valley network morphology, in particular the lack of dissection between consecutive drainage basins, and the invariant widths along the valleys (Grau Galofre et al., 2020b).

Grau Galofre et al. (2020b) studied the origin of a subset of the global database of valley networks and classified them by their most likely origin based on statistical morphometry. The study found that, whereas rivers and sapping valleys explain a large percentage of valley network morphologies, a fraction of them is most consistent with channels formed under ancient ice sheets, akin to the Finger Lakes in New York, or the subglacial channel networks found on Devon Island (Canadian Arctic Archipelago, Grau Galofre et al., 2020b; Lee et al., 1999). Morphological similarities between valley networks and subglacial channels, noted also regionally in Bouquety et al. (2019, 2020), as well as the spatial correlation of valley network locations with the near-marginal regions of a proposed late Noachian icy highlands ice sheet provide support for the existence of such an ice sheet with the presence of localized basal melt (Bouquety et al., 2019, 2020; Fastook & Head, 2015; Grau Galofre et al., 2020b). Whereas the exact origin of the Martian valley networks is still disputed, there is a general consensus that they were not active for a long period of time, and that they did not achieve a large degree of landscape modification (Fassett & Head, 2008a; Grau Galofre et al., 2020a; Penido et al., 2013; Som et al., 2009).

Fretted terrain, located in the transitional zone between the southern highlands and the northern lowlands, and in particular Aeolis Mensae and Deuteronilus-Protonilus Mensae (133°–147°E and 15°–55°E), display enigmatic terrains consisting of mesas of complex planimetric configuration and sinuous valleys (Carr, 2001; Head et al., 2006b; Parker et al., 1989). Whereas the exact origin of this terrain is under debate, some hypotheses have brought forward the idea of Hesperian-aged wet-based glaciation. In particular, fretted valleys such as Mamers, Auqakuh, or Clasia Vallis display characteristics typical of integrated valley glacial systems on Earth, including multiple theater-headed, alcove-like accumulation areas, U-shaped or box-shaped cross sections, sharp arete-like ridges (see glacial valleys and analogs shown in Fig. 3, bottom panels), converging patterns of downslope valley flow, valley lineation patterns typical of ice deformation, streamlines indicative of flow around obstacles, and broad piedmont-like lobes in the valley termination (Head et al., 2006a, 2006b). Whereas some of these characteristics, such as present-day valley infill and lineated patterns, are Amazonian in age, erosion by wet-based glaciation is required to sculpt cirques, aretes, and U-shaped valleys (e.g., Benn et al., 2003; Brook et al., 2006; Davila et al., 2013; Eyles, 1983; Harbor et al., 1988). Substantial ice covers, ranging from 1000 to 2500 m thick, and dated from the Late Noachian to the Early Hesperian, have been suggested to explain the morphology, high flow lines, orientation, and extent of the fretted valleys (Davila et al., 2013; Dickson et al., 2008; Morgan et al., 2009).

A key missing piece of geological evidence for a late Noachian icy highlands ice sheet is the existence of scouring marks and lineated features, such as drumlins, mega-scale glacial lineations, or ribbed moraines, which map the direction of glacial sliding, which are common indicators of the extent of terrestrial Quaternary ice sheets (e.g., Fastook & Head, 2015; Fowler & Chapwanya, 2014; Greenwood & Clark, 2009; Ramirez & Craddock, 2018; Svendsen et al., 2004; Wordsworth, 2016).

Three explanations for this lack of features have been proposed. First, the late Noachian icy highlands ice sheet did not exist (Ramirez & Craddock, 2018), requiring an unknown greenhouse gas that would maintain surface temperatures above freezing and inhibit the nucleation of significant portions of surface snow, preventing the ice-albedo feedback. Second, a late Noachian icy highlands ice sheet existed but was fundamentally cold-based, preventing glacial sliding except in very localized areas, and thus the development of any characteristic lineated landform and the areal scouring of the substrate (Fastook & Head, 2015; Wordsworth, 2016). This scenario hints at surface temperatures at or below freezing, an ice cover up to 2 km thick, and a water inventory under 5× present global equivalent layer

(GEL). The third possibility involves the role of surface gravity in setting the dominant mode of glacial drainage. Even if water accumulated under a late Noachian icy highlands ice sheet, the development of high-efficiency drainage channels under the ice could prevent significant glacial sliding from developing on Mars in comparison to Earth, producing a geological record that would differ from that characteristic of Quaternary glaciation (Grau Galofre et al., 2022). The third hypothesis is also consistent with the lack of sliding recorded in the DAF.

4 Question 3: How did the subsurface water inventory and cryosphere evolve with time, and what is its effect on the global hydrologic cycle?

The onset of a Noachian cryosphere and its impact on the hydrologic cycle. As discussed in the previous sections, strong evidence points to the fact that liquid water must have been more abundant on Mars' surface during the Noachian than at any time since. Therefore, the early Mars hydrology may have been similar to relatively arid environments on present-day Earth, where precipitation, surface runoff, and infiltration led to the existence of standing bodies of water, lakes and possibly oceans (e.g., Guzewich et al., 2021; Palumbo & Head, 2019; Schmidt et al., 2022), and an associated underlying groundwater system (Clifford, 1993; Clifford & Parker, 2001). While early Mars may have been warm and wet, the less degraded nature of post-Noachian terrains suggests a geologically rapid transition to less erosive conditions, similar to those that exist on Mars today (Carr, 1999; Kreslavsky & Head, 2018). As the planet transitioned to a colder climate, a freezing front developed from the planet's surface through its crust, creating a growing cold-trap for both atmospheric and subsurface H_2O—a region known as the cryosphere as illustrated in Fig. 6. The downward propagation of this freezing front, in response to the planet's long-term decline in geothermal heat flux, had two important consequences for the nature of the planet's hydrologic cycle and the state and distribution of subsurface water.

First, ice condensed in the pores of the near-surface regolith would seal away any groundwater from the atmosphere. This would result in a ground-ice depth distribution primarily governed by the thermal structure of the crust, in effect mirroring to the first order the local surface topography as illustrated in Fig. 6. Taking into account reasonable values of crustal permeability (i.e., column-averaged values of ~10^{-15} m^2 within the top 10 km, Clifford, 1993; Clifford & Parker, 2001; Manning & Ingebritsen, 1999; Mustard et al., 2001), the elimination of atmospheric recharge would rapidly lead to the decay of any precipitation-driven influence on the shape of the global water table in 10^6–10^8 years. In the absence of major seismic or thermal disturbances of the crust (such as those caused by large impacts, earthquakes, and volcanic activity), the liquid water at depth would have constituted an aquifer in hydrostatic equilibrium, saturating the lowermost porous regions of the crust. Due to the large range in surface altimetry on Mars (Aharonson et al., 2001), the vertical distance between the base of the cryosphere and the groundwater table may have varied considerably, creating an intervening unsaturated zone (also called vadose zone), whose thickness was maximized in regions of high elevation and minimized (or absent) at low elevation.

Second, as the cryosphere extent deepened with time (by decreasing geothermal flux and surface equilibrium temperatures), the condensation of ice behind the advancing freezing front would have created a growing sink for the planet's inventory of groundwater. Where the cryosphere and water table were in direct contact, the groundwater would have frozen to the base of the cryosphere as the freezing front propagated downward with time. However, in many locations throughout the highlands, the vertical distances separating the base of the cryosphere from the water table may have been several

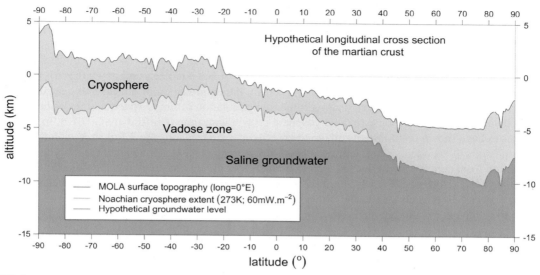

FIG. 6

Hypothetical meridian cross section of the Noachian Martian crust (along 0°E, a longitude representative of the global North-South dichotomy). The figure illustrates the potential relationship between surface topography, ground ice, and groundwater. The vadose zone is a potential subsurface zone where saline groundwater exchanges with the bottom part of the cryosphere by unsaturated vapor circulation. The lower level of the cryosphere is calculated from the subsurface heat propagation model with a geothermal flux of $60\,mW\,m^{-2}$ and a 273 K melting point of water (Surface topography from MOLA (Smith et al., 2001); cryosphere depth from Schwenzer et al., 2012).

Figure adapted from Clifford (1993).

kilometers or more. Under such conditions, the depletion of groundwater is expected to have occurred via the thermally induced diffusion of vapor from warmer depths to the colder pores under the advancing freezing-front (Clifford, 1991, 1993; Clifford & Parker, 2001).

Potential extent of the Late Noachian cryosphere. We use a 1D finite difference thermal model of the subsurface (Clifford et al., 2010; Schwenzer et al., 2012) to calculate Martian mean annual surface temperatures and the insolation heat wave propagation through the Late Noachian cryosphere. The calculations assume that: (1) the thermal conductivity of ice-saturated basalt varies with temperature following a simple inverse law (Clifford et al., 2010; Hobbs, 1974); (2) the mean value of Late Noachian geothermal heat flux can be deduced from estimates of the rheologic properties of the Martian lithosphere necessary to support the relief of the Noachian topography (Grott et al., 2005; McGovern et al., 2002, 2004); (3) the Martian atmosphere at that time did not induce a significant greenhouse effect as compared to current surface conditions; and (4) the variations in Martian obliquity and orbital eccentricity were comparable to those of today. This last assumption's validity is largely dependent on the early developmental state of Tharsis at that time which would have led to true polar wander of the orientation of the planet. A late growth of Tharsis would lead to significant shifts in the position of the deep cryosphere border, while not influencing majorly the overall results regarding the global cryosphere extent and trapped water inventory (Bouley et al., 2016; Ward et al., 1979).

Fig. 7 illustrates the variation in the depth of the cryosphere as a function of latitude based on the average and extreme obliquity and orbital eccentricity values of Laskar et al. (2004), combined with an assumed Noachian geothermal heat flux of $60\,mW\,m^{-2}$ (Grott et al., 2005; McGovern et al., 2004) and three groundwater freezing temperatures depending on its salt content. The brine models considered correspond to pure water (273 K) and eutectic brines of NaCl (252 K) and of $Mg(ClO_4)_2$ (203 K). One expects salty brines as groundwater on Mars from the combination of crustal rock leaching over hundreds of million years, vadose zone hydrothermal circulation of liquid water, and the reduction of groundwater volume in response to the cryosphere growth with time (Clifford et al., 2010). In situ observations have shown that Martian soil could contain up to 25 wt% of salts (Clark & Van Hart, 1981). The study of salt abundances in fractures of Nakhlite meteorites has also indicated the evaporative salt deposition from concentrated brines as well as dilute brines giving constraints on the fluid salt concentration on Mars (Bridges & Schwenzer, 2012; Rao et al., 2005, 2008). NaCl has been detected on Mars and is expected to be a common salt, with a eutectic freezing temperature of 252 K. This corresponds to the first brine considered in our model.

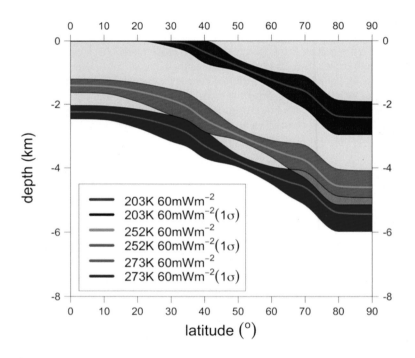

FIG. 7

Depth of the Martian cryosphere as a function of latitude assuming a Noachian geothermal heat flux of $60\,mW\,m^{-2}$. The three colored lines correspond to the average cryosphere depth for different brine salt contents: pure water (273 K), NaCl brine (252 K), and perchlorate brine (203 K). The colored area around the average depth corresponds to statistical variations due to the orbital parameters of Mars with time (Laskar et al., 2004; Schwenzer et al., 2012).

Chloride- and sulfur-rich salts are also expected with varying freezing depression points typically lower than the one for NaCl brines, for example, it is 238 K for a eutectic solution of $MgCl_2$ and 218 K for a eutectic solution of $CaCl_2$ (Burt & Knauth, 2003; Fairén et al., 2009). However, the concentrations of the different salt species in Martian underground water are not well constrained. The detection of perchlorate in the regolith of Mars (Clark et al., 2021; Hecht et al., 2009) and its ability to be readily dissolved suggest it may have infiltrated in the underground water system of Mars (Clifford et al., 2010). The strong freezing depression point of perchlorates (as low as 203 K for $Mg(ClO_4)_2$) is used as our second brine model for the cryosphere, as an extreme case to determine the limits of validity of the cryosphere extent. While it is unlikely that brines fully saturated in magnesium perchlorate salts exist globally on the planet (Glavin et al., 2013; Hecht et al., 2009; Kounaves et al., 2014), the simulations using such a low freezing depression point gives a lower limit to the depth of the cryosphere that can be reached by groundwater brines with mixed salts located in the underground.

The calculated equilibrium depths of a pure water cryosphere are represented in Fig. 6. The cryosphere maximum depth ranges from 2.3 km at the equator to 5.6 km near the poles (assuming a pure water depression freezing point of 273 K). This latitudinal variation is due to the colder average annual temperatures near the poles resulting from both lower insolation and the higher albedo of the seasonal and perennial polar ice deposits, compared to Mars' typical regolith. The polar deposits are assumed to be similar in extent to today's polar caps. The depth of the cryosphere would be much reduced under the conditions relevant to brines. For example, in the case of a solution saturated in magnesium perchlorate (i.e., 203 K), the cryosphere extent would reach its maximum of 2.5 km around the poles and continuously decrease with lower latitudes until it completely disappears at latitudes of less than about 30°–35°, where mean annual temperatures exceed the assumed perchlorate-depressed freezing point of groundwater.

In Fig. 7, the colored thickness of the lines corresponds to simulations of the high and low extreme 1σ Mars orbital parameters (obliquity and eccentricity) assuming a 20 Ma obliquity history equivalent to the nominal current model of Laskar et al. (2004). Such variations in orbital parameters translate into a 0.5 km depth change at the equator and a 1 km depth change at the poles, these two changes compensating on average one another due to the change in the obliquity of the planet.

Water content in the Noachian cryosphere. Once established and stabilized, the global planet cryosphere would act as an underground cold-trap in the pore volume of the Martian crust for the available water. Assuming a surface porosity of 0.2 and a typical exponential depth decay constant of 2.82 km (Clifford, 1993; Clifford & Parker, 2001), we can calculate the total pore volume available for this cold-trap over the entire planet at the depth where the cryosphere reaches its maximum extent. The pure water Noachian cryosphere model described in Fig. 7 could have cold trapped about 446 m (±8 m) GEL. As freezing-point depression salts are included, the depth of the cryosphere decreases as well as the potential volume of water trapped. In the case of a NaCl-saturated aquifer ($T = 252$ K), the cryosphere could trap 390 m (±15 m) GEL, and for the more extreme perchlorate-saturated brine ($T = 203$ K), a volume corresponding to 104 m (±13 m) GEL. As we do not expect the aquifer water to be entirely saturated in salts, this is certainly enough to trap a significant fraction or even the totality of the available liquid water inventory expected at the end of the Noachian, as discussed in other studies (Carr & Head, 2015, 2019; Palumbo & Head, 2020).

Local variations in the cryosphere thickness and state. The calculations presented above correspond to the ideal case of a global cryosphere in equilibrium with the planet's surface and interior conditions. It is expected that local variations in geothermal heat flow, and heat produced by volcanic activity as well as impacts, would locally temporarily affect the extent of the cryosphere, or melt it entirely. Such heat events could generate specific hydrothermal systems and have been suggested as a possible source for some of the fluvial features detected on Mars (Gulick, 1998, 2001). Even with a global cryosphere extent as shown in Fig. 7, it would have been thermally penetrated by relatively small impactors, the heat generated by the impact being sufficient to melt the cryosphere over its whole thickness. Calculations based on a model developed by Abramov and Kring (2005) and applied to a two-layer Mars subsurface have shown that a 2-km thick cryosphere, like around the equator, can be thermally penetrated by an impactor as small as 0.6 km in diameter, while the thicker 6-km cryosphere below the poles would be thermally penetrated by projectiles as small as 1.8 km in diameter (Schwenzer et al., 2012). Such impacts would also occur more frequently during the Noachian era and lead to the generation of local hydrothermal systems with liquid water lasting up to 10 million years due to the time necessary for the dissipation of heat in the system (Schwenzer et al., 2012). This would also further enable exchanges between the deep underground aquifers and the atmosphere.

Local changes in the depth of the cryosphere are expected due to local changes in the geothermal heat flux of the planet. To this day, the average heat flux is around $25\,\mathrm{mW\,m^{-2}}$ but can still increase to $50\,\mathrm{mW\,m^{-2}}$ in some locations (Plesa et al., 2016). Such high heat flux may have resulted in geothermally induced melting at the base of the cryosphere in the past. Such events can be linked with surficial meltwater features such as valley networks and the location of hydrous minerals on Mars (Ojha et al., 2020). There is also evidence that such processes have continued in the relatively recent past as candidate eskers have been identified in Amazonian terrains, Phlegra Montes, and Tempe Terra, for which an origin from basal melting due to enhanced geothermal heat flux was suggested (Butcher et al., 2017; Gallagher & Balme, 2015; Koutnik et al., 2024, pp. 101–142). Under such conditions, it could even be envisioned that liquid water could accumulate as a northern ocean fed by groundwater under a subfreezing Martian surface climate (Palumbo & Head, 2019). The existence of an ocean with a frozen surface would be consistent with the relatively low level of erosion of Hesperian age terrains but would be inconsistent with reported geological evidence for tsunami emplacement (Costard et al., 2017, 2019; Rodriguez et al., 2016).

As the cryosphere would continue to extend, and conditions for melting became rarer, it is possible that the ice-cemented cryosphere (the part of the cryosphere where pores are completely saturated by ice) would seal away liquid groundwater at a depth that could remain there to this day, without the possibility of escape (Grimm et al., 2017). However, there has been no detected evidence of such a global reservoir of ice and liquid water at depth to date, either from radars (Clifford et al., 2010) or from seismometry (Manga & Wright, 2021). The morphology of the Hesperian-Amazonian-aged single-layered ejecta crater indicates a smaller extent of cemented cryosphere at the Hesperian to Amazonian transition, with an estimated maximum depth of 1.3 km at the equator and 2.3 km at the poles (Weiss & Head, 2017a, 2017b). The volume considered corresponds to 200 m GEL assuming ~20% porosity at the surface. Comparing it to the results given in Fig. 7, it would mean that the available pore volume that could be filled by ice was supply limited in the cases of the pure water ice (446 m GEL), or saturated NaCl brine (390 m GEL) cryospheres which never reached saturation. The single-layered ejecta crater observations can be consistent with a perchlorate salts brine (104 m GEL) fully saturated cryosphere; though this remains an extreme case of salt saturation that is certainly less likely. This would mean that the majority of the subsurface water inventory of Mars during the Hesperian, up to 200 m GEL from the single-layered ejecta crater observations, would have been fully trapped in the expected cryosphere volumes (Weiss & Head, 2017a, 2017b). As a consequence, no underground extensive liquid water aquifers probably exist today on the planet while about 30 m GEL of water still exists above the surface to form the currently visible polar caps (Carr & Head, 2015; Lasue et al., 2013; Villanueva et al., 2015). This is also consistent with the seismic velocity investigations made by the InSight mission on Mars, indicating that there is no pore ice saturation of the upper 10 km of the Martian crust at Elysium Planitia (Manga & Wright, 2021).

5 Conclusions

While the extent of surface ice on early Mars remains as uncertain as the rest of the climate, the use of 3D GCMs to investigate possible climates continues to generate hypotheses for geologists to test. The glacial geomorphological record corresponding to early Mars is notoriously difficult to interpret, as many features that have possible glacial interpretations are ambiguous, are heavily degraded, and/ or are covered in later deposits (e.g., the latitude-dependent mantle—LDM). In particular, even very large cold-based Noachian or Hesperian ice sheets may never have left evidence capable of surviving to the present day, save for possible effects on volcanic flows that encountered them or on the impact cratering record. It is a worthwhile exercise to consider what any given observed Amazonian glacial landform (Koutnik et al., 2024, pp. 101–142) might look like after 3 Ga of impact cratering, aeolian erosion, and subsequent glacial cycles. Still, case studies of regional landform assemblages lend increasing support to the presence of Noachian and Hesperian ice sheets in regions such as the Isidis basin and the high southern latitudes, and global statistical studies of valley morphometry and impact crater degradation are a promising avenue for greater clarity on the prevalence of ice across space and time. Furthermore, recent work has highlighted the discrepancies that should be expected between terrestrial and Martian glacial landscapes, warranting a fresh examination of Martian glacial geology, both ancient and modern.

During the Noachian to Hesperian transition, it is expected that the decreasing geothermal heat flux and global surface temperatures induced the onset of a global subsurface cryosphere formation, which would have cold-trapped subsurface Martian water in the crustal pore system. The extant surface water ice of modern Mars, having decreased in abundance throughout the Hesperian and into the Amazonian by this and other mechanisms, provides another window to how water ice interacts with volcanic, atmospheric, impact, and aeolian processes to affect the Martian geologic record (Koutnik et al., 2024, pp. 101–142).

Acknowledgments

The authors would like to kindly thank the time and effort of the three reviewers that considered this manuscript, whose queries led to significant improvements, as well as insightful discussions with F. Butcher and comments by J.P. Williams. A.G.G. acknowledges the funding received from the European Union's Horizon 2020 research and innovation program under the Marie Skłokodowska-Curie grant agreement MGFR 101027900, and from the School of Earth and Space Exploration (ASU) through the Exploration Fellowship. K.E.S. gratefully acknowledges support from the NASA Postdoctoral Program at Goddard Space Flight Center, administered by ORAU under contract with NASA.

References

Abramov, O., & Kring, D. A. (2005). Impact-induced hydrothermal activity on early Mars. *Journal of Geophysical Research, Planets*, *110*(E12).

Ackiss, S., Horgan, B., Seelos, F., Farrand, W., & Wray, J. (2018). Mineralogic evidence for subglacial volcanism in the Sisyphi Montes region of Mars. *Icarus*, *311*, 357–370.

Ackiss, S. E., & Wray, J. J. (2014). Occurrences of possible hydrated sulfates in the southern high latitudes of Mars. *Icarus*, *243*, 311–324.

Aharonson, O., Zuber, M. T., & Rothman, D. H. (2001). Statistics of Mars' topography from the Mars Orbiter Laser Altimeter: Slopes, correlations, and physical models. *Journal of Geophysical Research, Planets*, *106*(E10), 23723–23735.

Allen, C. C. (1979). Volcano-ice interactions on Mars. *Journal of Geophysical Research. Solid Earth*, *84*(B14), 8048–8059.

Ann Hodges, C., & Moore, H. J. (1979). The subglacial birth of Olympus Mons and its aureoles. *Journal of Geophysical Research. Solid Earth*, *84*(B14), 8061–8074.

Archer, B. R., Hynek, B. M., & Robbins, S. J. (2021). Crater evidence does not support the late Noachian Icy Highlands for Mars. In *52nd lunar and planetary science conference (no. 2548)* (p. 2220).

Baker, V. R. (2001). Water and the Martian landscape. *Nature*, *412*(6843), 228–236.

Baker, V. R., Strom, R. G., Gulick, V. C., Kargel, J. S., Komatsu, G., & Kale, V. S. (1991). Ancient oceans, ice sheets and the hydrological cycle on Mars. *Nature*, *352*(6336), 589–594.

Balco, G. (2011). Contributions and unrealized potential contributions of cosmogenic-nuclide exposure dating to glacier chronology, 1990–2010. *Quaternary Science Reviews*, *30*(1–2), 3–27.

Banks, M. E., Lang, N. P., Kargel, J. S., McEwen, A. S., Baker, V. R., Grant, J. A., … Strom, R. G. (2009). An analysis of sinuous ridges in the southern Argyre Planitia, Mars using HiRISE and CTX images and MOLA data. *Journal of Geophysical Research, Planets*, *114*(E9).

Benn, D. I., Kirkbride, M. P., Owen, L. A., & Brazier, V. (2003). Glaciated valley land systems. *Glacial Land Systems*, 372–406.

Bernhardt, H., Hiesinger, H., Reiss, D., Ivanov, M., & Erkeling, G. (2013). Putative eskers and new insights into glacio-fluvial depositional settings in southern Argyre Planitia, Mars. *Planetary and Space Science*, *85*, 261–278.

Bernhardt, H., Reiss, D., Hiesinger, H., & Ivanov, M. A. (2016). The honeycomb terrain on the Hellas basin floor, Mars: A case for salt or ice diapirism. *Journal of Geophysical Research, Planets*, *121*(4), 714–738.

Bernhardt, H., Reiss, D., Ivanov, M., Hauber, E., Hiesinger, H., Clark, J. D., & Orosei, R. (2019). The banded terrain on northwestern Hellas Planitia: New observations and insights into its possible formation. *Icarus*, *321*, 171–188.

Black, B. A., Perron, J. T., Hemingway, D., Bailey, E., Nimmo, F., & Zebker, H. (2017). Global drainage patterns and the origins of topographic relief on Earth, Mars, and Titan. *Science*, *356*(6339), 727–731.

Boatwright, B. D., & Head, J. W. (2021). A Noachian proglacial paleolake on Mars: Fluvial activity and lake formation within a closed-source drainage basin crater and implications for early Mars climate. *The Planetary Science Journal*, *2*(2), 52.

Bouley, S., Baratoux, D., Matsuyama, I., Forget, F., Séjourné, A., Turbet, M., & Costard, F. (2016). Late Tharsis formation and implications for early Mars. *Nature*, *531*(7594), 344–347.

Bouquety, A., Sejourné, A., Costard, F., Bouley, S., & Leyguarda, E. (2020). Glacial landscape and paleoglaciation in Terra Sabaea: Evidence for a 3.6 Ga polythermal plateau ice cap. *Geomorphology*, *350*, 106858.

Bouquety, A., Sejourné, A., Costard, F., Mercier, D., & Bouley, S. (2019). Morphometric evidence of 3.6 Ga glacial valleys and glacial cirques in Martian highlands: South of Terra Sabaea. *Geomorphology*, *334*, 91–111.

Bridges, J. C., & Schwenzer, S. P. (2012). The nakhlite hydrothermal brine on Mars. *Earth and Planetary Science Letters*, *359*, 117–123.

Bristow, T. F., Haberle, R. M., Blake, D. F., Des Marais, D. J., Eigenbrode, J. L., Fairén, A. G., ... Vasavada, A. R. (2017). Low Hesperian P CO2 constrained from in situ mineralogical analysis at Gale Crater, Mars. *Proceedings of the National Academy of Sciences, 114*(9), 2166–2170.

Brook, M. S., Kirkbride, M. P., & Brock, B. W. (2006). Cirque development in a steadily uplifting range: rates of erosion and long-term morphometric change in alpine cirques in the Ben Ohau Range, New Zealand. *Earth Surface Processes and Landforms: The Journal of the British Geomorphological Research Group, 31*(9), 1167–1175.

Burt, D. M., & Knauth, L. P. (2003). Electrically conducting, Ca-rich brines, rather than water, expected in the Martian subsurface. *Journal of Geophysical Research, Planets, 108*(E4).

Butcher, F. E., Balme, M. R., Gallagher, C., Arnold, N. S., Conway, S. J., Hagermann, A., & Lewis, S. R. (2017). Recent basal melting of a mid-latitude glacier on Mars. *Journal of Geophysical Research, Planets, 122*(12), 2445–2468.

Butcher, F. E., Conway, S. J., & Arnold, N. S. (2016). Are the dorsa argentea on mars eskers? *Icarus, 275*, 65–84.

Carr, M. H. (1995). The Martian drainage system and the origin of valley networks and fretted channels. *Journal of Geophysical Research, Planets, 100*(E4), 7479–7507.

Carr, M. H. (1999). Retention of an atmosphere on early Mars. *Journal of Geophysical Research, Planets, 104*(E9), 21897–21909.

Carr, M. H. (2001). Mars global surveyor observations of Martian fretted terrain. *Journal of Geophysical Research, Planets, 106*(E10), 23571–23593.

Carr, M. H. (2007). *The surface of Mars. Vol. 6.* Cambridge University Press.

Carr, M. H., & Head, J. W. (2003). Basal melting of snow on early Mars: A possible origin of some valley networks. *Geophysical Research Letters, 30*(24).

Carr, M. H., & Head, J. W. (2015). Martian surface/near-surface water inventory: Sources, sinks, and changes with time. *Geophysical Research Letters, 42*(3), 726–732.

Carr, M., & Head, J. (2019). Mars: Formation and fate of a frozen Hesperian Ocean. *Icarus, 319*, 433–443.

Cassanelli, J. P., & Head, J. W. (2019). Glaciovolcanism in the Tharsis volcanic province of Mars: Implications for regional geology and hydrology. *Planetary and Space Science, 169*, 45–69.

Cassanelli, J. P., Head, J. W., & Fastook, J. L. (2015). Sources of water for the outflow channels on Mars: Implications of the late Noachian "icy highlands" model for melting and groundwater recharge on the Tharsis rise. *Planetary and Space Science, 108*, 54–65.

Chapman, M. G., Neukum, G., Dumke, A., Michael, G., Van Gasselt, S., Kneissl, T., ... Mangold, N. (2010a). Amazonian geologic history of the Echus Chasma and Kasei Valles system on Mars: New data and interpretations. *Earth and Planetary Science Letters, 294*(3–4), 238–255.

Chapman, M. G., Neukum, G., Dumke, A., Michael, G., Van Gasselt, S., Kneissl, T., ... Masson, P. (2010b). Noachian–Hesperian geologic history of the Echus Chasma and Kasei Valles system on Mars: New data and interpretations. *Earth and Planetary Science Letters, 294*(3–4), 256–271.

Citron, R. I., Manga, M., & Tan, E. (2018). A hybrid origin of the Martian crustal dichotomy: Degree-1 convection antipodal to a giant impact. *Earth and Planetary Science Letters, 491*, 58–66.

Clark, J., Sutter, B., Archer, P. D., Jr., Ming, D., Rampe, E., McAdam, A., ... Mahaffy, P. (2021). A review of sample analysis at Mars-Evolved Gas Analysis Laboratory analog work supporting the presence of perchlorates and chlorates in Gale Crater, Mars. *Minerals, 11*(5), 475.

Clark, B. C., & Van Hart, D. C. (1981). The salts of Mars. *Icarus, 45*(2), 370–378.

Clifford, S. M. (1991). The role of thermal vapor diffusion in the subsurface hydrologic evolution of Mars. *Geophysical Research Letters, 18*(11), 2055–2058.

Clifford, S. M. (1993). A model for the hydrologic and climatic behavior of water on Mars. *Journal of Geophysical Research, Planets, 98*(E6), 10973–11016.

Clifford, S. M., Lasue, J., Heggy, E., Boisson, J., McGovern, P., & Max, M. D. (2010). Depth of the Martian cryosphere: Revised estimates and implications for the existence and detection of subpermafrost groundwater. *Journal of Geophysical Research, Planets, 115*(E7).

Clifford, S. M., & Parker, T. J. (2001). The evolution of the Martian hydrosphere: Implications for the fate of a primordial ocean and the current state of the northern plains. *Icarus, 154*(1), 40–79.

Costard, F., Séjourné, A., Kelfoun, K., Clifford, S., Lavigne, F., Di Pietro, I., & Bouley, S. (2017). Modeling tsunami propagation and the emplacement of thumbprint terrain in an early Mars Ocean. *Journal of Geophysical Research, Planets, 122*(3), 633–649.

Costard, F., Séjourné, A., Lagain, A., Ormö, J., Rodriguez, J. A. P., Clifford, S., … Lavigne, F. (2019). The Lomonosov crater impact event: A possible mega-tsunami source on Mars. *Journal of Geophysical Research, Planets, 124*(7), 1840–1851.

Craddock, R. A., & Howard, A. D. (2002). The case for rainfall on a warm, wet early Mars. *Journal of Geophysical Research, Planets, 107*(E11). 21-1.

Craddock, R. A., & Maxwell, T. A. (1993). Geomorphic evolution of the Martian highlands through ancient fluvial processes. *Journal of Geophysical Research, Planets, 98*(E2), 3453–3468.

Cuffey, K. M., & Paterson, W. S. B. (2010). *The physics of glaciers*. Academic Press.

Davila, A. F., Fairén, A. G., Stokes, C. R., Platz, T., Rodriguez, A. P., Lacelle, D., … Pollard, W. (2013). Evidence for Hesperian glaciation along the Martian dichotomy boundary. *Geology, 41*(7), 755–758.

De Blasio, F. V. (2018). The pristine shape of Olympus Mons on Mars and the subaqueous origin of its aureole deposits. *Icarus, 302*, 44–61.

De Blasio, F. V. (2020). Frontal aureole deposit on Acheron fossae ridge as evidence for landslide-generated tsunami on Mars. *Planetary and Space Science, 187*, 104911.

Di Achille, G., & Hynek, B. M. (2010). Ancient Ocean on Mars supported by global distribution of deltas and valleys. *Nature Geoscience, 3*(7), 459–463.

Dickson, J., & Head, J. W. (2006). Evidence for an Hesperian-aged south circum-polar lake margin environment on Mars. *Planetary and Space Science, 54*(3), 251–272.

Dickson, J. L., Head, J. W., & Marchant, D. R. (2008). Late Amazonian glaciation at the dichotomy boundary on Mars: Evidence for glacial thickness maxima and multiple glacial phases. *Geology, 36*(5), 411–414.

Dohm, J. M., Hare, T. M., Robbins, S. J., Williams, J. P., Soare, R. J., El-Maarry, M. R., … Maruyama, S. (2015). Geological and hydrological histories of the Argyre province, Mars. *Icarus, 253*, 66–98.

Edwards, C. S., Nowicki, K. J., Christensen, P. R., Hill, J., Gorelick, N., & Murray, K. (2011). Mosaicking of global planetary image datasets: 1. Techniques and data processing for Thermal Emission Imaging System (THEMIS) multi-spectral data. *Journal of Geophysical Research, Planets, 116*(E10).

Erkeling, G., Reiss, D., Hiesinger, H., Ivanov, M. A., Hauber, E., & Bernhardt, H. (2014). Landscape formation at the Deuteronilus contact in southern Isidis Planitia, Mars: Implications for an Isidis Sea? *Icarus, 242*, 329–351.

Eyles, N. (1983). Modern Icelandic glaciers as depositional models for «hummocky moraine» in the Scottish Highlands. In *INQUA symposia on the genesis and lithology of quaternary deposits* (pp. 47–59).

Fairén, A. G., Davila, A. F., Gago-Duport, L., Amils, R., & McKay, C. P. (2009). Stability against freezing of aqueous solutions on early Mars. *Nature, 459*(7245), 401–404.

Fan, B. M., Jansen, M. F., Mischna, M. A., & Kite, E. S. (2022). Why are mountain-tops cold?—The decorrelation of surface temperature and topography due to the decline of the greenhouse effect on early Mars. *LPI Contributions, 2678*, 1698.

Fassett, C. I., & Head, J. W., III. (2008a). The timing of Martian valley network activity: Constraints from buffered crater counting. *Icarus, 195*(1), 61–89.

Fassett, C. I., & Head, J. W., III. (2008b). Valley network-fed, open-basin lakes on Mars: Distribution and implications for Noachian surface and subsurface hydrology. *Icarus, 198*(1), 37–56.

Fastook, J. L., & Head, J. W. (2015). Glaciation in the Late Noachian Icy Highlands: Ice accumulation, distribution, flow rates, basal melting, and top-down melting rates and patterns. *Planetary and Space Science, 106*, 82–98.

Fastook, J. L., Head, J. W., Scanlon, K. E., Weiss, D. K., & Palumbo, A. M. (2021). Hellas Basin, Mars: A model of rim and wall glaciation in the late Noachian and predictions for enhanced flow, basal melting, wet-based glaciation and Erosion, and generation and fate of meltwater. In *52nd lunar and planetary science conference (no. 2548)* (p. 1528).

Forget, F., Wordsworth, R., Millour, E., Madeleine, J. B., Kerber, L., Leconte, J., … Haberle, R. M. (2013). 3D modelling of the early martian climate under a denser CO2 atmosphere: Temperatures and CO2 ice clouds. *Icarus, 222*(1), 81–99.

Fowler, A. C., & Chapwanya, M. (2014). An instability theory for the formation of ribbed moraine, drumlins and mega-scale glacial lineations. *Proceedings of the Royal Society A: Mathematical, Physical and Engineering Sciences, 470*(2171). 20140185.

Gallagher, C., & Balme, M. (2015). Eskers in a complete, wet-based glacial system in the Phlegra Montes region, Mars. *Earth and Planetary Science Letters, 431*, 96–109.

Ghatan, G. J., & Head, J. W., III. (2002). Candidate subglacial volcanoes in the south polar region of Mars: Morphology, morphometry, and eruption conditions. *Journal of Geophysical Research, Planets, 107*(E7). 2-1.

Ghatan, G. J., & Head, J. W., III. (2004). Regional drainage of meltwater beneath a Hesperian-aged south circumpolar ice sheet on Mars. *Journal of Geophysical Research, Planets, 109*(E7).

Ghatan, G. J., Head, J. W., III, & Pratt, S. (2003). Cavi Angusti, Mars: Characterization and assessment of possible formation mechanisms. *Journal of Geophysical Research, Planets, 108*(E5).

Glavin, D. P., Freissinet, C., Miller, K. E., Eigenbrode, J. L., Brunner, A. E., Buch, A., … Mahaffy, P. R. (2013). Evidence for perchlorates and the origin of chlorinated hydrocarbons detected by SAM at the Rocknest aeolian deposit in Gale Crater. *Journal of Geophysical Research, Planets, 118*(10), 1955–1973.

Godin, P. J., Ramirez, R. M., Campbell, C. L., Wizenberg, T., Nguyen, T. G., Strong, K., & Moores, J. E. (2020). Collision-induced absorption of CH_4-CO_2 and H_2-CO_2 complexes and their effect on the ancient Martian atmosphere. *Journal of Geophysical Research: Planets, 125*(12). e2019JE006357.

Goudge, T. A., Aureli, K. L., Head, J. W., Fassett, C. I., & Mustard, J. F. (2015). Classification and analysis of candidate impact crater-hosted closed-basin lakes on Mars. *Icarus, 260*, 346–367.

Grau Galofre, A., Bahia, R. S., Jellinek, A. M., Whipple, K. X., & Gallo, R. (2020a). Did Martian valley networks substantially modify the landscape? *Earth and Planetary Science Letters, 547*, 116482.

Grau Galofre, A., Jellinek, A. M., & Osinski, G. R. (2020b). Valley formation on early Mars by subglacial and fluvial erosion. *Nature Geoscience, 13*(10), 663–668.

Grau Galofre, A., Whipple, K. X., Christensen, P. R., & Conway, S. J. (2022). Valley networks and the record of glaciation on ancient Mars. *Geophysical Research Letters, 49*(14). e2022GL097974.

Greenwood, S. L., & Clark, C. D. (2009). Reconstructing the last Irish Ice Sheet 2: A geomorphologically-driven model of ice sheet growth, retreat and dynamics. *Quaternary Science Reviews, 28*(27–28), 3101–3123.

Grimm, R. E., Harrison, K. P., Stillman, D. E., & Kirchoff, M. R. (2017). On the secular retention of ground water and ice on Mars. *Journal of Geophysical Research, Planets, 122*(1), 94–109.

Grott, M., Hauber, E., Werner, S. C., Kronberg, P., & Neukum, G. (2005). High heat flux on ancient Mars: Evidence from rift flank uplift at Coracis Fossae. *Geophysical Research Letters, 32*(21).

Guidat, T., Pochat, S., Bourgeois, O., & Souček, O. (2015). Landform assemblage in Isidis Planitia, Mars: Evidence for a 3 Ga old polythermal ice sheet. *Earth and Planetary Science Letters, 411*, 253–267.

Gulick, V. C. (1998). Magmatic intrusions and a hydrothermal origin for fluvial valleys on Mars. *Journal of Geophysical Research, Planets, 103*(E8), 19365–19387.

Gulick, V. C. (2001). Origin of the valley networks on Mars: A hydrological perspective. *Geomorphology, 37*(3–4), 241–268.

Guzewich, S. D., Way, M. J., Aleinov, I., Wolf, E. T., Del Genio, A., Wordsworth, R., & Tsigaridis, K. (2021). 3D simulations of the early Martian hydrological cycle mediated by a H2-CO2 greenhouse. *Journal of Geophysical Research, Planets, 126*(7). e2021JE006825.

Haberle, R. M., Zahnle, K., Barlow, N. G., & Steakley, K. E. (2019). Impact degassing of H2 on early Mars and its effect on the climate system. *Geophysical Research Letters, 46*(22), 13355–13362.

Halevy, I., Zuber, M. T., & Schrag, D. P. (2007). A sulfur dioxide climate feedback on early Mars. *Science*, *318*(5858), 1903–1907.

Harbor, J. M., Hallet, B., & Raymond, C. F. (1988). A numerical model of landform development by glacial erosion. *Nature*, *333*(6171), 347–349.

Hartmann, W. K., & Neukum, G. (2001). Cratering chronology and the evolution of Mars. *Chronology and Evolution of Mars*, 165–194.

Head, J. W., & Hallet, B. (2001). Origin of sinuous ridges in the Dorsa Argentea formation: New observations and tests of the esker hypothesis. In: *Lunar and Planetary Science Conference (abstract 1373)*.

Head, J. W., III, & Pratt, S. (2001). Extensive Hesperian-aged south polar ice sheet on Mars: Evidence for massive melting and retreat, and lateral flow and ponding of meltwater. *Journal of Geophysical Research, Planets*, *106*(E6), 12275–12299.

Head, J. W., Marchant, D. R., Agnew, M. C., Fassett, C. I., & Kreslavsky, M. A. (2006a). Extensive valley glacier deposits in the northern mid-latitudes of Mars: Evidence for late Amazonian obliquity-driven climate change. *Earth and Planetary Science Letters*, *241*(3–4), 663–671.

Head, J. W., Nahm, A. L., Marchant, D. R., & Neukum, G. (2006b). Modification of the dichotomy boundary on Mars by Amazonian mid-latitude regional glaciation. *Geophysical Research Letters*, *33*(8).

Hecht, M. H., Kounaves, S. P., Quinn, R. C., West, S. J., Young, S. M., Ming, D. W., … Smith, P. H. (2009). Detection of perchlorate and the soluble chemistry of Martian soil at the Phoenix lander site. *Science*, *325*(5936), 64–67.

Helgason, J. (1999). Formation of Olympus Mons and the aureole-escarpment problem on Mars. *Geology*, *27*(3), 231–234.

Hiesinger, H., & Head, J. W., III. (2002). Topography and morphology of the Argyre Basin, Mars: Implications for its geologic and hydrologic history. *Planetary and Space Science*, *50*(10−11), 939–981.

Hobbs, P. V. (1974). *Ice physics* (pp. 837–853). Clarendon.

Howard, A. D. (1981). Etched plains and braided ridges of the south polar region of Mars: Features produced by basal melting of ground ice? *Reports of Planetary Geology Program*, 286–288.

Howard, A. D., Moore, J. M., & Irwin, R. P., III. (2005). An intense terminal epoch of widespread fluvial activity on early Mars: 1. Valley network incision and associated deposits. *Journal of Geophysical Research, Planets*, *110*(E12).

Hynek, B. M., Beach, M., & Hoke, M. R. (2010). Updated global map of Martian valley networks and implications for climate and hydrologic processes. *Journal of Geophysical Research, Planets*, *115*(E9).

Irwin, R. P., III, Craddock, R. A., Howard, A. D., & Flemming, H. L. (2011). Topographic influences on development of Martian valley networks. *Journal of Geophysical Research, Planets*, *116*(E2).

Irwin, R. P., III, Howard, A. D., Craddock, R. A., & Moore, J. M. (2005). An intense terminal epoch of widespread fluvial activity on early Mars: 2. Increased runoff and paleolake development. *Journal of Geophysical Research, Planets*, *110*(E12).

Irwin, R. P., III, Wray, J. J., Mest, S. C., & Maxwell, T. A. (2018). Wind-eroded crater floors and intercrater plains, Terra Sabaea, Mars. *Journal of Geophysical Research, Planets*, *123*(2), 445–467.

Ivanov, M. A., Hiesinger, H., Erkeling, G., Hielscher, F. J., & Reiss, D. (2012). Major episodes of geologic history of Isidis Planitia on Mars. *Icarus*, *218*(1), 24–46.

Jakosky, B. M. (2021). Atmospheric loss to space and the history of water on Mars. *Annual Review of Earth and Planetary Sciences*, *49*, 71–93.

Johnson, S. S., Mischna, M. A., Grove, T. L., & Zuber, M. T. (2008). Sulfur-induced greenhouse warming on early Mars. *Journal of Geophysical Research, Planets*, *113*(E8).

Kamada, A., Kuroda, T., Kasaba, Y., Terada, N., Nakagawa, H., & Toriumi, K. (2020). A coupled atmosphere–hydrosphere global climate model of early Mars: A 'cool and wet' scenario for the formation of water channels. *Icarus*, *338*, 113567.

Kargel, J. S., & Strom, R. G. (1992). Ancient glaciation on Mars. *Geology*, *20*(1), 3–7.

Kerber, L., Forget, F., & Wordsworth, R. (2015). Sulfur in the early Martian atmosphere revisited: Experiments with a 3-D global climate model. *Icarus*, *261*, 133–148.

Kite, E. S. (2019). Geologic constraints on early Mars climate. *Space Science Reviews*, *215*(1), 1–47.

Kite, E. S., & Daswani, M. M. (2019). Geochemistry constrains global hydrology on early Mars. *Earth and Planetary Science Letters*, *524*, 115718.

Kite, E. S., & Hindmarsh, R. C. (2007). Did ice streams shape the largest channels on Mars? *Geophysical Research Letters*, *34*(19).

Kite, E. S., Matsuyama, I., Manga, M., Perron, J. T., & Mitrovica, J. X. (2009). True polar wander driven by late-stage volcanism and the distribution of paleopolar deposits on Mars. *Earth and Planetary Science Letters*, *280*(1–4), 254–267.

Kite, E. S., Williams, J. P., Lucas, A., & Aharonson, O. (2014). Low palaeopressure of the Martian atmosphere estimated from the size distribution of ancient craters. *Nature Geoscience*, *7*(5), 335–339.

Kounaves, S. P., Carrier, B. L., O'Neil, G. D., Stroble, S. T., & Claire, M. W. (2014). Evidence of Martian perchlorate, chlorate, and nitrate in Mars meteorite EETA79001: Implications for oxidants and organics. *Icarus*, *229*, 206–213.

Koutnik, M., Butcher, F., Soare, R., Hepburn, A., Hubbard, B., Brough, B., Gallagher, C., McKeown, L., & Pathare, A. (2024). *Glacial deposits, remnants, and landscapes on Amazonian Mars: Using setting, structure, and stratigraphy to understand ice evolution and climate history*. In R. J. Soare, J.-P. Williams, C. Ahrens, F. E. G. Butcher, & M. R. El-Maarry (Eds.), *Ices in the solar system, a volatile-driven journey from the inner solar system to its far reaches*. Elsevier Books.

Kreslavsky, M. A., & Head, J. W. (2018). Mars climate history: Insights from impact crater wall slope statistics. *Geophysical Research Letters*, *45*(4), 1751–1758.

Kress, A. M., & Head, J. W. (2015). Late Noachian and early Hesperian ridge systems in the south circumpolar Dorsa Argentea Formation, Mars: Evidence for two stages of melting of an extensive late Noachian ice sheet. *Planetary and Space Science*, *109*, 1–20.

Lamb, M. P., Howard, A. D., Johnson, J., Whipple, K. X., Dietrich, W. E., & Perron, J. T. (2006). Can springs cut canyons into rock? *Journal of Geophysical Research, Planets*, *111*(E7).

Laskar, J., Correia, A. C. M., Gastineau, M., Joutel, F., Levrard, B., & Robutel, P. (2004). Long term evolution and chaotic diffusion of the insolation quantities of Mars. *Icarus*, *170*(2), 343–364.

Lasue, J., Mangold, N., Hauber, E., Clifford, S., Feldman, W., Gasnault, O., … Mousis, O. (2013). Quantitative assessments of the Martian hydrosphere. *Space Science Reviews*, *174*(1), 155–212.

Lee, P., Rice, J. W., Jr., Bunch, T. E., Grieve, R. A. F., McKay, C. P., Schutt, J. W., & Zent, A. P. (1999). Possible analogs for small valleys on Mars at the Haughton impact crater site, Devon Island, Canadian High Arctic. In *Lunar and Planetary Science*.

Lopes, R., Guest, J. E., & Wilson, C. J. (1980). Origin of the Olympus Mons aureole and perimeter scarp. *The Moon and the Planets*, *22*(2), 221–234.

Lucchitta, B. K. (1982). Ice sculpture in the Martian outflow channels. *Journal of Geophysical Research: Solid Earth*, *87*(B12), 9951–9973.

Lucchitta, B. K. (2001). Antarctic ice streams and outflow channels on Mars. *Geophysical Research Letters*, *28*(3), 403–406.

Lucchitta, B. K., & Ferguson, H. M. (1983). Chryse Basin channels: Low-gradients and ponded flows. *Journal of Geophysical Research. Solid Earth*, *88*(S02), A553–A568.

Malin, M. C., & Carr, M. H. (1999). Groundwater formation of Martian valleys. *Nature*, *397*(6720), 589–591.

Manga, M., Patel, A., Dufek, J., & Kite, E. S. (2012). Wet surface and dense atmosphere on early Mars suggested by the bomb sag at Home Plate, Mars. *Geophysical Research Letters*, *39*(1).

Manga, M., & Wright, V. (2021). No cryosphere-confined aquifer below InSight on Mars. *Geophysical Research Letters*, *48*(8). e2021GL093127.

Manning, C. E., & Ingebritsen, S. E. (1999). Permeability of the continental crust: Implications of geothermal data and metamorphic systems. *Reviews of Geophysics*, *37*(1), 127–150.

Marinova, M. M., Aharonson, O., & Asphaug, E. (2008). Mega-impact formation of the Mars hemispheric dichotomy. *Nature*, *453*(7199), 1216–1219.

Martínez-Alonso, S., Mellon, M. T., Banks, M. E., Keszthelyi, L. P., McEwen, A. S., & Team, T. H. (2011). Evidence of volcanic and glacial activity in Chryse and Acidalia Planitiae, Mars. *Icarus, 212*(2), 597–621.

Matherne, C., Skok, J. R., Mustard, J. F., Karunatillake, S., & Doran, P. (2020). Multistage ice-damming of volcanic flows and fluvial systems in northeast Syrtis Major. *Icarus, 340*, 113608.

McGovern, P. J., Solomon, S. C., Smith, D. E., Zuber, M. T., Simons, M., Wieczorek, M. A., … Head, J. W. (2002). Localized gravity/topography admittance and correlation spectra on Mars: Implications for regional and global evolution. *Journal of Geophysical Research, Planets, 107*(E12). 19-1.

McGovern, P. J., Solomon, S. C., Smith, D. E., Zuber, M. T., Simons, M., Wieczorek, M. A., … Head, J. W. (2004). Correction to "Localized gravity/topography admittance and correlation spectra on Mars: Implications for regional and global evolution". *Journal of Geophysical Research, Planets, 109*(E7).

Milkovich, S. M., Head, J. W., III, & Pratt, S. (2002). Meltback of Hesperian-aged ice-rich deposits near the south pole of Mars: Evidence for drainage channels and lakes. *Journal of Geophysical Research, Planets, 107*(E6), 10–11.

Mischna, M. A., Baker, V., Milliken, R., Richardson, M., & Lee, C. (2013). Effects of obliquity and water vapor/trace gas greenhouses in the early Martian climate. *Journal of Geophysical Research, Planets, 118*(3), 560–576.

Montgomery, D. R., Som, S. M., Jackson, M. P., Schreiber, B. C., Gillespie, A. R., & Adams, J. B. (2009). Continental-scale salt tectonics on Mars and the origin of Valles Marineris and associated outflow channels. *Geological Society of America Bulletin, 121*(1–2), 117–133.

Morgan, G. A., Head, J. W., III, & Marchant, D. R. (2009). Lineated valley fill (LVF) and lobate debris aprons (LDA) in the Deuteronilus Mensae northern dichotomy boundary region, Mars: Constraints on the extent, age and episodicity of Amazonian glacial events. *Icarus, 202*(1), 22–38.

Mustard, J. F., Cooper, C. D., & Rifkin, M. K. (2001). Evidence for recent climate change on Mars from the identification of youthful near-surface ground ice. *Nature, 412*(6845), 411–414.

Neukum, G., Basilevsky, A. T., Kneissl, T., Chapman, M. G., Van Gasselt, S., Michael, G., … Lanz, J. K. (2010). The geologic evolution of Mars: Episodicity of resurfacing events and ages from cratering analysis of image data and correlation with radiometric ages of Martian meteorites. *Earth and Planetary Science Letters, 294*(3–4), 204–222.

Ojha, L., Buffo, J., Karunatillake, S., & Siegler, M. (2020). Groundwater production from geothermal heating on early Mars and implication for early Martian habitability. *Science Advances, 6*(49), eabb1669.

Palumbo, A. M., & Head, J. W. (2019). Oceans on Mars: The possibility of a Noachian groundwater-fed ocean in a sub-freezing Martian climate. *Icarus, 331*, 209–225.

Palumbo, A. M., & Head, J. W. (2020). Groundwater release on early Mars: Utilizing models and proposed evidence for groundwater release to estimate the required climate and subsurface water budget. *Geophysical Research Letters, 47*(8). e2020GL087230.

Pan, L., Ehlmann, B. L., Carter, J., & Ernst, C. M. (2017). The stratigraphy and history of Mars' northern lowlands through mineralogy of impact craters: A comprehensive survey. *Journal of Geophysical Research, Planets, 122*(9), 1824–1854.

Parker, T. J., Saunders, R. S., & Schneeberger, D. M. (1989). Transitional morphology in west Deuteronilus Mensae, Mars: Implications for modification of the lowland/upland boundary. *Icarus, 82*(1), 111–145.

Penido, J. C., Fassett, C. I., & Som, S. M. (2013). Scaling relationships and concavity of small valley networks on Mars. *Planetary and Space Science, 75*, 105–116.

Perron, J. T., Mitrovica, J. X., Manga, M., Matsuyama, I., & Richards, M. A. (2007). Evidence for an ancient Martian ocean in the topography of deformed shorelines. *Nature, 447*(7146), 840–843.

Phillips, R. J., Zuber, M. T., Solomon, S. C., Golombek, M. P., Jakosky, B. M., Banerdt, W. B., … Hauck, S. A., II. (2001). Ancient geodynamics and global-scale hydrology on Mars. *Science, 291*(5513), 2587–2591.

Plesa, A. C., Grott, M., Tosi, N., Breuer, D., Spohn, T., & Wieczorek, M. A. (2016). How large are present-day heat flux variations across the surface of Mars? *Journal of Geophysical Research, Planets, 121*(12), 2386–2403.

Ramirez, R. M., & Craddock, R. A. (2018). The geological and climatological case for a warmer and wetter early Mars. *Nature Geoscience, 11*(4), 230–237.

Ramirez, R. M., Kopparapu, R., Zugger, M. E., Robinson, T. D., Freedman, R., & Kasting, J. F. (2014). Warming early Mars with CO_2 and H_2. *Nature Geoscience, 7*(1), 59–63.

Rao, M. N., Nyquist, L. E., Wentworth, S. J., Sutton, S. R., & Garrison, D. H. (2008). The nature of Martian fluids based on mobile element studies in salt-assemblages from Martian meteorites. *Journal of Geophysical Research, Planets, 113*(E6).

Rao, M. N., Sutton, S. R., McKay, D. S., & Dreibus, G. (2005). Clues to Martian brines based on halogens in salts from nakhlites and MER samples. *Journal of Geophysical Research, Planets, 110*(E12).

Read, P. L., & Lewis, S. R. (2004). *The Martian climate revisited: Atmosphere and environment of a desert planet.* Springer Science & Business Media.

Robbins, S. J., Hynek, B. M., Lillis, R. J., & Bottke, W. F. (2013). Large impact crater histories of Mars: The effect of different model crater age techniques. *Icarus, 225*(1), 173–184.

Rodriguez, J. A. P., Fairén, A. G., Tanaka, K. L., Zarroca, M., Linares, R., Platz, T., … Glines, N. (2016). Tsunami waves extensively resurfaced the shorelines of an early Martian Ocean. *Scientific Reports, 6*(1), 1–8.

Scanlon, K. E. (2016). *Ice sheet melting throughout Mars climate history: Mechanisms, rates, and implications.* PhD thesis.

Scanlon, K. E., Head, J. W., Fastook, J. L., & Wordsworth, R. D. (2018). The dorsa Argentea formation and the Noachian-Hesperian climate transition. *Icarus, 299*, 339–363.

Schmidt, F., Way, M. J., Costard, F., Bouley, S., Séjourné, A., & Aleinov, I. (2022). Circumpolar Ocean stability on Mars 3 Gy ago. *Proceedings of the National Academy of Sciences, 119*(4). e2112930118.

Schwenzer, S. P., Abramov, O., Allen, C. C., Clifford, S. M., Cockell, C. S., Filiberto, J., … Wiens, R. C. (2012). Puncturing Mars: How impact craters interact with the Martian cryosphere. *Earth and Planetary Science Letters, 335*, 9–17.

Smith, D., Neumann, G., Arvidson, R. E., Guinness, E. A., & Slavney, S. (2003). *Mars Global Surveyor laser altimeter mission experiment gridded data record.* NASA Planetary Data System.

Smith, D. E., Zuber, M. T., Frey, H. V., Garvin, J. B., Head, J. W., Muhleman, D. O., … Sun, X. (2001). Mars orbiter laser altimeter: Experiment summary after the first year of global mapping of Mars. *Journal of Geophysical Research, Planets, 106*(E10), 23689–23722.

Som, S. M., Montgomery, D. R., & Greenberg, H. M. (2009). Scaling relations for large Martian valleys. *Journal of Geophysical Research, Planets, 114*(E2).

Souček, O., Bourgeois, O., Pochat, S., & Guidat, T. (2015). A 3 Ga old polythermal ice sheet in Isidis Planitia, Mars: Dynamics and thermal regime inferred from numerical modeling. *Earth and Planetary Science Letters, 426*, 176–190.

Šrámek, O., & Zhong, S. (2012). Martian crustal dichotomy and Tharsis formation by partial melting coupled to early plume migration. *Journal of Geophysical Research, Planets, 117*(E1).

Storrar, R. D., Evans, D. J., Stokes, C. R., & Ewertowski, M. (2015). Controls on the location, morphology and evolution of complex esker systems at decadal timescales, Breiðamerkurjökull, Southeast Iceland. *Earth Surface Processes and Landforms, 40*(11), 1421–1438.

Storrar, R. D., Stokes, C. R., & Evans, D. J. (2013). A map of large Canadian eskers from Landsat satellite imagery. *Journal of Maps, 9*(3), 456–473.

Storrar, R. D., Stokes, C. R., & Evans, D. J. (2014). Morphometry and pattern of a large sample (> 20,000) of Canadian eskers and implications for subglacial drainage beneath ice sheets. *Quaternary Science Reviews, 105*, 1–25.

Svendsen, J. I., Alexanderson, H., Astakhov, V. I., Demidov, I., Dowdeswell, J. A., Funder, S., … Stein, R. (2004). Late quaternary ice sheet history of northern Eurasia. *Quaternary Science Reviews, 23*(11–13), 1229–1271.

Tanaka, K. L. (1985). Ice-lubricated gravity spreading of the Olympus Mons aureole deposits. *Icarus, 62*(2), 191–206.

Tanaka, K. L., Robbins, S. J., Fortezzo, C. M., Skinner, J. A., Jr., & Hare, T. M. (2014). The digital global geologic map of Mars: Chronostratigraphic ages, topographic and crater morphologic characteristics, and updated resurfacing history. *Planetary and Space Science, 95*, 11–24.

Turbet, M., Boulet, C., & Karman, T. (2020). Measurements and semi-empirical calculations of $CO_2 + CH_4$ and $CO_2 + H_2$ collision-induced absorption across a wide range of wavelengths and temperatures. Application for the prediction of early Mars surface temperature. *Icarus, 346*, 113762.

Turbet, M., Forget, F., Head, J. W., & Wordsworth, R. (2017). 3D modelling of the climatic impact of outflow channel formation events on early Mars. *Icarus, 288*, 10–36.

Turbet, M., Tran, H., Pirali, O., Forget, F., Boulet, C., & Hartmann, J. M. (2019). Far infrared measurements of absorptions by $CH_4 + CO_2$ and $H_2 + CO_2$ mixtures and implications for greenhouse warming on early Mars. *Icarus, 321*, 189–199.

Urata, R. A., & Toon, O. B. (2013). Simulations of the Martian hydrologic cycle with a general circulation model: Implications for the ancient Martian climate. *Icarus, 226*(1), 229–250.

Villanueva, G. L., Mumma, M. J., Novak, R. E., Käufl, H. U., Hartogh, P., Encrenaz, T., … Smith, M. D. (2015). Strong water isotopic anomalies in the Martian atmosphere: Probing current and ancient reservoirs. *Science, 348*(6231), 218–221.

Ward, W. R., Burns, J. A., & Toon, O. B. (1979). Past obliquity oscillations of Mars: The role of the Tharsis uplift. *Journal of Geophysical Research. Solid Earth, 84*(B1), 243–259.

Warren, A. O., Kite, E. S., Williams, J. P., & Horgan, B. (2019). Through the thick and thin: New constraints on Mars paleopressure history 3.8–4 Ga from small exhumed craters. *Journal of Geophysical Research, Planets, 124*(11), 2793–2818.

Weiss, D. K., & Head, J. W. (2015). Crater degradation in the Noachian highlands of Mars: Assessing the hypothesis of regional snow and ice deposits on a cold and icy early Mars. *Planetary and Space Science, 117*, 401–420.

Weiss, D. K., & Head, J. W. (2017a). Evidence for stabilization of the ice-cemented cryosphere in earlier Martian history: Implications for the current abundance of groundwater at depth on Mars. *Icarus, 288*, 120–147.

Weiss, D. K., & Head, J. W. (2017b). Salt or ice diapirism origin for the honeycomb terrain in Hellas basin, Mars?: Implications for the early Martian climate. *Icarus, 284*, 249–263.

Werner, S. C. (2009). The global Martian volcanic evolutionary history. *Icarus, 201*(1), 44–68.

Wordsworth, R. D. (2016). The climate of early Mars. *Annual Review of Earth and Planetary Sciences, 44*, 381–408.

Wordsworth, R., Ehlmann, B., Forget, F., Haberle, R., Head, J., & Kerber, L. (2018). Healthy debate on early Mars. *Nature Geoscience, 11*(12), 888.

Wordsworth, R., Forget, F., Millour, E., Head, J. W., Madeleine, J. B., & Charnay, B. (2013). Global modelling of the early martian climate under a denser CO2 atmosphere: Water cycle and ice evolution. *Icarus, 222*(1), 1–19.

Wordsworth, R., Kalugina, Y., Lokshtanov, S., Vigasin, A., Ehlmann, B., Head, J., … Wang, H. (2017). Transient reducing greenhouse warming on early Mars. *Geophysical Research Letters, 44*(2), 665–671.

Wordsworth, R. D., Kerber, L., Pierrehumbert, R. T., Forget, F., & Head, J. W. (2015). Comparison of "warm and wet" and "cold and icy" scenarios for early Mars in a 3-D climate model. *Journal of Geophysical Research, Planets, 120*(6), 1201–1219.

Yin, A., Moon, S., & Day, M. (2021). Landform evolution of Oudemans crater and its bounding plateau plains on Mars: Geomorphological constraints on the Tharsis ice-cap hypothesis. *Icarus, 360*, 114332.

Zealey, W. J. (2009). Glacial, periglacial and glacio-volcanic structures on the Echus Plateau, upper Kasei Valles. *Planetary and Space Science, 57*(5–6), 699–710.

Glacial deposits, remnants, and landscapes on Amazonian Mars: Using setting, structure, and stratigraphy to understand ice evolution and climate history

Michelle Koutnik[a], Frances E.G. Butcher[b], Richard J. Soare[c], Adam J. Hepburn[d], Bryn Hubbard[e], Stephen Brough[f], Colman Gallagher[g], Lauren E. Mc Keown[h] and Asmin Pathare[i]

[a]Department of Earth and Space Sciences, University of Washington, Seattle, WA, United States, [b]Department of Geography, University of Sheffield, Sheffield, United Kingdom, [c]Department of Geography, Dawson College, Montreal, QC, Canada, [d]European Space Astronomy Centre, European Space Agency, Madrid, Spain, [e]Department of Geography and Earth Sciences, Aberystwyth University, Aberystwyth, United Kingdom, [f]School of Environmental Sciences, University of Liverpool, Liverpool, United Kingdom, [g]UCD School of Geography, University College Dublin, Dublin, Ireland, [h]NASA Jet Propulsion Laboratory, California Institute of Technology, Pasadena, CA, United States, [i]Planetary Science Institute, Tucson, AZ, United States

Abstract

Significant amounts of ice are located on the surface and in the subsurface of Mars. These polar and non-polar deposits are primarily water ice but, at the poles, carbon dioxide (CO_2) ice exists on the surface where it exchanges seasonally with the atmosphere, while buried CO_2 ice deposits have also been found. Analogous to Earth, Martian glacial ice deposits, as well as glacial remnants and landscapes from past glaciations, record how volatiles and components in the atmosphere, surface, and subsurface have interacted over time. Surface and subsurface expressions of past glaciations and deglaciations are critical to our understanding of the past climate on Mars, which is one of the highest priority goals in Mars science.

Mars' ice and climate record is constrained by the glacial record that extends over the last ~1 billion years of the Amazonian Period. Imagery, elevation models, radar, and spectral data have revealed aspects of the setting and structure of glacial deposits, glacial remnants, and geomorphological signatures of receded glaciers. The stratigraphy of these landforms has the capacity to provide the most highly resolved record available of past climate conditions on Mars. We discuss three key questions, leading with: what history of the Late Amazonian Epoch climate is recorded in the Polar Layered Deposits? Then, what sequence of glaciation and deglaciation developed non-polar glacial remnants? Related to interpreting glacial landscapes, we discuss: how widespread were past warm-based conditions among extant Amazonian-aged buried glaciers? Addressing these questions is necessary as part of continued efforts to advance our understanding of ice and climate histories on Mars.

1 Introduction

Glacial ice on Mars originates from the atmospheric precipitation of snow and ice onto the surface. Polar *glacial deposits* are predominantly composed of ice and exchange seasonally with the atmosphere. Polar and non-polar *glacial remnants* comprise ice, or at least likely contain ice, but most of the ice is buried and is not in exchange with the modern atmosphere. There are also *glaciated landscapes* that were shaped by past glaciations, and where only geomorphological signatures of receded glaciers remain. Analogous to Earth, these Martian glacial ice deposits, as well as glacial remnants and landscapes, record how volatiles and components in the atmosphere, surface, and subsurface have interacted over time (see also Gallagher, 2024, pp. 31–72).

It has been over 50 years since the multiple Mariner spacecraft validated early telescopic observations of significant polar-ice deposits on Mars (e.g., Herschel, 1784), and hypotheses about the latitude-dependent seasonal exchange with the atmosphere (e.g., Leighton et al., 1969; Murray et al., 1972; Soderblom et al., 1973). Since then, high-resolution visual imagery (e.g., Malin et al., 2007, 2010; Malin & Edgett, 2001; McEwen et al., 2007) has revealed fascinating structures that we will discuss here, including seasonally dynamic carbon dioxide (CO_2) ice structures, detailed stratigraphy within glacial deposits, exposures of subsurface massive ice, surface expressions of glacial remnants, and landforms that relate to past glaciations. Neutron spectrometer data revealed the distribution of water equivalent hydrogen in the upper 1 m of the subsurface, and especially poleward of 60°N and S where the shallow subsurface is rich in hydrogen (e.g., Feldman et al., 2002, 2004). Additionally, radar soundings (e.g., Holt et al., 2008; Orosei et al., 2015; Plaut et al., 2009; Seu et al., 2007) showed that significant deposits of water ice are also sequestered more deeply beneath the surface regolith in Mars' mid-latitudes (e.g., Bramson et al., 2015; Dundas et al., 2018, 2021b). Subsurface ice in the mid-to-high latitudes has been validated directly by the Phoenix lander (Mellon et al., 2009) and by mapping ice-exposing impact craters (Dundas et al., 2014, 2021b). Indirectly, radar sounding has also revealed CO_2 ice deposits buried within the predominantly water-ice polar deposits near Mars' south pole (e.g., Phillips et al., 2011).

Mars lacks the gravitationally stabilizing effect of a large moon and the planet's orbital parameters are therefore highly variable over time. Mars' spin-axis obliquity is particularly important for long-term climatic variability; currently, the obliquity is about 25° but it may have exceeded 65° in the last 250 Myr, and even higher obliquities were possible further back in time (e.g., Holo et al., 2018; Laskar et al., 2004). There is general agreement that surface accumulation of ice on Mars was episodic and cyclical and was tied to variations in obliquity and orbital eccentricity (e.g., Laskar et al., 2004; also Baker & Head, 2015; Dickson et al., 2008; Forget et al., 2006; Head et al., 2003; Madeleine et al., 2009; Mellon & Jakosky, 1995; Milliken et al., 2003; Mustard et al., 2001; Souness & Hubbard, 2012).

The surface and subsurface expressions of past glaciations and deglaciations are critical to our understanding of the past climate on Mars, which is one of the highest priority goals in Mars science. Recent summaries and reviews emphasize this priority, as well as introduce polar and non-polar deposits on Mars (e.g., Becerra et al., 2021; Bramson et al., 2020; Butcher, 2022; Diniega & Smith, 2020;

Koutnik & Pathare, 2021; Lasue et al., 2019; MEPAG, 2020; MEPAG ICE-SAG, 2019; Smith et al., 2018, 2020; Thomas et al., 2021).

In contrast to Early Mars (see Grau Galofre et al., 2024, pp. 73–100), significant polar and non-polar ice from more recent glaciations remain on the surface and in the near subsurface. Here, the focus is on the last ~1 billion years of Mars' history, which covers the Middle-to-Late Amazonian Epochs (Michael, 2013; Tanaka et al., 2014). This focus is based on the available age constraints for glacial deposits and glacial remnants, as well as associated landscapes with geomorphological signatures of receded glaciers in the non-polar regions. We explore three fundamental questions where recent observations and interpretations have advanced our understanding and revealed the complexity about ice and climate history, as well as challenged paradigms.

2 Key questions about Mars' Amazonian ice and climate history

The three questions that we address span polar ice, non-polar ice, glacial remnants, and past glacial processes. First, *what history of the Late Amazonian Epoch climate is archived in the Polar Layered Deposits?* Mars' polar deposits are the largest reservoirs of surface water ice on the planet and their structure and stratigraphy may provide the most highly resolved record of climate conditions over time. Deposits of CO_2 ice are also found at the poles, and their distribution and behavior may constrain past climate conditions. Second, *what sequence of glaciation and deglaciation developed non-polar glacial remnants?* Numerous glacial remnants, including extant buried glaciers, of different scales and in different settings have been mapped across the mid-latitudes (~30°–60°N and S) and also in select equatorial regions; these are referred to generally as "viscous flow features." They provide a record of glacial cycles, where evidence for ice accumulation, ice flow, and ice retreat is widespread. Glaciation, deglaciation, and ice flow on Mars have changed through time, and it is possible that the thermal regimes of glaciers have changed accordingly. Supraglacial and subglacial melting—whether liquid water existed on the surface or in the near subsurface of Mars, even if only ephemerally during the Amazonian—has far-reaching implications. Third, we address how glacial landscapes are beginning to shed light on the question: *How widespread were warm-based viscous flow features during the Amazonian?* In particular, the difference between cold-based and warm-based subglacial conditions has significant implications for the rate and nature of ice movement (and consequent landscape evolution), as well as for providing wet subglacial environments that were potentially conducive for life.

2.1 What history of the Late Amazonian Epoch climate is archived in the North and South polar layered deposits?

Mars' polar layered deposits (*PLD*) comprise the North polar layered deposits (*NPLD*) and the South polar layered deposits (*SPLD*), both multikilometer thick surface deposits of ice and dust (see Fig. 1). Layer sequences, unconformities, buried structures, and the large-scale morphology of the *PLD* reflect how climate patterns and climate changes control the deposition and erosion of polar ice (e.g., Smith et al., 2020). At the scale of individual layers, bright-to-dark visual variations in polar stratigraphy could be due to insolation-driven climate variations, where variations in obliquity and eccentricity have been hypothesized to control the flux of ice and dust to the poles that, over time, leads to a stack of polar layers with different ice/dust ratios (e.g., Cutts, 1973; Cutts & Lewis, 1982; Howard et al., 1982; Laskar et al., 2002, 2004; Murray et al., 1973; Smith et al., 2016; Toon et al., 1980).

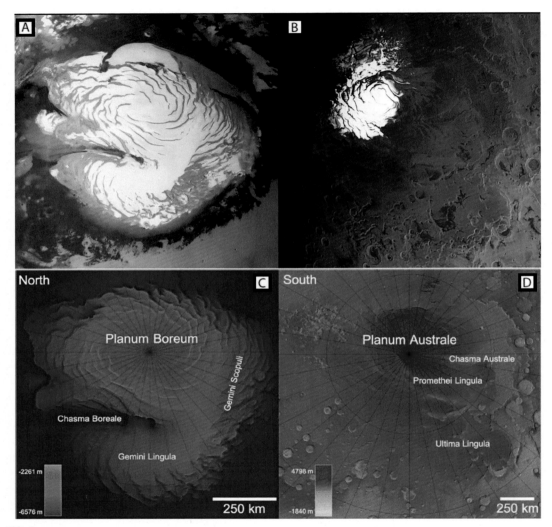

FIG. 1

(A) High-resolution stereo camera (*HRSC*; Neukum and Jaumann, 2004; Jaumann et al., 2007) imagery of the north polar layered deposits (*NPLD)* in late spring, without the seasonal CO_2 frost. The water-ice north polar residual cap appears bright white. (B) *HRSC* imagery of the south polar layered deposits (*SPLD*) in late summer, where the CO_2 ice of the south polar residual cap appears in bright white. The width of images in (A) and (B) is ~1250 km. Panels (C) and (D) are the topography of the *NPLDs* (panel C) and *SPLDs* (panel D) from Fergason et al. (2018) blended *HRSC* with Mars Orbiter Laser Altimeter (*MOLA*) data (Smith et al., 2001).

Panel A: Data credit: ESA/DLR/FU Berlin, adapted from Smith et al. (2020); panel D: Data credit: MOLA-NASA, adapted from Thomas et al. (2021).

The *NPLD* is approximately 1100 km across and up to 2 km thick (Phillips et al., 2008; Tanaka, 2005). In the south, the heavily cratered highlands surrounding the south polar dome of Planum Australe are approximately 6 km higher in elevation than the northern plains (Smith et al., 1999). The *SPLD* are approximately 1500 km across and up to 3.7-km thick (Plaut et al., 2009). For the *NPLD*, Shallow Subsurface Radar (*SHARAD*) reflectance data suggest a bulk average dielectric constant consistent with water ice and that the bulk dust concentration is a few percent or less (Grima et al., 2009; Phillips et al., 2008; Picardi et al., 2005). Internal layers are exposed along marginal scarps and along spiral sequences of troughs incised up to ~500 m deep into the *NPLD* (Figs. 1A and 2C). The origin of these troughs has been debated for decades, and recent hypotheses include formation involving trough migration (Bramson et al., 2019; Howard, 1978; Smith & Holt, 2010; Smith et al., 2013) and in situ erosion (e.g., Rodriguez et al., 2021). The different hypotheses have important implications for

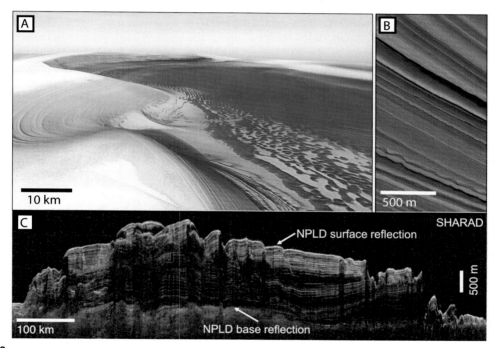

FIG. 2

(A) *NPLD* perspective view with a vertical exaggeration of 2.5. Head of Chasma Boreale, which is a canyon reaching 570 km from the *NPLD* edge. Canyon walls extend ~1400 m above the chasma floor. (B) Color portion of high-resolution imaging experiment (*HiRISE*; McEwen et al., 2007) ESP_018636_2745 showing fine-scale layering near the *NPLD* margin at 85.71 N, 179.37 E; image width is ~1 km. (C) Shallow radar (*SHARAD*; Seu et al., 2007) radargram of the *NPLD* interior showing subsurface layering that is extensive both in the horizontal direction and in the vertical, especially in the upper ~500 m but also evident to a few km depth.

Panel A: Credit: NASA/JPL/Arizona State University, R. Luk; panel B: Data credit: NASA/JPL/University of Arizona; panel C: Credit: adapted from Thomas et al. (2021).

the amount of ice that was available to cycle from the poles to the mid-latitudes as climate changes. Visually distinct layers and layer packets have been traced across both *PLD* (Figs. 2 and 3), especially in the north, where fine-scale layering is continuous over tens to hundreds of kilometers from trough to trough (Becerra et al., 2016; Fishbaugh & Hvidberg, 2006; Milkovich & Head, 2005).

There are at least four layered packets detected in *SHARAD* data that make up the *NPLD* stack and are separated by inter-packet regions with few radar reflections (Phillips et al., 2008). It is proposed that the inter-packet regions contain less dust within the ice compared to the packet regions, which in turn, are comprised of layers with varying proportions of ice and dust. High-resolution imagery and *SHARAD* data have been used to advance the hypothesis of a "widespread, recent accumulation package" (Smith et al., 2016), which is a near-surface unit identified at both poles. Across the *NPLD* this unit is up to ~320 m thick and overlies a notable unconformity (Putzig et al., 2018; Smith et al., 2016). The uppermost unit of the *NPLD* is the North Residual Cap, defined as the ~1-m thick portion of the *NPLD* surface where water-ice frost remains at the surface through the summer (e.g., Thomas et al., 2000).

Underlying much of the *NPLD* is a sand and ice deposit termed the "basal unit" (e.g., Brothers et al., 2015; Byrne & Murray, 2002); underlying the basal unit is the relatively flat Vastitas Borealis interior unit (Tanaka et al., 2005). However, the *NPLD* does not simply drape the basal unit. While the initial accumulation of the *NPLD* is inferred to be relatively uniform, it was punctuated by multiple episodes of erosion, evidenced by large-scale unconformities (Holt et al., 2010) in the form of isolated buried domes and deep valleys (e.g., Holt et al., 2010; Nerozzi & Holt, 2018; Putzig et al., 2009). Subsequent *NPLD* units formed on top of these earlier, eroded *NPLD* units in a complex manner; this included non-uniform accumulation, solar ablation, katabatic wind sculpting, as well as potential feedbacks between ice accumulation, ice loss, and overall topography (e.g., Byrne, 2009). For example, there is evidence of multiple *NPLD* ice depocenters (Brothers et al., 2015; Nerozzi et al., 2022). Chasma Boreale, a ~500-km long and up to 100-km wide canyon that cuts across the *NPLD* is proposed to have formed because of non-uniform accumulation (Holt et al., 2010; Fig. 2A).

Depocenter changes over time have also been proposed for the *SPLD* (Whitten et al., 2017; Whitten & Campbell, 2018). If layers detected by the radar were truncated by erosion (forming an unconformity), it is expected that they would intersect the surface, and if the unconformity is buried then it is expected that erosion was followed by subsequent deposition (Fig. 3B; Whitten & Campbell, 2018). By mapping such unconformities, it has been proposed that there were at least three major periods of deposition and two major periods of erosion (e.g., Kolb & Tanaka, 2006; Milkovich & Plaut, 2008). From visual imagery, Rodriguez et al. (2015) detailed evidence of plateau retreats that may have occurred in the last 10 Myr and that they attribute to wind erosion.

To what extent can PLD structure and stratigraphy constrain the Late Amazonian Epoch climate? The multiple major layer packets and units mapped within the *NPLD* and *SPLD* may relate to major changes in past climate. Global circulation modeling supports the hypothesis that the abundance of water and dust, atmospheric transport, and stability of surface ice deposits at the poles are sensitive to variations in insolation driven largely by variations in obliquity and eccentricity (Emmett et al., 2020; Forget et al., 2006; Levrard et al., 2007; Madeleine et al., 2009, 2014; Mischna et al., 2003; Newman et al., 2005; Toigo et al., 2020). Simpler climate models and/or process models have also progressed in relating ice and dust deposition to the formation of the *PLDs* (e.g., Cutts & Lewis, 1982; Greve et al., 2010; Hvidberg et al., 2012; Laskar et al., 2002; Toon et al., 1980). Fig. 4 indicates some of the major changes in volatile distribution in response to obliquity-driven climate change (see, e.g., Forget et al., 2017).

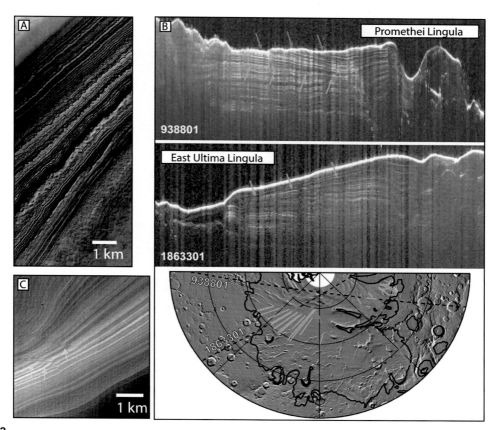

FIG. 3

(A) Fine-scale *SPLD* layers from *HiRISE* PSP_005011_0885, including non-horizontal layers that were likely distorted, possibly by ice flow. Fine-scale layering observed in high-resolution imagery has revealed individual layers that are decimeters to tens of meters in thickness and appear to have different textures (Becerra et al., 2017, 2019; Fishbaugh et al., 2010a, 2010b; Fishbaugh & Hvidberg, 2006; Limaye et al., 2012; Milkovich & Head, 2005). (B) The top and middle panels show *SHARAD* data of unconformities in the *SPLD* from Promethei Lingula and East Ultima Lingula. Locations of these traces are indicated by black dashed lines and the corresponding trace number in the bottom panel. The locations of mapped unconformities are shown with *light and dark green lines* in the bottom panel. *Dark green arrows and lines* indicate where surface layers truncate due to erosion. *Light green arrows and lines* indicate where subsurface layers truncate due to erosion and subsequent deposition. (C) *HiRISE* ESP_012934_1070 showing fine-scale layering and fault traces (indicated by *yellow arrows*) in an *SPLD* outcrop.

Panel A: Data credit: NASA/JPL/University of Arizona; panel B: Credit: adapted from Whitten and Campbell (2018); panel C: Data credit: NASA/JPL/University of Arizona.

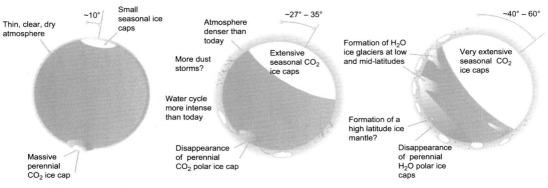

FIG. 4

Obliquity varies significantly on Mars, which affects Mars' climate, H_2O ice deposits, and seasonal CO_2 ice. The cartoons diagram ice deposits in three obliquity configurations: (A) low obliquity (0–10°); (B) high obliquity (~27–35°); and (C) very-high obliquity (40–60°). Generally, lower obliquity drives perennial polar-ice accumulation, higher obliquity drives polar-ice loss, and mid-to-low latitude H_2O ice deposits are proposed to form during high to very-high obliquity. Dust-storm activity is expected to be reduced with lower obliquity, and enhanced with higher obliquity due to changes in surface pressure and surface winds.

Credit: Forget et al. (2017; fig. 16.19).

Attempts have been made to interpret layer packets and individual layers directly, where "reading" the climate record of the *PLDs* typically refers to correlating a time series of quasi-periodic orbital variations with a depth series of layer-brightness, layer-thickness, and/or layer-morphology variations derived from stratigraphy. Time-series, spectral, and structural analyses have been applied to imagery (Becerra et al., 2017, 2019; Fishbaugh & Hvidberg, 2006; Laskar et al., 2002; Limaye et al., 2012; Milkovich & Head, 2005; Sori et al., 2014, 2022) and radar data (Putzig et al., 2009; Smith et al., 2016; Whitten et al., 2017) to detect periodicities from layer patterns that correspond to insolation oscillations. This is challenging because the system response for Mars may be nonlinear, absolute ages are unknown, annual-layer thicknesses of the ice record are unknown, and it is inherently challenging to distinguish a periodic orbital signal of climate change from stochastic variability in this type of stratigraphic record (e.g., Perron & Huybers, 2009; Sori et al., 2014). Even so, going beyond inferences from the *PLDs*, Sori et al. (2022) detected a signal of orbital forcing in the stratigraphy of an outlier ice deposit found in a South Polar crater. In addition, Sinha and Horgan (2022) found sediments in the *NPLD* that may relate to Late Amazonian impacts and volcanism, and could provide age constraints, if sampled.

An important question is whether the darker *PLD* layers (e.g., Courville et al., 2021; Lalich et al., 2019; Putzig et al., 2009) formed when atmospheric conditions were different (i.e., atmospheric dust was more abundant or the ratio of dust deposition to ice deposition increased), or formed as sublimation lag deposits. Global models are necessary to relate atmospheric conditions to layer formation and to address interactions between the dust, water, and CO_2 cycles (e.g., Kahre et al., 2017). Related to this question, global circulation modeling by Toigo et al. (2020) suggests that dust transport may be a primary driver of layer composition. Dust transport may vary non-monotonically with obliquity, which could help explain why there are more observed layers than obliquity cycles if the *NPLD* formed in the past ~4–5 Myr (Toigo et al., 2020).

Can present-day CO_2 ice distribution and behavior constrain past climate conditions?
The *SPLD* are composed primarily of water ice with a bulk dust concentration likely less than ~10% (Plaut et al., 2007), but also contain buried CO_2 ice deposits that appear to hold a similar mass of CO_2 to that of Mars' modern atmosphere (Bierson et al., 2016; Buhler et al., 2020; Phillips et al., 2011; Putzig et al., 2018). This deposit is referred to as the South Polar Massive CO_2 Ice Deposit (*MCID*; Buhler et al., 2020; Bierson et al., 2016; Phillips et al., 2011; Putzig et al., 2018), and in some locations, the CO_2 deposit is up to 1 km thick and may have undergone viscous flow (Smith et al., 2022). Although more CO_2 ice is expected to be deposited at times of lower obliquity (e.g., Forget et al., 2017; Fig. 4), the timing of *MCID* unit formation and whether these units were (or currently are) in exchange with the atmosphere are open questions. One hypothesis is that overlying densified water-ice layers fully sequester the accumulations of CO_2 ice, and that sequestration is necessary for the buried CO_2 to survive higher obliquity states (Bierson et al., 2016; Manning et al., 2019). Alternatively, the water-ice layer may be sufficiently permeable for buried CO_2 ice to exchange with the atmosphere, allowing gradual equilibration during higher obliquity states (Buhler et al., 2020; Phillips et al., 2011).

Seasonally, more than 25% of Mars' atmospheric CO_2 condenses in the north and south circumpolar regions (e.g., Forget et al., 1995; Kieffer & Titus, 2001; Leighton & Murray, 1966). The resulting deposited layer ranges in thickness from 1 to 2 m at the poles (Aharonson et al., 2004; Nier & McElroy, 1977), down to a few millimeters toward the mid-latitudes (Vincendon, 2015). As insolation increases in spring, the CO_2 layer sublimates, and this seasonal exchange is now recognized as a cardinal driver of present-day surface modification on Mars (Diniega et al., 2021; Dundas et al., 2021a). Martian features formed by the seasonal CO_2 cycle have no Earth analogs, and these forms include linear gullies and their pits (Diniega et al., 2013; Dundas et al., 2012; Pasquon et al., 2016), sand furrows (Bourke & Cranford, 2011), and the dendritic araneiform terrain near Mars' South pole (Hansen et al., 2010; Malin & Edgett, 2001; Piqueux et al., 2003).

Araneiforms, often referred to as "spiders," are large (up to 1 km wide), dendritic, tortuous negative topography features eroded in the Martian south polar substrate (Hansen et al., 2010; Piqueux et al., 2003; Fig. 5A and B). They have been observed to show "activity" in the form of dark fans and spot appearance within their locales in spring and summer (Kieffer et al., 2006; Malin et al., 1998). A combined effort by Piqueux et al. (2003), who identified and described spiders, and Kieffer et al. (2006), who developed a formation hypothesis, linked the formation of spiders with the seasonal appearance of fans and spots. Based on this activity and the finding that fans and spots were not at the same temperature as soil, but rather just below that of CO_2 ice, it was proposed by Kieffer et al. (2006) that spiders are formed by a type of "Solid State Greenhouse Effect" (Matson & Brown, 1989) (see Fig. 5C). This process involves insolation penetrating translucent CO_2 slab ice, which is opaque to thermal radiation, which then gets trapped, warming the regolith and causing the ice to sublimate from its base (Kieffer et al., 2006). As pressure increases, the ice ruptures and forms a vent. Gas escapes through this vent, entraining unconsolidated fines from beneath the ice in the form of a plume, and depositing the material as low albedo fans and spots at the surface.

Elements of the process that develops spiders have been demonstrated in the laboratory (e.g., Kaufmann & Hagermann, 2017; Mc Keown et al., 2017, 2021; Portyankina et al., 2019; Fig. 5D). However, spiders near the South Pole have never been observed to grow in the present day, despite ongoing fan and spot activity. This suggests that spiders either (i) are growing too slowly to have been detected by *HiRISE* or (ii) developed in a different climate, under which sublimation was much more energetic and capable of eroding large swaths of terrain. While many unknowns remain about the

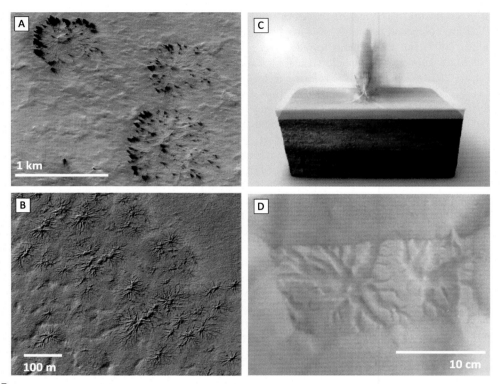

FIG. 5

(A) "Starburst" spiders showing dark fans strewn from within troughs. These spiders are on the scale of ~1 km in diameter and often have >1000 branches; *HiRISE* ESP_011697_0980. (B) "Classic" spiders that have slightly wider centers and dendritic, tortuous troughs; *HiRISE* ESP_014413_0930. (C) Cartoon of the hypothesis for spider formation (following Kieffer et al., 2006). (D) Spider patterns were observed when CO_2 ice-containing vents were allowed to sublimate in contact with granular substrate under Mars conditions in the laboratory.

Panel A: Data credit: NASA/JPL/University of Arizona; panel B: Data credit: NASA/JPL/University of Arizona; panel C: Image credit: Featherwax retouching/CGI; panel D: Credit: Mc Keown et al. (2021).

modern CO_2 cycle on Mars (see, e.g., Titus et al., 2017), observations of interactions between CO_2, water ice, and dust are important for polar-ice evolution, in particular how CO_2 ice may influence the polar surface composition, albedo, and emissivity, as well as the accumulation, sublimation, and potential sequestration of CO_2 ice within water-ice units.

2.2 What sequence of Late Amazonian Epoch glaciation and deglaciation developed non-polar glacial remnants?

On Earth, landscapes in both polar and non-polar regions have been shaped by glaciers that have waxed and waned in response to external climate changes and internal dynamics. Evidence from the ice, rock, and sediment records of glacial systems can be used to constrain the extent, timing, and style of past glaciations (see Gallagher, 2024, pp. 31–72; also, e.g., Benn & Evans, 2010).

On Mars, glacial remnants that are presently icy (as opposed to ice-rich permafrost, comprising frozen ground, see Soare et al., 2024, pp. 143–192) are evidence for past glaciations during the Amazonian, and especially during the Late Amazonian Epoch. Numerous lines of morphological evidence for mid-latitude (~30°–60°N and S) icy glacial remnants have been presented over the past decades (e.g., Head et al., 2006; Holt et al., 2008; Karlsson et al., 2015; Lucchitta, 1984; Milliken et al., 2003; Sharp, 1973; Squyres, 1978; Squyres & Carr, 1986). Fig. 6 shows how geomorphological mapping using available data can constrain the spatial distribution of subsurface ice across the mid-latitudes (e.g., Morgan et al., 2021).

Many mid-latitude features exhibit detailed surface patterns consistent with the viscous deformation of flowing ice, and collectively, have become known as viscous flow features (*VFFs*; Milliken et al., 2003). Forming a complex and coalescing system, *VFFs* are subcategorized as a function of their size and topographic setting. The global distribution of *VFFs* is shown in Fig. 7.

Although *VFF* morphology can vary widely, even over short spatial scales, four predominant sub-types have been identified and described based on size and topographic setting. *Lobate debris aprons* (*LDAs*) are wide and often radially flowing lobes of icy material that extend from, or in some cases, completely encircle isolated mesas and escarpments (Fig. 8B) (Head et al., 2010; Holt et al., 2008; Lucchitta, 1984; Squyres, 1978). *LDAs* are often isolated, and radar sounding (e.g., Gallagher et al., 2021; Holt et al., 2008; Parsons et al., 2011) indicates these deposits are thick (up to 100s of meters)

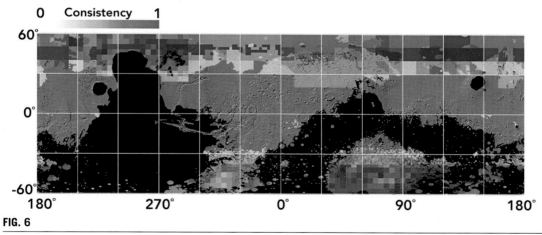

FIG. 6

Subsurface water-ice mapping (*SWIM*) geomorphology ice-consistency map for all depths that was derived from a grid-mapping approach of periglacial and glacial features between 60°S and 60°N. The product uses Context Camera (*CTX*; Malin et al., 2007) mosaic tiles (beta01 from the CalTech Murray Lab; Dickson et al., 2023), and represents grid sizes of 1 × 1 degrees, and 4 × 4 degrees used for the southern hemisphere and Tharsis Montes. For the rest of the northern hemisphere, the grid mapping was extrapolated to geologic units (Tanaka et al., 2005, USGS SIM 2888). Previously mapped viscous flow features (*VFFs*) were given an ice consistency value of 1, previously mapped pedestal craters a value of 0.75, detailed mapping of scalloped terrain a value of 0.75, and the number of newly mapped landforms was normalized with equal weighting to give ice consistency between 0 and 1; areas with no features were assigned a value of −1. *SWIM* products (e.g., Morgan et al., 2021) delineate estimates of where subsurface ice is likely to exist.

Credit: https://swim.psi.edu/SWIM2Products.php.

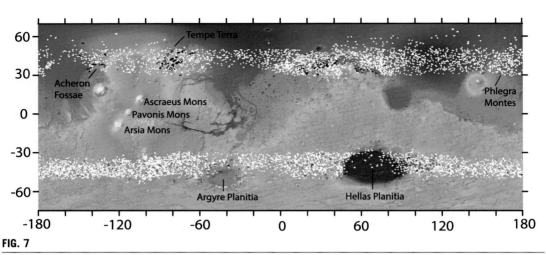

FIG. 7

Global distribution of the four primary classifications of non-polar glacial remnants: lobate debris aprons (*LDAs—yellow polygons*), lineated valley fill (*LVF—black polygons*), and concentric crater fill (*CCF—white polygons*) from Levy et al. (2014), and glacier-like forms (*GLFs—magenta polygons*) from Brough et al. (2019). In total, over 12,000 *VFFs* have been identified and mapped predominantly between 30° and 50° latitude in both hemispheres (e.g., Brough et al., 2019; Dickson et al., 2012; Levy et al., 2014; Souness et al., 2012), with ~9500 mapped as *CCF* (e.g., Dickson et al., 2012; Levy et al., 2014). *CCFs* are found at all longitudes. *LDAs* are primarily concentrated in the fretted terrain of the northern hemisphere and surrounding the Hellas and Argyre impact basins in the southern hemisphere (Levy et al., 2014; Squyres, 1979; Squyres & Carr, 1986). Major locations mentioned in the text are noted.

and comprise relatively pure water ice (less than 10% lithic content). *LDA* surfaces are often heavily pitted, consistent with sublimation-driven ice loss (Head et al., 2010), and convex outward ridges are frequently observed at *LDA* margins (Baker et al., 2010; Levy et al., 2007). *Lineated valley fill (LVF)* is located within constrained valleys and is likely formed through the convergence or coalescing of two or more *LDA*-type flows and subsequent along-valley flow (Lucchitta, 1984; Squyres, 1978, 1979). This convergence of material can produce highly integrated and anastomosing flows that can extend as continuous valley systems for hundreds of kilometers (e.g., Head et al., 2006, 2010). *Concentric crater fill (CCF)* has concentric quasi-circular surface patterns reminiscent of their host craters (e.g., Squyres, 1979; Squyres & Carr, 1986). However, true "concentric" flow is only observed in craters poleward of ~45°. At lower latitudes (~30°–45°), *CCF* deposits have a poleward flow preference suggesting that flow occurred preferentially—or at least evidence of flow is best preserved—on steep, cold, polar-facing slopes in these latitudes (Dickson et al., 2012), indicating a sensitivity to insolation. Finally, *glacier-like forms (GLFs)* are the lowest order form of *VFF* and are identified by their lobate termini often consisting of raised moraine-like ridges, the presence of longitudinal curvilinear ridges on their surfaces, and their characteristic occupation of small alcoves with amphitheater-like headwalls (Hubbard et al., 2011; Souness et al., 2012; Fig. 8A). *GLFs* commonly coalesce with larger *VFFs*, though they may also occur in isolation (Brough et al., 2019). *GLFs*, along with other *VFFs*, appear similar to terrestrial valley glaciers (e.g., Arfstrom & Hartmann, 2005; Hubbard et al., 2011; Milliken et al., 2003) and share characteristics of terrestrial debris-covered glaciers (see, e.g., Koutnik & Pathare, 2021). The survival of ice deposits that may remain in the *VFFs* is attributed to surface de-

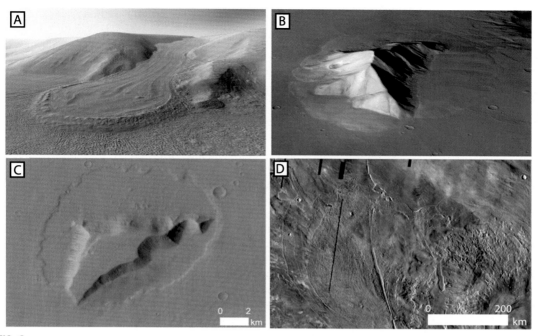

FIG. 8

(A) Glacier-like form (*GLF*) in *HiRISE* PSP_009455_2215 draped over topography. (B) Lobate debris apron (*LDA*) shown in perspective false color from *HRSC* image 0451. *LDAs* are characterized by sharply defined flow fronts, have convex down-slope profiles, and generally have visible perpendicular or parallel lineations and compressional ridges on their surface, reflecting viscous internal deformation and down-slope advection of icy material (e.g., Head et al., 2010; Li et al., 2005; Mangold et al., 2003; Pierce and Crown, 2003; Squyres, 1978). (C) Ghost LDA depression (*GLDA*; Hepburn et al., 2020a) in Kasei Valles. These radial depressions surrounding mesas formed at the boundary of interaction between a former *LDA* and lava flow ~1.3 Ga (Hauber et al., 2008; figure adapted from Hepburn et al., 2020a). (D) Fan-shaped deposit (*FSD*) west of Arsia Mons (e.g., De Blasio, 2014).

    Panel A: Credit: NASA/JPL/University of Arizona and rendered by S. Doran; panel B: Credit: ESA/DLR/FU Berlin, adapted from Helbert et al. (2015); panel D: Data credit: THEMIS mosaic, NASA/JPL/Arizona State University, adapted from De Blasio (2014).

bris protecting the underlying ice from sublimation (e.g., Baker & Carter, 2019; Fastook et al., 2014; Gallagher et al., 2021; Head et al., 2005).

    The output of global climate models run at a high spin-axis obliquity (>35° to 40°) supports the zonal redistribution of volatiles between Mars' polar regions and the mid-latitudes in particular, and where significant precipitation can be generated in areas of Mars where *VFFs* are abundant (Forget et al., 2006, Madeleine et al., 2009; see Fig. 4). However, long-term obliquity (>20 Ma) is chaotic and cannot be modeled definitively, only statistically bound (Laskar et al., 2004). Without a deterministic understanding of astronomically driven climate change, constructing a sequence of glacial activity from landform evidence is a tractable step by which we can interrogate past variability in Mars' climate (e.g., Hepburn et al., 2020a, 2020b). To do so, we need to constrain the origin, spatiotemporal evolution, and preservation of the mid-latitude glacial remnants, especially considering the volume of ice remaining (e.g., Karlsson et al., 2015), primarily spanning the 30°–60° latitude bands of both hemispheres (Fig. 7).

There is evidence that suggests that *LDAs* have formed over multiple glacial cycles during the past ~300–800 Ma (e.g., Dickson et al., 2008; Levy et al., 2021), and evidence for at least one glacial recession (e.g., Brough et al., 2016; Dickson et al., 2008; Levy et al., 2007; Morgan et al., 2009) as well as multiple glacial cycles in the past few million to tens of millions of years (Hepburn et al., 2020b; Soare et al., 2021a). Approximately one-third of *GLFs* appear recessed or depressed, which is predominantly evidenced by forefield arcuate moraines isolated from the buried ice margin and sequential latero-terminal moraine ridges (Brough et al., 2016; see Fig. 8A). Also, surface deflation relative to putative lateral moraines (Dickson et al., 2008) and steep scarps observed at the contact between *LVF* surfaces and abutting talus deposits indicate substantial *LVF* surface lowering since talus formation (Levy et al., 2007).

Changes in the thickness and extent of Mars' ice masses would likely be accompanied by corresponding variations in their dynamism. Driving stresses are reduced as ice surfaces are thin and lower, resulting in less dynamic ice flow regimes and eventual flow stagnation. While the surface morphologies of *VFFs* indicate past flow, no evidence has been identified to date for the present-day flow of ice in these features. Thus, they are either flowing so slowly that their motion is undetectable at current imaging resolutions and observing timescales, or they have stagnated.

Multiple *GLFs* have been identified, which debouch from alcoves and appear to superpose onto underlying *VFFs* (e.g., Brough et al., 2019; Dickson et al., 2008; Head et al., 2005; Hepburn et al., 2020b; Levy et al., 2007). Rather than coalescing with lower elevation and larger-sized *LDAs* and *LVFs*, these features—termed superposed glacier-like forms (*SGLFs*; e.g., Hepburn et al., 2020b)—terminate abruptly at a sharp contact and have surface lineations that are orientated obliquely to those on underlying flows (Hepburn et al., 2020b). This stratigraphic relationship implies that *SGLFs* are younger than the *VFFs* onto which they flow, indicating distinct phases: an early phase of glaciation emplacing large-scale ice sheets/caps that receded to the remnant *LDAs*, *LVFs*, and *CCFs*, and then a later alpine-style phase of glaciation that formed the now-remnant *SGLFs* (Baker et al., 2010; Brough et al., 2016; Dickson et al., 2008; Levy et al., 2007; Morgan et al., 2009). Global mapping of *SGLFs* and dating of both *SGLFs* and underlying *VFFs* confirms this apparent relative age difference in absolute terms, suggesting that while *VFFs* formed diffusively over the last 300 Ma (consistent with previous work), *SGLFs* formed in two distinct clusters of 2–20 Ma and 45–65 Ma (Hepburn et al., 2020b). These clusters are evidence of two growth-recession cycles in the Martian mid-latitude glaciation record within the period of reconstructable astronomical forcing.

At lower latitudes, a series of circumferential depressions (termed ghost lobate debris aprons, or *GLDAs*; Hepburn et al., 2020a; Fig. 8C)—considered the landscape record of former *LDAs* embayed by lava (Hauber et al., 2008)—suggest spatially variable accumulation and ablation may have taken place over the last ~1.3 Ga (Hepburn et al., 2020a). Additionally, numerous large (160,000 km^2) fan-shaped deposits, interpreted as sequences of moraines, extending from the northwest flanks of the equatorial Arsia, Pavonis, and Ascraeus Mons volcanoes (e.g., Fig. 8D), suggest formerly extensive ice at these locations (Fastook et al., 2008; Head & Marchant, 2003; Kadish et al., 2014; Scanlon et al., 2015; Shean et al., 2005, 2007). These fan-shaped deposits are thought to have formed during periods of high obliquity ($\gtrsim$40°) in the Amazonian, when ice was stable at equatorial latitudes, as moisture-laden air cooled adiabatically as it encountered these volcanoes, driving local precipitation (Fastook et al., 2008).

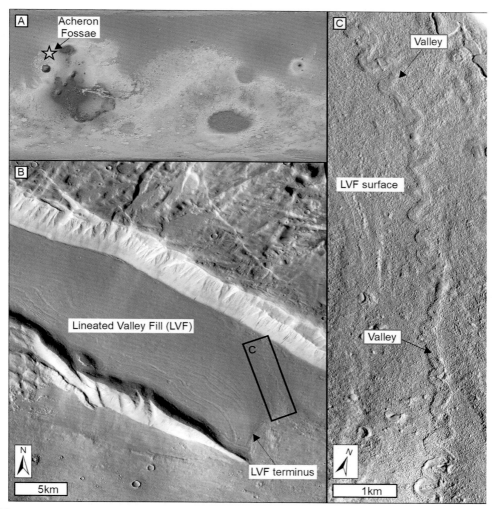

FIG. 10

Small, kilometer-scale valley on the surface of a lineated valley fill in Acheron Fossae (~230°E, 35.9°N), identified by Fassett et al. (2010). (A) Location of Acheron Fossae (indicated by the *yellow star*) on the *MOLA* elevation map of Mars. (B) *CTX* image P02_001933_2174_XN_37N130W showing the context of the lineated valley fill occupying a steep-sided NW-SE oriented valley. The *black box* shows the extent of panel C. (C) Enlarged view of the valley (extent in panel B) on the surface of the lineated valley fill shown in panel B. The valley incises a surface with a best-fit crater retention age estimate of ~80 to 110 Myr (Fassett et al., 2010).

Panel A: Data credit: MOLA—NASA/USGS; panel B and C: Data credits: CTX—NASA/JPL/Malin Space Science Systems/Arizona State University.

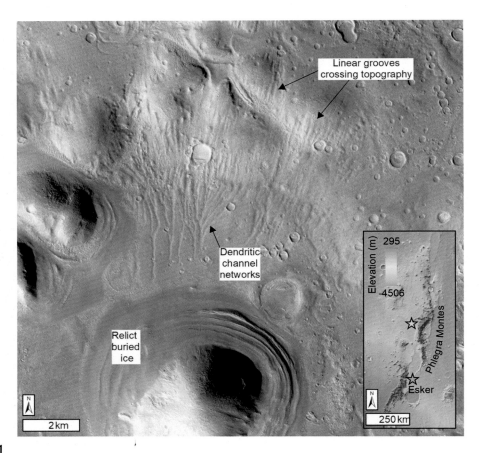

FIG. 11

Linear grooves in Phlegra Montes, which transition downslope into dendritic channel networks. Gallagher et al. (2021) interpreted these features as having formed by abrasion and meltwater flow beneath a wet-based ice mass, and they interpret various landforms across the region as having been eroded by wet-based glacial ice and/or glacial meltwater. *CTX* image G07_020778_2193_XI_39N197W. Inset *MOLA* elevation map shows the location of this site (*yellow star*) relative to the esker (*blue star*; see esker in Fig. 12C) identified by Gallagher and Balme (2015).

Data credit: CTX—NASA/JPL/Malin Space Science Systems/Arizona State University. Inset: MOLA—NASA/USGS.

In Phlegra Montes, there is also evidence for localized, transient, basal melting of ice in the form of candidate eskers associated with a late Amazonian-aged (~150 Ma) *LVF*, shown in Fig. 12 (Gallagher & Balme, 2015). Eskers on Earth are sinuous ridges of glaciofluvial sediment that trace the former locations of meltwater channels within or beneath glacial ice, forming key elements of some landsystem models (e.g., Fig. 9). Confidently identified eskers indicate past glacial melting and do not form beneath entirely cold-based glaciers; however, not all warm-based or polythermal glaciers on Earth form eskers. Fig. 12 also shows landforms potentially formed by proglacial runoff of glacial meltwater in the foreland of the *LVF* that formed the esker (Gallagher & Balme, 2015).

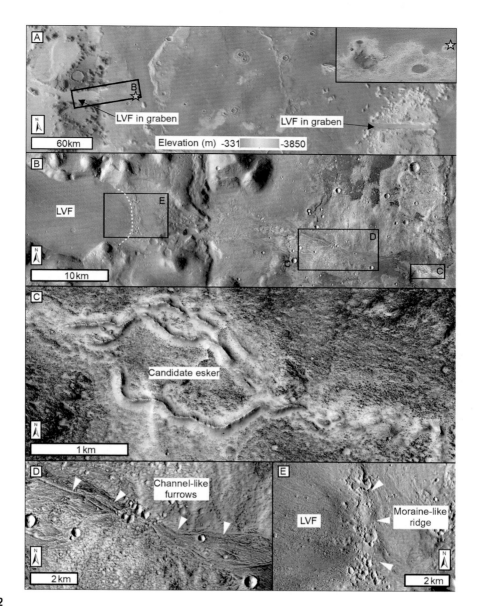

FIG. 12

A candidate esker and selected examples of other proglacial landforms identified by Gallagher and Balme (2015) in association with an Amazonian-aged *LVF* in Phlegra Montes. The candidate esker is separated from the present-day glacier terminus by a ~40km wide foreland hosting evidence for incision and reworking by glacial meltwater. The presence of an extant ice mass, a self-consistent system of candidate glacial landforms, and an absence of associated evidence for alternative sinuous ridge-forming processes lead to high confidence in the interpretation of this landform as an esker. (A) *MOLA* elevation map overlain on a *THEMIS* daytime IR image mosaic showing the location (*blue star*) of the candidate esker. The *black box* shows the extent of panel B. *Yellow star* in the inset global elevation map shows the location of Phlegra Montes. (B) *CTX* image mosaic showing the LVF (terminus indicated by *white dashed line*) and its foreland. *Black boxes* show the extent of panels C–E. (C) *HiRISE* image ESP_044316_2130 of the candidate esker. The extent shown in panel B. (D) *CTX* image P22_009583_2132_XN_33N197W of channel-like furrows possibly formed by glacial meltwater, which are located between the candidate esker and the present-day terminus of the *LVF*. Extent shown in panel B. (E) *CTX* image P18_007935_2132_XN_33N197W of a moraine-like ridge at the terminus of the *LVF*.

Panel A: Data credit: MOLA—NASA/USGS, THEMIS—NASA/USGS/Arizona State University; panel C: Data credit: NASA/JPL/University of Arizona; panel E: Data credit: NASA/JPL/Malin Space Science Systems/Arizona State University. Modified from Butcher (2019).

The existence of a Late Amazonian Epoch ice sheet in Phlegra Montes has been hypothesized on the basis of global climate modeling experiments (Madeleine et al., 2009) and separately based on geomorphic evidence (Dickson et al., 2010). However, even in the locations where an Amazonian-aged esker has been identified, it may represent transient, localized episodes of basal melting (perhaps under polythermal regimes) before their parent glaciers returned to a cold-based state (Butcher et al., 2017, 2020, 2021; Gallagher & Balme, 2015; Woodley et al., 2022). In addition, depending on the environment, sinuous ridges may have alternative explanations (e.g., Butcher et al., 2021; Ramsdale et al., 2015). However, the spatial association between extant Amazonian ice deposits, the identified candidate esker associated with a *VFF* (Fig. 12C; Gallagher & Balme, 2015), and the regionally extensive assemblage of glacial erosional landforms at Phelgra Montes (Gallagher et al., 2021), means that it is a distinct possibility that Amazonian wet-based glaciation played a central role in shaping this landscape.

In northwest Tempe Terra, Butcher et al. (2017) identified an esker-like sinuous ridge emerging from an *LVF* (Fig. 13A), with an impact crater retention age of ~110 Myr (of magnitude similar to that in Phlegra Montes). Woodley et al. (2022) also identified two additional eskers emerging from VFFs nearby with a combined crater retention age of 220 Ma. The host glaciers in Phlegra Montes and Tempe Terra have receded from the positions of the candidate eskers; hence, Late Amazonian Epoch age estimates from impact craters on the surfaces of those glaciers provide absolute minimum esker ages. Esker-like sinuous ridges have also been identified as emerging from an *LDA* with an impact crater retention age of ~330 Ma in the Chukhung crater in central Tempe Terra (Butcher et al., 2021) (Fig. 13C). However, the esker interpretation in this location is more ambiguous (the ridges could alternatively be older fluvial features), although at least one morphological subtype of sinuous ridges in Chukhung crater is currently best explained by the esker hypothesis (Butcher et al., 2021). It has also been proposed that ridges in the Lyot crater are Amazonian-aged eskers formed beneath a wet-based ice cap, which has since retreated (Hobley et al., 2014), but this interpretation, and the broader hypothesis of wet-based glaciation in the Lyot crater, have been challenged (Brooker, 2019).

Has there been recent polythermal glaciation of impact crater walls? Arcuate moraine-like ridges (e.g., Arfstrom & Hartmann, 2005; Hubbard et al., 2011; Milliken et al., 2003) are commonly bound extant glacier-like forms (Section 2.2) on Mars and are abundant within mid-latitude impact craters. Moraines are widespread on Earth and, while most commonly formed by warm-based and polythermal ice, they can be formed (more rarely) by cold-based glaciers (Fitzsimons & Howarth, 2020). Thus, moraines cannot necessarily be taken alone to indicate the former presence of basal meltwater and must be considered in their wider landscape context. There are, however, some locations on Mars where detailed studies provide evidence for the formation of arcuate ridges by polythermal glaciers.

The Greg crater, east of Hellas Planitia, hosts classic examples of moraine-like arcuate ridges bounding *GLFs*, which accumulated within the last 50 Myr (Fig. 14; e.g., Hartmann et al., 2014; Tsibulskaya et al., 2020). In a detailed characterization of one of these *GLFs*, Hubbard et al. (2011) identified streamlined landforms upslope of the arcuate ridges and downslope of remnant buried ice that are reminiscent of (but smaller in scale than) streamlined bedforms produced by wet-based ice on Earth. They considered the glaciotectonic deformation of frozen sediments to be the most likely explanation for the bounding arcuate ridges. Glaciotectonic deformation is most often associated with polythermal glaciers where it arises from stresses caused by differential flow between wet-based and cold-based ice and/or the pressurization of subglacial meltwater trapped by cold-based ice and frozen sediments. Thus, Hubbard et al. (2011) suggested that the *GLF* in the Greg crater may previously have had a polythermal regime with a wet-based interior zone (evidenced by streamlined landforms) and a cold-based periphery (evidenced by candidate glaciotectonic moraines).

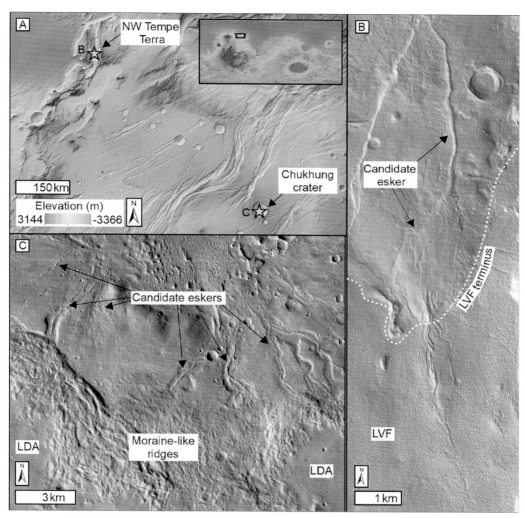

FIG. 13

Candidate eskers associated with extant ice deposits in Tempe Terra, identified by Butcher et al. (2017, 2021). (A) *MOLA* elevation map overlain on a shaded relief map of the Tempe Terra region of Mars, showing the locations (*yellow stars*) of candidate eskers identified in association with viscous flow features. The location is shown by a *black box* in the *inset* global elevation map of Mars. (B) *CTX* image P05_002907_2258_ XN_45N083W of the candidate esker emerging from a lineated valley fill in NW Tempe Terra identified by Butcher et al. (2017). See also Butcher et al. (2020). Location shown in panel A. (C) *CTX* image P04_002577_2186_XN_38N072W of the southern floor of Chukhung crater (location shown in panel A) where sinuous ridges emerge from moraine-like ridges bounding LDA. Butcher et al. (2021) interpreted these ridges as either eskers or older, topographically inverted fluvial landforms (or a mixture of both).

Panel A: Data credit: NASA/USGS; panel C: Data credit: NASA/JPL/Malin Space Science Systems/Arizona State University.

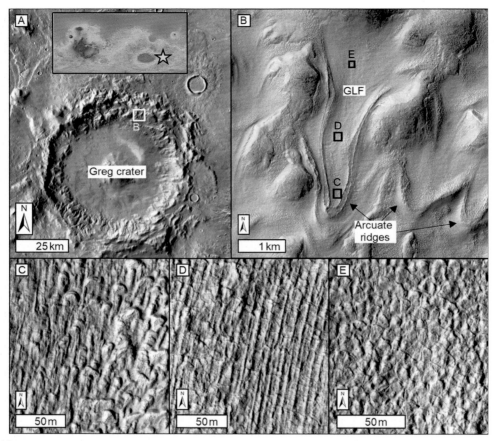

FIG. 14

Arcuate moraine-like ridges in the Greg crater are associated with streamlined landforms on an exposed portion of the bed of a glacier-like form (*GLF*). (A) *THEMIS* daytime IR image mosaic of the Greg crater. The *white box* shows the extent of panel B. *Inset*: *MOLA* Global Elevation Map shows the location of the Greg crater (*yellow star*). (B) *CTX* image G12_022745_1415_XN_38S246W shows the glacier-like form that was characterized by Hubbard et al. (2011). The glacier-like form has pronounced bounding arcuate ridges. *Black boxes* show the locations of panels C–E. (C) *HiRISE* image PSP_003243_1415 of streamlined mound-and-tail landforms proximal to the arcuate ridges, interpreted by Hubbard et al. (2011) as drumlins exposed by ice retreat. Location shown in panel B. (D) Elongate streamlined landforms located upslope of the mound-and-tail terrain and interpreted by Hubbard et al. (2011) as small-scale analogs of mega-scale glacial lineations exposed by ice retreat. From the same image product as panel C. Location shown in panel B. (E) Polygonized terrain located upslope of the exposed bedforms in panels C and D and interpreted by Hubbard et al. (2011) as indicators of extant buried ice in the upslope reaches of the glacier-like form. From the same image product as panel C. Location shown in panel B.

Panel A: Data credits: MOLA—NASA/USGS, THEMIS—NASA/USGS/Arizona State University; panel B: Data credit: CTX—NASA/JPL/Malin Space Science Systems/Arizona State University; panel E: Data credit: HiRISE—NASA/JPL/University of Arizona.

Analyses of the walls of 11 glaciated impact craters across Mars' mid-latitudes by Conway et al. (2018) also revealed past erosion rates similar to those in warm-based glacial environments on Earth, perhaps as recently as 5–10 Ma. This challenges the more generalized cold-based erosion rates estimated by Levy et al. (2016) for Amazonian glacial landscapes. Landform assemblages on the impact craters studied by Conway et al. (2018) showed evidence of wall erosion, a reworking of basal sediments by glacial ice, and glaciotectonic deformation at former ice margins. As in Hubbard et al. (2011), Conway et al. (2018) invoked past polythermal glaciation for small crater-wall glaciers, whereby subglacial meltwater, produced in relatively small volumes, refroze subglacially as it approached cold-based ice at the ice margin.

Warm-based subglacial conditions likely require exogenic and endogenic heat sources. While candidate eskers identified in association with Amazonian-aged glaciers are rare compared to the abundance of those glaciers, those described above (e.g., Butcher et al., 2017, 2021; Gallagher & Balme, 2015) are located within tectonic rifts or grabens (or, in the case of Chukhung crater, on plains between major branches of a large rift system; see Figs. 12A and 13A). These settings have been used to invoke locally elevated geothermal flux as a possible explanation for esker-forming meltwater production, perhaps supplemented by strain heating enhanced by the convergence of ice flow in steep-sided topographic settings (Butcher et al., 2017, 2021; Gallagher & Balme, 2015).

While major geothermal activity waned early in Mars' geological history, the possibility of late-stage, transient, and localized geothermal-flux anomalies cannot be ruled out, based on current data. Indeed, evidence for localized, wet-based conditions associated with subglacial volcanism has been suggested in association with the large Middle-to-Late-Amazonian-aged moraine deposits on the flank of the equatorial Arsia Mons volcano (Scanlon et al., 2015). However, while localized, late-stage, endogenic heating could explain a handful of isolated eskers, and may be required for their formation, evidence for more regional, warm-based glaciation during the Amazonian could be difficult to explain with geothermal heating alone. Evidence for regional warm-based glaciation in Phlegra Montes (Gallagher et al., 2021), and polythermal glacial erosion of the walls of numerous, spatially disparate, impact craters (Conway et al., 2018; Hubbard et al., 2011) suggested that more widespread heat sources, such as climate changes sufficient to raise the temperature of glacial ice masses, might have been required.

Constraints on past thermal conditions of Mars' *VFFs* can come from the landscape record, but large areas of the mid-latitudes have yet to be evaluated for signatures of past warm-based glaciation. In addition, the extent to which the ice melting point may have been depressed by salts is unknown (e.g., Conway et al., 2018), as is whether climate changes and endogenic heat sources operated (together or separately) to drive warm-based/polythermal glaciation on Mars. Complementary approaches to geomorphology, such as modeling and mineralogical observations, are currently underutilized. Improving our understanding of the spatial distribution of past, warm-based glaciation in Mars' mid-latitudes could provide important insights into the nature and magnitude of environmental change(s), as well as endogenic heat sources, on Amazonian Mars.

3 Constraints on the timing of Martian glaciations

Glacial deposits, remnants, and landscapes are dated by crater size-frequency distributions (*CSFDs*), relative stratigraphy, and/or cross-cutting relationships. Age estimates inform how far back in time we can directly constrain past glaciations on Mars, which is critical because this evidence is used to reconstruct past climates, as well as orbital history.

Few craters remain on the surface of the *NPLD*, and therefore the uppermost surface of the *NPLD* has a very young exposure age (or resurfacing age, the theoretical time since a surface was last free of craters) that could be less than tens of thousands of years (Banks et al., 2010). In contrast, *CSFD* dating suggests the *SPLD*'s surface is 10–100 Ma (Herkenhoff & Plaut, 2000; Koutnik et al., 2002; Plaut et al., 1988), or perhaps older depending on crater removal by viscous relaxation (Pathare et al., 2005). The bulk accumulation of the ~2-km thick *NPLD* is hypothesized to have taken place ~4–5 Ma (Montmessin, 2006; Levrard et al., 2007; Hvidberg et al., 2012) based on modeled variations in orbital parameters over the past 10 Myr (Laskar et al., 2004). However, the actual age of the *NPLD* is poorly constrained, and model ages appear to disagree with geologic data that instead indicate the *NPLD* may contain layers deposited during the past 1 Gyr (Tanaka et al., 2005, 2008; Tanaka & Fortezzo, 2012). Rodriguez et al. (2021) emphasized this age uncertainty, where the uppermost few hundred meters of the *NPLD* have been proposed to form in the past ~0.37 Ma (Smith et al., 2016) or in the past ~3.6 ± 2.5 Ma (Tanaka et al., 2005).

Mid-latitude *VFFs* can also be dated using *CSFDs*, which provide a minimum stagnation date. However, constructing an absolute chronology is difficult due to significant uncertainties. Care must be exercised when interpreting ages based on small-crater/area chronometry and the subject of small craters, in particular, is controversial (e.g., Hartmann, 2005; Hartmann & Daubar, 2017; Warner et al., 2015; Williams et al., 2018). In addition, there is a general lack of flow-deformed craters, as craters may become buried as ice accumulates, or ice flow may remove craters. Ice strength also influences the size-frequency distribution and long-term preservation of impact craters (e.g., Landis et al., 2016; Senft & Stewart, 2008), an effect that is poorly constrained on *VFFs* (Berman et al., 2015; Hepburn et al., 2020b).

CSFD dating of Mars' mid-latitude glacial remnants extends from ~1 Ma back to ~1 Ga. This includes studies that span approximately 1 Ma to ≤100 Ma (e.g., Brough et al., 2016; Gallagher & Balme, 2015; Head et al., 2003; Hepburn et al., 2020b; Hubbard et al., 2014; Morgan et al., 2009; Sinha & Murty, 2015; Souness & Hubbard, 2013), from approximately ≥100 Ma to ≤1 Ga (e.g., Baker et al., 2010; Butcher et al., 2017, 2021; Gallagher & Balme, 2015; Hepburn et al., 2020b; Morgan et al., 2009; Sinha & Murty, 2015; Woodley et al., 2022) and, perhaps, even older at >1 Ga (Levrard et al., 2004; Hepburn et al., 2020a).

In comparison, crater-based dating of possible *periglacial* landscapes around the Mars dichotomy or at the mid-to-high northern latitudes exhibit relatively young ages: <0.1 Ma (Mustard et al., 2001; also, Milliken et al., 2003); ~0.1–1 Ma (Levy et al., 2009; Mangold, 2005); approximately ≤3.0 Ma (Kostama et al., 2006); ~0.4–2.1 Ma (Head et al., 2003); 0.5–2 Ma (Gallagher et al., 2011); and no older than 2–8 Ma (Balme & Gallagher, 2009; Balme et al., 2009). However, Soare et al. (2020, 2021b) identified a notable exception: polygonized and thermokarst-like terrain at the mid-latitudes of Utopia Planitia and north of the Moreux impact crater with an age estimated to be >100 Ma. Landform evidence and landscape features in Protonilus Mensae (near the dichotomy boundary) indicate cycling between periods of periglacial and glacial activity over at least the past 1 billion years (Soare et al., 2022).

4 Conclusions and outlook

Observations of glacial deposits and glacial remnants on Mars provide direct constraints on polar and non-polar deposition and erosion of ice during the last ~1 billion years of the Amazonian, and glacial landscapes provide indirect constraints on past glacial conditions. Integrating high-resolution imagery and topography with ice-penetrating radar has transformed our understanding of some of the fundamental aspects of the distribution and structure of surface and subsurface water ice and CO_2 ice on Mars. Recent observations have advanced our understanding and inquiry around three key questions discussed here that involve polar glacial deposits, non-polar glacial remnants, and glacial landscapes. Related to these questions, some of the key targets and open questions for ongoing and future research, in order to understand ice evolution and climate history on Mars, include:

- Variations in Mars' orbit and spin axis provide a framework for assessing climate history on Mars; but, we need to improve our understanding of the atmospheric processes that drive the deposition and erosion of ice and dust under the present climate and especially during different climate states. Questions also remain about whether Mars' climate experienced stochastic variability, including the extent to which climate changes resulted from internal forcing as opposed to astronomical forcing.
- The nature of the thinnest *PLD* layers, as well as fine layers that mantle regions of the mid-to-high latitudes, in relation to global climate, local climate, and planetary sources and sinks of ice and dust is largely unknown. An improved understanding of past and present fluxes of water ice, CO_2 ice, and dust that form and modify layers is a recognized priority.
- Adding to the debate about liquid water at the base of the *SPLD*, Arnold et al. (2022) demonstrated that local *SPLD* surface topography could be consistent with subglacial water in a way similar to the impact of subglacial lakes on an ice flow and surface topography on Earth, and that ice-surface topography may offer supplemental information about subglacial conditions. With respect to advances in radar signal interpretation, analyses of multiple sets of radar data at different frequencies, together with an improved understanding of chemical, geological, and thermal processes, are needed in order to better constrain *SPLD* subglacial conditions (Schroeder & Steinbrügge, 2021).
- Determining the degree to which buried CO_2 deposits exchange with the atmosphere has important implications for the evolution of atmospheric pressure and the state of the Late Amazonian Epoch climate, as well as the potential for liquid water stability over time.
- Related to CO_2 ice in the modern climate, extended surveys and further experiments collecting empirical data on the solid-state greenhouse hypothesis may elucidate whether araneiforms are representative of seasonal dynamics during past climates.
- Age uncertainties of *PLDs* and *VFFs* make it difficult to reconstruct a robust glacial chronology from dating alone, including the identification of multiple glacial-interglacial cycles. It is not clear if the ages for *VFFs* imply recent glaciation in the mid-latitudes, enhanced resurfacing upon *VFF* surfaces, latitude-dependent mantle deposition, or some combination thereof.
- The mass-balance relationship of Martian ice masses is a fundamental unknown that drives ice flow and their volumetric response to climate forcing. However, it is not known for Mars if mass accumulates at high elevations and ablates at low elevations as occurs on the Earth. Present-day Mars has a significantly lower atmospheric pressure compared to the Earth, but how this changes back in time is not well constrained. Empirical evidence is needed to constrain the nature of the spatial distribution of mass gain and loss across Mars' *VFFs*. In addition, the next level of analysis for *VFFs* includes further development and application of glaciological models to investigate how *VFFs* accumulate and ablate, and how debris gets entrained.
- While *VFFs* have been mapped over an extensive portion of the mid-latitudes, limited depositional and erosional landforms consistent with more extensive ice-sheet scale glaciation have been identified to date. Whether this means that large-scale advances were rare and/or large-scale ice

masses were cold based and thus left minimal landscape signature is not well constrained. If ice-sheet scale glaciation occurred, as supported by climate models, the role of expanded ice cover in relation to planetary geochemical processes and cycles is not well constrained.

- A regionally extensive assemblage of glacial erosional landforms at Phlegra Montes means that the influence of regional wet-based glaciation in this landscape, which currently hosts cold-based *VFFs*, is a distinct possibility. In general, mid-latitude *VFFs* are glacial remnants that sequester ice in the modern climate, but *VFFs* were larger in the past and climate conditions may have been very different (from modern Mars) during and following their emplacement. It is thought that geothermal heating could have driven basal melting of *VFFs* to form in Phlegra Montes and in Tempe Terra, but the potential for climate change as an additional or alternative heat source is poorly constrained.
- It is not confirmed, which terrestrial glacial landsystem analogs may have been active on Mars in the past; but, there is potential to interpret Mars' landform assemblages more holistically using a landsystem approach to better understand the style of past glaciations, the subglacial thermal environment, and how deglaciated landscapes preserve information relating to past glacial processes.

We have focused primarily on how observations of glacial deposits, as well as glacial-geomorphological remnants and landscapes, may be used to directly constrain ice and climate histories. While models largely agree with latitudinal trends in the distribution of these landforms, model-based evaluations of past climate are primarily theoretical (see, e.g., Forget et al., 2017). New observations are critical (see, e.g., Garvin et al., 2024, pp. 193–220). Future missions will target surface and subsurface glacial deposits for their scientific value and as potential water resources for future human missions to Mars (e.g., Starr & Muscatello, 2020). These efforts will most likely focus on the mid-latitudes, which is less challenging for landing than at the poles, and where near-surface ice may be accessible (e.g., Morgan et al., 2021). Separate from the mid-latitude targets, the *PLDs* are a distinct stratigraphic archive and target for future investigations (e.g., MEPAG ICE-SAG, 2019; Smith et al., 2020, 2021; Thomas et al., 2021). Outlier polar-ice deposits (e.g., Conway et al., 2012; Sori et al., 2022) exist at lower circumpolar latitudes and may also retain stratigraphic archives. Regarding new measurements, time-resolved global atmospheric conditions and dynamics (including winds, clouds, and dust content) are priorities (e.g., MEPAG, 2020; MEPAG ICE-SAG, 2019), and could provide necessary constraints for process-based models and global circulation models. Sample-return missions could be transformative to our understanding, and, in particular, recovering an ice sample would enable entirely new branches of Martian science (e.g., Mars Ice Core Working Group, 2021).

Studies of glacial ice on Mars benefit from and contribute to comparative planetology. Our understanding of glacial ice on Earth distinctly informs the interpretation of glacial landforms and past glacial processes on Mars, and studies of glacial ice on Mars contribute to solar-system-wide investigations of surface processes involving water ice and/or carbon dioxide ice. Bridging interdisciplinary expertise related to ice evolution and climate can significantly advance our understanding of icy planetary bodies.

Acknowledgments

A portion of this chapter was written at the Jet Propulsion Laboratory, California Institute of Technology, under a contract with the National Aeronautics and Space Administration. M.R.K. acknowledges research support from NASA Solar System Workings. F.E.G.B. is funded by a Leverhulme Trust Early Career Fellowship. We thank Gareth Morgan for thoughtful comments.

References

Aharonson, O., Zuber, M. T., Smith, D. E., Neumann, G. A., Feldman, W. C., & Prettyman, T. H. (2004). Depth, distribution, and density of CO2 deposition on Mars. *Journal of Geophysical Research, Planets, 109*(E5). https://doi.org/10.1029/2003JE002223.

Andrews, J. T. (2000). Dating glacial events and correlation to global climate change. In *Quaternary geochronology* (pp. 447–455). American Geophysical Union (AGU). https://doi.org/10.1029/RF004p0447.

Arfstrom, J., & Hartmann, W. K. (2005). Martian flow features, moraine-like ridges, and gullies: Terrestrial analogs and interrelationships. *Icarus, 174*(2), 321–335. https://doi.org/10.1016/j.icarus.2004.05.026.

Arnold, N. S., Butcher, F. E. G., Conway, S. J., Gallagher, C., & Balme, M. R. (2022). Surface topographic impact of subglacial water beneath the south polar ice cap of Mars. *Nature Astronomy, 1–7*. https://doi.org/10.1038/s41550-022-01782-0.

Baker, D. M. H., & Carter, L. M. (2019). Probing supraglacial debris on Mars 1: Sources, thickness, and stratigraphy. *Icarus, 319*, 745–769. https://doi.org/10.1016/j.icarus.2018.09.001.

Baker, D. M. H., & Head, J. W. (2015). Extensive Middle Amazonian mantling of debris aprons and plains in Deuteronilus Mensae, Mars: Implications for the record of mid-latitude glaciation. *Icarus, 260*, 269–288. https://doi.org/10.1016/j.icarus.2015.06.036.

Baker, D. M. H., Head, J. W., & Marchant, D. R. (2010). Flow patterns of lobate debris aprons and lineated valley fill north of Ismeniae Fossae, Mars: Evidence for extensive mid-latitude glaciation in the Late Amazonian. *Icarus, 207*(1), 186–209. https://doi.org/10.1016/j.icarus.2009.11.017.

Balme, M. R., & Gallagher, C. (2009). An equatorial periglacial landscape on Mars. *Earth and Planetary Science Letters, 285*(1–2), 1–15.

Balme, M. R., Gallagher, C. J., Page, D. P., Murray, J. B., & Muller, J.-P. (2009). Sorted stone circles in Elysium Planitia, Mars: Implications for recent Martian climate. *Icarus, 200*(1), 30–38.

Banks, M. E., Byrne, S., Galla, K., McEwen, A. S., Bray, V. J., Dundas, C. M., … Murray, B. C. (2010). Crater population and resurfacing of the Martian north polar layered deposits. *Journal of Geophysical Research, Planets, 115*(E8). https://doi.org/10.1029/2009JE003523.

Becerra, P., Byrne, S., Sori, M. M., Sutton, S., & Herkenhoff, K. E. (2016). Stratigraphy of the north polar layered deposits of Mars from high-resolution topography. *Journal of Geophysical Research, Planets, 121*(8), 1445–1471. https://doi.org/10.1002/2015JE004992.

Becerra, P., Smith, I. B., Hibbard, S., Andres, C., Bapst, J., Bramson, A. M., … Yoldi, Z. (2021). Past, present, and future of Mars polar science: Outcomes and outlook from the 7th international conference on Mars polar science and exploration. *The Planetary Science Journal, 2*(5), 209. https://doi.org/10.3847/PSJ/ac19a5.

Becerra, P., Sori, M. M., & Byrne, S. (2017). Signals of astronomical climate forcing in the exposure topography of the north polar layered deposits of Mars. *Geophysical Research Letters, 44*(1), 62–70. https://doi.org/10.1002/2016GL071197.

Becerra, P., Sori, M. M., Thomas, N., Pommerol, A., Simioni, E., Sutton, S. S., … Cremonese, G. (2019). Timescales of the climate record in the south polar ice cap of Mars. *Geophysical Research Letters, 46*(13), 7268–7277. https://doi.org/10.1029/2019GL083588.

Benn, D., & Evans, D. J. A. (2010). *Glaciers and Glaciation* (2nd ed.). London: Routledge. https://doi.org/10.4324/9780203785010.

Berman, D. C., Crown, D. A., & Joseph, E. C. S. (2015). Formation and mantling ages of lobate debris aprons on Mars: Insights from categorized crater counts. *Planetary and Space Science, 111*, 83–99. https://doi.org/10.1016/j.pss.2015.03.013.

Bickerdike, H. L., Cofaigh, C.Ó., Evans, D. J. A., & Stokes, C. R. (2018). Glacial landsystems, retreat dynamics and controls on Loch Lomond stadial (younger dryas) glaciation in Britain. *Boreas, 47*(1), 202–224. https://doi.org/10.1111/bor.12259.

Bierson, C. J., Phillips, R. J., Smith, I. B., Wood, S. E., Putzig, N. E., Nunes, D., & Byrne, S. (2016). Stratigraphy and evolution of the buried CO2 deposit in the Martian south polar cap. *Geophysical Research Letters, 43*(9), 4172–4179. https://doi.org/10.1002/2016GL068457.

Bierson, C. J., Tulaczyk, S., Courville, S. W., & Putzig, N. E. (2021). Strong MARSIS radar reflections from the base of Martian south polar cap may be due to conductive ice or minerals. *Geophysical Research Letters, 48*(13). https://doi.org/10.1029/2021GL093880. e2021GL093880.

Bourke, M. C., & Cranford, A. (2011). *Seasonal formation of furrows on polar dunes.* Fifth Mars polar science conference.

Bramson, A. M., Andres, C., Bapst, J., Becerra, P., Courville, S. W., Dundas, C., ... Wooster, P. (2020). Mid-Latitude ice on Mars: A science target for planetary climate histories and an exploration target for in situ resources. White Paper #115 Submitted to the Planetary Science and Astrobiology Decadal Survey 2023–2032. *Bulletin of the AAS, 53*(4). https://doi.org/10.3847/25c2cfeb.cc90422d.

Bramson, A. M., Byrne, S., Bapst, J., Smith, I. B., & McClintock, T. (2019). A migration model for the polar spiral troughs of Mars. *Journal of Geophysical Research, Planets, 124*(4), 1020–1043. https://doi.org/10.1029/2018JE005806.

Bramson, A. M., Byrne, S., Putzig, N. E., Sutton, S., Plaut, J. J., Brothers, T. C., & Holt, J. W. (2015). Widespread excess ice in Arcadia Planitia, Mars. *Geophysical Research Letters, 42*(16), 6566–6574. https://doi.org/10.1002/2015GL064844.

Brooker, L. (2019). *The history of water at Lyot Crater, Mars: Possible surface manifestations of ancient groundwater and/or recent climate change.* PhD thesis The Open University.

Brothers, T. C., Holt, J. W., & Spiga, A. (2015). Planum Boreum basal unit topography, Mars: Irregularities and insights from SHARD. *Journal of Geophysical Research: Planets, 120*(7), 1357–1375.

Brough, S., Hubbard, B., & Hubbard, A. (2016). Former extent of glacier-like forms on Mars. *Icarus, 274,* 37–49. https://doi.org/10.1016/j.icarus.2016.03.006.

Brough, S., Hubbard, B., & Hubbard, A. (2019). Area and volume of mid-latitude glacier-like forms on Mars. *Earth and Planetary Science Letters, 507,* 10–20. https://doi.org/10.1016/j.epsl.2018.11.031.

Buhler, P. B., Ingersoll, A. P., Piqueux, S., Ehlmann, B. L., & Hayne, P. O. (2020). Coevolution of Mars's atmosphere and massive south polar CO_2 ice deposit. *Nature Astronomy, 4*(4), 364–371. https://doi.org/10.1038/s41550-019-0976-8.

Butcher, F. E. G. (2019). *Wet-based glaciation on Mars.* The Open University.

Butcher, F. E. G. (2022). Water ice at mid-latitudes on Mars. *Planetary Science.* https://doi.org/10.1093/acrefore/9780190647926.013.239.

Butcher, F. E. G., Balme, M. R., Conway, S. J., Gallagher, C., Arnold, N. S., Storrar, R. D., ... Hagermann, A. (2020). Morphometry of a glacier-linked esker in NW Tempe Terra, Mars, and implications for sediment-discharge dynamics of subglacial drainage. *Earth and Planetary Science Letters, 542,* 116325. https://doi.org/10.1016/j.epsl.2020.116325.

Butcher, F. E. G., Balme, M. R., Conway, S. J., Gallagher, C., Arnold, N. S., Storrar, R. D., ... Davis, J. M. (2021). Sinuous ridges in Chukhung crater, Tempe Terra, Mars: Implications for fluvial, glacial, and glaciofluvial activity. *Icarus, 357,* 114131. https://doi.org/10.1016/j.icarus.2020.114131.

Butcher, F. E. G., Balme, M. R., Gallagher, C., Arnold, N. S., Conway, S. J., Hagermann, A., & Lewis, S. R. (2017). Recent basal melting of a mid-latitude glacier on Mars. *Journal of Geophysical Research, Planets, 122*(12), 2445–2468. https://doi.org/10.1002/2017JE005434.

Byrne, S. (2009). The polar deposits of Mars. *Annual Review of Earth and Planetary Sciences, 37,* 535–560.

Byrne, S., & Murray, B. C. (2002). North polar stratigraphy and the paleo-erg of Mars. *Journal of Geophysical Research: Planets, 107*(E6), 11-1–11-12. https://doi.org/10.1029/2001JE001615.

Conway, S. J., & Balme, M. R. (2014). Decameter thick remnant glacial ice deposits on Mars. *Geophysical Research Letters, 41*(15), 5402–5409. https://doi.org/10.1002/2014GL060314.

Conway, S. J., Butcher, F. E. G., de Haas, T., Deijns, A. A. J., Grindrod, P. M., & Davis, J. M. (2018). Glacial and gully erosion on Mars: A terrestrial perspective. *Geomorphology, 318*, 26–57. https://doi.org/10.1016/j.geomorph.2018.05.019.

Conway, S. J., Hovius, N., Barnie, T., Besserer, J., Le Mouélic, S., Orosei, R., & Read, N. A. (2012). Climate-driven deposition of water ice and the formation of mounds in craters in Mars' north polar region. *Icarus, 220*(1), 174–193. https://doi.org/10.1016/j.icarus.2012.04.021.

Courville, S. W., Perry, M. R., & Putzig, N. E. (2021). Lower bounds on the thickness and dust content of layers within the north polar layered deposits of Mars from radar forward modeling. *The Planetary Science Journal, 2*(1), 28. https://doi.org/10.3847/PSJ/abda50.

Cuffey, K. M., Conway, H., Gades, A. M., Hallet, B., Lorrain, R., Severinghaus, J. P., … White, J. W. C. (2000). Entrainment at cold glacier beds. *Geology, 28*(4), 351–354. https://doi.org/10.1130/0091-7613(2000)28<351:EACGB>2.0.CO;2.

Cuffey, K. M., & Paterson, W. S. B. (2010). *The physics of glaciers* (4th ed.). Retrieved from https://www.elsevier.com/books/the-physics-of-glaciers/cuffey/978-0-12-369461-4.

Cutts, J. A. (1973). Nature and origin of layered deposits of the Martian polar regions. *Journal of Geophysical Research (1896-1977), 78*(20), 4231–4249. https://doi.org/10.1029/JB078i020p04231.

Cutts, J. A., & Lewis, B. H. (1982). Models of climate cycles recorded in Martian polar layered deposits. *Icarus, 50*(2), 216–244. https://doi.org/10.1016/0019-1035(82)90124-5.

De Blasio, F. V. (2014). Fan-shaped deposit (Tharsis, Mars). In *Encyclopedia of planetary landforms*. New York, NY: Springer.

Dickson, J. L., Ehlmann, B. L., Kerber, L. H., & Fassett, C. I. (2023). Release of the global CTX mosaic of Mars: An experiment in information-preserving image data processing. In *Proceedings of the 54th Lunar and Planetary Science Conference #2353*.

Dickson, J. L., Head, J. W., & Fassett, C. I. (2012). Patterns of accumulation and flow of ice in the mid-latitudes of Mars during the Amazonian. *Icarus, 219*(2), 723–732. https://doi.org/10.1016/j.icarus.2012.03.010.

Dickson, J. L., Head, J. W., & Marchant, D. R. (2008). Late Amazonian glaciation at the dichotomy boundary on Mars: Evidence for glacial thickness maxima and multiple glacial phases. *Geology, 36*(5), 411–414. https://doi.org/10.1130/G24382A.1.

Dickson, J. L., Head, J. W., & Marchant, D. R. (2010). Kilometer-thick ice accumulation and glaciation in the northern mid-latitudes of Mars: Evidence for crater-filling events in the Late Amazonian at the Phlegra Montes. *Earth and Planetary Science Letters, 294*(3), 332–342. https://doi.org/10.1016/j.epsl.2009.08.031.

Diniega, S., Bramson, A. M., Buratti, B., Buhler, P., Burr, D. M., Chojnacki, M., … Widmer, J. M. (2021). Modern Mars' geomorphological activity, driven by wind, frost, and gravity. *Geomorphology, 380*, 107627. https://doi.org/10.1016/j.geomorph.2021.107627.

Diniega, S., Hansen, C. J., McElwaine, J. N., Hugenholtz, C. H., Dundas, C. M., McEwen, A. S., & Bourke, M. C. (2013). A new dry hypothesis for the formation of Martian linear gullies. *Icarus, 225*(1), 526–537. https://doi.org/10.1016/j.icarus.2013.04.006.

Diniega, S., & Smith, I. B. (2020). High-priority science questions identified at the Mars workshop on Amazonian and present-day climate. *Planetary and Space Science, 182*, 104813. https://doi.org/10.1016/j.pss.2019.104813.

Dundas, C. M., Becerra, P., Byrne, S., Chojnacki, M., Daubar, I. J., Diniega, S., … Valantinas, A. (2021a). Active Mars: A dynamic world. *Journal of Geophysical Research, Planets, 126*(8). https://doi.org/10.1029/2021JE006876. e2021JE006876.

Dundas, C. M., Bramson, A. M., Ojha, L., Wray, J. J., Mellon, M. T., Byrne, S., … Holt, J. W. (2018). Exposed subsurface ice sheets in the Martian mid-latitudes. *Science, 359*(6372), 199–201. https://doi.org/10.1126/science.aao1619.

Dundas, C. M., Byrne, S., McEwen, A. S., Mellon, M. T., Kennedy, M. R., Daubar, I. J., & Saper, L. (2014). HiRISE observations of new impact craters exposing Martian ground ice. *Journal of Geophysical Research, Planets, 119*(1), 109–127. https://doi.org/10.1002/2013JE004482.

Dundas, C. M., Diniega, S., Hansen, C. J., Byrne, S., & McEwen, A. S. (2012). Seasonal activity and morphological changes in Martian gullies. *Icarus*, *220*(1), 124–143. https://doi.org/10.1016/j.icarus.2012.04.005.

Dundas, C. M., Mellon, M. T., Conway, S. J., Daubar, I. J., Williams, K. E., Ojha, L., … Pathare, A. V. (2021b). Widespread exposures of extensive clean shallow ice in the midlatitudes of Mars. *Journal of Geophysical Research, Planets*, *126*(3). https://doi.org/10.1029/2020JE006617. e2020JE006617.

Emmett, J. A., Murphy, J. R., & Kahre, M. A. (2020). Obliquity dependence of the formation of the Martian polar layered deposits. *Planetary and Space Science*, *193*, 105047. https://doi.org/10.1016/j.pss.2020.105047.

Evans, D. (2003). Introduction to glacial landsystems. In *Glacial landsystems* (1st ed., pp. 1–11). London: Arnold.

Eyles, N. (1983). The glaciated valley landsystem. In *Glacial geology* (pp. 91–110). Oxford: Pergamon.

Fassett, C. I., Dickson, J. L., Head, J. W., Levy, J. S., & Marchant, D. R. (2010). Supraglacial and proglacial valleys on Amazonian Mars. *Icarus*, *208*(1), 86–100. https://doi.org/10.1016/j.icarus.2010.02.021.

Fastook, J. L., Head, J. W., & Marchant, D. R. (2014). Formation of lobate debris aprons on Mars: Assessment of regional ice sheet collapse and debris-cover armoring. *Icarus*, *228*, 54–63. https://doi.org/10.1016/j.icarus.2013.09.025.

Fastook, J. L., Head, J. W., Marchant, D. R., & Forget, F. (2008). Tropical mountain glaciers on Mars: Altitude-dependence of ice accumulation, accumulation conditions, formation times, glacier dynamics, and implications for planetary spin-axis/orbital history. *Icarus*, *198*(2), 305–317. https://doi.org/10.1016/j.icarus.2008.08.008.

Feldman, W. C., Boynton, W. V., Tokar, R. L., Prettyman, T. H., Gasnault, O., Squyres, S. W., … Reedy, R. C. (2002). Global distribution of neutrons from Mars: Results from Mars Odyssey. *Science*, *297*(5578), 75–78. https://doi.org/10.1126/science.1073541.

Feldman, W. C., Prettyman, T. H., Maurice, S., Plaut, J. J., Bish, D. L., Vaniman, D. T., … Tokar, R. L. (2004). Global distribution of near-surface hydrogen on Mars. *Journal of Geophysical Research, Planets*, *109*(E9). https://doi.org/10.1029/2003JE002160.

Fergason, R. L., Hare, T. M., & Laura, J. (2018). HRSC and MOLA blended digital elevation model at 200m v2. Astrogeology PDS Annex, U.S. Geological Survey. http://bit.ly/HRSC_MOLA_Blend_v0.

Fishbaugh, K. E., Byrne, S., Herkenhoff, K. E., Kirk, R. L., Fortezzo, C., Russell, P. S., & McEwen, A. (2010a). Evaluating the meaning of "layer" in the Martian north polar layered deposits and the impact on the climate connection. *Icarus*, *205*(1), 269–282. https://doi.org/10.1016/j.icarus.2009.04.011.

Fishbaugh, K. E., & Hvidberg, C. S. (2006). Martian north polar layered deposits stratigraphy: Implications for accumulation rates and flow. *Journal of Geophysical Research, Planets*, *111*(E6). https://doi.org/10.1029/2005JE002571.

Fishbaugh, K. E., Hvidberg, C. S., Byrne, S., Russell, P. S., Herkenhoff, K. E., Winstrup, M., & Kirk, R. (2010b). First high-resolution stratigraphic column of the Martian north polar layered deposits. *Geophysical Research Letters*, *37*(7). https://doi.org/10.1029/2009GL041642.

Fitzsimons, S., & Howarth, J. (2020). Development of push moraines in deeply frozen sediment adjacent to a cold-based glacier in the McMurdo Dry Valleys, Antarctica. *Earth Surface Processes and Landforms*, *45*(3), 622–637. https://doi.org/10.1002/esp.4759.

Forget, F., Byrne, S., Head, J. W., Mischna, M. A., Schörghofer, N., Forget, F., Smith, M. D., & Clancy, R. T. (2017). Recent climate variations. In R. W. Zurek, & R. M. Haberle (Eds.), *The atmosphere and climate of Mars* (pp. 497–525). Cambridge: Cambridge University Press. https://doi.org/10.1017/9781139060172.016.

Forget, F., Haberle, R. M., Montmessin, F., Levrard, B., & Head, J. W. (2006). Formation of glaciers on Mars by atmospheric precipitation at high obliquity. *Science*, *311*(5759), 368–371. https://doi.org/10.1126/science.1120335.

Forget, F., Hansen, G. B., & Pollack, J. B. (1995). Low brightness temperatures of Martian polar caps: CO2 clouds or low surface emissivity? *Journal of Geophysical Research, Planets*, *100*(E10), 21219–21234. https://doi.org/10.1029/95JE02378.

Gallagher, C. (2024). *Glaciation and glacigenic geomorphology on Earth in the Quaternary Period*. In R. J. Soare, J.-P. Williams, C. Ahrens, F. E. G. Butcher, & M. R. El-Maarry (Eds.), *Ices in the solar system, a volatile-driven journey from the inner solar system to its far reaches*. Elsevier Books.

Gallagher, C., & Balme, M. (2015). Eskers in a complete, wet-based glacial system in the Phlegra Montes region, Mars. *Earth and Planetary Science Letters*, *431*, 96–109. https://doi.org/10.1016/j.epsl.2015.09.023.

Gallagher, C., Balme, M. R., Conway, S. J., & Grindrod, P. M. (2011). Sorted clastic stripes, lobes and associated gullies in high-latitude craters on Mars: Landforms indicative of very recent, polycyclic ground-ice thaw and liquid flows. *Icarus*, *211*(1), 458–471.

Gallagher, C., Butcher, F. E. G., Balme, M., Smith, I., & Arnold, N. (2021). Landforms indicative of regional warm based glaciation, Phlegra Montes, Mars. *Icarus*, *355*, 114173. https://doi.org/10.1016/j.icarus.2020.114173.

Garvin, J. B., Soare, R. J., Hepburn, A. J., Koutnik, M., & Godin, E. (2024). *Ice Exploration on Mars: Whereto and when?* In R. J. Soare, J.-P. Williams, C. Ahrens, F. E. G. Butcher, & M. R. El-Maarry (Eds.), *Ices in the solar system, a volatile-driven journey from the inner solar system to its far reaches*. Elsevier Books.

Grau Galofre, A., Lasue, J., & Scanlon, K. (2024). *Ice on Noachian and Hesperian Mars: Atmospheric, surface, and subsurface processes*. In R. J. Soare, J.-P. Williams, C. Ahrens, F. E. G. Butcher, & M. R. El-Maarry (Eds.), *Ices in the solar system, a volatile-driven journey from the inner solar system to its far reaches*. Elsevier Books.

Grau Galofre, A., Whipple, K. X., Christensen, P. R., & Conway, S. J. (2022). Valley networks and the record of glaciation on ancient Mars. *Geophysical Research Letters*, *49*(14). https://doi.org/10.1029/2022GL097974. e2022GL097974.

Greve, R., Grieger, B., & Stenzel, O. J. (2010). MAIC-2, a latitudinal model for the Martian surface temperature, atmospheric water transport and surface glaciation. *Planetary and Space Science*, *58*(6), 931–940. https://doi.org/10.1016/j.pss.2010.03.002.

Grima, C., Kofman, W., Mouginot, J., Phillips, R. J., Herique, A., Biccari, D., … Cutigni, M. (2009). North polar deposits of Mars: Extreme purity of the water ice. *Geophysical Research Letters*, *36*, L03203. https://doi.org/10.1029/2008GL036326.

Guidat, T., Pochat, S., Bourgeois, O., & Souček, O. (2015). Landform assemblage in Isidis Planitia, Mars: Evidence for a 3 Ga old polythermal ice sheet. *Earth and Planetary Science Letters*, *411*, 253–267. https://doi.org/10.1016/j.epsl.2014.12.002.

Hansen, C. J., Thomas, N., Portyankina, G., McEwen, A., Becker, T., Byrne, S., … Mellon, M. (2010). HiRISE observations of gas sublimation-driven activity in Mars' southern polar regions: I. Erosion of the surface. *Icarus*, *205*(1), 283–295. https://doi.org/10.1016/j.icarus.2009.07.021.

Hartmann, W. K. (2005). Martian cratering 8: Isochron refinement and the chronology of Mars. *Icarus*, *174*(2), 294–320. https://doi.org/10.1016/j.icarus.2004.11.023.

Hartmann, W. K., Ansan, V., Berman, D. C., Mangold, N., & Forget, F. (2014). Comprehensive analysis of glaciated Martian crater Greg. *Icarus*, *228*, 96–120. https://doi.org/10.1016/j.icarus.2013.09.016.

Hartmann, W. K., & Daubar, I. J. (2017). Martian cratering 11. Utilizing decameter scale crater populations to study Martian history. *Meteoritics & Planetary Science*, *52*(3), 493–510. https://doi.org/10.1111/maps.12807.

Hauber, E., van Gasselt, S., Chapman, M. G., & Neukum, G. (2008). Geomorphic evidence for former lobate debris aprons at low latitudes on Mars: Indicators of the Martian paleoclimate. *Journal of Geophysical Research, Planets*, *113*(E2). https://doi.org/10.1029/2007JE002897.

Head, J. W., & Marchant, D. R. (2003). Cold-based mountain glaciers on Mars: Western Arsia Mons. *Geology*, *31*(7), 641–644. https://doi.org/10.1130/0091-7613(2003)031<0641:CMGOMW>2.0.CO;2.

Head, J. W., Marchant, D. R., Dickson, J. L., Kress, A. M., & Baker, D. M. (2010). Northern mid-latitude glaciation in the late Amazonian period of Mars: Criteria for the recognition of debris-covered glacier and valley glacier landsystem deposits. *Earth and Planetary Science Letters*, *294*(3), 306–320. https://doi.org/10.1016/j.epsl.2009.06.041.

Head, J. W., Mustard, J. F., Kreslavsky, M. A., Milliken, R. E., & Marchant, D. R. (2003). Recent ice ages on Mars. *Nature*, *426*(6968), 797–802. https://doi.org/10.1038/nature02114.

Head, J. W., Nahm, A. L., Marchant, D. R., & Neukum, G. (2006). Modification of the dichotomy boundary on Mars by Amazonian mid-latitude regional glaciation. *Geophysical Research Letters*, *33*(8). https://doi.org/10.1029/2005GL024360.

Head, J. W., Neukum, G., Jaumann, R., Hiesinger, H., Hauber, E., Carr, M., … van Gasselt, S. (2005). Tropical to mid-latitude snow and ice accumulation, flow and glaciation on Mars. *Nature, 434*(7031), 346–351. https://doi.org/10.1038/nature03359.

Hecht, M. H. (2002). Metastability of liquid water on Mars. *Icarus, 156*(2), 373–386. https://doi.org/10.1006/icar.2001.6794.

Hecht, M. H., Kounaves, S. P., Quinn, R. C., West, S. J., Young, S. M. M., Ming, D. W., … Smith, P. H. (2009). Detection of perchlorate and the soluble chemistry of Martian soil at the phoenix lander site. *Science, 325*(5936), 64–67. https://doi.org/10.1126/science.1172466.

Helbert, J., Hauber, E., & Reiss, D. (2015). Water on the terrestrial planets. In *Treatise on geophysics* (pp. 367–409). https://doi.org/10.1016/B978-0-444-53802-4.00174-3.

Hepburn, A. J., Ng, F. S. L., Holt, T. O., & Hubbard, B. (2020a). Late Amazonian ice survival in Kasei Valles, Mars. *Journal of Geophysical Research, Planets, 125*(11). https://doi.org/10.1029/2020JE006531. e2020JE006531.

Hepburn, A. J., Ng, F. S. L., Livingstone, S. J., Holt, T. O., & Hubbard, B. (2020b). Polyphase mid-latitude glaciation on Mars: Chronology of the formation of superposed glacier-like forms from crater-count dating. *Journal of Geophysical Research, Planets, 125*(2). https://doi.org/10.1029/2019JE006102. e2019JE006102.

Herkenhoff, K. E., & Plaut, J. J. (2000). Surface ages and resurfacing rates of the polar layered deposits on Mars. *Icarus, 144*(2), 243–253. https://doi.org/10.1006/icar.1999.6287.

Herschel, W. (1784). XIX. On the remarkable appearances at the polar regions of the planet Mars, and its spheroidical figure; with a few hints relating to its real diameter and atmosphere. *Philosophical Transactions of the Royal Society of London, 74*, 233–273. https://doi.org/10.1098/rstl.1784.0020.

Hobley, D. E. J., Howard, A. D., & Moore, J. M. (2014). Fresh shallow valleys in the Martian midlatitudes as features formed by meltwater flow beneath ice. *Journal of Geophysical Research, Planets, 119*(1), 128–153. https://doi.org/10.1002/2013JE004396.

Holo, S. J., Kite, E. S., & Robbins, S. J. (2018). Mars obliquity history constrained by elliptic crater orientations. *Earth and Planetary Science Letters, 496*, 206–214. https://doi.org/10.1016/j.epsl.2018.05.046.

Holt, J. W., Fishbaugh, K. E., Byrne, S., Christian, S., Tanaka, K., Russell, P. S., … Phillips, R. J. (2010). The construction of Chasma Boreale on Mars. *Nature, 465*, 446–449.

Holt, J. W., Safaeinili, A., Plaut, J. J., Head, J. W., Phillips, R. J., Seu, R., … Gim, Y. (2008). Radar sounding evidence for buried glaciers in the southern mid-latitudes of Mars. *Science, 322*(5905), 1235–1238. https://doi.org/10.1126/science.1164246.

Howard, A. D. (1978). Origin of the stepped topography of the Martian poles. *Icarus, 34*(3), 581–599.

Howard, A. D., Cutts, J. A., & Blasius, K. R. (1982). Stratigraphic relationships within Martian polar cap deposits. *Icarus, 50*(2), 161–215. https://doi.org/10.1016/0019-1035(82)90123-3.

Hubbard, B., Milliken, R. E., Kargel, J. S., Limaye, A., & Souness, C. (2011). Geomorphological characterisation and interpretation of a mid-latitude glacier-like form: Hellas Planitia, Mars. *Icarus, 211*(1), 330–346. https://doi.org/10.1016/j.icarus.2010.10.021.

Hubbard, B., Souness, C., & Brough, S. (2014). Glacier-like forms on Mars. *The Cryosphere, 8*(6), 2047–2061. https://doi.org/10.5194/tc-8-2047-2014.

Hughes, A. L. C., Clark, C. D., & Jordan, C. J. (2014). Flow-pattern evolution of the last British ice sheet. *Quaternary Science Reviews, 89*, 148–168. https://doi.org/10.1016/j.quascirev.2014.02.002.

Hvidberg, C. S., Fishbaugh, K. E., Winstrup, M., Svensson, A., Byrne, S., & Herkenhoff, K. E. (2012). Reading the climate record of the Martian polar layered deposits. *Icarus, 221*(1), 405–419. https://doi.org/10.1016/j.icarus.2012.08.009.

Jaumann, R., Neukum, G., Behnke, T., Duxbury, T. C., Eichentopf, K., Flohrer, J., … Wählisch, M. (2007). The high-resolution stereo camera (HRSC) experiment on Mars express: Instrument aspects and experiment conduct from interplanetary cruise through the nominal mission. *Planetary and Space Science, 55*(7), 928–952. https://doi.org/10.1016/j.pss.2006.12.003.

Kadish, S. J., Head, J. W., Fastook, J. L., & Marchant, D. R. (2014). Middle to late Amazonian tropical mountain glaciers on Mars: The ages of the Tharsis Montes fan-shaped deposits. *Planetary and Space Science*, *91*, 52–59. https://doi.org/10.1016/j.pss.2013.12.005.

Kahre, M. A., Murphy, J. R., Newman, C. E., Wilson, R. J., Cantor, B. A., Lemmon, M. T., & Wolff, M. J. (2017). The Mars dust cycle. In F. Forget, M. D. Smith, R. T. Clancy, R. W. Zurek, & R. M. Haberle (Eds.), *The atmosphere and climate of Mars* (pp. 295–337). Cambridge: Cambridge University Press. https://doi.org/10.1017/9781139060172.010.

Kargel, J. S., Baker, V. R., Begét, J. E., Lockwood, J. F., Péwé, T. L., Shaw, J. S., & Strom, R. G. (1995). Evidence of ancient continental glaciation in the Martian northern plains. *Journal of Geophysical Research, Planets*, *100*(E3), 5351–5368. https://doi.org/10.1029/94JE02447.

Karlsson, N. B., Schmidt, L. S., & Hvidberg, C. S. (2015). Volume of Martian midlatitude glaciers from radar observations and ice flow modeling. *Geophysical Research Letters*, *42*(8), 2627–2633. https://doi.org/10.1002/2015GL063219.

Kaufmann, E., & Hagermann, A. (2017). Experimental investigation of insolation-driven dust ejection from Mars' CO_2 ice caps. *Icarus*, *282*, 118–126. https://doi.org/10.1016/j.icarus.2016.09.039.

Khuller, A. R., & Plaut, J. J. (2021). Characteristics of the basal interface of the Martian south polar layered deposits. *Geophysical Research Letters*, *48*(13). https://doi.org/10.1029/2021GL093631. e2021GL093631.

Kieffer, H. H., Christensen, P. R., & Titus, T. N. (2006). CO_2 jets formed by sublimation beneath translucent slab ice in Mars' seasonal south polar ice cap. *Nature*, *442*(7104), 793–796. https://doi.org/10.1038/nature04945.

Kieffer, H. H., & Titus, T. N. (2001). TES mapping of Mars' north seasonal cap. *Icarus*, *154*(1), 162–180. https://doi.org/10.1006/icar.2001.6670.

Kolb, E. J., & Tanaka, K. L. (2006). *Accumulation and erosion of south polar layered deposits in the Promethei Lingula region, Planum Australe, Mars*. MARS. https://doi.org/10.1555/mars.2006.0001.

Kostama, V.-P., Kreslavsky, M. A., & Head, J. W. (2006). Recent high-latitude icy mantle in the northern plains of Mars: Characteristics and ages of emplacement. *Geophysical Research Letters*, *33*(11). https://doi.org/10.1029/2006GL025946.

Koutnik, M., Byrne, S., & Murray, B. (2002). South polar layered deposits of Mars: The cratering record. *Journal of Geophysical Research, Planets*, *107*(E11), 10-1–10-10. https://doi.org/10.1029/2001JE001805.

Koutnik, M. R., & Pathare, A. V. (2021). Contextualizing lobate debris aprons and glacier-like forms on Mars with debris-covered glaciers on Earth. *Progress in Physical Geography: Earth and Environment*, *45*(2), 130–186. https://doi.org/10.1177/0309133320986902.

Kreslavsky, M. A., & Head, J. W. (2002). Mars: Nature and evolution of young latitude-dependent water-ice-rich mantle. *Geophysical Research Letters*, *29*(15), 14-1–14-4. https://doi.org/10.1029/2002GL015392.

Lalich, D. E., Hayes, A. G., & Poggiali, V. (2022). Explaining bright radar reflections below the south pole of Mars without liquid water. *Nature Astronomy*, 1–5. https://doi.org/10.1038/s41550-022-01775-z.

Lalich, D. E., Holt, J. W., & Smith, I. B. (2019). Radar reflectivity as a proxy for the dust content of individual layers in the Martian north polar layered deposits. *Journal of Geophysical Research, Planets*, *124*(7), 1690–1703. https://doi.org/10.1029/2018JE005787.

Landis, M. E., Byrne, S., Daubar, I. J., Herkenhoff, K. E., & Dundas, C. M. (2016). A revised surface age for the north polar layered deposits of Mars. *Geophysical Research Letters*, *43*(7), 3060–3068. https://doi.org/10.1002/2016GL068434.

Langevin, Y., Poulet, F., Bibring, J.-P., & Gondet, B. (2005). Sulfates in the north polar region of mars detected by OMEGA/Mars express. *Science*, *307*(5715), 1584–1586. https://doi.org/10.1126/science.1109091.

Laskar, J., Correia, A. C. M., Gastineau, M., Joutel, F., Levrard, B., & Robutel, P. (2004). Long term evolution and chaotic diffusion of the insolation quantities of Mars. *Icarus*, *170*(2), 343–364. https://doi.org/10.1016/j.icarus.2004.04.005.

Laskar, J., Levrard, B., & Mustard, J. F. (2002). Orbital forcing of the Martian polar layered deposits. *Nature*, *419*(6905), 375–377. https://doi.org/10.1038/nature01066.

Lasue, J., Clifford, S. M., Conway, S. J., Mangold, N., & Butcher, F. E. G. (2019). Chapter 7—The hydrology of Mars including a potential cryosphere. In J. Filiberto, & S. P. Schwenzer (Eds.), *Volatiles in the Martian crust* (pp. 185–246). Elsevier. https://doi.org/10.1016/B978-0-12-804191-8.00007-6.

Lauro, S. E., Pettinelli, E., Caprarelli, G., Guallini, L., Rossi, A. P., Mattei, E., … Orosei, R. (2021). Multiple subglacial water bodies below the south pole of Mars unveiled by new MARSIS data. *Nature Astronomy*, *5*(1), 63–70. https://doi.org/10.1038/s41550-020-1200-6.

Leighton, R. B., Horowitz, N. H., Murray, B. C., Sharp, R. P., Herriman, A. G., Young, A. T., … Leovy, C. B. (1969). Mariner 6 television pictures: First report. *Science*, *165*(3894), 685–690. https://doi.org/10.1126/science.165.3894.685.

Leighton, R. B., & Murray, B. C. (1966). Behavior of carbon dioxide and other volatiles on Mars. *Science*, *153*(3732), 136–144. https://doi.org/10.1126/science.153.3732.136.

Levrard, B., Forget, F., Montmessin, F., & Laskar, J. (2004). Recent ice-rich deposits formed at high latitudes on Mars by sublimation of unstable equatorial ice during low obliquity. *Nature*, *431*, 1072–1075.

Levrard, B., Forget, F., Montmessin, F., & Laskar, J. (2007). Recent formation and evolution of northern Martian polar layered deposits as inferred from a global climate model. *Journal of Geophysical Research, Planets*, *112*(E6). https://doi.org/10.1029/2006JE002772.

Levy, J. S., Fassett, C. I., & Head, J. W. (2016). Enhanced erosion rates on Mars during Amazonian glaciation. *Icarus*, *264*, 213–219. https://doi.org/10.1016/j.icarus.2015.09.037.

Levy, J. S., Fassett, C. I., Head, J. W., Schwartz, C., & Watters, J. L. (2014). Sequestered glacial ice contribution to the global Martian water budget: Geometric constraints on the volume of remnant, midlatitude debris-covered glaciers. *Journal of Geophysical Research, Planets*, *119*(10), 2188–2196. https://doi.org/10.1002/2014JE004685.

Levy, J. S., Fassett, C. I., Holt, J. W., Parsons, R., Cipolli, W., Goudge, T. A., … Armstrong, I. (2021). Surface boulder banding indicates Martian debris-covered glaciers formed over multiple glaciations. *Proceedings of the National Academy of Sciences*, *118*(4), e2015971118. https://doi.org/10.1073/pnas.2015971118.

Levy, J. S., Head, J. W., & Marchant, D. R. (2007). Lineated valley fill and lobate debris apron stratigraphy in Nilosyrtis Mensae, Mars: Evidence for phases of glacial modification of the dichotomy boundary. *Journal of Geophysical Research, Planets*, *112*(E8). https://doi.org/10.1029/2006JE002852.

Levy, J., Head, J., & Marchant, D. (2009). Thermal contraction crack polygons on Mars: Classification, distribution, and climate implications from HiRISE observations. *Journal of Geophysical Research, Planets*, *114*(E1). https://doi.org/10.1029/2008JE003273.

Li, H., Robinson, M. S., & Jurdy, D. M. (2005). Origin of Martian northern hemisphere mid-latitude lobate debris aprons. *Icarus*, *176*(2), 382–394. https://doi.org/10.1016/j.icarus.2005.02.011.

Limaye, A. B. S., Aharonson, O., & Perron, J. T. (2012). Detailed stratigraphy and bed thickness of the Mars north and south polar layered deposits. *Journal of Geophysical Research, Planets*, *117*(E6). https://doi.org/10.1029/2011JE003961.

Lucchitta, B. K. (1984). Ice and debris in the Fretted Terrain, Mars. *Journal of Geophysical Research. Solid Earth*, *89*(S02), B409–B418. https://doi.org/10.1029/JB089iS02p0B409.

Madeleine, J.-B., Forget, F., Head, J. W., Levrard, B., Montmessin, F., & Millour, E. (2009). Amazonian northern mid-latitude glaciation on Mars: A proposed climate scenario. *Icarus*, *203*(2), 390–405. https://doi.org/10.1016/j.icarus.2009.04.037.

Madeleine, J.-B., Head, J. W., Forget, F., Navarro, T., Millour, E., Spiga, A., … Dickson, J. L. (2014). Recent ice ages on Mars: The role of radiatively active clouds and cloud microphysics. *Geophysical Research Letters*, *41*(14), 4873–4879. https://doi.org/10.1002/2014GL059861.

Malin, M. C., Bell, J. F., III, Cantor, B. A., Caplinger, M. A., Calvin, W. M., Clancy, R. T., … Wolff, M. J. (2007). Context camera investigation on board the Mars reconnaissance orbiter. *Journal of Geophysical Research, Planets*, *112*(E5). https://doi.org/10.1029/2006JE002808.

Malin, M. C., Carr, M. H., Danielson, G. E., Davies, M. E., Hartmann, W. K., Ingersoll, A. P., … Warren, J. L. (1998). Early views of the Martian surface from the Mars orbiter camera of Mars global surveyor. *Science*, *279*(5357), 1681–1685. https://doi.org/10.1126/science.279.5357.1681.

Malin, M. C., & Edgett, K. S. (2001). Mars global surveyor mars orbiter camera: Interplanetary cruise through primary mission. *Journal of Geophysical Research, Planets*, *106*(E10), 23429–23570. https://doi.org/10.1029/2000JE001455.

Malin, M. C., Edgett, K. S., Cantor, B. A., Caplinger, M. A., Danielson, G. E., Jensen, E. H., Ravine, M. A., Sandoval, J. L., & Supulver, K. D. (2010). An overview of the 1985-2006 Mars orbiter camera science investigation. *The Mars Journal*, *5*, 1–60. https://doi.org/10.1555/mars.2010.0001.

Mangold, N. (2005). High latitude patterned grounds on Mars: Classification, distribution and climatic control. *Icarus*, *174*(2), 336–359. https://doi.org/10.1016/j.icarus.2004.07.030.

Mangold, N., Costard, F., & Forget, F. (2003). Debris flows over sand dunes on Mars: Evidence for liquid water. *Journal of Geophysical Research, Planets*, *108*(E4). https://doi.org/10.1029/2002JE001958.

Manning, C. V., Bierson, C., Putzig, N. E., & McKay, C. P. (2019). The formation and stability of buried polar CO_2 deposits on Mars. *Icarus*, *317*, 509–517. https://doi.org/10.1016/j.icarus.2018.07.021.

Mars Ice Core Working Group. (2021). *First ice cores from Mars*. co-chairs: M.R. Albert and M. Koutnik. Retrieved from https://www.nasa.gov/sites/default/files/atoms/files/mars_ice_core_working_group_report_final_feb2021.pdf.

Massé, M., Beck, P., Schmitt, B., Pommerol, A., McEwen, A., Chevrier, V., … Séjourné, A. (2014). Spectroscopy and detectability of liquid brines on mars. *Planetary and Space Science*, *92*, 136–149. https://doi.org/10.1016/j.pss.2014.01.018.

Matson, D. L., & Brown, R. H. (1989). Solid-state greenhouse and their implications for icy satellites. *Icarus*, *77*(1), 67–81. https://doi.org/10.1016/0019-1035(89)90007-9.

Mc Keown, L. E., Bourke, M. C., & McElwaine, J. N. (2017). Experiments on sublimating carbon dioxide ice and implications for contemporary surface processes on Mars. *Scientific Reports*, *7*(1), 14181. https://doi.org/10.1038/s41598-017-14132-2.

Mc Keown, L., McElwaine, J. N., Bourke, M. C., Sylvest, M. E., & Patel, M. R. (2021). The formation of Araneiforms by carbon dioxide venting and vigorous sublimation dynamics under Martian atmospheric pressure. *Scientific Reports*, *11*(1), 6445. https://doi.org/10.1038/s41598-021-82763-7.

McEwen, A. S., Eliason, E. M., Bergstrom, J. W., Bridges, N. T., Hansen, C. J., Delamere, W. A., … Weitz, C. M. (2007). Mars reconnaissance orbiter's high resolution imaging science experiment (HiRISE). *Journal of Geophysical Research, Planets*, *112*(E5). https://doi.org/10.1029/2005JE002605.

Mellon, M. T., Arvidson, R. E., Sizemore, H. G., Searls, M. L., Blaney, D. L., Cull, S., … Zent, A. P. (2009). Ground ice at the Phoenix Landing Site: Stability state and origin. *Journal of Geophysical Research: Planets*, *114*, E00E07. https://doi.org/10.1029/2009JE003417.

Mellon, M. T., & Jakosky, B. M. (1995). The distribution and behavior of Martian ground ice during past and present epochs. *Journal of Geophysical Research, Planets*, *100*(E6), 11781–11799. https://doi.org/10.1029/95JE01027.

MEPAG. (2020). *Mars scientific goals, objectives, investigations, and priorities* (p. 89).

MEPAG ICE-SAG. (2019). MEPAG ICE-SAG Final Report 2019, Report from the Ice and Climate Evolution Science Analysis group (ICE-SAG). Chaired by S. Diniega and N. E. Putzig, 157 pages posted 08 July 2019, by the Mars Exploration Program Analysis Group (MEPAG) at: http://mepag.nasa.gov/reports.cfm.

Michael, G. G. (2013). Planetary surface dating from crater size–frequency distribution measurements: Multiple resurfacing episodes and differential isochron fitting. *Icarus*, *226*(1), 885–890. https://doi.org/10.1016/j.icarus.2013.07.004.

Milkovich, S. M., & Head, J. W. (2005). North polar cap of Mars: Polar layered deposit characterization and identification of a fundamental climate signal. *Journal of Geophysical Research, Planets*, *110*(E1). https://doi.org/10.1029/2004JE002349.

Milkovich, S. M., & Plaut, J. J. (2008). Martian South Polar Layered Deposit stratigraphy and implications for accumulation history. *Journal of Geophysical Research: Planets*, *113*(E6). https://doi.org/10.1029/2007JE002987.

Milliken, R. E., & Mustard, J. F. (2003). *Erosional morphologies and characteristics of latitude-dependent surface mantles on Mars* (p. 3240). Retrieved from https://ui.adsabs.harvard.edu/abs/2003mars.conf.3240M.

Milliken, R. E., Mustard, J. F., & Goldsby, D. L. (2003). Viscous flow features on the surface of Mars: Observations from high-resolution Mars orbiter camera (MOC) images. *Journal of Geophysical Research, Planets*, *108*(E6). https://doi.org/10.1029/2002JE002005.

Mischna, M. A., Richardson, M. I., Wilson, R. J., & McCleese, D. J. (2003). On the orbital forcing of Martian water and CO2 cycles: A general circulation model study with simplified volatile schemes. *Journal of Geophysical Research, Planets*, *108*(E6). https://doi.org/10.1029/2003JE002051.

Montmessin, F. (2006). The orbital forcing of climate changes on Mars. *Space Science Reviews*, *125*(1), 457–472. https://doi.org/10.1007/s11214-006-9078-x.

Morgan, G. A., Head, J. W., & Marchant, D. R. (2009). Lineated valley fill (LVF) and lobate debris aprons (LDA) in the Deuteronilus Mensae northern dichotomy boundary region, Mars: Constraints on the extent, age and episodicity of Amazonian glacial events. *Icarus*, *202*(1), 22–38. https://doi.org/10.1016/j.icarus.2009.02.017.

Morgan, G. A., Putzig, N. E., Perry, M. R., Sizemore, H. G., Bramson, A. M., Petersen, E. I., … Campbell, B. A. (2021). Availability of subsurface water-ice resources in the northern mid-latitudes of Mars. *Nature Astronomy*, *5*(3), 230–236. https://doi.org/10.1038/s41550-020-01290-z.

Murray, B. C., Soderblom, L. A., Cutts, J. A., Sharp, R. P., Milton, D. J., & Leighton, R. B. (1972). Geological framework of the south polar region of Mars. *Icarus*, *17*(2), 328–345. https://doi.org/10.1016/0019-1035(72)90004-8.

Murray, B. C., Ward, W. R., & Yeung, S. C. (1973). Periodic insolation variations on Mars. *Science*, *180*(4086), 638–640. https://doi.org/10.1126/science.180.4086.638.

Mustard, J. F., Cooper, C. D., & Rifkin, M. K. (2001). Evidence for recent climate change on Mars from the identification of youthful near-surface ground ice. *Nature*, *412*(6845), 411–414. https://doi.org/10.1038/35086515.

Nerozzi, S., & Holt, J. W. (2018). Earliest accumulation history of the north polar layered deposits, Mars from SHARAD. *Icarus*, *308*, 128–137.

Nerozzi, S., Ortiz, M. R., & Holt, J. W. (2022). The north polar basal unit of Mars: An Amazonian record of surface processes and climate events. *Icarus*, *373*, 114716. https://doi.org/10.1016/j.icarus.2021.114716.

Neukum, G., & Jaumann, R. (2004). HRSC: The high resolution stereo camera of Mars express. In A. Wilson, & A. Chicarro (Eds.), *Mars express scientific payload* (vol. 1240, pp. 17–35). ESA Special Publication.

Newman, C. E., Lewis, S. R., & Read, P. L. (2005). The atmospheric circulation and dust activity in different orbital epochs on Mars. *Icarus*, *174*(1), 135–160. https://doi.org/10.1016/j.icarus.2004.10.023.

Nier, A. O., & McElroy, M. B. (1977). Composition and structure of Mars' upper atmosphere: Results from the neutral mass spectrometers on Viking 1 and 2. *Journal of Geophysical Research (1896-1977)*, *82*(28), 4341–4349. https://doi.org/10.1029/JS082i028p04341.

Orosei, R., Jordan, R. L., Morgan, D. D., Cartacci, M., Cicchetti, A., Duru, F., … Picardi, G. (2015). Mars advanced radar for subsurface and ionospheric sounding (MARSIS) after nine years of operation: A summary. *Planetary and Space Science*, *112*, 98–114. https://doi.org/10.1016/j.pss.2014.07.010.

Orosei, R., Lauro, S. E., Pettinelli, E., Cicchetti, A., Coradini, M., Cosciotti, B., … Seu, R. (2018). Radar evidence of subglacial liquid water on Mars. *Science*, *361*(6401), 490–493.

Parsons, R. A., Nimmo, F., & Miyamoto, H. (2011). Constraints on Martian lobate debris apron evolution and rheology from numerical modeling of ice flow. *Icarus*, *214*(1), 246–257. https://doi.org/10.1016/j.icarus.2011.04.014.

Pasquon, K., Gargani, J., Massé, M., & Conway, S. J. (2016). Present-day formation and seasonal evolution of linear dune gullies on Mars. *Icarus*, *274*, 195–210. https://doi.org/10.1016/j.icarus.2016.03.024.

Pathare, A. V., Paige, D. A., & Turtle, E. (2005). Viscous relaxation of craters within the Martian south polar layered deposits. *Icarus*, *174*(2), 396–418. https://doi.org/10.1016/j.icarus.2004.10.031.

Perron, J. T., & Huybers, P. (2009). Is there an orbital signal in the polar layered deposits on Mars? *Geology, 37*(2), 155–158. https://doi.org/10.1130/G25143A.1.

Phillips, R. J., Davis, B. J., Tanaka, K. L., Byrne, S., Mellon, M. T., Putzig, N. E., … Seu, R. (2011). Massive CO2 ice deposits sequestered in the south polar layered deposits of Mars. *Science, 332*(6031), 838–841. https://doi.org/10.1126/science.1203091.

Phillips, R. J., Zuber, M. T., Smrekar, S. E., Mellon, M. T., Head, J. W., Tanaka, K. L., … Marinangeli, L. (2008). Mars north polar deposits: Stratigraphy, age, and geodynamical response. *Science, 320*(5880), 1182–1185. https://doi.org/10.1126/science.1157546.

Picardi, G., Plaut, J. J., Biccari, D., Bombaci, O., Calabrese, D., Cartacci, M., … Zampolini, E. (2005). Radar soundings of the subsurface of Mars. *Science, 310*(5756), 1925–1928. https://doi.org/10.1126/science.1122165.

Pierce, T. L., & Crown, D. A. (2003). Morphologic and topographic analyses of debris aprons in the eastern Hellas region, Mars. *Icarus, 163*(1), 46–65.

Piqueux, S., Byrne, S., & Richardson, M. I. (2003). Sublimation of Mars's southern seasonal CO2 ice cap and the formation of spiders. *Journal of Geophysical Research, Planets, 108*(E8). https://doi.org/10.1029/2002JE002007.

Plaut, J. J., Kahn, R., Guinness, E. A., & Arvidson, R. E. (1988). Accumulation of sedimentary debris in the south polar region of Mars and implications for climate history. *Icarus, 76*(2), 357–377. https://doi.org/10.1016/0019-1035(88)90076-0.

Plaut, J. J., Picardi, G., Safaeinili, A., Ivanov, A. B., Milkovich, S. M., Cicchetti, A., … Edenhofer, P. (2007). Subsurface radar sounding of the south polar layered deposits of Mars. *Science, 316*(5821), 92–95. https://doi.org/10.1126/science.1139672.

Plaut, J. J., Safaeinili, A., Holt, J. W., Phillips, R. J., Head, J. W., Seu, R., … Frigeri, A. (2009). Radar evidence for ice in lobate debris aprons in the mid-northern latitudes of Mars. *Geophysical Research Letters, 36*(2). https://doi.org/10.1029/2008GL036379.

Portyankina, G., Merrison, J., Iversen, J. J., Yoldi, Z., Hansen, C. J., Aye, K.-M., … Thomas, N. (2019). Laboratory investigations of the physical state of CO2 ice in a simulated Martian environment. *Icarus, 322*, 210–220. https://doi.org/10.1016/j.icarus.2018.04.021.

Putzig, N. E., Phillips, R. J., Campbell, B. A., Holt, J. W., Plaut, J. J., Carter, L. M., … Seu, R. (2009). Subsurface structure of planum boreum from Mars reconnaissance orbiter shallow radar soundings. *Icarus, 204*(2), 443–457. https://doi.org/10.1016/j.icarus.2009.07.034.

Putzig, N. E., Smith, I. B., Perry, M. R., Foss, F. J., Campbell, B. A., Phillips, R. J., & Seu, R. (2018). Three-dimensional radar imaging of structures and craters in the Martian polar caps. *Icarus, 308*, 138–147. https://doi.org/10.1016/j.icarus.2017.09.023.

Ramsdale, J. D., Balme, M. R., Conway, S. J., & Gallagher, C. (2015). Ponding, draining and tilting of the Cerberus Plains; a cryolacustrine origin for the sinuous ridge and channel networks in Rahway Vallis, Mars. *Icarus, 253*, 256–270. https://doi.org/10.1016/j.icarus.2015.03.005.

Rodriguez, J. A. P., Leonard, G. J., Platz, T., Tanaka, K. L., Kargel, J. S., Fairen, A., … Oguma, M. (2015). New insights into the Late Amazonian zonal shrinkage of the Martian south polar plateau. *Icarus, 248*, 407–411.

Rodriguez, J. A. P., Tanaka, K. L., Bramson, A. M., Leonard, G. J., Baker, V. R., & Zarroca, M. (2021). North polar trough formation due to in-situ erosion as a source of young ice in mid-latitudinal mantles on Mars. *Scientific Reports, 11*(1), 6750. https://doi.org/10.1038/s41598-021-83329-3.

Scanlon, K. E., Head, J. W., & Marchant, D. R. (2015). Remnant buried ice in the equatorial regions of Mars: Morphological indicators associated with the Arsia Mons tropical mountain glacier deposits. *Planetary and Space Science, 111*, 144–154. https://doi.org/10.1016/j.pss.2015.03.024.

Schon, S. C., Head, J. W., & Fassett, C. I. (2012). Recent high-latitude resurfacing by a climate-related latitude-dependent mantle: Constraining age of emplacement from counts of small craters. *Planetary and Space Science, 69*(1), 49–61. https://doi.org/10.1016/j.pss.2012.03.015.

Schroeder, D. M., & Steinbrügge, G. (2021). Alternatives to liquid water beneath the south polar ice cap of Mars. *Geophysical Research Letters, 48*(19). https://doi.org/10.1029/2021GL095912. e2021GL095912.

Senft, L. E., & Stewart, S. T. (2008). Impact crater formation in icy layered terrains on Mars. *Meteoritics & Planetary Science*, *43*(12), 1993–2013. https://doi.org/10.1111/j.1945-5100.2008.tb00657.x.

Seu, R., Phillips, R. J., Biccari, D., Orosei, R., Masdea, A., Picardi, G., … Nunes, D. C. (2007). SHARAD sounding radar on the Mars reconnaissance orbiter. *Journal of Geophysical Research, Planets*, *112*(E5). https://doi.org/10.1029/2006JE002745.

Sharp, R. P. (1973). Mars: Fretted and chaotic terrains. *Journal of Geophysical Research (1896-1977)*, *78*(20), 4073–4083. https://doi.org/10.1029/JB078i020p04073.

Shean, D. E., Head, J. W., Fastook, J. L., & Marchant, D. R. (2007). Recent glaciation at high elevations on Arsia Mons, Mars: Implications for the formation and evolution of large tropical mountain glaciers. *Journal of Geophysical Research, Planets*, *112*(E3). https://doi.org/10.1029/2006JE002761.

Shean, D. E., Head, J. W., & Marchant, D. R. (2005). Origin and evolution of a cold-based tropical mountain glacier on Mars: The Pavonis Mons fan-shaped deposit. *Journal of Geophysical Research, Planets*, *110*(E5). https://doi.org/10.1029/2004JE002360.

Sinha, P., & Horgan, B. (2022). Sediments within the icy north polar deposits of mars record recent impacts and volcanism. *Geophysical Research Letters*, *49*(8). https://doi.org/10.1029/2022GL097758. e2022GL097758.

Sinha, R. K., & Murty, S. V. S. (2015). Amazonian modification of Moreux crater: Record of recent and episodic glaciation in the Protonilus Mensae region of Mars. *Icarus*, *245*, 122–144. https://doi.org/10.1016/j.icarus.2014.09.028.

Smith, I., Calvin, W. M., Smith, D. E., Hansen, C., Diniega, S., McEwen, A., … Dundas, C. M. (2021). Solar-system-wide significance of Mars polar science. *Bulletin of the American Astronomical Society*, *53*(4). Retrieved from https://eprints.whiterose.ac.uk/177196/.

Smith, I. B., Diniega, S., Beaty, D. W., Thorsteinsson, T., Becerra, P., Bramson, A. M., … Titus, T. N. (2018). 6th international conference on Mars polar science and exploration: Conference summary and five top questions. *Icarus*, *308*, 2–14. https://doi.org/10.1016/j.icarus.2017.06.027.

Smith, I. B., Hayne, P. O., Byrne, S., Becerra, P., Kahre, M., Calvin, W., … Siegler, M. (2020). The Holy Grail: A road map for unlocking the climate record stored within Mars' polar layered deposits. *Planetary and Space Science*, *184*, 104841. https://doi.org/10.1016/j.pss.2020.104841.

Smith, I. B., & Holt, J. W. (2010). Onset and migration of spiral troughs on Mars revealed by orbital radar. *Nature*, *465*, 450–453.

Smith, I. B., Holt, J. W., Spiga, A., Howard, A. D., & Parker, G. (2013). The spiral troughs of Mars as cyclic steps. *Journal of Geophysical Research: Planets*, *118*(9), 1835–1857.

Smith, I. B., Putzig, N. E., Holt, J. W., & Phillips, R. J. (2016). An ice age recorded in the polar deposits of Mars. *Science*, *352*(6289), 1075–1078. https://doi.org/10.1126/science.aad6968.

Smith, I. B., Schlegel, N.-J., Larour, E., Isola, I., Buhler, P. B., Putzig, N. E., & Greve, R. (2022). Carbon dioxide ice glaciers at the south pole of Mars. *Journal of Geophysical Research, Planets*, *127*(4). https://doi.org/10.1029/2022JE007193. e2022JE007193.

Smith, D. E., Zuber, M. T., Frey, H. V., Garvin, J. B., Head, J. W., Muhleman, D. O., … Sun, X. (2001). Mars orbiter laser altimeter: Experiment summary after the first year of global mapping of Mars. *Journal of Geophysical Research, Planets*, *106*(E10), 23689–23722. https://doi.org/10.1029/2000JE001364.

Smith, D. E., Zuber, M. T., Solomon, S. C., Phillips, R. J., Head, J. W., Garvin, J. B., … Duxbury, T. C. (1999). The global topography of Mars and implications for surface evolution. *Science*, *284*(5419), 1495–1503. https://doi.org/10.1126/science.284.5419.1495.

Soare, R. (2014a). Latitude-dependent mantle (in MOC) (with stratigraphically associated periglacial landforms). In *Encyclopedia of planetary landforms*. New York, NY: Springer.

Soare, R. (2014b). Latitude-dependent mantle (in HiRISE) (with no stratigraphically associated periglacial landforms). In *Encyclopedia of planetary landforms*. New York, NY: Springer.

Soare, R. J., Conway, S. J., Williams, J.-P., Gallagher, C., & Keown, L. E. M. (2020). Possible (closed system) pingo and ice-wedge/thermokarst complexes at the mid latitudes of Utopia Planitia, Mars. *Icarus, 342*, 113233. https://doi.org/10.1016/j.icarus.2019.03.010.

Soare, R. J., Conway, S. J., Williams, J.-P., & Hepburn, A. J. (2021a). Possible polyphase periglaciation and glaciation adjacent to the Moreux impact-crater, Mars. *Icarus, 362*, 114401. https://doi.org/10.1016/j.icarus.2021.114401.

Soare, R. J., Conway, S. J., Williams, J.-P., Philippe, M., Mc Keown, L. E., Godin, E., & Hawkswell, J. (2021b). Possible ice-wedge polygonisation in Utopia Planitia, Mars and its latitudinal gradient of distribution. *Icarus, 358*, 114208. https://doi.org/10.1016/j.icarus.2020.114208.

Soare, R. J., Costard, F., Pearce, G. D., & Séjourné, A. (2012). A re-interpretation of the recent stratigraphical history of Utopia Planitia, Mars: Implications for late-Amazonian periglacial and ice-rich terrain. *Planetary and Space Science, 60*(1), 131–139. https://doi.org/10.1016/j.pss.2011.07.007.

Soare, R. J., Williams, J.-P., Hepburn, A. J., & Butcher, F. E. G. (2022). A billion or more years of possible periglacial/glacial cycling in Protonilus Mensae, Mars. *Icarus, 385*, 115115. https://doi.org/10.1016/j.icarus.2022.115115.

Soare, R. J., Costard, F., Williams, J.-P., Gallagher, C., Hepburn, A. J., Stillman, D., … Godin, E. (2024). *Evidence, arguments, and cold-climate geomorphology that favour periglacial cycling at the Martian mid-to-high latitudes in the Late Amazonian Epoch*. In R. J. Soare, J.-P. Williams, C. Ahrens, F. E. G. Butcher, & M. R. El-Maarry (Eds.), *Ices in the solar system, a volatile-driven journey from the inner solar system to its far reaches*. Elsevier Books.

Soderblom, L. A., Malin, M. C., Cutts, J. A., & Murray, B. C. (1973). Mariner 9 observations of the surface of Mars in the north polar region. *Journal of Geophysical Research (1896-1977), 78*(20), 4197–4210. https://doi.org/10.1029/JB078i020p04197.

Sori, M. M., Becerra, P., Bapst, J., Byrne, S., & McGlasson, R. A. (2022). Orbital forcing of Martian climate revealed in a south polar outlier ice deposit. *Geophysical Research Letters, 49*(6). https://doi.org/10.1029/2021GL097450. e2021GL097450.

Sori, M. M., & Bramson, A. M. (2019). Water on Mars, with a grain of salt: Local heat anomalies are required for basal melting of ice at the south pole today. *Geophysical Research Letters, 46*(3), 1222–1231. https://doi.org/10.1029/2018GL080985.

Sori, M. M., Perron, J. T., Huybers, P., & Aharonson, O. (2014). A procedure for testing the significance of orbital tuning of the Martian polar layered deposits. *Icarus, 235*, 136–146. https://doi.org/10.1016/j.icarus.2014.03.009.

Souness, C., & Hubbard, B. (2012). Mid-latitude glaciation on Mars. *Progress in Physical Geography: Earth and Environment, 36*(2), 238–261. https://doi.org/10.1177/0309133312436570.

Souness, C. J., & Hubbard, B. (2013). An alternative interpretation of late Amazonian ice flow: Protonilus Mensae, Mars. *Icarus, 225*(1), 495–505. https://doi.org/10.1016/j.icarus.2013.03.030.

Souness, C., Hubbard, B., Milliken, R. E., & Quincey, D. (2012). An inventory and population-scale analysis of Martian glacier-like forms. *Icarus, 217*(1), 243–255. https://doi.org/10.1016/j.icarus.2011.10.020.

Squyres, S. W. (1978). Martian fretted terrain: Flow of erosional debris. *Icarus, 34*(3), 600–613. https://doi.org/10.1016/0019-1035(78)90048-9.

Squyres, S. W. (1979). The distribution of lobate debris aprons and similar flows on Mars. *Journal of Geophysical Research. Solid Earth, 84*(B14), 8087–8096. https://doi.org/10.1029/JB084iB14p08087.

Squyres, S. W., & Carr, M. H. (1986). Geomorphic evidence for the distribution of ground ice on Mars. *Science, 231*(4735), 249–252. https://doi.org/10.1126/science.231.4735.249.

Starr, S. O., & Muscatello, A. C. (2020). Mars in situ resource utilization: A review. *Planetary and Space Science, 182*, 104824. https://doi.org/10.1016/j.pss.2019.104824.

Tanaka, K. L. (2005). Geology and insolation-driven climatic history of Amazonian north polar materials on Mars. *Nature, 437*, 991–994.

Tanaka, K. L., & Fortezzo, C. M. (2012). Geological Map of the North Polar Region of Mars. Scientific Investigations Map 3177. United States Geological Survey.

Tanaka, K. L., Rodriguez, J. A. P., Skinner, J. A., Jr., Bourke, M. C., Fortezzo, C. M., Herkenhoff, K. E., … Okubo, C. H. (2008). North polar region of Mars: Advances in stratigraphy, structure, and erosional modification. *Icarus, 196*(2), 318–358.

Tanaka, K. L., Skinner, J. A., Dohm, J. M., Irwin, R. P., III, Kolb, E. J., Fortezzo, C. M., … Hare, T. M. (2014). Geologic map of Mars. In *Geologic map of Mars (USGS numbered series)*. Reston, VA: U.S. Geological Survey. https://doi.org/10.3133/sim3292. No. 3292; p. 48.

Tanaka, K. L., Skinner, J. A., & Hare, T. M. (2005). Geologic map of the northern plains of Mars. In *U.S. Geological Survey Scientific Investigations Map 2888*. Retrieved from https://pubs.usgs.gov/sim/2005/2888/.

Thomas, N., Becerra, P., & Smith, I. B. (2021). Mars and the ESA science programme—The case for Mars polar science. *Experimental Astronomy*. https://doi.org/10.1007/s10686-021-09760-6.

Thomas, P. C., Malin, M. C., Edgett, K. S., Carr, M. H., Hartmann, W. K., Ingersoll, A. P., & Sullivan, R. (2000). North-South geological differences between the residual polar caps on Mars. *Nature, 404*, 161–164.

Titus, T. N., Byrne, S., Colaprete, A., Forget, F., Michaels, T. I., & Prettyman, T. H. (2017). The CO2 cycle. In F. Forget, M. D. Smith, R. T. Clancy, R. W. Zurek, & R. M. Haberle (Eds.), *The Atmosphere and Climate of Mars* (pp. 374–404). Cambridge: Cambridge University Press. https://doi.org/10.1017/9781139060172.012.

Toigo, A. D., Waugh, D. W., & Guzewich, S. D. (2020). Atmospheric transport into polar regions on Mars in different orbital epochs. *Icarus, 347*, 113816. https://doi.org/10.1016/j.icarus.2020.113816.

Toon, O. B., Pollack, J. B., Ward, W., Burns, J. A., & Bilski, K. (1980). The astronomical theory of climatic change on Mars. *Icarus, 44*(3), 552–607. https://doi.org/10.1016/0019-1035(80)90130-X.

Tsibulskaya, V., Hepburn, A. J., Hubbard, B., & Holt, T. (2020). Surficial geology and geomorphology of Greg crater, Promethei Terra, Mars. *Journal of Maps, 16*(2), 524–533. https://doi.org/10.1080/17445647.2020.1785343.

Vincendon, M. (2015). Identification of Mars gully activity types associated with ice composition. *Journal of Geophysical Research, Planets, 120*(11), 1859–1879. https://doi.org/10.1002/2015JE004909.

Warner, N. H., Gupta, S., Calef, F., Grindrod, P., Boll, N., & Goddard, K. (2015). Minimum effective area for high resolution crater counting of Martian terrains. *Icarus, 245*, 198–240. https://doi.org/10.1016/j.icarus.2014.09.024.

Whitten, J. L., & Campbell, B. A. (2018). Lateral continuity of layering in the Mars south polar layered deposits from SHARAD sounding data. *Journal of Geophysical Research, Planets, 123*(6), 1541–1554. https://doi.org/10.1029/2018JE005578.

Whitten, J. L., Campbell, B. A., & Morgan, G. A. (2017). A subsurface depocenter in the south polar layered deposits of Mars. *Geophysical Research Letters, 44*(16), 8188–8195. https://doi.org/10.1002/2017GL074069.

Williams, J.-P., van der Bogert, C. H., Pathare, A. V., Michael, G. G., Kirchoff, M. R., & Hiesinger, H. (2018). Dating very young planetary surfaces from crater statistics: A review of issues and challenges. *Meteoritics & Planetary Science, 53*(4), 554–582. https://doi.org/10.1111/maps.12924.

Woodley, S. Z., Butcher, F. E. G., Fawdon, P., Clark, C. D., Ng, F. S. L., Davis, J. M., & Gallagher, C. (2022). Multiple sites of recent wet-based glaciation identified from eskers in western Tempe Terra, Mars. *Icarus, 386*, 115147. https://doi.org/10.1016/j.icarus.2022.115147.

Evidence, arguments, and cold-climate geomorphology that favour periglacial cycling at the Martian mid-to-high latitudes in the Late Amazonian Epoch

Richard J. Soare[a], F. Costard[b], Jean-Pierre Williams[c], Colman Gallagher[d], Adam J. Hepburn[e], D. Stillman[f], Michelle Koutnik[g], S.J. Conway[h], M. Philippe[h], Frances E.G. Butcher[i], Lauren E. Mc Keown[j], and E. Godin[k]

[a]Department of Geography, Dawson College, Montreal, QC, Canada, [b]Geosciences Paris Saclay, University of Paris-Saclay, Orsay, France, [c]Department of Earth, Planetary, and Space Sciences, University of California, Los Angeles, CA, United States, [d]UCD School of Geography, University College Dublin, Dublin, Ireland, [e]European Space Astronomy Centre, European Space Agency, Madrid, Spain, [f]Department of Space Studies, Southwest Research Institute, Boulder, CO, United States, [g]Department of Earth and Space Sciences, University of Washington, Seattle, WA, United States, [h]CNRS, UMR 6112 Planetology and Geodynamics Laboratory, University of Nantes, Nantes, France, [i]Department of Geography, University of Sheffield, Sheffield, United Kingdom, [j]NASA Jet Propulsion Laboratory, California Institute of Technology, Pasadena, CA, United States, [k]Northern Studies Centre, Laval University, Quebec City, QC, Canada

Abstract

Absent of a humanly wielded pick and shovel digging beneath the surface, the identification of ground (interstitial) ice on Mars formed by the freeze–thaw cycling of water has largely been inferred from presumed periglacial analogues on Earth.

Here, we reach beyond the *looks-like therefore-must-be* paradigm and seek to validate this presumption by two means:

(1) Presenting diverse data sets, tools, scales (temporal and spatial), and case studies that point, collectively, to the plausibility of periglacial processes having occurred at the northern mid-latitudes of Mars through the Late Amazonian Epoch.

(2) Suggesting that and showing why the litmus test of periglacial plausibility ought not to be derived from current boundary conditions on Mars or models derived therefrom.

Towards these twinned ends, we draw upon an admixture of recently published and new work by this chapter's authors.

Ices in the Solar System. https://doi.org/10.1016/B978-0-323-99324-1.00008-0

1 Introduction

Mean surface temperatures on Mars at non-polar latitudes are extremely low, as is the mean atmospheric vapour pressure. Areas where water is metastable are relatively few and the temporal span of metastability is small (e.g. Haberle et al., 2001; Hecht, 2002). This suggests that the opportunities are meagre for landscape-scale revisions to occur by *wet* periglacial processes, i.e. the freeze–thaw cycling of water (e.g. Barrett et al., 2018; Costard & Kargel, 1995; Costard et al., 2016; Gallagher et al., 2011; Hauber et al., 2011; Seibert & Kargel, 2001; Séjourné et al., 2011; Soare et al., 2008, 2011) compared to non-wet or *dry* periglacial processes, i.e. sublimation and/or diffusion/adsorption cycles (e.g. Dundas et al., 2015; Dundas, 2017; Lefort et al., 2009; Levy et al., 2009b, 2010; Mellon & Jakosky, 1993, 1995; Morgenstern et al., 2007). By this token, hypotheses driven by apparent morphological similarities between possible *wet* periglacial landscapes on Mars and presumed analogues on Earth typically have been discounted.

Contrarily, the idea that glaciation has occurred cyclically on Mars throughout the Late Amazonian Epoch (e.g. Hepburn et al., 2020; Hubbard et al., 2014; Koutnik et al., 2024, pp. 101–142), if not much earlier (e.g. Grau-Galofre et al., 2020, 2024, pp. 73–100), is generally accepted. This presumption is underpinned by a suite of analytical and observational/empirical pillars:

(a) *G*eneral *c*irculation *m*odels (*GCMs*) (e.g. Forget et al., 2006; Madeleine et al., 2009, 2014) that tie glacial-deglacial cycles to shifts of obliquity and eccentricity.

(b) Wide-ranging observations of glacier- and moraine-like landforms, possible viscous-flow features and lobate-debris aprons (e.g. Baker, 2003; Baker & Carter, 2019; Forget et al., 2006; Head et al., 2003; Hepburn et al., 2020; Hubbard et al., 2014; Kargel & Strom, 1992; Lucchitta, 1981; Squyres, 1978).

(c) Cross-sectional exposures of *icy* scarps (e.g. Dundas et al., 2018, 2021; Harish et al., 2020).

(d) Radar profiles at the mid-latitudes pointing to the possible presence of massive and near-surface tabular ice that is decametres thick (e.g. Bramson et al., 2015; Stuurman et al., 2016).

Here, we reach beyond the *looks-like therefore-must-be* paradigm and propose that wet-based periglacial processes are part and parcel of the geological history of Mars in the Late Amazonian Epoch, even if only ephemerally and episodically. We do this on the basis of geomorphological/geological consilience. This is the idea that form feeds process and that understanding landscape evolution requires bundling data and information through as many convergent lenses, i.e. geological, geomorphological, climatic, hydrological, etc., and analytical pathways, i.e. photogeology, statistics, ground-penetrating radar, geophysical modelling, etc., as possible (derived from Baker, 2003, 2014; also, e.g. Hauber et al., 2011; Soare et al., 2008, 2021a, 2021b). Coeval with this idea is the suggestion that the *relative* plausibility or implausibility of competing hypotheses should be determined by the greater or lesser robustness of each hypothesis in deriving and explaining landscape-scale processes from form.

The geological consilience proposed by us comprises:

(a) **Data Sets**—*HiRISE* (*h*igh-*r*esolution *i*maging *s*cience *e*xperiment)/*CTX* (*c*ontext) images, *HRSC-based DEMs* (high-resolution science camera, *d*igital *e*levation *m*odels), *MOLA* (*M*ars *O*rbiter *L*aser *A*ltimeter), *THEMIS* (*th*ermal *em*ission *i*maging *s*ystem), and *TES* (*t*hermal *e*mission *s*pectrometer) data.

(b) **Tools**—field-based observations; statistics; lab experiments and modelling; geochemistry; and production functions for crater-size frequency distributions.

(c) **Scales** (spatial/geographical)—*HiRISE* (local)/*CTX* (regional) imagery, *MOLA*-based *DEMs,* (local/regional) and associated maps (regional/local), crater-size frequency distributions (local/regional), and lab chambers (small).

(d) **Scales** (temporal) – $\leq 10^0$ years (lab experiments); $\sim 10^1$–$^{\mathsf{c}}10^2$ years (field observations on Earth), $\sim 10^5$–10^7 years (obliquity cycles on Mars).

(e) **Case studies** – Utopia Planitia, Galaxias Chaos, the Elysium/Athabasca Valles, and Protonilus Mensae.

2 Ice-rich landscapes on Earth and on Mars
2.1 Ice complexes on Earth

In areas of continuous permafrost such as the Tuktoyaktuk Coastlands of northern Canada (e.g. Mackay, 1998; Murton, 2001; Murton et al., 2005; Rampton, 1988; Rampton & Bouchard, 1975) and the Yamal Peninsula coastlands of eastern Russia (e.g. Grosse et al., 2007; Morgenstern et al., 2013; Schirrmeister et al., 2002, 2013) genetically associated assemblages of *wet* periglacial landforms, or *ice complexes*, are commonplace (Fig. 1). These complexes, in part, comprise metre- to decametres-thick sequences of silts and clays. The abundant pore space and relatively fine grains of these sediments facilitate (**1**) the migration of pore water to a freezing front or fringe (cryosuction) and (**2**) the formation of segregation (excess) ice, i.e. ice lenses at that front or fringe.

Excess ice describes the volume of interstitial ice in the ground that exceeds the total pore volume that the ground would have were it not frozen (e.g. Harris et al., 1988; also, see French, 2017; Penner, 1959; Rampton, 1988; Rampton & Mackay, 1971; Taber, 1930). Permafrost that contains excess ice is deemed to be *ice rich* (e.g. Harris et al., 1988).

Segregation ice is a type of excess ice that varies in thickness from hairline to tabular masses, decametres deep at some locations, and formed by the iterative freeze–thaw cycling of water (e.g. Black, 1954; French, 2017; Harris et al., 1988; Penner, 1959; Rampton, 1988; Rampton & Mackay, 1971; Taber, 1930). As the volume of segregation ice increases or decreases, so does the sensitivity of the ground to heave (by aggradation) and deflation (by degradation and thaw/melt) (e.g. Grosse et al., 2007; Morgenstern et al., 2013; Penner, 1959; Rampton, 1988; Rampton & Bouchard, 1975; Schirrmeister et al., 2002).

Thermokarst is a periglacial term about form and process. It refers to permafrost landscapes and landscape features whose ice-rich composition makes them particularly susceptible to the volumetric loss of water by thaw or melt, forming depressions (French, 2017; Harris et al., 1988). Commonly, this loss is driven by region-wide or global rises in mean temperature (e.g. Czudek & Demek, 1970; Grosse et al., 2007; Murton, 2001; Osterkamp et al., 2009; Péwé, 1959; Schirrmeister et al., 2013; Taber, 1930; Wetterich et al., 2014).

Commonplace features of coastal ice-rich landscapes such as in northern Canada and in eastern Russia are (Fig. 1): hummocky terrain where the presence or absence of near-surface excess ice generates positive or negative topography, i.e. thermokarst; polygonised alases, i.e. thermokarst depressions absent of water and fractured by thermal-contraction cracking; low/high-centred polygons

FIG. 1

(a) Ice-rich and continuous permafrost (Tuktoyaktuk Coastlands [*TC*], Northwest Territories [*NWT*], Canada. Noteworthy landscape features: hummocky thermokarst; thermokarst lakes comprised of pooled meltwater; polygonised alases (i.e. drained or evaporated thermokarst lakes); and, beaded streams linking thaw pools of degraded ice-wedge polygons (Husky Lakes [*HL*], *NWT*). (b) Panoramic and cross-sectional view of an ice complex comprised of ice wedges (on the right-hand side) and massive ice (on the left-hand side) [Peninsula Point (*PP*), *NWT*, ~6 km southwest of Tuktoyaktuk]. The small, surface depressions flanking the largest ice wedge (on the right-hand side, *black arrows*) mark the location of bilateral polygonal troughs associated with a field of high-centred polygons. In a wet permafrost-environment high-centred, ice-wedge polygons form when and if local or regional (mean) temperatures rise to the point of inducing thaw and melt at or near the surface where the ice wedges occur. Greyish-beige material on the right and throughout the floor of the terrace comprises silts and clays that have slumped to the floor from the near-surface as retrogressive thaw slumping has eroded this Beaufort Sea headland. The oblong block on the right-hand side of the terrace floor points to the recentness of thaw-based detachment as there is a gap in the overhang of the terrain immediately above the block location.

Image credits: R. Soare.

whose morphologies delineate the aggradation (low-centredness) or degradation (high-centredness) of near-surface ice at the margins of thermal-contraction polygons; polygon-corner pits, often filled by icy-derived meltwater and formed at the confluence of polygon margins by the thermal and mechanical erosion of the latter; retrogressive thaw slumps that exhume and cross-sectionally expose ice-rich permafrost and massive tabular ice as well as ice-wedge and sand-wedge polygons; and, perennially ice-cored mounds, i.e. closed or hydrostatic pingos (e.g. Mackay, 1974, 1998; Rampton, 1988; Rampton & Mackay, 1971).

2.2 Possible ice-complexes and ice-rich landscapes on Mars

Landforms whose morphologies, scales, traits, surface textures, and close spatial association are synonymous with ice complexes on Earth are ubiquitous on Mars at the mid-to-high latitudes of both hemispheres (Costard & Kargel, 1995; Balme & Gallagher, 2009; Barrett et al., 2018; Costard et al., 2016; Lefort et al., 2009; Morgenstern et al., 2007; Séjourné et al., 2011, 2012, 2019 ; Soare et al., 2005, 2007, 2008, 2011, 2020; Ulrich et al., 2010; Wan Bun Tseung & Soare, 2006). Examples of these landforms and landscape features, are:

(a) Thermokarst-like and rimless depressions formed not by thaw-driven ice loss, as they might be on Earth, but by sublimation; contrarily, ice enrichment, which is the precursive state of a thermokarst land-form, would be best explained on Mars as on Earth by water undergoing iterative freeze-thaw cycling. This is especially true if the thermokarst is decametres deep.
(b) Low- and high-centred polygons possibly underlain at the margins by aggraded or degraded ice wedges, respectively.
(c) Polygon-margin pits formed, perhaps, by sublimation-driven ice-wedge degradation; however, as with the thermokarst-like depressions, the formation and aggradation of ice wedges are explained more easily by the iterative freeze–thaw cycling of water than by sublimation.
(d) Possible retrogressive thaw slumps.
(e) Mounds whose form, features, and scale are similar to that of closed-system (hydrostatic) pingos (Fig. 2).

2.3 Excess ice on Mars

The distribution of water-equivalent hydrogen has been identified by the Mars Odyssey's neutron spectrometer to a depth of ~1 m throughout the mid-latitudes of the northern plains (e.g. Feldman et al., 2002; Mitrofanov et al., 2002) and, broadly speaking, wherever the possible ice complexes are located (e.g. Ramsdale et al., 2017, 2018; Séjourné et al., 2019).

Some relatively recent radar tracks generated by the *Sha*llow *Rad*ar (*SHARAD)* sounding experiment show near-surface reflectors in Arcadia and Utopia Planitae, for example, that are consistent with decametres-thick tabular and possibly excess ice. The near-surface reflectors coincide spatially with the presence of thermokarst-like depressions and polygonised terrain at the surface (Bramson et al., 2015, 2017; Stuurman et al., 2016).

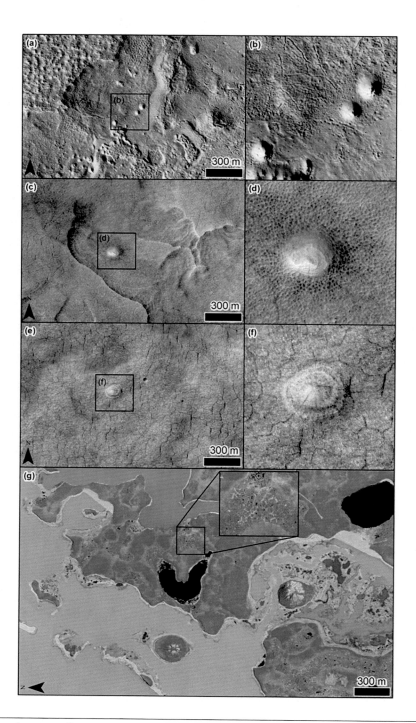

FIG. 2

See figure legend on opposite page.

3 Low to high-centred polygon ratios in Utopia Planitia: Statistical support of an ice-wedge origin

3.1 Ice-wedge polygons (Earth)

As noted above, clastically non-sorted polygons are commonplace in ice-rich permafrost landscapes on Earth (e.g. French, 2017; Lachenbruch, 1962; Mackay, 1974; Rampton, 1988; Rampton & Bouchard, 1975; Washburn, 1973) (Fig. 3). Typically ≤25 m in diameter, the polygons are produced by the tensile-induced *thermal-contraction cracking* of frozen sediment cemented, even if only minimally, by an ice fraction (Lachenbruch, 1962). This cracking or fracturing occurs when frozen icy sediments undergo a sharp drop in sub-zero (Celsius) temperatures (de Leffingwell, 1915; Lachenbruch, 1962). In-filling, by water or sediments, prevents the cracked ground from relaxing and returning to a seamless pre-cracked state as mean temperatures rise, diurnally or seasonally. Iterative infilling by meltwater transforms shallow and narrow vertical veins into metres-wide and decametre-deep (vertically-foliated) wedges (e.g. French, 2017; Lachenbruch, 1962; Mackay, 1974; Washburn, 1973).

Fill types vary. They are constrained by local or regional boundary conditions and the availability of (a) meltwater derived from thawed snow or ice, vs (b) winter hoarfrost, to a much lesser degree, or (c) windblown sand, mineral soil, or a mixture of the two (e.g. de Leffingwell, 1915; Hallet et al., 2011; Lachenbruch, 1962; Péwé, 1959; Sletten et al., 2003; Washburn, 1973).

As wedge cracks become increasingly dense in their distribution, they intercept one another and form a polygonal pattern at the surface. Some polygon networks are expansive, covering tens if not hundreds of km^2 in places like the Tuktoyaktuk Coastlands, and are the work of countless seasonal/annual iterations of cracking and filling.

Wedge growth, regardless of the fill type, is vertical and horizontal. As wedges aggrade at the polygon margins, the sedimentary overburden rises above the elevation datum of the polygon centres; this forms *low-centred polygons* (*LCPs*) (French, 2017; Harris et al., 1988; Péwé, 1959; Rampton,

FIG. 2

Possible closed-system pingos (*CSPs*) on Mars. (a) Three candidates on the floor of a thermokarst-like and polygonised depression. The adjacent terrain is mantled and not polygonised (*HiRISE* image ESP_055038_2250). (b) Close-up of the mounds in panel (a). Note the irregularly shaped summit depressions and geometries. (c) Candidate *CSP* on the floor of a thermokarst-like and polygonised depression. The adjacent terrain is polygonised but has no apparent mantling (*HiRISE* image ESP_044042_2240). (d) Close-up of the mound in panel (c). Note the irregular topography of the summit and adjacent slopes as well as the small-scale pitting of the terrain surrounding terrain. (e) Candidate mound in topographically uneven and polygonised terrain (*HiRISE* image ESP_026556_2245). (f) The same candidate *CSP* exhibits a mound summit and a diametrical crack that intercepts the adjacent polygonised terrain. North is up in panels (a–f). (g) Split (bottom-centre) and Ibyuk (right) pingos in ice-rich terrain populated by thermokarst/ice-wedge complexes, i.e. the Pingo Canadian Landmark (*TC, NWT*). Each of these pingos, nested within a thermokarst lake, hosts summit depressions and radial fracturing. Note the benched and unnamed peninsula-shaped pingo at the top left-hand corner of the image, as well as the similarity of scale amongst the *CSPs* in the Tuktoyaktuk Coastlands and the candidate *CSPs* in Utopia Planitia.

Image credits (a–f): NASA/JPL/University of Arizona. Panel (g) Image credits: DigitalGlobe, GoogleEarth.

FIG. 3

(a) Early stage polygonisation of continuous permafrost by thermal-contraction cracking (*TC, NWT*). (b) Planimetric view of permafrost coverage by small-sized (and presumed ice-wedge) polygons (*HL, NWT*). (c) Low-centred (presumed ice-wedge) polygons with marginal and positively elevated troughs filled with ice-wedge meltwater. (d) High-centred (presumed ice-wedge) polygons with marginal and negatively elevated troughs filled with ice-wedge meltwater. (e) Cross-sectional view of ice-wedges exposed by retrogressive thaw slumping on the Beaufort seacoast (*PP, NWT*). Note the surface depressions above the large ice wedge on the right and above the smaller ice-wedges to its left. The depressions, were they observed planimetrically, would be seen to comprise the marginal troughs of high-centred (degradational) polygons such as the ones observed in panels (b) and (d). The brownish sediments in whose midst the ice wedges lie are silty clays/clayey silts. (f) Exposed sand wedge (*TC*) (Murton et al., 2000).

Image credits, for all panels other than (f): R. Soare.

1988; Washburn, 1973). Degradation, most often by thaw in the case of ice wedges or aeolian erosion in the case of sand or mineral wedges, depletes the wedge volume and mass and deflates the marginal overburden. **H**igh-**c**entred **p**olygons (**HCPs**) develop if and when this depletion or deflation lowers the polygon margins below the elevation of the centres (French, 2017; Harris et al., 1988; Péwé, 1959; Rampton, 1988; Washburn, 1973).

Some polygons show neither elevated nor deflated margins. As such, there is no topographical variance between margins and centres. This could be due to: (1) wedge nascency, whereby marginal wedges have evolved insufficiently to show overburden uplift; (2) truncated or stagnated growth, the result of thermal-contraction cycles having ended; or (3) a transitional stage between aggradation and degradation with the latter being insufficiently evolved for the margins to fall below the elevation of the centres.

The thermal flux of a periglacial system on a local or regional scale can be identified by the concurrence and proximity of *LCPs* and *HCPs* in that system (e.g. French, 2017; Washburn, 1973; also, Grosse et al., 2007; Morgenstern et al., 2013; Rampton, 1988; Rampton & Bouchard, 1975). A similar admixture of low and high-centred polygons with sand-wedge margins would not be expected on a local or regional scale as sand-wedge polygons are insensitive to thermal stress. As noted above, the latter aggrade sedimentarily and degrade erosionally largely by aeolian processes.

3.2 Possible ice-wedge polygons (Mars)

The distribution of small-sized polygons is wide ranging and dense across the mid to high latitudes of the northern plains and to the north of the Mars dichotomy, i.e. Utopia Planitia, Elysium Planitia, Arcadia Planitia, Protonilus Mensae, etc. (e.g. Levy et al., 2009b, 2010; Mangold, 2005; Seibert & Kargel, 2001; Soare et al., 2022). The localised distribution of these small-sized polygons (~10–50 m in diameter) may comprise *LCPs*, *HCPs* or both types, concurrently. Where thermokarst-like depressions are observed, it is not unusual for the terrain also to be punctuated by these polygons (e.g. Burr et al., 2005; Costard & Kargel, 1995; Lefort et al., 2009; Levy et al., 2009a, 2009b, 2010; Morgenstern et al., 2007; Seibert & Kargel, 2001; Séjourné et al., 2011; Soare et al., 2008, 2011, 2021b) (Fig. 4). Historically, the planimetric similarities between ice and sand-wedge polygons on Earth let alone on Mars have obviated the (observational) possibility of validating or invalidating assumptions concerning the possible presence on Mars of near-surface ice wedges at polygon margins.

3.3 Testing the ice-wedge hypothesis statistically

Ground ice is stable where the annual mean surface and near-surface temperatures are below the atmospheric frost point; typically, this occurs at and polewardly of the mid-Martian latitudes in each Martian hemisphere (Mellon & Jakosky, 1993, 1995). This suggests that the thermal and, consequently, the structural integrity of near-surface ice wedges, were they to be/have been present at polygon margins, varies latitudinally. As such, the ratio, i.e. presence, of low centred (aggradational) to high-centred (degradational) polygons would be expected to increase polewardly and decrease equatorially. Conversely, changes in surface and near-surface temperatures (in and of themselves) would not be expected to exercise any influence on the aggradation or degradation of sand-wedge polygons. As such, the ratio of low to high-centred polygons, were the margins underlain by sand wedges, ought not to vary with latitude.

Here, we present and summarise the results of a test that recently evaluated these presumptions (Soare et al., 2021b). The study region comprised a quadrat at the mid-latitudes of *UP* (40–50° N; 100–125° E) (Fig. 5). Using a grid-mapping approach (e.g. Ramsdale et al., 2017, 2018; Séjourné et al., 2019) (Fig. 6), all of the *HiRISE* images in this region ($N=135$) were surveyed for the presence of *LCPs* or *HCPs*. Eleven images were too blurry to be evaluated; 16 images were excluded because other images with higher resolution either overlapped them completely or they exhibited equal resolution but greater spatial extent. The remaining sample comprised 108 images, divided into two groups:

(1) Images whose resolution and/or quality support the unambiguous identification of *LCPs* or *HCPs*, if and when they are observed (67 images, of which 41 have polygons).
(2) Images whose resolution or quality is of a lesser order, making the identification of *LCPs* and *HCPs* more difficult (41 images, of which 35 have polygons).

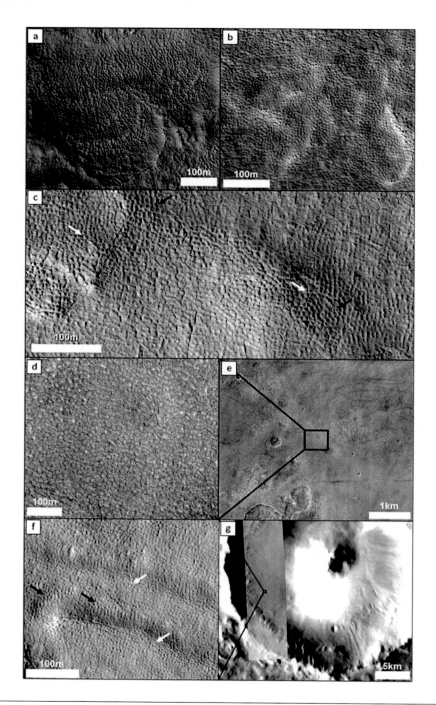

FIG. 4

(a) Thermokarst-like depressions overprinted by *LCPs* (*HiRISE* image PSP_006908). (b) Thermokarst-like depressions overprinted by *HCPs* (*HiRISE* image ESP_046467_2280). (c) *LCPs* (black arrows) and *HCPs* (white arrows) (*HiRISE* image ESP_055038_2250). (d) Polygons with no observable elevation difference between centres and margins (*HiRISE* image ESP_026450_2270). (e) Larger-scale view of panel (d) within the inter-crater plains. (f) *LCPs* (*black arrows*) and *HCPs* (*white arrows*) in close proximity (*HiRISE* image ESP_036366_2235). (g) Panel (f) at a larger scale. Note: these polygons are in Type 2 terrain (background: *THEMIS* day-*IR* controlled mosaic). North is up in all figures.

HiRISE image credits: NASA/JPL/University of Arizona. THEMIS image credit: NASA/JPL-Caltech/Arizona State University.

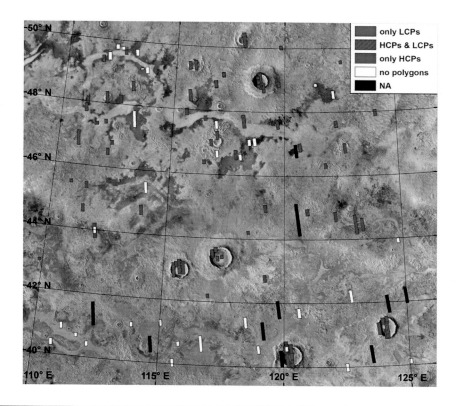

FIG. 5

Our study region is in Utopia Planitia (Lambert conformal conic-projection). The footprints of the 119 *HiRISE* images scanned for this study displayed with different colours according to the type(s) of polygons observed therein: *blue*, low-centred polygons (*LCPs*); *red*, high-centred polygons (*HCPs*); *blue/red strips*: *LCPs* and *HCPs*; *white*: no polygons; and, *black*: non-analysable image. Background: *MOLA* elevation data overlapping *THEMIS* daytime-*IR* mosaics; *MOLA* data credit: *MOLA* Science Team, *ASU*.

THEMIS image credit: NASA/JPL-Caltech/Arizona State University.

The *HiRISE* image footprints were gridded into reference squares of 250,000 m^2 (37,639 squares). Some squares were not fully within the *HiRISE* footprints. We excluded 4237 of those squares from our data base since they comprised less than half the area of a reference square, i.e. 125,000 m^2. A further 111 squares were discounted because they overlapped with data gaps in the *HiRISE* images. As such, 33,291 squares were evaluated.

If five or more candidates, *LCPs* or *HCPs* were observed within each square they were recorded. Grid squares also were recorded even if the polygons showed neither low nor high margins relative to their centres. We are aware of and acknowledge that our five-unit (count/no count) threshold is arbitrary but no more so than any other small unit threshold that could have been used to differentiate meaningful samples from meaningless ones.

For each *HiRISE* image, the number of grid squares containing *LCP* data points was divided by the number of grid squares containing *HCP* data points to calculate the *LCP/HCP* ratio of each

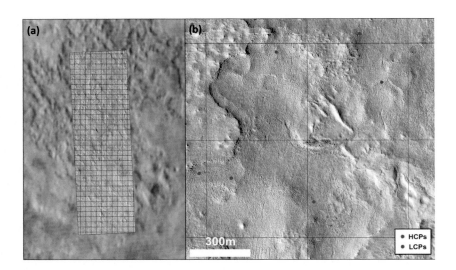

FIG. 6

(a) *HiRISE* image ESP_055038_2250 gridded by 500 × 500 m² (background: *MOLA* elevation data overlapping a *THEMIS* daytime-*IR* mosaic). (b) Zoom on four 500 × 500 m² that illustrate the mapping method. *Red* and *blue dots* respectively locate *LCPs* and *HCPs*. North is up in all figures.

HiRISE image credit: NASA/JPL/University of Arizona. MOLA data credit: MOLA Science Team, ASU. THEMIS image credit: NASA/JPL-Caltech/Arizona State University. Figure after Soare et al. (2021b).

image. Also, each scanned grid square was related to (a) its centre latitude; (b) the presence/absence of *LCPs* or *HCPs*; and (c) one of two terrain types (*Type 1* and *Type 2*) by the location of their centre point. *Type 1* comprises the whole study region exclusive of *Type 2*. The latter comprises the walls, rims, and interiors of craters ≥2 km (in diameter). These outputs were used to evaluate the possible correlation between polygon type (i.e. the *LCP/HCP* ratio) and latitude/terrain type using regression analysis. All of the *LCPs* or *HCPs* evaluated by us show similar changes of morphology with their poleward latitude.

3.4 Results and interpretation

The study area comprised 23,380 grid squares classified as terrain Type 1 and 9911 classified as terrain Type 2. We identified 313 squares with *LCPs* and no *HCPs*, 3325 squares with *HCPs* and no *LCPs*, 663 squares with both *LCPs* and *HCPs*, and 10,349 squares with polygons that display neither high nor low centres. This generates a total of 14,650 polygonised grid squares.

Overall, the ratio of low- to high-centred polygons does not show any correlation with increasing latitude ($R^2 \sim 0.0143$), despite the frequency of polygonised squares increasing with latitude. However, when polygons in Type 2 terrain are excluded from the sample data, a statistically significant linear correlation appears between the *LCP*-to-*HCP* ratio and latitude ($R^2 > 0.87$, *P-value* < 0.05) (Fig. 7). This exclusion is based on the assumption that crater interiors, often in shadow, are cold traps. The relatively

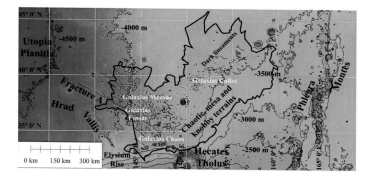

FIG. 9

Location of Galaxias Chaos, Galaxias Mensae, and Galaxias Colles in relation to Hecates Tholus, Hrad Vallis, Utopia Planitia, and Phlegra Montes (*THEMIS* Day IR Controlled Mosaic Cebrenia 30 N 120E 100 mpp with *MOLA MEGDR* 128 ppd. colour-shaded relief overlay, contours shown at 500 m intervals). *Heavy black line* bounding the cross-hatched area shows the extent of chaotic, mesa, and knobby terrains. Line work within this area shows locations of fractures, pitted cones, and broad ridges (a region characterised by dark lineaments is labelled). North is up.

THEMIS image credit: NASA/JPL-Caltech/Arizona State University. MOLA data credit: MOLA Science Team, ASU. Figure after Gallagher et al. (2015).

the pitted-cone assemblages and crosscutting fractures are focused topographically into a dominant axial fracture that occupies the lowest regional elevation between the Galaxias regions (Fig. 12). Longitudinally, the axial fracture develops into a distal channel with morphologies consistent with having been formed by liquid discharges. The spatial and topographic relationship between the pitted cone assemblages and the axial fracture-channel system suggest that groundwater or low-viscosity lava effluxes emerged from the pitted cone chains and cone ridges and were discharged through the axial fracture and, ultimately, along the distal channel.

On the basis of morphology alone, it is difficult to know whether the discharged liquid was water or lava. However, spectroscopic data indicate that the fractured mesa-like platforms in Galaxias Mensae, and those directly bounding Galaxias Fossae, contain phyllosilicate minerals produced by weathering in an aqueous environment (Fig. 13A) and outcrops of water ice (Fig. 13B). In addition, it is evident that material rich in phyllosilicates, and including some water ice, accumulated around the NW-SE striking fracture shown in Fig. 12 (Mustard et al., 2008; Pelkey, 2007). This indication of a water-rich eruption (that later froze) of liquefied material through the fracture from the substrate suggests strongly that the dissection of the Galaxias Mensae/Fossae region into its characteristic terrain of mesas crosscut by fractures was associated with the production of subsurface liquid water. Most probably this was in association with the geothermal disturbance of ground ice (Gallagher et al., 2018). The resulting scale of material loss was high ($\sim$22,000 km^3, average thickness, 77 m, across $\sim$285,000 km^2; Gallagher et al., 2018), particularly across the Galaxias Mensae-Colles surface unit that was inferred to be icy by Pedersen and Head (2010, 2011). Additionally, the loss of surface materials occurred within a geometric configuration that follows the directional fabric of pitted cones in the underlying unit. As such, it is reasonable to hypothesise that significant quantities of water would have been released as a paravolcanic by-product of the disturbance of ground ice, as suggested by the evidence in Fig. 13.

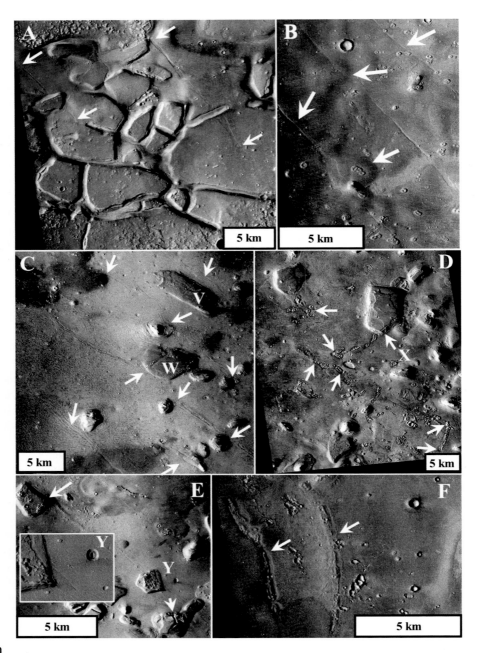

FIG. 10

See figure legend on opposite page.

The fractures in Galaxias Chaos and Galaxias Mensae-Colles are coherent in both strike and length with super-regional fracture systems associated with Elysium Mons' volcanism. This suggests that cryo-volcanic processes contributed to the production of the groundwater consonant with chaos-terrain formation across the region (Gallagher et al., 2018). The thermal effects of the same fracture system can be seen as far afield as Phlegra Montes. Here, the uplands are traversed by grabens that can be traced northwestward to fractures originating in the Galaxias region; they also extend for hundreds of kilometres southeastward, along a strike, beyond Phlegra Montes (Gallagher & Balme, 2015).

An assemblage of sinuous ridges occurs within one of the Phlegra Montes grabens, within a dead-ice zone close to the margins of an extant glacier, which also occupies the graben. The ridges could comprise an esker (Fig. 14). Were this the case it would be indicative of a warm/wet-based glacial thermal regime, rooted in the Amazonian Epoch and possibly associated with an elevated geothermal-heat flux along the graben (Gallagher & Balme, 2015; Gallagher et al., 2021). Although there is more widespread evidence of warm/wet-based processes in Phlegra Montes (Gallagher et al., 2021), the esker appears to be unique in the region. This reinforces the conclusion that its location could reflect an elevated geothermal-heat flux along the graben.

Overall, these observations are consistent with the effects of endogenic ground heating in regional-wide landscape development across Galaxias Chaos, Galaxias Mensae-Colles, and further afield in the same volcanic province during the Amazonian Epoch (Pedersen & Head, 2010). For Galaxias Chaos

FIG. 10

(a) Galaxias Chaos; chaotic terrain is pervaded by parallel sets of SE-NW fractures (*arrowed*) (*CTX* image B22_018260_2159_XN_35N213W). The fractures are thought to be the result of uplift and tensional adjustment across all of the Galaxias regions and could be associated with the volcanic up-doming of either Tharsis or Elysium Mons, possibly both (Carr, 1974; Hall et al., 1986; Wilson & Head, 2002). (b) In Galaxias Mensae, the plains terrain between mesas is patterned by regularly spaced, parallel, SE-NW fractures. Pitted cones chains (>2 cones in sequence; examples *arrowed*) are generally aligned parallel or orthogonal to the SE-NW fractures. Rounded hummocks are degraded mesas, often pierced by short chains of pitted cones (*CTX* image P17_007751_2150_XN_ 35N214W). (c) Galaxias Mensae; the parallel arrangement of SE_NW fractures is mirrored by the arrangement of free-standing mesas; arrowed mesa edges are located alongside SE-NW fractures; crosscutting (SW–NE) fractures pervade mesas at **V** and **W** (*CTX* image B17_016335_2187_XN_38N215W). (d) Galaxias Mensae; pitted cone chains with bilateral and orthogonal SE-NW/SW–NE arrangements associated with the regional crosscutting fracture sets; where a pitted cone chain terminates at a mesa, a fracture pervades the mesa directly along strike from the cone chain arrowed at **X**) (from *CTX* image B22_018260_2159_XN_35N213W). (e) Galaxias Mensae; a mesa edge directly constrained by an adjacent fracture (*arrowed* upper left corner of image) contrasts with a mesa (at **Y**) that is fractured along a strike offset from the fracture in the substrate; indicative of partial rotation of the mesa at **Y**, perhaps as a consequence of ground disturbance; a deeply pitted crater near **Y** is indicative of voids in the subsurface, perhaps contributory to local ground disturbance (*CTX* image B22_018260_2159_XN_35N213W). (f) Three broad ridges (aligned approximately N–S) showing variable emergence of summital pitted cones. Eastern ridge, no cones evident but summital thinning is apparent. The central ridge is fractured along its summit. Western ridge, a summital sequence of pitted cones fully emerged—the ridge is partially onlapped from the west by a darker cap of softened terrain (*CTX* image B18_016625_2179_XN_37N212W). North is up in all figures and panels.

Image credits: NASA/JPL/Arizona State University. Figure after Gallagher et al. (2018).

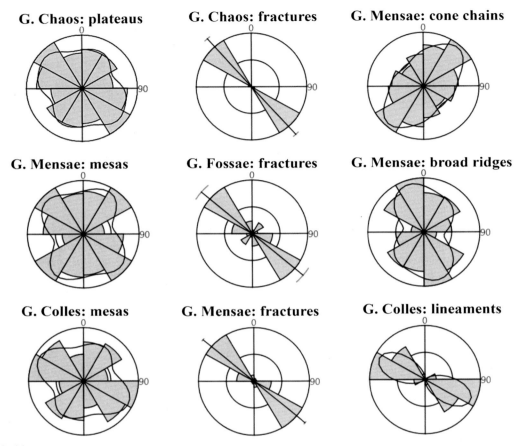

FIG. 11

Direction roses, showing orientation characteristics of plateaus/mesas, fractures, and pitted cone chains in the Galaxias regions. Left stack; chaos plateau and plateau/mesa edge axial orientation planes. Centre stack; primary fracture axial strike planes. Right stack; pitted cone chain, broad ridge, and dark lineament axial orientation planes. External arcs are the bootstrapped 95% confidence interval. The circular-kernel density estimate is shown as a *black* outline within the circles (not required for the fracture strike distributions, which are tightly unimodal).

Figure after Gallagher et al. (2018).

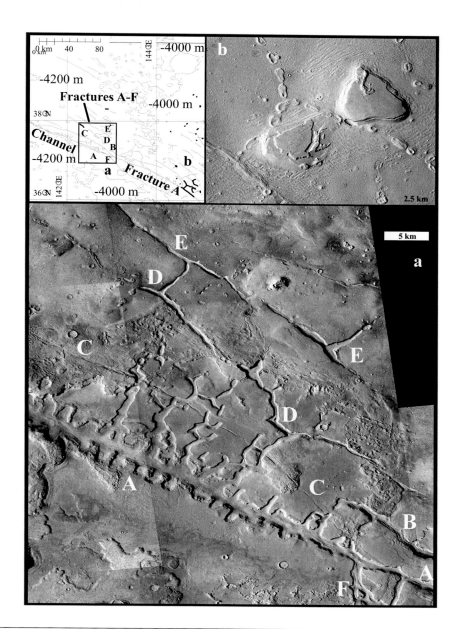

FIG. 12

Left upper panel (map); the spatial and topographic context of primary fractures (*grey lines*) in Galaxias Fossae, including Fracture **A** and associated fractures **B–F**. (a) Fractures **A–F**, detail showing the trellis patterned (orthogonal) dissection of terrain and consequent chasm-channel network development between fractures **A–E** (CTX B19_017179_2177_ XN_37N217W, P18_008173_2195_XN _39N217W and F20_043565_2156 _XN_35N21 7W). (b) Pitted cone chains and overlying mesas developed in the upper reaches of Fracture A. The pitted cones compartment the mesas and converge step-wise towards Fracture **A** by developing curving transitions from primary (SE-NW) to secondary (SW–NE) strike. Erosion of overlying mesas reveals variably mantled pitted cones; cone tops are visible as pits where mesas have been thinned, rather than completely eroded (*CTX* B21_017680_2179_XN_37N215W). North is up in all figures and panels.

Image credits NASA/JPL/Arizona State University. Figure after Gallagher et al. (2018).

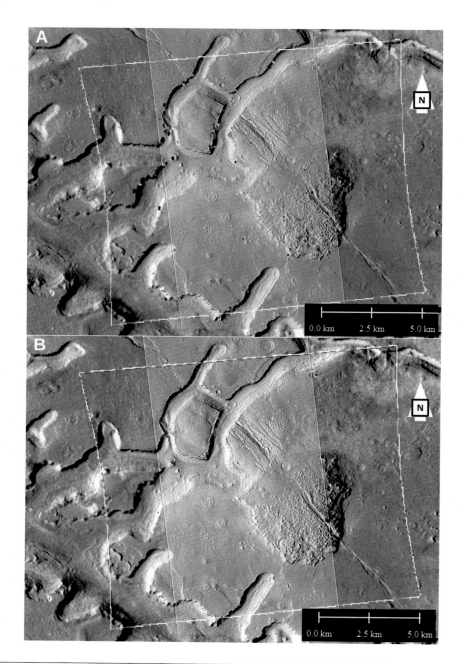

FIG. 13

Spectroscopic indications of water and water ice in Galaxias Fossae. (a) *CRISM* data from image HRS00018B25_07_IF176L (within the irregular *white* border): Fe/Mg phyllosilicates indicated in *red*; Al phyllosilicate or hydrated glass indicated in *green*; and hydrated sulphates, clays, glass, or water ice indicated in *blue*. (b) *CRISM* data from image HRS00018B25_07_IF176L (within the irregular *white* border): water ice or hydrated sulphates, clays, or glass indicated in *red*; water ice indicated in *green*; and CO_2 ice indicated in *blue*. In both panels, spectroscopic indications of water or water ice are located on steep faces and disturbed terrain, particularly the accumulation surrounding the fracture striking NW-SE across the image scene. Main background image: *CTX* B21_017746_2172_XN_37N 217W. Central image: *HiRISE* ESP_017746_2175.

CRISM image credit: NASA/JPL-Caltech/University of Arizona/JHUAPL. CTX image credit: NASA/JPL/Arizona State University. HiRISE image credit: NASA/JPL/University of Arizona.

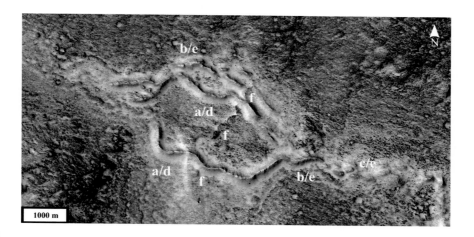

FIG. 14

Sinuous ridges revealed by the degradation of a glacial terminus in Phlegra Montes. The ridges may be continuous (a), discontinuous (b) or segmented (c), with crests that are sharp (d) or rounded (e). A possible transverse ice-margin remnant crosses the sinuous ridges (f). The contextual landform association of the ridges, and their morphological consilience with terrestrial analogues, suggest that the ridges are an esker complex. *HiRISE* ESP_044316_2130.

Image credit: NASA/JPL/University of Arizona. Figure after Gallagher and Balme (2015).

and Galaxias Mensae-Colles, there is no unequivocal answer to whether the channel-forming fluid involved was groundwater, generated by the thermal destabilisation of the icy deposits, or low-viscosity lava. However, it is likely that the degradation of the icy surface units, in which chaotic terrain, mesas, and knobs of Galaxias Chaos and Galaxias Mensae-Colles developed, was not a consequence of sublimation alone (Pedersen & Head, 2010, 2011).

Rather, the evident role of fracturing and cone emergence, both from the unit underlying the icy chaotic, mesa, and knobby surface, underscores two things. First, the potential of geothermally induced destabilisation of icy surface units to modify the landscape; and, second, the significance of cryovolcanic interactions in the cycling of water between the Martian surface and the atmosphere.

6.2 Endogenesis (II)

The margins of Lethe Valles in Elysium Planitia show polygonal to circular patterns of size-sorted clasts surrounding low centres, relative to the clasts (Fig. 15). The origin and development of these landforms are thought to be relatively recent, i.e. $\leq\sim 10$ Ma (Balme et al., 2009) and, as is the case with the evolution of the mesa-knobby terrains discussed above, need not be connected with variances in the spin axis of Mars or with the eccentricity of its orbit. Possible cold-climate analogues on Earth form from the differential rates of disturbance and movement between mixed coarse and fine clasts, over many cycles of liquid-water-based cryoturbation (Kessler & Werner, 2003).

The precursor moisture of the clastically sorted circles could be attributed to the large-scale (catastrophic) or mega-flooding in Lethe Vallis; for the sorted circles formed on the edge of a fluvial channel immediately upstream of a system of distributary channels and islands (Balme et al., 2011). This

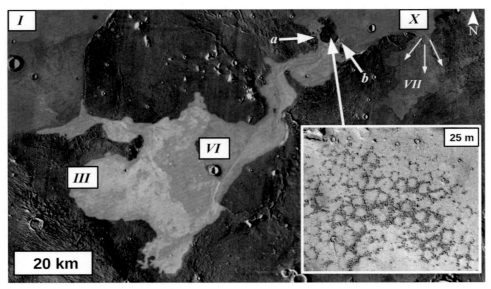

FIG. 15

Main image (*THEMIS* daytime thermal infrared): main sub-basins (by surface area) in the fill and spill sequence of the Lethe Vallis system, showing how the high stand of the floods through the system evolved over time (sub-basins *I*, *III*, *VI*, and *VII* shown). Sub-basin *X*, at the north-eastern corner of the scene, was filled before Lethe formed but overspilled from its southern margin during into sub-basin *VII*. Inset: *HiRISE* image sub-scene showing sorted clastic circles that developed at the margin of Lethe Vallis in deposits <10 Ma, based on crater counts *HiRISE* PSP-004072_1845. *THEMIS* image credit: *NASA/JPL*-Caltech/ Arizona State University.

HiRISE image credit: NASA/JPL/University of Arizona. Figure after Balme et al. (2011).

hydrological system seemingly evolved as a series of linked basins that filled, over-spilled, and then drained catastrophically into the next, topographically lower, basin. At least three such *fill and spill* events could have occurred (Balme et al., 2011) (Fig. 15).

In the head regions of Athabasca Vallis, near Mars' equator (Fig. 16), Balme and Gallagher (2009) observed morphological analogues of thermokarst basins (alases) on Earth and high-centred polygons (Fig. 17). The edges of the polygonised terrain are indented by amphitheatre-shaped retrogressive basins. Planimetrically, rather than having a continuously curving break of slope, the margins of these basins are facetted escarpments. Each escarpment facet is a free edge of a polygon exposed by the wastage of former bounding polygons. Hence, as adjacent polygons were wasted, the basins expanded, recording the loss of polygons through the faceting of each basin margin. Significantly, channels were incised into debris slumped at the mouths of the retrogressive basins, becoming extensive contributary networks that converge in shallow basins crosscut by the Cerberus Fossae fracture system. On Earth, basins that develop from the retrogressive wastage of polygonised terrain and also involve both the production of thaw water and the slumping of liquefied debris are definitive indicators of periglacial conditions characterised by freeze–thaw oscillations.

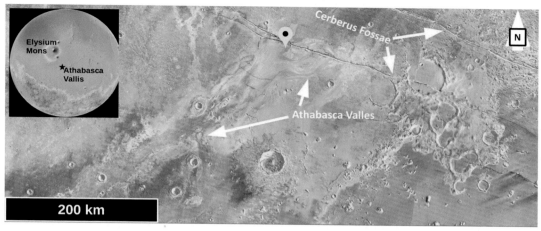

FIG. 16

Athabasca Valles and Cerberus Fossae, context. The location of the periglacial landform assemblage is marked by the map pin symbol. *THEMIS* Daytime-*IR* image with *USGS MOLA-HRSC* 200 m elevation model *DEM*.

THEMIS image credit: NASA/JPL-Caltech/Arizona State University. MOLA-HRSC DEM credit NASA/JPL/USGS/ESA.

The Cerberus Fossae are thought to have formed when an aquifer was breached by faulting and dike emplacement. This resulted in the catastrophic discharge of flood water and the formation of streamlined, fluvial-like landforms downstream (e.g. Burr et al., 2002, 2005; Hanna & Phillips, 2006; Vetterlein & Roberts, 2009), as well as providing the precursor moisture for the emplacement of ground ice and, ultimately, the development of the periglacial landforms.

In the same small region, the margins of the alas-like basins also appear to have been eroded by retrogressive thaw slumps, dependent on the production of meltwater (Fig. 18). These retrogressive slumps are overprinted by high-centred (possibly ice wedge) polygons and pingo-like mounds (Balme & Gallagher, 2009; also, Burr et al., 2005 in the case of the latter) (Fig. 18).

The high-centred polygons on the slumps must represent a second generation of polygonisation, contingent on the liquefaction, and then freezing of debris produced by the wastage of the earlier generation of polygonised terrain in which the alases developed. The presence of pingo-like mounds supports this interpretation.

On Earth, pingos originate and develop as permafrost aggradation entraps and constrains pore water beneath recently frozen lakebeds (e.g. Mackay, 1998). This leads to the positive deformation of the overburden and, eventually, as the permafrost grades through the trapped porewater, to the congealing of a permanent ice core (e.g. Mackay, 1998).

Collectively, the spatial association of the possible periglacial landforms points to a geomorphological regime that requires the highly iterative freeze–thaw cycling of liquid water and surface/near-surface conditions sufficiently warm to maintain these cycles, even if only intermittently. Modelled climate predictions for the Elysium and Athabasca regions are inconsistent with these boundary conditions (Balme et al., 2009) and, as with the Galaxias terrains discussed antecedently, warming boundaries

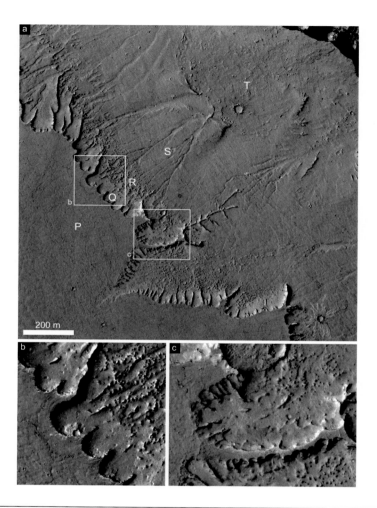

FIG. 17

Retrogressive scarps with cuspate niches, long *tuning fork* spurs and associated fluvial-like channels that form distinctive dendritic networks in Athabasca Valles. (a) Surface types and sequence: **P**, the upper surface, characterised by domed polygons occurring throughout this area; **Q**, a smooth-textured surface, pitted by small (up to 5 m) circular depressions, fronting the retrogressively indented scarp bounding surface P (detail in panel b); **R**, dominated by blocks, up to a few metres wide, and dendritic networks of channels that emerge from within the blocky regions; **S**, a rutted, polygonally patterned surface, incised by networks of channels that emerged from the blocky regions and widen beyond tributary junctions; **T**, the shallow terminal basin of the channels has a hummocky floor and is bounded by discontinuous curving ridges. The two white boxes indicate the extent of (b) and (c). (b) Close-up view of cuspate niches showing smooth, pitted terrain at the base of the scarp, trending into blocky material further downslope. The tributary channel system originates at the boundary between the smooth pitted terrain and the blocky material. (c) Close-up view of a medium-sized retrogressive embayment containing higher-order niches. North is up in all figures. HiRISE image PSP_009280_ 1905.

Image credit: NASA/JPL/University of Arizona. Figure after Balme and Gallagher (2009).

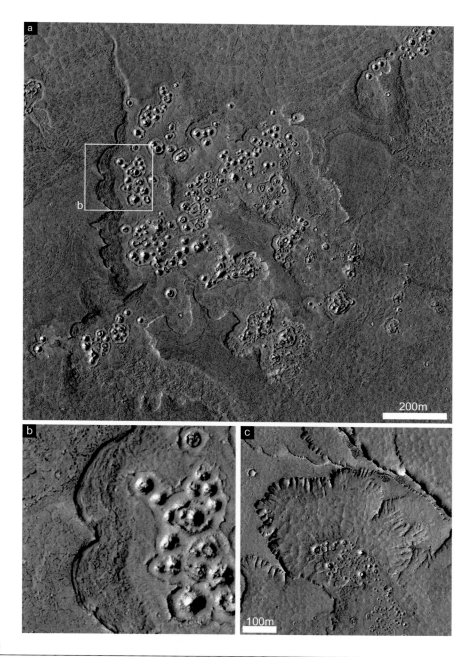

FIG. 18

Athabasca Valles; shallow basins with retrogressive margins and containing mound and cone landforms with morphologies and contextual relationships consistent with pingos. (a) A large (~1 km), multi-part basin in polygonally patterned ground. (b) Close-up view, marked by a white box, of subsidence failure of part of the basin wall. And basin floor is pierced by pingo-like landforms; the latter is characterised by mound and cone structures with surrounding moats. (c) Smaller, open basin containing mound and cone landforms. The basin periphery is itself indented by smaller basins with the same scale as the polygons on the upper surface. The population of domed polygons inside the basin postdate the retrogressive erosion of the basin into the upper polygonised surface. Image sub-scenes: panels (a) and (b) (*HiRISE* PSP_009280_1905); panel (c) (*HiRISE* PSP_007843_1905). North is up in all figures.

Image credits: NASA/JPL/University of Arizona. Figure after Balme and Gallagher (2009).

induced by volcanism could have been at work, here. They also may be responsible for the origin and development of the proposed periglacial complexes.

7 Periglacial landscapes: relatively recent, ancient, or both?

Typically, the crater-based age dating of possible near-surface ground ice or of polygonised terrain at the Martian mid to high latitudes returns ages that are extremely young ($\leq$ ~0.1 Ma, Mustard et al., 2001; ~0.1–1 Ma, Levy et al., 2009b; ~0.5–2 Ma, Gallagher et al., 2011; ~1–2 Ma, Levy et al., 2009a; ~0.4–2.1 Ma, Sinha & Murty, 2015; ~1.8 Ma, Viola, 2021).

Interpreting *crater-size frequency distributions* (*CSFDs*) on such terrains is challenging. Modification of the surface can obfuscate the true production crater population upon which model ages are derived. Derived model ages from crater count on terrains that have experienced complex histories, as many of the periglacial landscapes on Mars have, may not represent a formation age of a geologic unit or feature. Rather, the crater population represents a temporal marker identifying the period of time since an episode of erosion, deposition, or exhumation (Hartmann, 1971; Öpik, 1966). Thus, the derived age is commonly referred to as a *crater-retention age* and represents the total time that craters have accumulated and been retained for a given size (Hartmann, 1966, 2005). The rate of crater destruction is typically size dependent and thus will alter the power-law slope of the *CSFD* on a log–log plot with the preferential loss of smaller diameter craters resulting in a shallower slope (Michael & Neukum, 2010; Rubanenko et al., 2021). Higher obliteration rates of craters result in larger uncertainties, especially for count areas of limited extent (Palucis et al., 2020).

In addition to crater loss and destruction, other factors complicate the interpretation of *CSFDs*. There is an ongoing debate on whether the impact rate has been relatively constant over the last ~3 Gyr (e.g. Williams et al., 2018; Mazrouei et al., 2019), as is assumed in the absolute model ages (Hartmann, 2005; Neukum et al., 2001), or whether the recently observed new impacts on the Moon and Mars are representative of past rates (e.g. Daubar et al., 2013; Speyerer et al., 2016). Additionally, the role that secondary cratering plays in shaping the production function is not entirely clear (e.g. Bierhaus et al., 2018; Hartmann & Daubar, 2017; Powell et al., 2021; Williams, 2018).

In spite of these challenges, crater counts still have the potential to provide insightful information on ages and often can generate at least lower bounds on possible crater-retention ages.

Crater counts of possible periglacial terrain populated with polygons and polygonised thermokarst-like depressions at the mid-latitudes of Utopia Planitia as well as at/near the Mars dichotomy in Protonilus Mensae and the Moreux impact crater, have generated minimum retention ages that span three or more orders of magnitude (Soare et al., 2020, 2021a, 2021c) (Fig. 19).

For example, crater counts of terrain in Utopia Planitia dotted with polygonised but clastically non-sorted terrain and thermokarst-like depressions showed a shallower power-law slope than predicted by production functions. Though an absolute age could not be established, the number of craters larger than 50 m suggested the material was emplaced and exposed to impact cratering for at least ~10 Myr. Crater counts of similar terrain in western Utopia Planitia, based on observed ~100 m diameter craters suggested that the terrain was no younger than ~100 Ma (Soare et al., 2020). The heavily degraded appearance of many of the craters and the shallow power-law slope of the *CSFD* were consistent with a substantial loss of smaller craters. Rather than reflect a production function, the *CSFD* likely was the

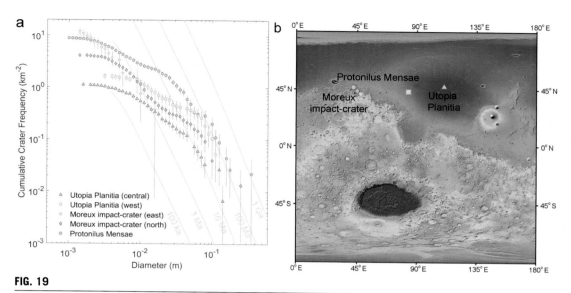

FIG. 19

(a) Compilation of *CSFDs* from crater counts conducted on the polygonised, periglacial terrains in Utopia Planitia (Soare et al., 2020, 2021b), around the Moreux impact crater (Soare et al., 2021a, 2021c), and within Protonilius Mensae (Soare et al., 2022). The Moreux impact-crater (north) *CSFD* represents the combined counts of the three areas from Soare et al. (2021c). Labelled absolute model age isochrons are from Hartmann (2005). (b) *MOLA* shaded relief of the eastern hemisphere with markers showing the location of the crater count studies shown in panel (a).

MOLA data credit: MOLA Science Team, Arizona State University.

result of a crater population that is in an equilibrium regulated by a balance of formation and destruction. Thus, an absolute model age could not be determined, however, an age of ≤100 Ma was deemed unlikely.

The crater counts of the polygonised (and clastically non-sorted) terrain in the vicinity of the Moreux impact crater also show a wide range of ages. For example, on the eastern margin of the crater, the *CSFD* overlapped a model age of ~1 Ma (Soare et al., 2021c). Crater counts conducted by Soare et al. (2021a) north of the Moreux impact crater yielded *CSFDs* that displayed a broader range of model ages. *CSFDs* overlapped isochrons ~1 Ma at the small diameters while extending up to the ~500 Ma isochron at the largest diameters. More recently, counted craters amidst (clastically sorted) polygonised terrain in Protonilius Mensae showed that the terrain was unlikely to have been exposed to impact cratering more recently than ~100 Ma and, possibly up to ~1 Ga (Soare et al., 2022).

Collectively, these crater counts comprise two important findings. First, the temporal span of possible periglaciation may be much broader than had been thought, reaching back at least into the mid-Amazonian Epoch and, maybe, even further than that. Second, each of the zones studied by us exhibited a stratigraphical alternation or intertwining of possible glacial and periglacial landforms, features, and traits. This cyclicity would not be unexpected on Earth and, perhaps, ought not to be unexpected on Mars.

8 Cold-room (permafrost/active layer) experiments

8.1 Background

Cold rooms comprise temperature-controlled (experimental) chambers or labs. They are used by planetary scientists to replicate or constrain cold-climate conditions thought to be relevant to the study of landscape-feature evolution, including off-Earth landscapes, and the geomorphological/geological processes that initiate or underlie these landscapes. In two cold rooms in France—the *GEOPS* laboratory (University of Paris-Saclay) and the M2C laboratory (University of Caen) (Fig. 20), the possible formation mechanisms of gullies observed on large sand dunes on Mars such as the ones observed at the Russell Crater (Figs 21 and 22) were explored by way of debris-flow simulations (Védie et al., 2008; Jouannic et al., 2015) (Figs 23 and 24).

FIG. 20

The cold room (University of Caen/*CNRS* France). Close-up of the experimental setup used for the physical modelling of the debris flows. Experiments were carried out using silty materials that had been saturated in water and frozen in a cold room. A controlled active layer was simulated by thawing the upper layer just before the experiment. A controlled flow rate was simulated by originating the latter at the top of the inclined substrate.

Image credit: E. Vedié.

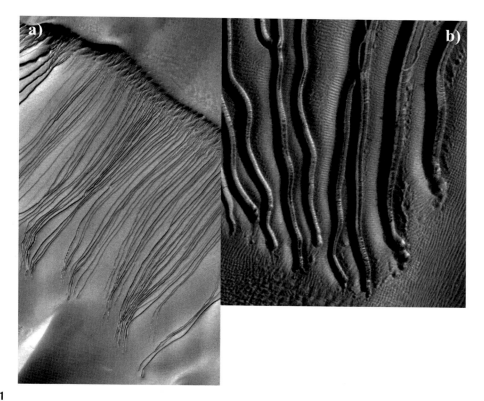

FIG. 21

(a) Linear gullies over a sand dune in the Russell impact crater (*HiRISE* image PSP_002904_1255). The image is 2 km wide. (b) Levees and sinuosities along linear gullies in their distal parts. Terminal deposits do not show terminal lobes, but rather a concentration of small pits of unknown origin (*MOC* E02–00893). The image is 0.5 km wide. North is up in each of the figures.

HiRISE image credit: NASA/JPL/University of Arizona. MOC image credit: NASA/JPL/MSSS.

8.2 Experiments and results

In order to replicate a cold-climate (permafrost) environment on Mars that mirrors the Earth, air temperatures in the cold room were kept below 0 °C and sediment temperatures were − 10 °C during the multiple experiments. Atmospheric pressure was not varied from current terrestrial boundaries, as water is assumed to have been stable or metastable for a sufficiently long period of time on Mars where debris flows of the type being simulated are observed (Védie et al., 2008; Jouannic et al., 2015). The spatial variance between the modelled sedimentary environment (discussed below) and Martian terrain is acknowledged and largely unavoidable in all scaled-down simulations. With regard to the reduced gravity conditions on Mars, the effect of water content in the thawed layer on the flow geometry has been adjusted (gravity on Earth = 9.81 m s^{-2}; gravity on Mars = 3.71 m s^{-2}) (Jouannic et al., 2015).

Field studies in Greenland have shown that some debris flows may be enhanced by the thawing of surface snow, near-surface ground ice, or the active layer (Costard et al., 2002, 2007; Védie et al., 2008;

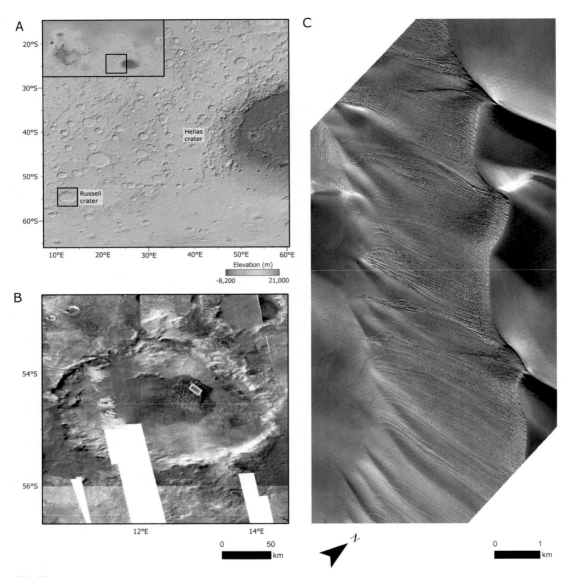

FIG. 22

(a) The location of the Russell impact crater, west of the Hellas impact basin. The extent of panel B is shown by a block box. Background colour represents *MOLA* global elevation (Zuber et al., 1992) in an equirectangular projection. *MOLA* data credit: *MOLA* Science Team, Arizona State University. (b) *CTX* mosaic (see Dickson et al., 2018) of the Russell crater; the dune field is visible in the centre of the image. The *yellow box* shows the extent of panel (c). The coverage of *CTX* images used to generate this mosaic is incomplete in this area; data gaps are shown in *white*. (c) *HiRISE* image ESP_018872_1255 highlights the large SSW-facing mega dune slope within the Russell crater field. This slope also is a monitor site for seasonal changes (Reiss et al., 2010). North is up in figures (a) and (b).

CTX image credits NASA/JPL/Arizona State University. HiRISE image credit: NASA/JPL/University of Arizona.

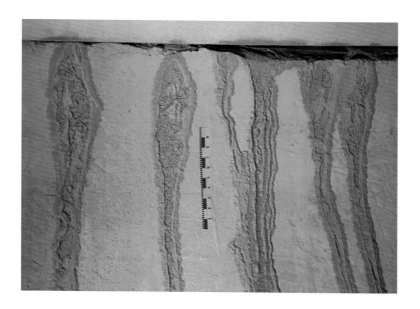

FIG. 23

Frozen ground with silty materials saturated with water but overlain by a thin active layer (few millimetres). Note narrow gullies with lateral levees. This morphology resembles the gullies on the Russell impact-crate dunes.

Image credit: E. Vedié.

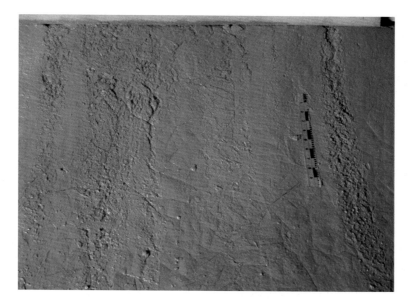

FIG. 24

Same conditions and simulation type as in the experiment exhibited in Fig. 20, absent of an active layer, however. The absence of an active layer induces a sheet flood without any incision.

Image credit: E. Vedié.

see French & Koster, 1988, for the originating hypothesis). Between 2007 and 2015, more than 80 cold-room experiments were carried out to study whether active-layer thaw could contribute water meaningfully to debris flows. The experiments comprised a rectangular box/bed 2.5 m by 0.5 m wide and 0.2 m deep. Disparate flow morphologies were tested, i.e. flows with a thin snow cover, or not, over the surface, as were different slope angles and disparate sediments whose grain size varied from silts (Védie et al., 2008) to fine sands (Jouannic et al., 2015). With regard to the latter, the median slope gradient was 15°; the top and bottom slope gradients were constant (8° and 50° respectively). The same volume of water supply was used regardless of the sediment type. Air and sediment temperatures also varied, to allow for the presence of an active layer.

Regardless of the sediment type, each test sediment was saturated with water and frozen. This created ice-rich permafrost for the experiments. Once frozen, the upper layer of the bedded permafrost was defrosted progressively (from a few mm to 2 cm of depth), creating disparate active-layer depths. Water was then introduced from an external reservoir at the top of the frozen bed to simulate a point source. On Earth, debris flows may occur and/or be enhanced if and when near-surface sediments become saturated with water derived of snow melt and/or of thawed ground ice.

The experiments were carried out by depositing a measured amount of water at a controlled flow rate from a point source at the top of the inclined substrate. This created a small debris flow that propagates downslope and terminates shortly after the flow of water at the top of the substrate is halted. The result of each experiment is a well-defined debris-flow landform that can include an eroded channel, levees, and a blunt lobe-like debris apron. The final morphology generated by each run was precisely measured using a laser scanner. During the experimental program, the following parameters were varied to explore the parameter space: discharge, slope angle, water content of the thawed layer, and thickness of the thawed layer.

The first experiments attempted to model debris flows in ice-rich permafrost, i.e. frozen ground saturated with water, with a thin active layer (Figs 23 and 24). The presence of an active layer was assumed to facilitate erosion and generate meltwater that would promote the debris-flow process. Active-layer formation took place by means of a controlled rise of air and surface temperatures to and above 0°C immediately before the experiments began. This created a frozen bed with a moist surface layer that can be inclined at various angles to produce an analogue for the slip face of a Martian or terrestrial dune.

Preliminary laboratory simulations showed that linear debris flows are best reproduced with the thawing of near-surface ground ice, mostly with silty materials and a low slope angle (10°–20°) (Védie et al., 2008). In fact, an active layer of less than 5 mm depth was sufficient to simulate narrow gullies with lateral levees and relatively small terminal lobes (Védie et al., 2008).

Later simulations showed the influence of active-layer water, previously frozen within a frozen substrate made of relatively fine-grained sand, on flow length (Jouannic et al., 2015). The water from the saturated active layer is incorporated progressively into the flow over time. This reduces flow viscosity and increases the runout length for a given initial volume of water. For example, an increase in the water content of the active layer by a factor of 4 resulted in an increase in the length of the flow by a factor of 3 (Jouannic et al., 2015). Furthermore, the presence of high water content in the thawed layer reduces infiltration of the flow into it and possibly reduces friction at the base of the flow.

When associated with the presence of silty debris material, a water saturated and relatively thin active layer could be helpful in explaining the long, linear morphology of some Martian dune gullies. Moreover, the cold-room simulations suggest that ice-rich albeit relatively thin active layers—thin active layers being a proxy for low mean temperatures—would require neither substantial nor sustained rises of temperature above 0°C to form, especially were brines also present at or near the surface.

9 The periodicity of periglacial/glacial cycles, on Mars as on Earth?

The cyclical connectivity of glacial and interglacial or deglacial periods on Earth, particularly in the Quaternary period, is fundamental to understanding the geological record of Earth through the last ~2.5 million years; no less so, perhaps, with regard to much earlier cycles (e.g. French, 2017; Rampton, 1988; Washburn, 1973; also, Gallagher, 2024, pp. 31–72). In particular, as the Earth periodically has swung away from glaciation, the opportunities for meltwater and thaw to influence the evolution of deglaciated surfaces and near-surfaces by the work of freeze–thaw cycling have been substantial at poleward (e.g. French, 2017; Murton, 2021; Rampton, 1988; Washburn, 1973) and non-poleward latitudes (e.g. Ewing et al., 2014; Williams, 1975, 2008).

Here, in this section, Protonilus Mensae (*PM*) (43–49° N, 37–59° E) is the focus of our attention. *PM* lies in the Ismenius Lacus quadrangle. *PM* is wedged between the Lyot impact crater to the northwest, the Moreux impact crater to the southeast, and the Mars crustal dichotomy to the south (Fig. 25). The latter is a global geological boundary that separates the ancient southern highlands from the younger northern lowlands.

We use *PM* as a geochronological test bed for evaluating the possible periodicity of periglacial/glacial cycles on Mars from close to the present day through too much earlier periods of the Late

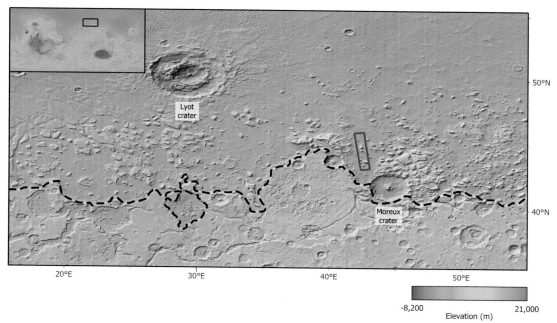

FIG. 25

The geographical footprint of our study area in the Protonilus-Deuteronilus Mensae region of Mars. The *black serrated line* highlights the Mars crustal dichotomy and the proximity of our footprint to it. Background colour comprises *MOLA* global elevation (Zuber et al., 1992) in an equirectangular projection.

MOLA data credit: MOLA Science Team, Arizona State University.

Amazonian Epoch (also see Soare et al., 2021a, 2021c). In particular, we study a sub-region within *PM* that straddles the geological contact between *[eHT]*, an early Hesperian Epoch transition unit and *[HNt]*, and a Noachian-Hesperian Epoch transition unit (Tanaka et al., 2014) (Fig. 26). The two units are distinguished by differences of surface tone and texture (Figs 26 and 28a,c, and e) as well as topography (Fig. 27).

Unit *eHt* exhibits a relatively dark tone (Figs 26 and 28a and e). As seen in *HiRISE* image ESP_028457_2255, the unit is punctuated continuously and densely by landforms whose shape, size, apparent size sorting of clasts, and metre-scale elevation differences between margins and centres (Fig. 28a) are akin to clastically sorted polygons/circles (*CSPs*) on Earth (e.g. Barrett et al., 2017, 2018; Gallagher et al., 2011; Soare et al., 2016) (Fig. 28b). In terrestrial permafrost environments, clastically sorted polygons/circles form if (e.g. French, 2017; Washburn, 1973):

(1) liquid water is proximal to the surface and available, even if only briefly, for geomorphological work;

(2) an admixture of relatively coarse and fine-grained clasts are present to metres of depth; and

(3) the freeze–thaw cycling of water, even if only periodically and for relatively short periods of time, occurs to that depth.

Unit *eHt* is relatively flat albeit broken up and discontinuous, particularly on the northern and eastern margins (Figs 26 and 27a). At one location, the elevation of the units drops by ~70 m (Fig. 27a). This could be the result of volatile-based degradation and erosion and may be a marker of unit *HNt*'s floor (Fig. 27a and b).

The surface of unit *HNt* is relatively light toned and fragmented, with the latter possibly being the work of devolatilisation and erosion, as is suggested for unit *eHt* (Soare et al., 2022). The distribution of the surface material is constrained by the contours of a massifs-centred basin (Figs 26–28a). Massif summits rise ~2.3 km above the surface material (Fig. 27).

Landscape features and landforms that show similarities of shape, size, and scale, as well as closely knit spatial association, all of which are akin to periglacial (Fig. 28c, **also see** Fig. 1a) and glacial landscapes (Fig. 28e, **also see** Fig. 28d) on Earth, are discerned within and amidst the surface material/terrain of this unit.

Age estimates of unit *eHt* derived of the crater-size frequency distribution are a minimum of ~100 Ma and a maximum of ~1 Ga (**see** Fig. 19 **and Section 7**). If the age of the candidate clastically sorted polygons is comparable to that of the terrain incised by them, then they would be one and perhaps two orders of magnitude older than any of the other Martian polygons whose ages have been estimated by *CSFDs* (e.g. Gallagher et al., 2011; Levy et al., 2009a, 2009b; Milliken et al., 2003; Mustard et al., 2001; Sinha & Murty, 2015).

Age estimates of unit *HNt*'s surface derived of crater-size frequency distribution are a minimum of ~10 Ma and a maximum of ~100 Ma (Soare et al., 2022). Depending upon whether the clastically non-sorted polygons and polygonised thermokarst-like assemblages are exhumed or not, they pre- or postdate the formation of unit *NHt*'s observed surface. If the latter, then this places their age well within the mean and youthful estimates of age for similar terrain at the mid to high latitudes of the northern plains (e.g. Mustard et al., 2001; Milliken et al., 2003; Levy et al., 2009a, 2009b; Gallagher et al., 2011; Sinha & Murty, 2015). If the former, then they too could be an order of magnitude older than any polygons of a similar type reported in the literature.

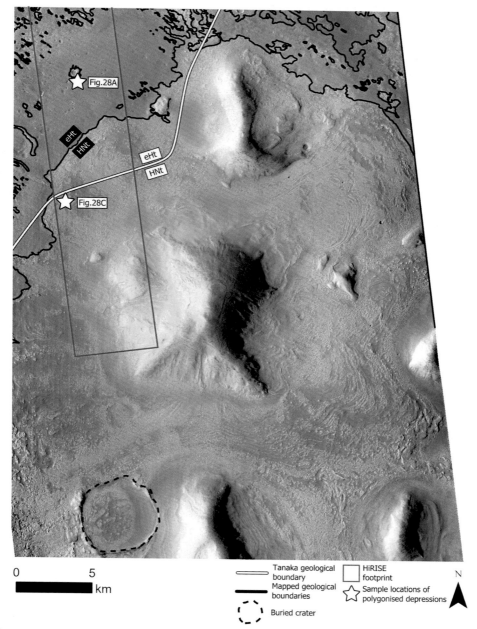

FIG. 26

Context image (*CTX* image F21_044083_2248_ XI_44N317W) of geological units *eHt* and *HNt* in the *PM* region. The two units are separated by a geological contact first identified by Tanaka et al. (2014) and refined, here, using *CTX* imagery. The *white line* coincides with Tanaka's original boundary; the *black line* marks the updated contact. Age estimates of the large crater (*serrated circle*) suggest that it intercepts the floor of *HNt* at depth (*CTX* image F21_044083_2248_ XI_44N317W) (Soare et al., 2022). The *red rectangle* represents the footprint of *HiRISE* image ESP_028457_2255. *Stars* mark the location of some candidate clastically sorted polygons/circles in unit *eHt* (see Fig. 28a) and polygonised but not clastically sorted thermokarst-like depressions in unit *HNt* (see Fig. 28c). The two locations are mapped using *ArcGIS* pro with units distinguished according to the systemic variations in surface texture visible at a 1:10,000 scale.

CTX image credit: NASA/JPL/Arizona State University. HiRISE image credit: NASA/JPL/University of Arizona.

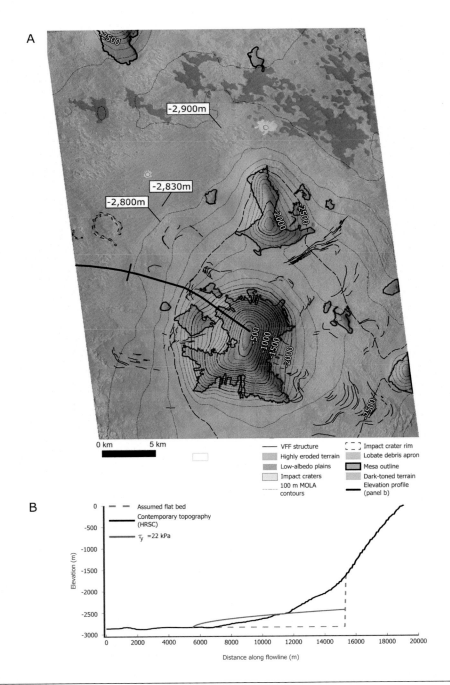

FIG. 27

See figure legend on opposite page.

Chevrier, V. F., Hanley, J., & Altheide, T. S. (2009). Stability of perchlorate hydrates and their liquid solutions at the Phoenix landing site, Mars. *Geophysical Research Letters*, *36*(L10202). https://doi.org/10.1029/2009GL037497.

Chevrier, V. F., Rivera-Valentin, E. G., Soto, A., & Altheide, T. S. (2020). Global temporal and geographic stability of brines on present-day Mars. *The Planetary Science Journal*, *1*, 64. 12pp https://doi.org/10.3847/PSJ/abbc14.

Costard, F., Forget, F., Jomelli, V., & Mangold, N. (2007). Debris flows in Greenland and on Mars. In M. Chapman (Ed.), *The geology of Mars: Evidence from earth-based analogs* (pp. 265–278). Cambridge: Cambridge University Press. https://doi.org/10.1017/CBO9780511536014. 011.

Costard, F., Forget, F., Mangold, N., & Peulvast, J. P. (2002). Formation of recent Martian debris flows by melting of near-surface ground ice at high obliquity. *Science*, *295*, 110–113. https://doi.org/10.1126/science.1066698.

Costard, F. M., & Kargel, J. S. (1995). Outwash plains and thermokarst on Mars. *Icarus*, *114*(1), 93–112. https://doi.org/10.1006/icar.1995.1046.

Costard, F., Séjourné, A., Kargel, J., & Godin, E. (2016). Modeling and observational occurrences of near-surface drainage in Utopia Planitia, Mars. *Geomorphology*, *275*, 80–89. https://doi.org/10.1016/j.geomorph.2016.09.034.

Czudek, T., & Demek, J. (1970). Thermokarst in Siberia and its influence on the development of lowland relief. *Quaternary Research*, *1*, 103–120.

Daubar, I. J., McEwen, A. S., Byrne, S., Kennedy, M. R., & Ivanov, B. (2013). The current Martian cratering rate. *Icarus*, *225*(1), 506–516. https://doi.org/10.1016/j.icarus.2013.04.009.

De Leffingwell, E. (1915). The dominant form of ground-ice on the north coast of Alaska. *Journal of Geology*, *23*(7), 635–654.

Dickson, J. L. J. L., Kerber, L. A., Fassett, C. I., & Ehlman, B. L. (2018). A global, blended CTX mosaic of Mars with vectorized seam mapping: A new mosaicking pipeline using principles of non-destructive image editing. In 49th Lunar and Planetary Science Conference, Houston, Texas.

Dundas, C. M. (2017). Effects of varying obliquity on Martian sublimation thermokarst landforms. *Icarus*, *281*(1), 115–120. https://doi.org/10.1016/j.icarus.2016.08.031.

Dundas, C. M., Bramson, A. M., Ojha, L., Wray, J. J., et al. (2018). Exposed subsurface ice sheets in the Martian mid-latitudes. *Science*, *359*, 199–201. https://doi.org/10.1126/science.aao1619.

Dundas, C. M., Byrne, S., & McEwen, A. S. (2015). Modeling the development of Martian sublimation thermokarst landforms. *Icarus*, *262*, 154–169. https://doi.org/10.1016/j.icarus.2015.07.033.

Dundas, C. M., Mellon, M. T., Conway, S. J., Daubar, I. J., et al. (2021). Widespread exposures of extensive clean shallow ice in the midlatitudes of Mars. *Journal of Geophysical Research*, *126*, e06617. https://doi.org/10.1029/2020JE006617.

Ewing, R. C., Eisenman, I., Lamb, M. P., Poppick, L., Maloof, A. C., & Fischer, W. W. (2014). New constraints on equatorial temperatures during a late Neoproterozoic snowball Earth glaciation. *Earth and Planetary Science Letters*, *406*, 110–122. https://doi.org/10.1016/j.epsl.2014.09.017.

Farquharson, L. M., Romanovsky, V. E., Cable, W. L., Walker, D. A., Kokelj, S. V., & Nicolsky, D. (2019). Climate change drives widespread and rapid thermokarst development in very cold permafrost in the Canadian high arctic. *Geophysical Research Letters*, 6681–6689. https://doi.org/10.1029/2019GL082187.

Feldman, W. C., et al. (2002). Global distribution of neutrons from Mars: Results from Mars odyssey. *Science*, *297*, 75. https://doi.org/10.1126/science.1073541.

Forget, F., Haberle, R. M., Montmessin, D., Levrard, B., & Head, J. W. (2006). Formation of glaciers on Mars by atmospheric precipitation at high obliquity. *Nature*, *311*(5759), 368–371. https://doi.org/10.1126/science.1120335.

French, H. M. (2017). *The periglacial environment* (4th ed., p. 544). West Sussex, England: J. Wiley & Sons.

French, H. M., & Koster, E. A. (1988). The periglacial phenomena: Ancient and modern. *Journal of Quaternary Science*, *3*(1), 110.

Gallagher, C. (2024). *Glaciation and glacigenic geomorphology on Earth in the Quaternary Period.* In R. J. Soare, J.-P. Williams, C. Ahrens, F. E. G. Butcher, & M. R. El-Maarry (Eds.), *Ices in the solar system, a volatile-driven journey from the inner solar system to its far reaches.* Elsevier Books.

Gallagher, C., & Balme, M. R. (2015). Eskers in a complete, wet-based glacial system in the Phlegra Montes region Mars. *Earth and Planetary Science Letters, 431,* 96–109. https://doi.org/10.1016/j.epsl.2015.09.023.

Gallagher, C. J., Balme, M. R., Soare, R. J., & Conway, S. J. (2018). Formation and degradation of chaotic terrain in the Galaxias regions of Mars: Implications for near-surface storage of ice. *Icarus, 309,* 69–83. https://doi.org/10.1016/j.icarus.2018.03.002.

Gallagher, C. J., Butcher, F. E. G., Balme, M., Smith, I., & Arnold, N. (2021). Landforms indicative of regional warm based glaciation, Phlegra Montes, Mars. *Icarus, 355,* 114173. https://doi.org/10.1016/j.icarus.2020.114173.

Gallagher, C., Balme, M. R., Conway, S. J., & Grindrod, P. M. (2011). Sorted clastic stripes, lobes and associated gullies in high-latitude craters on Mars: Landforms indicative of very recent, polycyclic ground-ice thaw and liquid flows. *Icarus, 211,* 458–471. https://doi.org/10.1016/j.icarus.2010.09.010.

Gough, R. V., Chevrier, V. F., Baustian, K. J., Wise, M. E., & Tolbert, M. A. (2011). Laboratory studies of perchlorate phase transitions: Support for metastable aqueous perchlorate solutions on Mars. *Earth and Planetary Science Letters, 312,* 371–377. https://doi.org/10.1016/j.epsl.2011.10.026.

Grau-Galofre, A., Jellinek, A. M., & Osinski, G. R. (2020). Valley formation on early Mars by subglacial and fluvial erosion. *Nature Geoscience.* https://doi.org/10.1038/s41561-020-0618-x.

Grau-Galofre, A., Lasue, J., & Scanlon, K. (2024). *Ice on Noachian and Hesperian Mars: Atmospheric, surface, and subsurface processes.* In R. J. Soare, J.-P. Williams, C. Ahrens, F. E. G. Butcher, & M. R. El-Maarry (Eds.), *Ices in the solar system, a volatile-driven journey from the inner solar system to its far reaches.* Elsevier Books.

Grimm, R. E., Stillman, D. E., Dec, S. F., & Bullock, M. (2008). Low frequency electrical properties of polycrystalline saline ice and salt hydrates. *The Journal of Physical Chemistry. B, 112*(15), 382. https://doi.org/10.1021/jp8055366.

Grosse, G., Schirrmeister, L., Siegert, C., Kunitsky, V. K., Slagoda, E. A., Andreev, A. A., & Dereviagyn, Y. (2007). Geological and geomorphological evolution of a sedimentary periglacial landscape in northeast Siberia during the late quaternary. *Geomorphology, 86,* 25–51. https://doi.org/10.1016/j.geomorph.2006.08.005.

Haberle, R. M., McKay, C. P., Schaeffer, J., Cabrol, N. A., Grin, E. A., Zent, A. P., & Quinn, R. (2001). On the possibility of liquid water on present-day Mars. *Journal of Geophysical Research, 106*(E10), 23317–23326.

Hall, J. L., Solomon, S. C., & Head, J. W. (1986). Elysium region, Mars: Test of lithospheric loading models for the formation of tectonic features. *Journal of Geophysical Research, 91*(B11), 11377–11392.

Hallet, B., Sletten, R., & Whilden, K. (2011). Micro-relief development in polygonal patterned ground in the dry valleys of Antarctica. *Quaternary Research, 75,* 347–355. https://doi.org/10.1016/j.yqres.2010.12.009.

Hanley, J., Chevrier, V. F., Berget, D. J., & Adams, R. D. (2012). Chlorate salts and solutions on Mars. *Geophysical Research Letters, 39*(L08201). https://doi.org/10.1029/2012GL051239.

Hanna, J. C., & Phillips, R. J. (2006). Tectonic pressurization of aquifers in the formation of Mangala and Athabasca Valles, Mars. *Journal of Geophysical Research, 111,* E03003. https://doi.org/10.1029/2005JE002546.

Harish, N., Vijayan, S., Mangold, N., & Bhardwaj, A. (2020). Water-ice exposing scarps within the northern mid-latitude craters on Mars. *Geophysical Research Letters, 47.* https://doi.org/10.1029/2020GL089057.

Harris, S. A., French, H. M., Heginbottom, J. A., Johnston, G. H., Ladanyi, B., Sego, D. C., & van Everdingen, R. O. (Eds.). (1988). *Technical memorandum: Vol. 142. Glossary of permafrost and related ground-ice terms.* National Research Council of Canada: Permafrost Subcommittee. 154 p.

Hartmann, W. K. (1966). Martian cratering. *Icarus, 5,* 565–576. https://doi.org/10.1016/00191035(66)90071-6.

Hartmann, W. K. (1971). Martian cratering III: Theory of crater obliteration. *Icarus, 15,* 410–428. https://doi.org/10.1016/0019-1035(71)90119-9.

Hartmann, W. K. (2005). Martian cratering 8: Isochron refinement and the chronology of Mars. *Icarus, 174,* 294–320. https://doi.org/10.1016/j.icarus.2004.11.023.

Hartmann, W. K., & Daubar, I. J. (2017). Martian cratering 11. Utilizing decameter scale crater populations to study Martian history. *Meteoritics and Planetary Science, 52*(3), 493–510. https://doi.org/10.1111/maps.12807.

Hauber, E., et al. (2011). Periglacial landscapes on Svalbard: Terrestrial analogs for cold-climate landforms on Mars. *Geological Society of America Special Paper, 483*, 177–201. https://doi:10.1130/2011.2483(12).

Head, J. W., Mustard, J. F., Kreslavsky, M. A., Milliken, R. E., & Marchant, D. R. (2003). Recent ice ages on Mars. *Nature, 426*, 797–802. https://doi.org/10.1038/nature02114.

Hecht, M. H. (2002). Metastability of liquid water on Mars. *Icarus, 156*, 373–386. https://doi.org/10.1006/icar.2001.6794.

Hepburn, A. J., Ng, F., Livingstone, S. J., Holt, T., & Hubbard, B. (2020). Polyphase mid-latitude glaciation on Mars: Chronology of the formation of superposed glacier-like forms from crater-count dating. *Journal of Geophysical Research, 125*(e2019JE006102), 102. https://doi.org/10.1029/2019JE006.

Hubbard, B., Souness, C., & Brough, S. (2014). Glacier-like forms on Mars. *The Cryosphere, 8*, 2047–2061. https://doi.org/10.5194/tc-8-2047-2014.

Jouannic, G., et al. (2015). Laboratory simulation of debris flows over sand dunes: Insights into gully-formation (Mars). *Geomorphology, 231*, 101–115. https://doi.org/10.1016/j.geomorph.2014.12.007.

Kargel, J. S., & Strom, R. G. (1992). Ancient glaciation on Mars. *Geology, 20*, 3–7. https://doi.org/10.1130/0091-7613(1992)020<0003:AGOM>2.3.CO;2.

Karlsson, N. B. B., Schmidt, L. S. S., & Hvidberg, C. S. S. (2015). Volume of Martian midlatitude glaciers from radar observations and ice flow modeling. *Geophysical Research Letters, 42*, 2627–2633. https://doi.org/10.1002/2015GL063219.

Kessler, M. A., & Werner, B. T. (2003). Self-organisation of sorted patterned ground. *Science, 299*(5605), 380–383. https://doi.org/10.1126/science.1077309.

Kounaves, S. P., Chaniotakis, N. A., Chevrier, V. F., Carrier, B. L., Folds, K. E., Hansen, V. M., McElhoney, K. M., O'Neil, G. D., & Weber, A. W. (2014). Identification of the perchlorate parent salts at the Phoenix Mars landing site and possible implications. *Icarus, 232*, 226–231. https://doi.org/10.1016/j.icarus.2014.01.016.

Koutnik, M., Butcher, F., Soare, R., Hepburn, A., Hubbard, B., Brough, S., Gallagher, C., Mc Keown, L., & Pathare, A. (2024). *Glacial deposits, remnants, and landscapes on Amazonian Mars: Using setting, structure, and stratigraphy to understand ice evolution and climate history*. In R. J. Soare, J.-P. Williams, C. Ahrens, F. E. G. Butcher, & M. R. El-Maarry (Eds.), *Ices in the solar system, a volatile-driven journey from the inner solar system to its far reaches*. Elsevier Books.

Lachenbruch, A. H. (1962). Mechanics of thermal contraction cracks and ice-wedge polygons in permafrost. In *vol. 69. GSA special paper 70*. New York: Geological Society of America.

Lefort, A., Russell, P. S., McEwen, A. S., Dundas, C. M., & Kirk, R. L. (2009). Observations of periglacial landforms in Utopia Planitia with the high resolution imaging science experiment (HiRISE). *Journal of Geophysical Research, 114*, E04005. https://doi.org/10.1029/2008JE003264.

Levy, J., Head, J., & Marchant, D. (2009a). Concentric crater fill in utopia Planitia: History and interaction between glacial "brain terrain" and periglacial mantle processes. *Icarus, 202*, 462–476. https://doi.org/10.1016/j.icarus.2009.02.018.

Levy, J., Head, J., & Marchant, D. (2009b). Thermal contraction crack polygons on Mars: Classification, distribution, and climate implications from HiRISE observations. *Journal of Geophysical Research, 114*(E01007). https://doi.org/10.1029/2008JE003273.

Levy, J. S., Marchant, D. R., & Head, J. W. (2010). Thermal contraction crack polygons on Mars. A synthesis from HiRISE, Phoenix and terrestrial analog studies. *Icarus, 206*, 229–252. https://doi.org/10.1016/j.icarus.2009.09.005.

Lucchitta, B. L. (1981). Mars and Earth: A comparison of cold-climate features. *Icarus, 45*, 264–303.

Mackay, J. R. (1974). Ice wedge cracks, Garry Island, Northwest Territories. *Canadian Journal of Earth Sciences, 11*, 1366–1383.

Mackay, J. R. (1997). A full-scale field experiment (1978-1995) on the growth of permafrost by means of lake drainage, western Arctic coast: A discussion of the method and some results. *Canadian Journal of Earth Sciences, 34*, 17–33.

Mackay, J. R. (1998). Pingo growth and collapse, Tuktoyaktuk peninsula areas, western arctic coast, Canada: A long-term field study. *Geographie Physique Et Quaternaire, 52*(3), 1–53.

Mackay, J. R., & Burn, C. R. (2002). The first 20 years (1978–1979 to 1998–1999) of active-layer development, Illisarvik experimental drained lake site, western Arctic coast. *The Canadian Journal of Earth Sciences, 39*, 1657–1674. https://doi.org/10.1139/E02-068.

Madeleine, J.-B., Forget, F., Head, J. W., Levrard, B., Montmessin, F., & Millour, E. (2009). Amazonian northern mid-latitude glaciation on Mars: A proposed climate scenario. *Icarus, 203*, 390–405. https://doi.org/10.1016/j.icarus.2009.04.037.

Madeleine, J. B., et al. (2014). Recent ice ages on Mars: The role of radiatively active clouds and cloud microphysics. *Geophysical Research Letters, 41*. https://doi:10.1002/2014GL059861.

Mangold, N. (2005). High latitude patterned grounds on Mars: Classification, distribution and climatic control. *Icarus, 174*, 336–359. https://doi.org/10.1016/j.icarus.2004.07.030.

Martín-Torres, F. J., et al. (2015). Transient liquid water and water activity at Gale crater on Mars. *Nature Geoscience*, 1–5. http://www.nature.com/doifinder/10.1038/ngeo2412.

Martinez, G. M., et al. (2017). The modern near-surface Martian climate: A review of in-situ meteorological data from viking to curiosity. *Space Science Reviews, 212*, 295–338. https://doi.org/10.1007/s11214-017-0360-x.

Mazrouei, S., Ghent, R. R., Bottke, W. F., Parker, A. H., & Gernon, T. M. (2019). Earth and moon impact flux increased at the end of the Paleozoic. *Science, 363*(6424), 253–257. https://doi.org/10.1126/science.aar4058.

Mellon, M. T., & Jakosky, B. M. (1993). Geographic variations in the thermal and diffusive stability of ground ice on Mars. *Journal of Geophysical Research, 98*(E2), 3345–3364.

Mellon, M. T., & Jakosky, B. M. (1995). The distribution and stability of ground ice during past and present epochs. *Journal of Geophysical Research, 100 (E6), 11*, 781–11799.

Michael, G., & Neukum, G. (2010). Planetary surface dating from crater size–frequency distribution measurements: Partial resurfacing events and statistical age uncertainty. *Earth and Planetary Science Letters, 294*, 223–229. https://doi.org/10.1016/j.epsl.2009.12.041.

Milliken, R. E., Mustard, J. F., & Goldsby, D. L. (2003). Viscous flow features on the surface of Mars: Observations from high-resolution Mars orbiter camera (MOC) images. *Journal of Geophysical Research, 108*(E6), 5057. https://doi.org/10.1029/2002JE002005.

Mitrofanov, I., et al. (2002). Maps of subsurface hydrogen from the high energy neutron detector, Mars Odyssey. *Science, 297*(5578), 78–81. https://doi.org/10.1126/science.1073616.

Morgenstern, A., Hauber, E., Reiss, D., van Gasselt, S., Grosse, G., & Schirrmeister, L. (2007). Deposition and degradation of a volatile-rich layer in Utopia Planitia, and implications for climate history on Mars. *Journal of Geophysical Research, 112*, E06010. https://doi.10.1029/2006JE002869.

Morgenstern, A., Ulrich, M., Günther, F., Roessler, S., Fedorova, I. V., Rudaya, N. A., Wetterich, S., Boike, J., & Schirrmeister, L. (2013). Evolution of thermokarst in east Siberian ice-rich permafrost: A case study. *Geomorphology, 201*, 363–379. https://doi.org/10.1016/j.geomorph.2013.07.011.

Murton, J. B. (2001). Thermokarst sediments and sedimentary structures, Tuktoyaktuk coastlands, western Arctic Canada. *Global and Planetary Change, 28*, 175–192. https://doi.org/10.1016/S0921-8181(00)00072-2.

Murton, J. B. (2021). What and where are periglacial landscapes? *Permafrost and Periglacial Processes, 32*, 186–212. https://doi.org/10.1002/ppp.2102.

Murton, J. B., Whiteman, C. A., Waller, R. I., Pollard, W. H., Clarke, I. D., & Dallimore, S. R. (2005). Basal ice facies and supraglacial melt-out till of the Laurentide ice sheet, Tuktoyaktuk Coastlands, western Arctic Canada. *Quaternary Science Reviews, 24*, 681–708. https://doi:10.1016/j. quascirev.2004.06.008.

Murton, J. B., Worsley, P., & Gozdsik, J. (2000). Sand veins and wedges in cold aeolian environments. *Quaternary Science Reviews, 19*, 899–922. https://doi.org/10.1016/S0277-3791(99)00045-1.

Mustard, J. F., Cooper, C. D., & Rifkin, M. R. (2001). Evidence for recent climate change on Mars from the identification of youthful near-surface ground ice. *Nature, 412*, 411–414. https://doi.org/10.1038/35086515.

Mustard, J. F., et al. (2008). Hydrated silicate minerals on Mars observed by the Mars Reconnaissance Orbiter *CRISM* instrument. *Nature, 454*(7202), 305–309. https://doi.org/10.1038/nature07097.

Neukum, G., Ivanov, B. A., & Hartmann, W. K. (2001). Cratering records in the inner solar system in relation to the lunar reference system. *Space Science Reviews, 96*, 55–86. https://doi.org/10.1007/978-94-017-1035-0_3.

Neukum, G., Jaumann, R., et al. (2004). HRSC: The high resolution stereo camera of Mars Express. In *The Scientific Payload, 1240*, 17–35.

Öpik, E. J. (1966). The Martian surface. *Science, 153*, 255–265. https://doi.org/10.1126/science.153.3733.255.

Osterkamp, T. E., Jorgenson, M. T., Schuur, E. A. G., Shur, Y. L., Kanevskiy, M. A., Vogel, J. G., & Tumskov, V. E. (2009). Physical and ecological changes associated with warming permafrost and thermokarst in interior Alaska. *Permafrost and Periglacial Processes, 20*(235), 256. https://doi.org/10.1002/ppp656.

Pál, B., & Kereszturi, A. (2017). Possibility of microscopic liquid water formation at landing sites on Mars and their observational potential. *Icarus, 282*, 84–92. https://doi.org/10.1016/j.icarus.2016.09.006.

Palucis, M. C., Jasper, J., Garczynski, B., & Dietrich, W. E. (2020). Quantitative assessment of uncertainties in modeled crater retention ages on Mars. *Icarus, 341*, 113623. https://doi.org/10.1016/j.icarus.2020.113623.

Pedersen, G. B. M., & Head, J. W. (2010). Evidence of widespread degraded Amazonian-aged ice-rich deposits in the transition between Elysium rise and Utopia Planitia, Mars: Guidelines for the recognition of degraded ice-rich materials. *Planetary and Space Science, 58*(14–15), 1953–1970. https://doi.org/10.1016/j.pss.2010.09.019.

Pedersen, G. B. M., & Head, J. W. (2011). Chaos formation by sublimation of volatile-rich substrate: Evidence from Galaxias Chaos, Mars. *Icarus, 211*(1), 316–329. https://doi.org/10.1016/j.icarus.2010.09.005.

Pelkey, S. M. (2007). *CRISM* multispectral summary products: Parameterizing mineral diversity on Mars from reflectance. *Journal of Geophysical Research, 112*(E08S14). https://doi.org/10.1029/2006JE002831.

Penner, E. (1959). The mechanism of frost heaving in soils. *Highway Research Board Bulletin, 225*, 1–22.

Péwé, T. (1959). Sand-wedge polygons (tessellations) in the McMurdo Sound region, Antarctica—A progress report. *American Journal of Science, 257*, 545–552.

Powell, T. M., Rubanenko, L., Williams, J.-P., & Paige, D. A. (2021). The role of secondary craters on Martian crater chronology. In R. Soare, S. Conway, J.-P. Williams, & D. Oehler (Eds.), *Mars Geological Enigmas: From the Late Noachian Epoch to the Present Day* (pp. 123–146). Amsterdam, the Netherlands: Elsevier. https://doi.org/10.1016/B978-0-12-820245-6.00006-9.

Primm, K. M., Gough, R. V., Wong, J., Rivera-Valentin, E. G., Martinez, G. M., Hogancamp, J. V., Archer, P. D., Ming, D. W., & Tolbert, M. A. (2018). The effect of Mars-relevant soil analogs on the water uptake of magnesium perchlorate and implications for the near-surface of Mars. *Journal of Geophysical Research, 123*, 2076–2088. https://doi.org/10.1029/2018JE005540.

Primm, K. M., Stillman, D. E., & Michaels, T. I. (2020). Investigating the hysteretic behavior of Mars-relevant chlorides. *Icarus, 342*, 113342. https://doi.org/10.1016/j.icarus.2019.06.003.

Rampton, V. N. (1988). Quaternary geology of the Tuktoyaktuk Coastlands, Northwest Territories, Geological Survey of Canada (GSC). *Memoir, 423*. 98 p.

Rampton, V. N., & Bouchard, M. (1975). *Surficial geology of Tuktoyaktuk, District of Mackenzie, Geological Survey of Canada, Paper 74-53*. 16 p.

Rampton, V. N., & Mackay, J. R. (1971). *Massive ice and icy sediments throughout the Tuktoyaktuk peninsula, Richards Island, and nearby areas, district of Mackenzie, Geological Survey of Canada, paper 71–21*. (16 p).

Ramsdale, J. D., et al. (2017). Grid-based mapping: A method for rapidly determining the spatial distributions of small features over very large areas. *Planetary and Space Science, 140*, 49–61. https://doi.org/10.1016/j.pss.2017.04.002.

Ramsdale, J. D., et al. (2018). Grid-mapping the northern plains of Mars: Geomorphological, radar and water-equivalent hydrogen results from Arcadia Planitia suggest possible fluvial and volcanic systems overlain by a ubiquitous and heavily degraded ice-rich latitude-dependent mantle. *Journal of Geophysical Research*, *123*. https://doi.org/10.1029/2018JE005663.

Reiss, D., Erkeling, G., Bauch, K. E., & Hiesinger, H. (2010). Evidence for present day gully activity on the Russell crater dune field, Mars. *Geophysical Research Letters*, *37*, 1–7. https://doi.org/10.1029/2009GL042192.

Rennó, N. O., et al. (2009). Possible physical and thermodynamical evidence for liquid water at the Phoenix landing site. *Journal of Geophysical Research*, *114*, E00E03. https://doi.org/10.1029/2009JE003362.

Rivera-Valentin, E. G., Chevrier, V. F., Soto, A., & Martinez, G. (2020). Distribution and habitability of (meta)stable brines on present-day Mars. *Nature Astronomy*, *4*, 756–761. https://doi.org/10.1038/s41550-020-1080-9.

Rubanenko, L., Powell, T. M., Williams, J.-P., Daubar, I., & Paige, D. A. (2021). Challenges in crater chronology on Mars as reflected in Jezero crater. In R. Soare, S. Conway, J.-P. Williams, & D. Oehler (Eds.), *Mars Geological Enigmas: From the Late Noachian Epoch to the Present Day* (pp. 97–122). Amsterdam, the Netherlands: Elsevier. https://doi.org/10.1016/B978-0-12-820245-6.00005-7.

Rummel, J. D., et al. (2014). A new analysis of Mars "special regions": Findings of the second *MEPAG* special regions science analysis group (SR-SAG2). *Astrobiology*, *14*. https://doi.org/10.1089/ast.2014.1227.

Schirrmeister, L., Froese, D., Tumskoy, V., Grosse, G., & Wetterich, S. (2013). Yedoma: Late Pleistocene ice-rich syngenetic permafrost of Beringia. In S. A. Elias, & C. J. Mock (Eds.), *vol. 3*. *Encyclopedia of quaternary science* (2nd ed., pp. 542–552).

Schirrmeister, L., Siegert, C., Kunitzky, V. V., Grootes, P. B., & Erlenkeuser, H. (2002). Late quaternary ice-rich permafrost sequences as a paleoenvironmental archive for the Laptev Sea region in northern Siberia. *International Journal of Earth Sciences*, *91*, 154–167. https://doi.org/10.1007/s005310100205.

Schmidt, L. S., Hvidberg, C. S. S., Kim, J. R., & Karlsson, N. B. B. (2019). Non-linear flow modelling of a Martian lobate debris apron. *Journal of Glaciology*, *65*, 889–899. https://doi.org/10.1017/jog.2019.54.

Seibert, N. M., & Kargel, J. S. (2001). Small-scale Martian polygonal terrain: Implications for liquid surface water. *Geophysical Research Letters*, *28*(5), 899–902. https://doi.org/10.1029/2000GL012093.

Séjourné, A., Costard, F., Gargani, J., Soare, R. J., Fedorov, A., & Marmo, C. (2011). Scalloped depressions and small-sized polygons in western Utopia Planitia: A new formation hypothesis. *Planetary and Space Science*, *59*, 412–422. https://doi:10.1016/j.pss.2011.01.007.

Séjourné, A., Costard, F., Gargani, J., Soare, R. J., & Marmo, C. (2012). Evidence of an eolian ice-rich and stratified permafrost in Utopia Planitia, Mars. *Planetary and Space Science*, *60*, 248–254. https://doi.org/10.1016/j.pss.2011.09.004.

Séjourné, A., et al. (2019). Grid mapping the Northern Plains of Mars: Using Morphotype and distribution of ice-related landforms to understand multiple ice-rich deposits in Utopia Planitia. *Journal of Geophysical Research*, *124*, 483–503. https://doi.org/10.1029/2018JE0 05665.

Sinha, R. K., & Murty, S. V. S. (2015). Amazonian modification of Moreux crater: Record of recent and episodic glaciation in the Protonilus Mensae region of Mars. *Icarus*, *245*, 122–144. https://doi.org/10.1016/j.icarus.2014.09.028.

Sizemore, H. G., Zent, A. P., & Rempel, A. W. (2015). Initiation and growth of Martian ice lenses. *Icarus*, *251*, 191–210. https://doi.org/10.1016/j.icarus.2014.04.013.

Sletten, R. S., Hallet, B., & Fletcher, R. C. (2003). Resurfacing time of terrestrial surfaces by the formation and maturation of polygonal patterned ground. *Journal of Geophysical Research*, *108*(E4), 8044. https://doi.org/10.1029/2002JE001914.

Soare, R. J., Burr, D. M., Tseung, W. B. J. M. (2005). Pingos and a possible periglacial landscape in northwest Utopia Planitia Mars. *Icarus*, *174*(2), 373–382. https://doi.org/10.1016/j.icarus.2004.11.013.

Soare, R. J., Conway, S. J., Gallagher, C., & Dohm, J. M. (2016). Sorted (clastic) polygons in the Argyre region, Mars, and possible evidence of pre- and post-glacial periglaciation in the Late Amazonian Epoch. *Icarus*, *264*, 184–197. https://doi.org/10.1016/j.icarus.2015.09.019.

Soare, R. J., Conway, S. J., Williams, J.-P., Gallagher, C., & McKeown, L. E. (2020). Possible (closed system) pingo-thermokarst complexes at the mid-latitudes of Utopia Planitia, Mars. *Icarus, 342*, 113233. https://doi.org/10.1016/j.icarus.2019.03.010.

Soare, R. J., Conway, S. J., Williams, J.-P., & Hepburn, A. J. (2021a). Possible polyphase periglaciation and glaciation adjacent to the Moreux impact-crater, Mars. *Icarus, 362*, 114401. https://doi.org/10.1016/j.icarus.2021.114401.

Soare, R. J., Conway, S. J., Williams, J.-P., Phillipe, M., Mc Keown, L. E., Godin, E., & Hawkswell, J. (2021b). Possible ice-wedge polygonisation in Utopia Planitia, Mars & its latitudinal gradient of distribution. *Icarus, 353*, 114208. https://doi.org/10.1016/j.icarus.2020.114208.

Soare, R. J., Williams, J.-P., Hepburn, A. J., & Butcher, F. E. G. (2022). One billion or more years of possible periglacial/glacial cycling in Protonilus Mensae, Mars. *Icarus, 385*(115115). https://doi.org/10.1016/j.icarus.2022.115115.

Soare, R. J., Kargel, J. S., Osinski, G. R., & Costard, F. (2007). Thermokarst processes and the origin of crater-rim gullies in utopia and western Elysium Planitia. *Icarus, 191*(1), 95–112. https://doi.org/10.1016/j.icarus.2007.04.018.

Soare, R. J., Osinski, G. R., & Roehm, C. L. (2008). Thermokarst lakes and ponds on Mars in the very recent (late Amazonian) past. *Earth and Planetary Science Letters, 272*(1–2), 382–393. https://doi.10.1016/j.epsl.2008.05.10.

Soare, R. J., Séjourné, A., Pearce, G., Costard, F., & Osinski, G. R. (2011). The Tuktoyaktuk Coastlands of northern Canada: A possible "wet" periglacial analogue of Utopia Planitia, Mars. *Geological Society of America, 483*, 203–218. https://doi.org/10.1130/2011.2483(13.

Soare, R. J., Williams, J.-P., Conway, S. J., & El-Maarry, M. R. (2021c). Pingo-like mounds and possible polyphase periglaciation/glaciation at/adjacent to the Moreux impact-crater. In R. Soare, S. Conway, J.-P. Williams, & D. Oehler (Eds.), *Mars Geological Enigmas: From the Late Noachian Epoch to the Present Day* (pp. 407–435). Amsterdam, the Netherlands: Elsevier. https://doi.org/10.1016/B978-0-12-820245-6.00014-8.

Speyerer, E. J., Povilaitis, R. Z., Robinson, M. S., Thomas, P. C., & Wagner, R. V. (2016). Quantifying crater production and regolith overturn on the moon with temporal imaging. *Nature, 538*, 215–218. https://doi.org/10.1038/nature19829.

Squyres, S. W. (1978). Martian fretted terrain: Flow of erosional debris. *Icarus, 34*, 600–613.

Stillman, D. E., & Grimm, R. E. (2011). Dielectric signatures of adsorbed and salty liquid water at the Phoenix landing site, Mars. *Journal of Geophysical Research, 116*, E09005. https://doi.org/10.1029/2011JE003838.

Stillman, D. E., Grimm, R. E., & Dec, S. F. (2010). Low-frequency electrical properties of ice–silicate mixtures. *The Journal of Physical Chemistry. B, 113*(18), 6065–6073.

Stuurman, C. M., Osinski, G. R., Holt, J. W., Levy, J. S., Brothers, T. C., Kerrigan, M., & Campbell, B. A. (2016). SHARAD detection and characterization of subsurface water ice deposits in Utopia Planitia, Mars. *Journal of Geophysical Research, 43*, 9484–9491. https://doi:10.1002/2016GL070138.

Taber, S., 1930. The mechanics of frost heaving, 9-26, in historical perspectives in frost heave research: The early works of S. Taber and G. Beskow. Special report 91-23, U.S. Army Corp of Engineers, (eds.) P.B. Black and M.J. Hardenburg, (1991, 159 pp.).

Tanaka, K. L., Skinner, J. A., Dohm, J. M., Irwin, R. P., Kolb, E. J., Fortezzo, C. M., Platz, T., Michael, G. G., & Hare, T. M. (2014). Geologic map of Mars: U.S.G.S. *Scientific Investigations Map 3292*. https://doi.org/10.3133/sim3292. scale 1:20,000,000.

Toner, J. D., Catling, D. C., & Light, B. (2014). The formation of supercooled brines, viscous liquids, and low temperature perchlorate glasses in aqueous solutions relevant to Mars. *Icarus, 233*, 36–47. https://doi.org/10.1016/j.icarus.2014.01.018.

Toner, J. D., Catling, D. C., & Light, B. (2015). A revised Pitzer model for low-temperature soluble salt assemblages at the Phoenix site, Mars. *Geochimica et Cosmochimica Acta, 166*, 327–343. https://doi.org/10.1016/j.gca.2015.06.011.

Wan Bun Tseung, J.-M., & Soare, R. J. (2006). Thermokarst and related landforms in western utopia Planitia, Mars: Implications for near-surface excess ice. In *37th Lunar and Planetary Science Conference*. Houston, Texas. (Abstract # 1414).

Ulrich, M., Morgenstern, A., Günther, F., Reiss, D., Bauch, K. E., Hauber, H., Rössler, S., & Schirrmeister, L. (2010). Thermokarst in Siberian ice-rich permafrost: Comparison to asymmetric scalloped depressions on Mars. *Journal of Geophysical Research*, *115*, E10009. https://doi.org/10.1029/2010JE003640.

Védie, E., Costard, F., Font, M., & Lagard, J. L. (2008). Laboratory simulations of Martian gullies on sand dunes. *Geophysical Research Letters*, *35*, L21501. https://doi.org/10.1029/2008GL035638.

Vetterlein, J., & Roberts, G. P. (2009). Postdating of flow in Athabasca Valles by faulting of the Cerberus Fossae, Elysium Planitia, Mars. *Journal of Geophysical Research*, *114*, E07003. https://doi.org/10.1029/2009JE003356.

Viola, D. (2021). Thermokarst on Mars: age constraints on ice degradation in Utopia Planitia. In R. Soare, S. Conway, J.-P. Williams, & D. Oehler (Eds.), *Mars Geological Enigmas: From the Late Noachian Epoch to the Present Day* (pp. 437–472). Amsterdam, the Netherlands: Elsevier. https://doi.org/10.1016/B978-0-12-820245-6.00015-X.

Washburn, A. L. (1973). *Periglacial processes and environment* (p. 320). New York, NY: St Martin's Press.

Wetterich, S., Tumskoy, V., Rudaya, N., Andreev, A. A., Opel, T., Meyer, H., Schirrmeister, L., & Hüls, M. (2014). Ice complex formation in arctic East Siberia during the MIS3 Interstadial. *Quaternary Science Reviews*, *84*, 39–55. https://doi.org/10.1016/j.quascirev.2013.11.009.

Williams, G. E. (1975). Late Precambrian glacial climate and the Earth's obliquity. *Geological Magazine*, *112*, 441–444.

Williams, G. E. (2008). Proterozoic (pre-Ediacaran) glaciation and the high obliquity, low-latitude ice, strong seasonality (*HOLIST*) hypothesis: Principles and tests. *Earth Science Reviews*, *87*, 61–93. https://doi.org/10.1016/j.earscirev.2007.11.002.

Williams, J.-P. (2018). Modification of the Martian surface by impact cratering. In R. Soare, S. Conway, & S. Clifford (Eds.), *Dynamic Mars: Recent and current landscape evolution of the red planet* (pp. 365–386). Amsterdam, The Netherlands: Elsevier Books. https://doi.org/10.1016/B978-0-12-813018-6.00012-1.

Williams, P. J., & Smith, M. W. (1989). *The frozen earth: Fundamentals of geocryology* (p. 329). Cambridge, UK: Cambridge University Press.

Williams, J.-P., van der Bogert, C. H., Pathare, A. V., Michael, G. G., Kirchoff, M. R., & Hiesinger, H. (2018). Dating very young planetary surfaces from crater statistics: A review of issues & challenges. *Meteoritics and Planetary Science*, *53*(4), 554–582. https://doi.org/10.1111/maps.12924.

Wilson, L., & Head, J. W. (2002). Tharsis-radial graben systems as the surface manifestation of plume-related dike intrusion complexes: Models and implications. *Journal of Geophysical Research*, *107*(8), 5057. https://doi.org/10.1029/2001JE001593.

Zuber, M. T., Smith, D., Solomon, S. C., Muhleman, D. O., Head, J. W., Garvin, J. B., Abshire, J. B., & Bufton, J. L. (1992). The Mars Observer Laser Altimeter investigation. *Journal of Geophysical Research*, *97*(E5), 7781–7797. https://doi.org/10.1029/92JE00341.

Ice Exploration on Mars: Whereto and when?

6

James B. Garvin[a], Richard J. Soare[b], Adam J. Hepburn[c], Michelle Koutnik[d], and E. Godin[e]

[a]*NASA Goddard Space Flight Center, Greenbelt, MD, United States,* [b]*Department of Geography, Dawson College, Montreal, QC, Canada,* [c]*European Space Astronomy Centre, European Space Agency, Madrid, Spain,* [d]*Department of Earth and Space Sciences, University of Washington, Seattle, WA, United States,* [e]*Northern Studies Centre, Laval University, Quebec City, QC, Canada*

Abstract

Identifying, characterizing, and inventorying the ground ice on Mars is an essential element of any Mars exploration strategy, with direct linkages to: (1) questions concerning Mars's paleo-climate and the stability or non-stability of liquid water at or near Mars's surface through its geological history; (2) concepts for astrobiological reconnaissance; and (3) support for human exploration needs tied in particular to water ice as a resource (Planetary and Astrobiology Sciences Decadal Survey, NASEM, 2022).

Here we summarize the electromagnetic sensing methods documented to have been the most effective thus far in following the story of ice on Mars, at least within the upper active surface layer (<10 m thick), and the path forward towards possibly understanding its sources, sinks, and, writ large, cycles. Building on the seminal experiences with ice reconnaissance on Earth and the exploration of Mars, we describe the in situ, orbital, and proposed ice-relevant sample-return steps constitutive of this path.

1 Introduction

Ground *p*enetrating *r*adar (*GPR*) is a microwave-based tool that has been used in airborne and ground-based campaigns to investigate glacial (e.g. Moorman & Michel, 2000; Robin et al., 1969; Schroeder et al., 2020; Weber & Andrieux, 1970) and periglacial (e.g. Angelopolous et al., 2013; Hinkel et al., 2001; Léger et al., 2017; Moorman et al., 1994; Schennen et al., 2022) landscapes and ice on Earth for over 50 years. These *ice* campaigns also have included:

1. surveys derived from *c*apacitively *c*oupled *r*esistivity (*CCR*); borehole data; and/or vertically exposed/exhumed icy or ice-rich (sedimentary) facies;
2. studies of geographical distribution; and
3. geochemical and (fine-grained/small-scaled) fabric analyses.

Section 2 summarizes the fieldwork and remote sensing carried out on Earth to constrain and shape our understanding of form and process in glacial and periglacial environments. Section 3 describes that which has been done, is being planned, and could be envisaged with regard to the remote detection and evaluation of surface (i.e. Polar Layered Deposits) (Koutnik et al., 2024, pp. 101–142), buried (i.e. viscous-flow features or putative buried glaciers) (Koutnik et al., 2024, pp. 101–142), and sub/near-surface ground or excess (Soare et al., 2024, pp. 143–192) ice on Mars.

2 What on Earth?

2.1 Glacial landscapes and ice (see Gallagher, 2024, pp. 31–72)

Glacial ice on Earth originates as airfall snow or icy precipitation (e.g. Hambrey, 1994). As a perennial mass, be it an ice sheet, ice cap, or mountain glacier, it forms if and when accumulation exceeds ablation. The ice itself comprises an aggregate of irregularly shaped, interlocking crystals that range in size from a few millimetres to several tens of centimetres. At its maximum coverage and depth, the distribution of glacial ice can be continental in scale and kilometres in thickness (i.e. the Antarctic ice sheet) (Jennings & Hambrey, 2021). Burial, compaction, and recrystallization generate the primary structure of glacial ice (Jennings & Hambrey, 2021).

Glaciologists use *GPR* to investigate: internal structure/stratigraphy and bed conditions (e.g. Schroeder et al., 2020); sub/englacial hydrology (e.g. Church et al., 2019; Egli et al., 2021; Irvine-Fynn et al., 2006); mass balance (e.g. Kohler et al., 1997; Sylvestre et al., 2013) volumetric changes over time (e.g. Ai et al., 2014; Fischer, 2009; Grab et al., 2018); and to characterize snow/firn layers and/or structures (e.g. Dunse et al., 2009; Machguth et al., 2006). The choice of radar frequency and data-collection strategies can be tuned according to the phenomena of interest. For example, higher-frequency (>1 GHz) signals allow closely spaced reflectors to be distinguished (e.g. firn layers); however, they lack the penetration depth of lower-frequency signals (<100 MHz). The latter, for example, can be used to map deep internal layers in the Greenland/Antarctic ice sheets. Combining these datasets allows layers to be identified and evaluated continuously over large regions of Greenland and Antarctica. The use of intersecting firn and/or ice cores enables cross-validation of layer interpretation and age dating as well as the identification of finer-scale features that fall below the resolution of radar (e.g. Cavitte et al., 2021; MacGregor et al., 2015; Winter et al., 2019).

Unconformities, indicative of past and/or recent changes in glacial ice dynamics (e.g. Holschuh et al., 2015), differences in the type and amount of internal debris (e.g. Forte et al., 2021; Petersen et al., 2020; Winter et al., 2019), and, writ large, paleo-climate conditions (e.g. Kassab et al., 2020; MacKay et al., 2014) also can be identified in *GPR* profiles (with associated radargrams) and spatially integrated geographical grids. For example, two neighbouring but otherwise distinct debris-covered glaciers in the McMurdo Dry Valleys of Antarctica were investigated by MacKay et al. (2014). Both glaciers, despite their geographical separation, exhibited a series of surface ridges with a consistent amplitude and spacing that suggested a shared origin. Radargrams showed that these ridges are the surface expression of a similarly consistent series of up-dipping or inclined debris layers to depth, pointing to the fact that climatologically the evolution of the two glaciers was determined by near-identical boundary conditions (MacKay et al., 2014).

Moreover, MacKay et al. (2014) surmised that the glaciers are cold-based. As such, the rates of basal entrainment are too low for subglacial accretion, which otherwise could be expected to form these debris layers (Hubbard et al., 2004). Using *GPR*, supplemented by ice-core analysis, geomorphic mapping, and numerical modelling, MacKay et al. (2014) proposed that the inclined debris layers ($\lesssim 3$ m in thickness) and arcuate surface ridges formed at the headwall by (climate-driven) temperature changes in net accumulation/ablation rates. They hypothesized that layers of englacial debris originate as supraglacial lags in ice-accumulation areas when net ice-accumulation rates are low (perhaps when mean temperatures are relatively high); when net ice-accumulation rates rise (perhaps when mean temperatures are relatively low), the lags are buried by snow and ice, flow englacially, and then intercept the ice surface in the ablation zone. Readily observable and distinct changes in the texture and topographic relief of supraglacial debris and in topographic relief are the result (MacKay et al., 2014). Accordingly, MacKay et al. (2014) suggest that the thickness and inclination of debris layers within the two glaciers comprise an insightful archive of climatic variances over time (Fig. 1).

Radar is also very sensitive to changes in liquid-water content within and beneath massive bodies of ice. For example, lakes have been widely identified beneath the Greenland and Antarctic ice sheets using radar (e.g. Goeller et al., 2016, and references contained therein). Diez et al. (2019) used airborne radar to map the presence of a large lake system beneath *Recovery Glacier*, located at the onset of fast ice flow along the main trunk. Similarly, macrostructures containing water within glaciers, including englacial conduits and crevasses, and water content have been mapped, as well as spatial variations in water content (Cuffey & Paterson, 2010).

Often, the thermal transition between temperate and cold ice is clearly visible in radargrams (e.g. Dowdeswell et al., 2017). By comparing *GPR* surveys to borehole measurements of glacier temperature, Pettersson et al. (2003) demonstrated that *GPR* could be used to constrain a thermal boundary to within ±1 m in Storglaciären, Sweden. Macrostructures containing water within or beneath glaciers, including englacial conduits, crevasses, and/or subglacial lakes, also appear as bright reflections (e.g. Christianson et al., 2012; Maguire et al., 2021; Williams et al., 2014) (Fig. 2).

The presence, location, and characteristics of these structures impose crucial controls on the dynamics of glacier flow (Cuffey & Paterson, 2010). In Svalbard, Bælum and Benn (2011) used *GPR* to track drainage within a small glacier, mapping the transition between supraglacial, englacial, and subglacial channels that form a downglacier. Even though the bed is cold at this glacier, these channels can store significant amounts of water and may alter glacier flow throughout the year (Bælum & Benn, 2011).

Boreholes, in addition to *GPR* surveys, were used to measure changes in the water level and sediment content within the 14,000 m^2 englacial network of the Rhone Glacier, Switzerland (Church et al., 2019) (Fig. 3). This combination of geophysical tools minimizes the impact of *GPR's* limited vertical resolution and maximizes the coverage area associated with borehole investigations alone.

Borehole samples of glacial ice also provide microscale information (10^{-1} to 1^{-6} m) about internal layering, crystallographic fabric, and impurities. Thin sections can be analysed for grain shape and size distribution, all of which influence ice flow (e.g. Gerbi et al., 2021). Additionally, deep cores drilled through ice sheets can provide information about the evolution of the ice microstructure (e.g. Faria et al., 2014; Kipfstuhl et al., 2006), including microstructural impurities (e.g. Gerbi et al., 2021; Stoll et al., 2021) (Fig. 4).

2.2 Periglacial landscapes and ice-rich permafrost

Unlike terrestrial glacial ice which incorporates sedimentary grains, the ice within the ice-rich periglacial sediment is intergranular or interstitial, i.e. it forms within the pore space of these sediments and does so by means of freeze-thaw cycling, cryosuction, and ice segregation (e.g. French, 2007; Harris et al., 1988). If and when the volume of ice exceeds the pore volume of the ground, were it not frozen, then the ice is referenced as excess ice (e.g. French, 2007; Harris et al., 1988). Landforms resulting from the loss of this ice, i.e. lakes, depressions (alases), etc., are known as thermokarst (e.g. French, 2007; Harris et al., 1988).

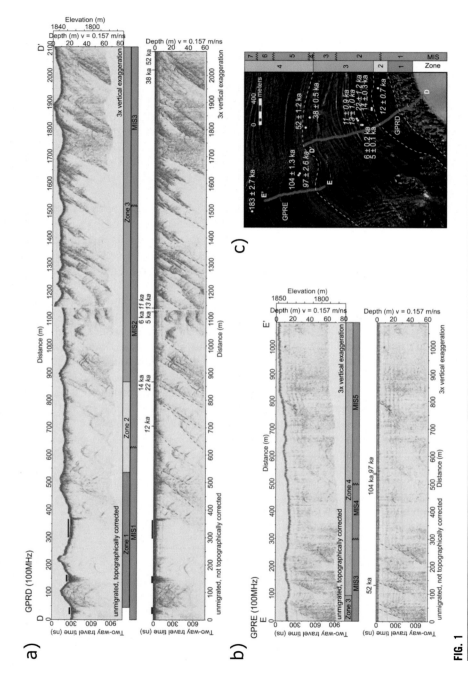

FIG. 1

GPR track along a longitudinal transect A_m–A'_m on the *Mullins Glacier*, Antarctica. (a) Aerial photograph (*USGS TMA* 3080V0276) showing the transect location, as well as a series of arcuate surface discontinuities (*ASD*). Glacial flow is from left to right. (b) Radar data (transformed to emphasize reflector brightness) outlining the presence of internal inclined debris layers (*IDL*) and their spatial correlation with surface *ASD*. (c) Sketch highlighting the major reflectors and a physical interpretation.

Panels reproduced from MacKay et al. (2014).

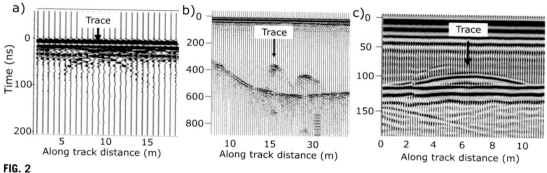

FIG. 2

GPR profiles of a changing hydrological system within a glacier. (a) Shallow (~2 m deep) meltwater channel at the glacier margin. The shape of the hyperbola indicates the radar signal is attenuated by a snow-filled channel. (b) Englacial channel ~2.5 m in diameter, partially filled with water. (c) The subglacial channel near the glacier terminus is again partially filled with water.

Figure credit: Bælum and Benn (2011).

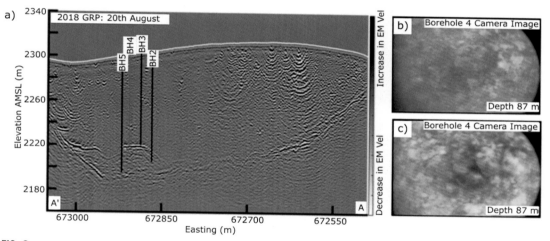

FIG. 3

Borehole camera investigations of an englacial feature. (a) *GPR* survey, with the englacial feature visible as a clear parabolic feature below the vertical lines that indicate the location of each borehole. (b and c) Borehole imagery shows sediment content within the englacial feature. The presence of sediment would be difficult to distinguish using *GPR* surveys alone.

Image credit: Church et al. (2019).

Freeze-thaw cycling facilitates the migration of meltwater through soil pores; cryosuction pulls the meltwater towards a freezing front and iterative ice segregation engenders the formation of ice lenses. Depending on the number and intensity of these iterations, the scale of ice lensing varies from microscopic to tabular (e.g. Black, 1954; French, 2007; Penner, 1959; Rampton, 1988; Taber, 1930) and massive, i.e. decametres thick and kilometres in diameter (e.g. Kanevskiy et al., 2011;

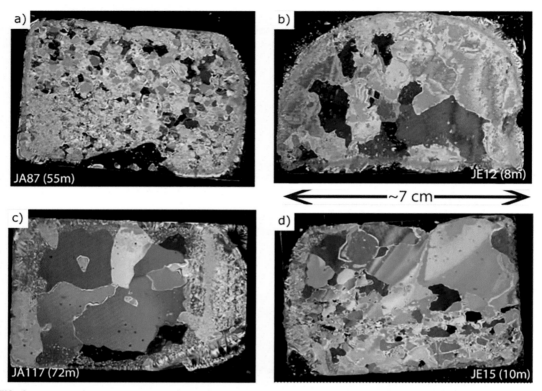

FIG. 4

Ice-core thin sections (a–d) from multiple cores on the *Jarvis Glacier*, Alaska. The section highlights grain shapes and sizes at different depths (bottom left of the text of each figure). Section margins display fine-grained areas where the water came under the sample.

Image credit: Gerbi et al. (2021).

MacKay & Dallimore, 1992). Ice-rich terrain, especially when populated by massive tabular ice, is diagnostic of relatively temperate boundary conditions at or near the surface having occurred iteratively/episodically (over time), intensely but perhaps less frequently, or a combination of the two options (e.g. Czudek & Demek, 1970; Grosse et al., 2007; Schirrmeister et al., 2013).

Using *GPR* and *CCR*, Angelopolous et al. (2013) surveyed, mapped (on a metre scale), and categorized near-surface vertical sections of thermokarst at Parson's Lake, near Tuktoyaktuk on the Beaufort seacoast in northern Canada (Fig. 5).

Boreholes were used to sample the terrain and, no less importantly, to cross-validate the *GPR/CCR* data on a relatively finer scale, i.e. sufficient to differentiate sediment types, boundaries, and horizons more clearly. This having been said, bulk periglacial landforms also can be constrained by *GPR*. For example, Léger et al. (2017) used *GPR* to locate and identify near-surface ice wedges, a commonplace feature amidst coastal or near-coastal permafrost landscapes (Fig. 6).

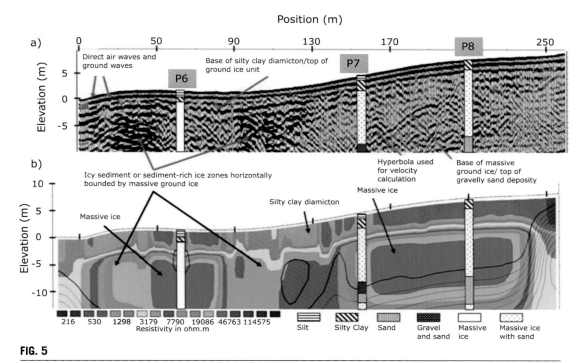

FIG. 5

(a) Radargram of disparate sedimentary and icy horizons to ~10 m of depth at Parson's Lake and (b) *CCR* (cross-sectional) profile that shows the location and distribution of ground ice/massive ice deposits with greater granularity.

Figure credit: Angelopolous et al. (2013).

Where a façade of ice is exposed, exhumed, and available for sampling and/or hands-on inspection, increasingly small-scale observations, interpretations, and age dating become possible. For example, the walls of the Cold Regions Research and Engineering Laboratory (*CRREL*) permafrost tunnel near Fairbanks, Alaska, show ice wedges and the (centimetre scale) vertical foliations of ice that comprise annual growth thereof (Fig. 7a) (Léger et al., 2017, also Douglas et al., 2011). On an even finer scale, ice lenses are observed (Fig. 7b). The ice lenses, as noted above, as well as the foliations of ice, are markers of iterative freeze-thaw cycling.

The cryostratigraphical layers and horizons of the *CRREL* tunnel wall comprise a regional cold-climate (non-glacial) time capsule of age, process, and form from the Wisconsin Glacial Period through to the Holocene Epoch and near the present day (Fig. 8). For example, the narrow layers (thin black lines) of segregation ice and horizons (pink) of ice-rich sediments have been age-dated and show that freeze-thaw cycling was particularly active between 31,970 and 28,380 ka (Fig. 8); the vertically elongated ice wedge (dark blue), for example, points to an origin and development that was syngenetic, i.e. upward growth and development tied to pulses, cycles, or episodes of permafrost aggradation and sedimentation (Fig. 8).

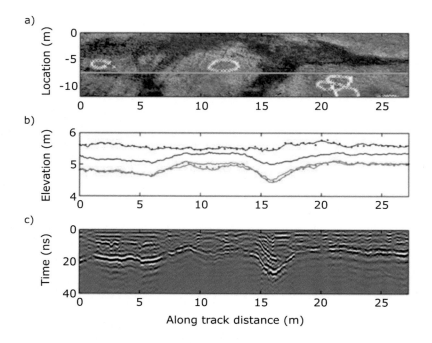

FIG. 6

(a) Plan view of the polygonized-ground surface with the location of *GPR* transect; non-sorted circles shown in white. (b) Elevations of the top of the snow from the probe (*blue dots*) and *GPR* (*blue line*), ground (*black line*), the base of the thaw layer from the probe (*red dots*) and *GPR* (*red line*) data in September 2012, the base of the active layer from *GPR* data (*pink line*) (May 2013). (c) Radargram with a base of the thaw layer from the probe (*dots*) and *GPR* (*line*) data (September 2012).

Image/figure credit: Léger et al. (2017).

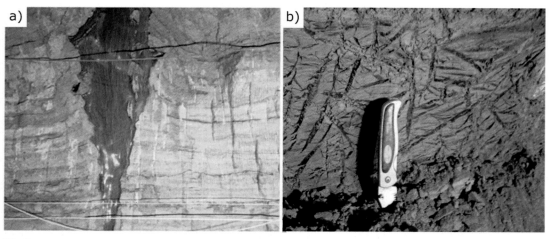

FIG. 7

(a) Partially exposed ice wedge in the wall of the *CRREL* tunnel, ~3 m in height. (b) Dense concentration of thin ice lenses (the handle of the knife is ~6 cm long).

Image credit: Panel (a): Image credit: Kanevskiy et al. (2022); panel (b): Fortier et al. (2008).

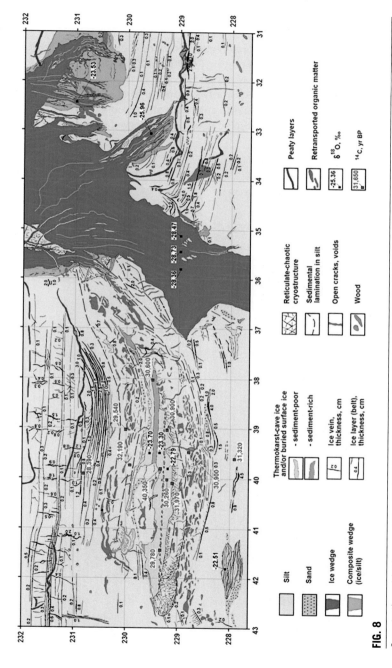

FIG. 8

Cryostratigraphical map of the *CRREL* tunnel wall section. Note the *vertical and horizontal insets* of age.

Figure credit: Kanevskiy et al. (2022).

As an analytic tool, nuclear magnetic resonance (*MRI*, a.k.a. "*NMR*") imagery constrains permafrost profiles by its ability to reach beyond naked-eye observations. There are two elements of particular interest in Fig. 9:

1) The fine-sand section of this permafrost core contains pore ice, i.e. ice that fills the pore space of the sediment but does not exceed it. This layer (86–90 cm) is shown as light-grey on the *MRI* and, unlike the visible image, does not show the pore ice.

2) The porosity and permeability of silt are especially conducive to the formation of ice veins and lenses, which is black in the *MRI*. Horizontal to sub-vertical ice veins are shown within the silt horizons in the *MRI* image (81–85 cm); these veins are not observed in the visible images.

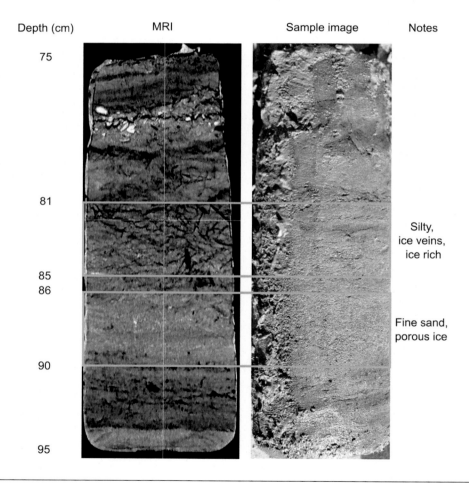

FIG. 9

Valley-floor permafrost core sample (Bylot Island, Nunavut, Canada). A magnetic resonance image (*MRI*) of the sample (*left*) and a non-*MRI* picture of the sample (*right*).

Photo credit: E. Godin, D. Fortier (Université de Montréal).

3 What on Mars?

3.1 Icy radargrams

An admixture of glacial (water) ice, lithic material, and dust comprises the *N*orth and *S*outh *P*olar *L*ayered *D*eposits (*NPLD/SPLD*) on Mars (e.g. Becerra et al., 2016; Byrne, 2009; Plaut et al., 2007; Sinha & Horgan, 2022; Whitten & Campbell, 2018). Additionally, CO_2 ice is present in portions of the *SPLD* as a ~10 m thick perennial surface deposit (e.g. Byrne & Ingersoll, 2003; Kieffer, 1979) and a massive subsurface deposit that is hundreds of metres to a kilometre thick (e.g. Phillips et al., 2011; also, see Koutnik et al., 2024, pp. 101–142).

Cross sections of the *PLD* (Fig. 10) and of (possible) near-surface massive ice distributed throughout the northern mid-latitudes (e.g. Bramson et al., 2015; Stuurman et al., 2016) have been generated by two orbital profiling MHz frequency radar-sounding instruments or *GPRs*, the *SHA*llow *RAD*ar *S*ounder (*SHARAD*, on the *NASA's MRO*) and the *M*ars *A*dvanced *R*adar for *S*ubsurface and *I*onosphere

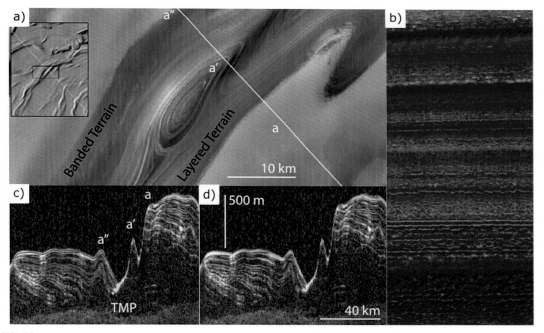

FIG. 10

Internal and exposed stratigraphy of the *NPLD*. (a) Layered and banded terrain within a trough. The line shows the transect in (c) and (d). Background image is *CTX* image B01_010014_2644. (b) A vertical exposure (~1 km) in a trough elsewhere on the *NPLD*. Differences in colouring indicate the relative concentration of dust compared to ice (*HiRISE* image PSP_001738_2670). (c and d) Internal layering at depth, below profile a to a' in panel (a) visible in *SHARAD* observation 556,602. These discontinuities—labelled as trough-migration pathways (*TMP*)—are associated with the (proposed) lateral migration of troughs across the *NPLD* over time, described in Smith and Holt (2015).

Panels (a), (c), and (d) from Smith and Holt (2015); panel (b) is reprinted from Wilcoski and Hayne (2020).

Sounding (*MARSIS*, on the *ESA's Mars Express*, *MEX*). The former has a horizontal (surface/planimetric) resolution of ~0.3–3 km, a vertical (depth) resolution of ~10 m (at best on the basis of the radar's RF bandwidth), and a maximum signal (depth) penetration of ~1 km (e.g. Seu et al., 2007). *SHARAD*'s resolution has been sufficient to detect mid-latitudinal glacier beds; by contrast, *MARSIS*'s resolution has been useful in penetrating the *PLD*s to depth and has a horizontal (surface/planimetric) resolution of ~5–9 km, a vertical (depth) resolution of ~150 m, and a maximum signal (depth) penetration of ~5 km (e.g. Jordan et al., 2009).

The higher-frequency *SHARAD* instrument (30 MHz) has identified internal reflectors (Fig. 10c and d), continuous over hundreds of kilometres in the *PLD*s, that could be indicative of vertical variability in the concentration/density of dust layers (Plaut et al., 2007). In some instances, and at some locations, a decametre-scale visible to near-infrared hyper-spectral data (i.e. from the *Compact Reconnaissance Imaging Spectrometer for Mars*, *CRISM*, *MRO*, e.g. Murchie et al., 2007) has driven the discrimination of water-ice from non-ice strata and/or horizons (e.g. Sinha & Horgan, 2022; also, Horgan et al., 2009; Massé et al., 2010, 2012).

Radar sounding of the *SPLD* using the lower-frequency (and higher penetration depth) *MARSIS* instrument has identified locations with a bright radar return that could comprise liquid water at the base of the ice (Lauro et al., 2021; Orosei et al., 2018) (Fig. 11). The possible identification of basal-ice water remains contentious and is inconsistent with our current geological understanding of basal heat-flux, as the latter is thought to be too low to generate meltwater at the bed of the *SPLD* (Sori & Bramson, 2019).

Bierson et al. (2021) posit that the reflectors could be hypersaline liquid-brine, saline ice, or conductive minerals, whereas Smith et al. (2021) suggest that they are partially hydrated smectite deposits. Regardless of their origin, these electromagnetic signature signals are enigmatic (Arnold et al., 2022; Schroeder & Steinbrügge, 2021).

The vertical (or subsurface depth-ranging) resolution of the *SHARAD/MARSIS* instruments is too low to resolve small (metre) scale internal layering, either in the *PLD*s or the mid-latitudinal Martian viscous-flow (glacial) features, if layering of this scale is present at all. While larger-scale internal layering and the basal topography are imaged for the PLDs (e.g. Putzig et al., 2009; Selvans et al., 2010), only the base of some mid-latitude lobate debris aprons is visible in radargrams (e.g. Berman et al., 2021; Holt et al., 2008; Karlsson et al., 2015; Petersen et al., 2018; Plaut et al., 2009). However, a putative internal structure has been identified within a small lobate debris apron (*LDA*) dissected by gully erosion (see Berman et al., 2021; Butcher et al., 2021) (Fig. 12).

This up-glacier (dipping) expression is consistent with a compressive flow regime that may be analogous to those within polythermal glaciers on Earth (also see Garvin et al., 2006). To date, this is the only observed location at which any internal structure is visible. The development and deployment of higher-frequency (GHz) radar instrumentation, e.g. *Hybrid Astronomical Radar Polarimeters* (*HARP*) systems (e.g. Raney, 2019, 2021), either in orbit or on the ground, would be extremely useful for widespread observation of internal layering at such metre scales. In this regard, the initial data and results delivered by the Mars *Perseverance* rover's *RIMFAX* radar (multi-frequency *GPR*) at the Jezero Crater are tantalizing (Hamran et al., 2022).

3.2 Icy exposures

At the mid-Martian latitudes of both hemispheres, glimpses of near-surface (possibly remnant glacial) ice have been observed within a group of geographically scattered, small, and freshly formed impact craters (e.g. Byrne et al., 2009; Dundas et al., 2014, 2021; Dundas and Byrne, 2010; and indirectly as

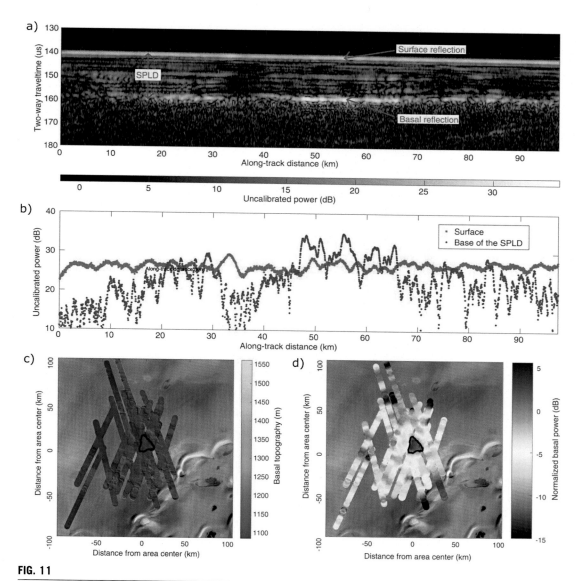

FIG. 11

(a) Bright basal reflection beneath the *SPLD*. (a) *MARSIS* radargram 10,737 across the *SPLD*. The bright reflection at 160 μs represents the *SPLD*/basal material interface; it may be related to the presence of water, hypersaline brine, reflective minerals, or some combination thereof. (b) Surface (*red dots*) and basal (*blue dots*) echo power, showing the echo associated with the bright basal interface in (a) is higher than that of the surface even after the signal has passed within the *SPLD*. (c) Basal topography surrounding the bright reflection. (d) Normalized echo power clearly shows the location of the main bright area; the value for each point represents the median from every radar footprint intersecting that location.

Figure adapted from Orosei et al. (2018).

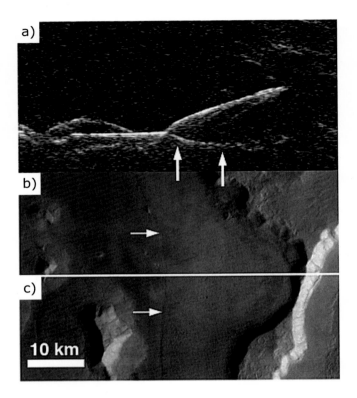

FIG. 12

(a) *MRO SHARAD*-based view of an *LDA*'s basal reflection in Deuteronilus Mensae. The vertical resolution of *SHARAD* is limited, and no internal layering is visible. In the upper panel (a and b), *arrows* indicate the location of the basal unconformity. In the lower panel (c), the *horizontal line* indicates the radar track, and *arrows* indicate the horizontal extent of the *LDA* (Plaut et al., 2009).

in Garvin et al., 2006) (Fig. 13) as well as on crater-based scarp faces (e.g. Dundas et al., 2018, 2021; Harish et al., 2020) in images acquired by the *H*igh *R*esolution *I*maging *S*cience *E*xperiment camera (*HiRISE*) on the *M*ars *R*econnaissance *O*rbiter (*MRO*), with a nominal resolution of ~25–50 cm/pixel (e.g. McEwen et al., 2007) (Fig. 14).

3.3 Icy scarps

At the mid-latitudes of both Martian hemispheres, some cross-sectional (*HiRISE*) images of scarps show massive and tabular water-ice deposits. The deposits are continuous, decametres deep, and buried within centimetres to metres of the surface (e.g. Dundas et al., 2018, 2021; Harish et al., 2020) (Fig. 14). The lateral reach or distribution of the ice beyond the observed exposures is unknown.

Typically, the ice overburden is incised by small-sized polygons (~10–15 m in diameter). At some locations, the polygonized terrain also comprises rimless and relatively shallow

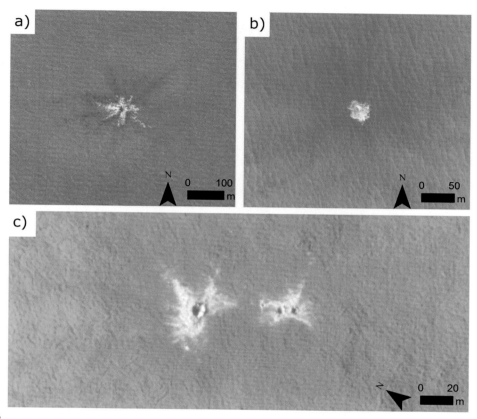

FIG. 13

Possible water ice exposed by the recent formation of impact craters; white colouration suggests that the ice is relatively pure. (a) *HiRISE* image ESP_017868_2440, Utopia Planitia. (b) *HiRISE* image ESP_025840_2240, Arcadia Planitia. (c) *HiRISE* image ESP_032340_1060, Aonia Terra.

Image credits: NASA/JPL/University of Arizona.

(thermokarst-like) depressions. It is presumed that the ice deposits formed by the airfall precipitation of snow or icy dust, its accumulation at the surface, and subsequently its burial by aeolian drift and a sublimation till (e.g. Bramson et al., 2015, 2017; Dundas et al., 2021; Stuurman et al., 2016). Layers exposed in these scarps can be oriented parallel to the surface and also oriented to intercept the surface (Fig. 14).

Similar (scarp-side) exposures of massive tabular ice, overlain or buried by a polygonized overburden dotted with thermokarstic depressions, occur in the continuous permafrost of the Tuktoyaktuk Coastlands in the Canadian arctic (e.g. MacKay, 1971; MacKay & Dallimore, 1992; Murton, 2001; Rampton, 1988; Soare et al., 2011; Vasil'chuk & Murton, 2016) and in the eastern Russian arctic (e.g. Grosse et al., 2007; Schirrmeister et al., 2013; Solomatin & Belova, 2012; Vasil'chuk & Murton, 2016).

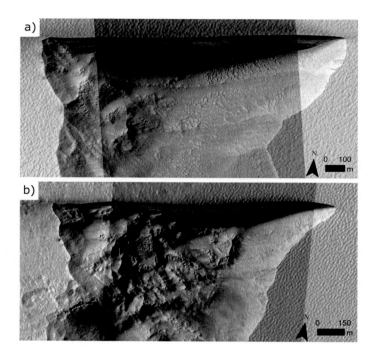

FIG. 14

Scarp exposures of water ice in *Promethei Terra*. (a) Cross-sectional view of an icy scarp-face and scarp pit in an enhanced-colour *HiRISE* image ESP_057466_1230 showing exposed water ice (*blue colouration*). The water-ice signature has been corroborated by spectral data (Dundas et al., 2018). The layers are oriented towards the surface and there is an apparent unconformity. (b) *HiRISE* image ESP_057321_1220 also shows exposed water ice. The layers are parallel to the surface.

Image credits: NASA/JPL/University of Arizona.

The origin of the massive tabular ice is glacial (never undergone freeze-thaw cycling at some locations) (e.g. French & Harry, 1990; Murton et al., 2004; Rampton, 1991), and periglacial or paraglacial (having undergone freeze-thaw cycling) at others (e.g. French & Harry, 1990; MacKay, 1971; Murton et al., 2004; Rampton, 1991). Where the overburden comprises a polygonized patterned ground, i.e. the work of thermal-contraction cracking, its origin postdates the formation and boundary conditions associated with the underlying ice and offers some insight on the variances of climate that shape this cold-climate landscape.

3.4 Ice-rich deposits

In the case of Martian subsurface deposits, structures, and features that could be the work of iterative freeze-thaw cycling, i.e. ice-rich sediments, ice wedges and lenticular cryo-structures, their relatively small scale means that the *SHARAD/MARSIS* instruments are ill-suited for their identification and volumetric characterization (Raney, 2019). However, glimpses of possible excess-ice deposits have

Bælum, K., & Benn, D. I. (2011). Thermal structure and drainage system of a small valley glacier (Tellbreen, Svalbard), investigated by ground penetrating radar. *The Cryosphere*, *5*(1), 139–149. https://doi.org/10.5194/tc-5-139-2011.

Bar-Cohen, Y., & Zacny, K. (2020). *Advances in terrestrial and extraterrestrial drilling: Ground, ice, and underwater*. Boca Raton, FL: CRC Press/Taylor & Francis Group LLC. https://www.routledge.com/Advances-in-Terrestrial-and-Extraterrestrial-Drilling-Ground-Ice-and/Bar-Cohen-Zacny/p/book/9781138341500.

Becerra, P., Byrne, S., Sori, M., Sutton, S., & Herkenhoff, K. E. (2016). Stratigraphy of the north polar layered deposits of Mars from high-resolution topography. *Journal of Geophysical Research*, *121*(8), 1445–1471. https://doi.org/10.1002/2015JE004992.

Berman, D. C., Chuang, F. C., Smith, I. B., & Crown, D. A. (2021). Ice-rich landforms of the southern mid-latitudes of Mars: A case study in Nereidum Montes. *Icarus*, *355*, 114170. https://doi.org/10.1016/j.icarus.2020.114170.

Bierson, C. J., Tulaczyk, S., Courville, S. W., & Putzig, N. E. (2021). Strong MARSIS radar reflections from the base of Martian south polar cap may be due to conductive ice or minerals. *Geophysical Research Letters*, *48*(13). https://doi.org/10.1029/2021GL093880. e2021GL093880.

Black, R. F. (1954). Permafrost—A review. *Bulletin Geological Society of America*, *85*, 839–856.

Bramson, A. M., Byrne, S., & Bapst, J. (2017). Preservation of midlatitude ice sheets on Mars. *Journal of Geophysical Research*, *122*, 2250–2266. https://doi.org/10.1002/2017JE005357.

Bramson, A. M., Byrne, S., Putzig, N. E., Sutton, S., Plaut, J. J., Brothers, T. C., & Holt, J. W. (2015). Widespread excess ice in Arcadia Planitia, Mars. *Geophys. Res. Lett.*, *42*(16), 6566–6574. https://doi.org/10.1002/2015GL064844.

Butcher, F. E. G., Arnold, N. S., Conway, S. J., Berman, D. C., Davis, M., & Balme, M. R. (2021). Potential for sampling of subglacial and englacial environments in Mars' mid latitudes, without deep drilling (*LPI* Contrib. No. 2548). In *52nd lunar and planetary science conference, Houston, Texas*.

Byrne, S. (2009). The polar deposits of Mars. *Annual Review of Earth and Planetary Sciences*, *37*(1), 535–560. https://doi.org/10.1146/annurev.earth.031208.100101.

Byrne, S., & Ingersoll, A. P. (2003). A sublimation model for Martian south polar ice features. *Science*, *299*, 1051–1053.

Byrne, S., et al. (2009). Distribution of mid-latitude ground ice on Mars from new impact craters. *Science*, *325*, 164. https://doi.org/10.1126/science.1175307.

Cavitte, M. G. P., et al. (2021). A detailed radiostratigraphic data set for the central East Antarctic Plateau spanning from the Holocene to the mid-Pleistocene. *Earth System Science Data*, *13*, 4759–4777. https://doi.org/10.5194/essd-13-4759-2021.

Christianson, K., Jacobel, R. W., Horgan, H. J., Anandakrishnan, S., & Alley, R. B. (2012). Subglacial Lake Whillans-ice-penetrating radar and GPS observations of a shallow active reservoir beneath a West Antarctic ice stream. *Earth and Planetary Science Letters*, *331*, 237–245. https://doi.org/10.1016/j.epsl.2012.03.013.

Christner, B. C., Mikucki, J. A., Foreman, C. M., Denson, J., & Priscu, J. C. (2005). Glacial ice cores: A model system for developing extraterrestrial decontamination protocols. *Icarus*, *174*(2), 572–584. https://doi.org/10.1016/j.icarus.2004.10.027.

Church, G., Bauder, A., Grab, M., Rabenstein, L., Singh, S., & Maurer, H. (2019). Detecting and characterising an englacial conduit network within a temperate Swiss glacier using active seismic, ground penetrating radar and borehole analysis. *Annals of Glaciology*, *60*(79), 193–205. https://doi.org/10.1017/aog.2019.19.

Cuffey, K. M., & Paterson, W. S. B. (2010). *The physics of glaciers*. Academic Press.

Cull, S., Arvidson, R. E., Mellon, M. T., Skemer, P., Shaw, A., & Morris, R. V. (2010). Compositions of subsurface ices at the Mars Phoenix landing site. *Geol. Res. Lett.*, *37*, L24203. https://doi.org/10.1029/2010GL045372.

Czudek, T., & Demek, J. (1970). Thermokarst in Siberia and its influence on the development of lowland relief. *Quaternary Research*, *1*, 103–120.

Diez, A., Matsuoka, K., Jordan, T. A., Kohler, J., Ferraccioli, F., Corr, H. F., Olesen, A. V., Forsberg, R., & Casal, T. G. (2019). Patchy lakes and topographic origin for fast flow in the recovery glacier system, East Antarctica. *J. Geophys. Res.*, *124*(2), 287–304. https://doi.org/10.1029/2018JF004799.

Douglas, T. A., Fortier, D., Shur, Y. L., Kanevskiy, M. Z., Guo, L., Cai, Y., & Bray, M. T. (2011). Biogeochemical and geocryological characteristics of wedge and thermokarst-cave ice in the CRREL permafrost tunnel, Alaska. *Permafrost Periglac. Process.*, *22*, 120–128. https://doi.org/10.1002/ppp.709.

Dowdeswell, J. A., Drewry, D. J., Liestøl, O., & Orheim, O. (2017). Radio echo-soundings of Spitsbergen glaciers: Problems in the interpretation of layer and bottom returns. *Journal of Glaciology*, *30*(104), 16–21. https://doi.org/10.3189/S0022143000008431.

Doyle, S., Dieser, M., Broemsen, E., & Christner, B. (2011). General characteristics of cold-adapted microorganisms, chapter 5. In R. V. Miller, & L. G. Whyte (Eds.), *Polar microbiology: Life in a deep freeze*. https://doi.org/10.1128/9781555817183.ch5.

Dundas, C. M., & Byrne, S. (2010). Modeling sublimation of ice exposed by new impacts in the Martian mid-latitudes. *Icarus*, *206*, 716–728. https://doi.org/10.1016/j.icarus.2009.09.007.

Dundas, C. M., et al. (2014). HiRISE observations of new impact craters exposing Martian ground ice. *Journal of Geophysical Research*, *119*, 109–127. https://doi.org/10.1002/2013JE004482.

Dundas, C. M., et al. (2018). Exposed subsurface ice sheets in the Martian mid-latitudes. *Science*, *359*, 199–201. https://doi.org/10.1126/science.aao1619.

Dundas, C. M., et al. (2021). Widespread exposures of extensive clean shallow ice in the mid-latitudes of Mars. *Journal of Geophysical Research*, *126*, e2020JE006617. https://doi.org/10.1029/2020JE.006617.

Dunse, T., Schuler, T., Hagen, J., Eiken, T., Brandt, O., & Høgda, K. (2009). Recent fluctuations in the extent of the firn area of Austfonna, Svalbard, inferred from GPR. *Annals of Glaciology*, *50*(50), 155–162. https://doi.org/10.3189/172756409787769780.

Egli, P. E., Belotti, B., Ouvry, B., Irving, J., & Lane, S. N. (2021). Subglacial channels, climate warming, and increasing frequency of alpine glacier snout collapse. *Geophysical Research Letters*, *48*(21). https://doi.org/10.1029/2021GL096031.

Faria, S., Wiekusat, I., & Azuma, N. (2014). The microstructure of polar ice. Part I: Highlights from ice core research. *Journal of Structural Geology*, *61*, 2–20. https://doi.org/10.1016/j.jsg.2013.09.010.

Farr, T. G., et al. (2002). Terrestrial analogs to Mars. In *The future of solar system exploration, 2003–2013. ASP conference series*.

Fischer, A. (2009). Calculation of glacier volume from sparse ice-thickness data, applied to Schaufelferner, Austria. *J. Glaciol.*, *55*(191), 453–460. https://doi.org/10.3189/002214309788816740.

Forte, E., Santin, I., Ponti, S., Colucci, R. R., Gutgesell, P., & Guglielmin, M. (2021). New insights in glaciers characterization by differential diagnosis integrating GPR and remote sensing techniques: A case study for the eastern gran Zebrù glacier (Central Alps). *Remote Sensing of Environment*, *267*, 112715. https://doi.org/10.1016/j.rse.2021.112715.

Fortier, D., Kanevskiy, M., & Shur, Y. (2008). Genesis of reticulate-chaotic cryostructure in permafrost. In D. L. Kane, & K. M. Hinkel (Eds.), *Vol. 1. Proceedings of the ninth international conference on permafrost, June 29–July 3, 2008* (pp. 451–456). Fairbanks, Alaska: Institute of Northern Engineering, University of Alaska.

French, H. M. (2007). *The periglacial environment* (3rd ed.). West Sussex, England: J. Wiley & Sons.

French, H. M., & Harry, D. G. (1990). Observations on buried glacial ice and massive segregated ice, western Arctic Coast, Canada. *Permafr. Perigl.*, *1*, 31–43.

Gallagher, C. (2024). *Glaciation and glacigenic geomorphology on Earth in the Quaternary Period*. In R. J. Soare, J.-P. Williams, C. Ahrens, F. E. G. Butcher, & M. R. El-Maarry (Eds.), *Ices in the solar system, a volatile-driven journey from the inner solar system to its far reaches*. Elsevier Books.

Garvin, J. B., Head, J. W., Marchant, D. R., & Kreslavsky, M. A. (2006). High-latitude cold-based glacial deposits on Mars: Multiple superposed drop moraines in a crater interior at 70° N latitude. *Meteoritics and Planetary Science*, *41*(10), 1659–1674. https://doi.org/10.1111/j.1945-5100.2006.tb00443.

Gerbi, C., et al. (2021). Microstructures in a shear margin: Jarvis Glacier, Alaska. *J. Glaciol.*, *67*(266), 1163–1176. https://doi.org/10.1017/jog.2021.62.

Goeller, S., Steinhage, D., Thoma, M., & Grosfeld, K. (2016). Assessing the subglacial lake coverage of Antarctica. *Annals of Glaciology*, *7*(72), 109–117.

Grab, M., Bauder, A., Ammann, F., Langhammer, L., Hellmann, S., Church, G. J., Schmid, L., Rabenstein, L., & Maurer, H. R. (2018). Ice volume estimates of Swiss glaciers using helicopter-borne GPR—An example from the glacier de la Plaine Morte. In *2018 17th international conference on ground penetrating radar (GPR)* (pp. 1–4). IEEE.

Grosse, G., Schirrmeister, L., Siegert, C., Kunitsky, V. K., Slagoda, E. A., Andreev, A. A., & Dereviagyn, Y. (2007). Geological and geomorphological evolution of a sedimentary periglacial landscape in Northeast Siberia during the late quaternary. *Geomorphology*, *86*, 25–51. https://doi.org/10.1016/j.geomorph.2006.08.005.

Hambrey, M. J. (1994). *Glacial environments*. UBC Press.

Hamran, S. E., et al. (2020). Radar imager for Mars' subsurface experiment-RIMFAX. *Space Science Reviews*, *216*, 128. https://doi.org/10.1007/s11214-020-00740-4.

Hamran, S.-E., et al. (2022). Ground penetrating radar observations of subsurface structures in the floor of Jezero crater, Mars. *Science Advances*, *8*(34). https://doi.org/10.1126/sciadv.abp8564.

Harish, Vijayan, S., Mangold, N., & Bhardwaj, A. (2020). Water-ice exposing scarps within the Northern Midlatitude Craters on Mars. *Geophysical Research Letters*, *47*. https://doi.org/10.1029/2020GL089057. e2020GL089057.

Harris, S. A., et al. (1988). Glossary of permafrost and related ground-ice terms. In S. A. Harris, H. M. French, J. A. Heginbottom, G. H. Johnston, B. Ladanyi, D. C. Sego, & R. O. van Everdingen (Eds.), *Technical Memorandum 142* (p. 154). National Research Council of Canada: Permafrost Subcommittee.

Hinkel, K. M., Doolittle, J. A., Bockheim, J. G., Nelson, F. E., Paetzold, R., Kimble, J. M., & Travis, R. (2001). Detection of subsurface permafrost features with ground-penetrating radar, Barrow, Alaska. *Permafr. Periglac. Process.*, *12*, 179–190. https://doi.org/10.1002/ppp.369.

Holschuh, N., Christianson, K., Conway, H., Jacobel, R. W., & Welch, B. C. (2015). Persistent tracers of historic ice flow in glacial stratigraphy near Kamb Ice Stream, West Antarctic. *The Cryosphere, 12*(9), 2821–2829. https://doi.org/10.5194/tc-12-2821-2018.

Holt, J. W., Safaeinilli, A., Plaut, J. J., Head, J. W., Phillips, R. J., Seu, R., … Gim, Y. (2008). Radar sounding evidence for buried glaciers in the Southern Mid-Latitudes of Mars. *Science, 322*, 1235–1238. https://doi.org/10.1126/science.1164246.

Horgan, B. H., Bell, J. F., III, Noe Dobrea, E. Z., Cloutis, E. A., Bailey, D. T., Craig, M. A., et al. (2009). Distribution of hydrated minerals in the north polar region of Mars. *Journal of Geophysical Research*, *114*(E1), E01005. https://doi.org/10.1029/2008je003187.

Hubbard, B., Glasser, N., Hambrey, M., & Etienne, J. (2004). A sedimentological and isotopic study of the origin of supraglacial debris bands: Kongsfjorden, Svalbard. *J. Glaciol.*, *50*(169), 157–170. https://doi.org/10.3189/172756504781830114.

I-MIM, M. D. T. (2022). *Final report of the International Mars Ice Mapper Reconnaissance*. Science Measurement Definition Team. 239 pp https://science.nasa.gov/researchers/ice-mapper-measurement-definition-team.

Irvine-Fynn, T. D. L., Moorman, B. J., Williams, J. L. M., & Walter, F. S. A. (2006). Seasonal changes in ground-penetrating radar signature observed at a polythermal glacier, Bylot Island, Canada. *Earth Surf. Process. Landf.*, *31*(7), 892–909. https://doi.org/10.1002/esp.1299.

Jennings, S. J. A., & Hambrey, M. J. (2021). Structures and deformation in glaciers and ice sheets. *Reviews of Geophysics*, *59*(3), 1–135. https://doi.org/10.1029/2021RG000743.

Jordan, R., et al. (2009). The Mars express MARSIS sounder instrument. *Planetary and Space Science*, *57*, 75–86. https://doi.org/10.1016/j.pss.2009.09.016.

Kanevskiy, M., Shur, Y., Fortier, D., Jorgenson, M. T., & Stephani, E. (2011). Cryostratigraphy of late Pleistocene syngenetic permafrost (yedoma) in northern Alaska, Itkillik River exposure. *Quatern. Res.*, *75*(3), 584–596. https://doi.org/10.1016/j.yqres.2010.12.003.

Kanevskiy, M., Yuri Shur, Y., Bigelow, N. H., Bjella, K. L., Douglas, T. A., Fortier, D., Jones, B. M., & Torre Jorgenson, M. (2022). Yedoma cryostratigraphy of recently excavated sections of the *CRREL* permafrost tunnel near fairbanks, Alaska. *Front. Earth Sci.*, *9*, 758800. https://doi.org/10.3389/feart.2021.758800.

Karlsson, N. B., Schmidt, L. S., & Hvidberg, C. S. (2015). Volume of Martian midlatitude glaciers from radar observations and ice flow modeling. *Geophysical Research Letters*, *42*(8), 2627–2633. https://doi.org/10.1002/2015GL063219.

Kassab, C. M., Licht, K. J., Petersson, R., Lindbäck, K., Graly, J. A., & Kaplan, M. R. (2020). Formation and evolution of an extensive blue ice moraine in central Transantarctic Mountains, Antarctica. *J. Glaciol.*, *66*(255), 49–60. https://doi.org/10.1017/jog.2019.83.

Kieffer, H. H. (1979). Mars south polar spring and summer temperatures: A residual CO_2 frost. *Journal of Geophysical Research*, *84*, 8263. https://doi.org/10.1029/JB084iB14p08263.

Kipfstuhl, S., Hamann, I., Lambrecht, A., Freitag, J., Faria, S. H., Grigoriev, D., & Azuma, N. (2006). Microstructure mapping: A new method for imaging deformation-induced microstructural features of ice on the grain scale. *Journal of Glaciology*, *52*(178), 398–406. https://doi.org/10.3189/172756506781828647.

Kohler, J., Moore, J. O. H. N., Kennett, M., Engeset, R., & Elvehøy, H. (1997). Using ground-penetrating radar to image previous years' summer surfaces for mass-balance measurements. *Annals of Glaciology*, *24*, 355–360. https://doi.org/10.3189/S0260305500012441.

Koutnik, M., Butcher, F., Soare, R., Hepburn, A., Hubbard, B., Brough, S., … Pathare, A. (2024). *Glacial deposits, remnants, and landscapes on Amazonian Mars: Using setting, structure, and stratigraphy to understand ice evolution and climate history*. In R. J. Soare, J.-P. Williams, C. Ahrens, F. E. G. Butcher, & M. R. El-Maarry (Eds.), *Ices in the solar system, a volatile-driven journey from the inner solar system to its far reaches*. Elsevier Books.

Lauro, S. E., et al. (2021). Multiple subglacial water bodies below the south pole of Mars unveiled by new MARSIS data. *Nature Astronomy*, *5*(1), 63–70. https://doi.org/10.1038/s41550-020-1200-6.

Léger, E., Dafflon, B., Soom, F., et al. (2017). Quantification of arctic soil and permafrost properties using ground-penetrating radar and electrical resistivity tomography datasets. *IEE. J.*, *10*(10). https://doi.org/10.1109/JSTARS.2017.2694447.

MacGregor, J. A., Fahnestock, M. A., Catania, G. A., Paden, J. D., Prasad Gogineni, S., Young, S. K., … Morlighem, M. (2015). Radiostratigraphy and age structure of the Greenland Ice Sheet. *Journal of Geophysical Research (Earth Surface)*. https://doi.org/10.1002/2014JF003215.

Machguth, H., Eisen, O., Paul, F., & Hoelzle, M. (2006). Strong spatial variability of snow accumulation observed with helicopter-borne GPR on two adjacent Alpine glaciers. *Geophysical Research Letters*, *33*(13). https://doi.org/10.1029/2006GL026576.

MacKay, J. R. (1971). The origin of massive icy beds in permafrost, Western Arctic Coast, Canada. *Can. J. Earth Sci.*, *8*(4), 397–422.

MacKay, J. R., & Dallimore, S. (1992). Massive ice of the Tuktoyaktuk area, western Arctic coast, Canada. *Can. J. Earth Sci.*, *29*(6), 1235–1249. https://doi.org/10.1139/e92-099.

MacKay, S. L., Marchant, D. R., Lamp, J. L., & Head, J. W. (2014). Cold-based debris-covered glaciers: evaluating their potential as climate archives through studies of ground-penetrating radar and surface morphology. *Journal of Geophysical Research*, *119*(11), 2505–2540. https://doi.org/10.1002/2014JF003178.

Maguire, R., Schmerr, N., Pettit, E., Riverman, K., Gardner, C., DellaGiustina, D. N., Avenson, B., Wagner, N., Marusiak, A. G., Habib, N., & Broadbeck, J. I. (2021). Geophysical constraints on the properties of a subglacial lake in Northwest Greenland. *The Cryosphere*, *15*(7), 3279–3291. https://doi.org/10.5194/tc-15-3279-2021.

Mars Ice Core Working Group. (2021). *First ice cores from Mars, co-chairs: M.R. Albert and M. Koutnik, 74p. White paper*. https://www.nasa.gov/sites/default/files/atoms/files/mars_ice_core_working_group_report_final_feb2021.pdf.

Massé, M., Bourgeois, O., le Mouélic, S., Verpoorter, C., le Deit, L., & Bibring, J.-P. (2010). Martian polar and circum-polar sulfate-bearing deposits: Sublimation tills derived from the north polar cap. *Icarus*, *209*(2), 434–451. https://doi.org/10.1016/j.icarus.2010.04.017.

Massé, M., Bourgeois, O., le Mouélic, S., Verpoorter, C., Spiga, A., & le Deit, L. (2012). Wide distribution and glacial origin of polar gypsum on Mars. *Earth and Planetary Science Letters, 317*, 44–55. https://doi.org/10.1016/j.epsl.2011.11.035.

McCubbin, F. M., et al. (2019). Advanced curation of astromaterials for planetary science. *Space Science Reviews, 215*(8), 48. https://doi.org/10.1007/s11214-019-0615-9.

McEwen, A. S., et al. (2007). Mars reconnaissance orbiter's high-resolution imaging science experiment (*HiRISE*). *Journal of Geophysical Research, 112*, E05S02. https://doi.org/10.1029/2005JE002.605.

Mellon, M. T., Malin, M. C., Arvidson, R. E., Searls, M. L., Sizemore, H. G., Heet, T. L., Lemmon, M. T., Keller, H. U., & Marshall, J. (2009a). The periglacial landscape at the Phoenix landing site. *Journal of Geophysical Research, 114*(E1). https://doi.org/10.1029/2009JE003418.

Mellon, M. T., et al. (2009b). Ground ice at the Phoenix landing site: Stability state and origin. *Journal of Geophysical Research, 114*, E00E07. https://doi.org/10.1029/2009JE003417.

MEPAG. (2020). In D. Banfield (Ed.), *Mars scientific goals, objectives, investigations, and priorities: 2020* The Mars Exploration Program Analysis Group (MEPAG). https://mepag.jpl.nasa.gov/reports.cfm.

MEPAG ICE-SAG Final Report, 2019. Report from the ICE and climate evolution science analysis group (ICE-SAG), chaired by S. Diniega and N. E. Putzig, the Mars exploration program analysis group (MEPAG) at: http://mepag.nasa.gov/reports.cfm.

Moorman, B. J., Judge, A. S., Burgess, M. M., & Fridel, T. W. (1994). Geotechnical investigations of insulated permafrost slopes along the Norman Wells pipeline using ground penetrating radar. In *GPR '94: Fifth international conference on ground penetrating radar, Kitchener, Ontario. Waterloo Centre for Groundwater Research, University of Waterloo* (pp. 477–491).

Moorman, B. J., & Michel, F. A. (2000). Glacial hydrological system characterization using ground-penetrating radar. *Hydrological Processes, 14*, 2645–2667.

Murchie, S., et al. (2007). Compact reconnaissance imaging spectrometer for Mars (CRISM) on Mars reconnaissance orbiter (MRO). *Journal of Geophysical Research, 112*, E05S03. https://doi.org/10.1029/2006JE002682.

Murton, J. B. (2001). Thermokarst sediments and sedimentary structures, Tuktoyaktuk Coastlands, western Arctic Canada. *Global and Planetary Change, 28*, 175–192. https://doi.org/10.1016/S0921-8181(00)00072-2.

Murton, J. B., Waller, R. I., Hart, J. K., Whitement, C. A., Pollard, W. H., & Clark, I. D. (2004). Stratigraphy and glaciotectonic structures of permafrost deformed beneath the northwest margin of the Laurentide ice sheet, Tuktoyaktuk Coastlands, Canada. *J. Glaciol., 50*(170), 399–411.

National Academies of Sciences, Engineering, and Medicine (NASEM). (2022). *Origins, worlds, and life: A decadal strategy for planetary science and astrobiology 2023–2032*. Washington, DC: The National Academies Press. https://doi.org/10.17226/26522.

Orosei, R., et al. (2018). Radar evidence of subglacial liquid water on Mars. *Science, 361*(6401), 490–493. https://doi.org/10.1126/science.aar72.

Penner, E. (1959). The mechanism of frost heaving in soils. *Highway Res. Board Bull., 225*, 1–22.

Petersen, E. I., Holt, J. W., & Levy, J. S. (2018). High ice purity of Martian lobate debris aprons at the regional scale: Evidence from an orbital radar sounding survey in Deuteronilus and Protonilus Mensae. *J. Geophys. Lett.*. https://doi.org/10.1029/2018GL079759.

Petersen, E. I., Levy, J. S., Holt, J. W., & Stuurman, C. M. (2020). New insights into ice accumulation at Galena Creek Rock Glacier from radar imaging of its internal structure. *Journal of Glaciology, 66*(255), 1–10. https://doi.org/10.1017/jog.2019.67.

Pettersson, R., Jansson, P., & Holmlund, P. (2003). Cold surface layer thinning on Storglaciären, Sweden, observed by repeated ground penetrating radar surveys. *Journal of Geophysical Research—Earth Surface, 108*(F1). https://doi.org/10.1029/2003JF000024.

Phillips, R. J., et al. (2011). Massive CO_2 ice deposits sequestered in the south polar layered deposits of Mars. *Science, 332*(6031), 838–841. https://doi.org/10.1126/science.1203091.

Plaut, J. J., Safaeinili, A., Holt, J. W., Phillips, R. J., Head, J. W., III, Seu, R., Putzig, N. E., & Frigeri, A. (2009). Radar evidence for ice in lobate debris aprons in the mid-northern latitudes of Mars. *Geophys. Res. Letters*, *36*(2). https://doi.org/10.1029/2008GL036379.

Plaut, J. J., et al. (2007). Subsurface radar sounding of the south polar layered deposits of Mars. *Science, 316*, 92–95. https://doi.org/10.1126/science.1139672.

Putzig, N. E., et al. (2009). Subsurface structure of Planum Boreum from Mars reconnaissance orbiter shallow radar soundings. *Icarus, 204*(2), 443–457. https://doi.org/10.1016/j.icarus.2009.07.034.

Rampton, V. N. (1988). *Quaternary geology of the Tuktoyaktuk Coastlands, Northwest Territories* (p. 423). Geological Survey of Canada, Memoir (98 p).

Rampton, V. N. (1991). Observations on buried glacial ice and massive segregated ice, Western Arctic Canada: Discussion. *Permafr. Periglac, 2*, 163–1991.

Raney, R. K. (2019). Hybrid dual-polarization synthetic aperture radar. *Remote Sensing, 11*(13), 1521. https://doi.org/10.3390/rs11131521.

Raney, R. K. (2021). Polarimetric portraits. *Earth Space Sci., 8*(7). https://doi.org/10.1029/2021EA001768. e2021EA001768.

Robin, G.d. Q., Evans, E., & Bailey, J. T. (1969). Interpretation of radio echo sounding in polar ice sheets. *Philosophical Transactions of the Royal Society of London Series A, 265*, 437–505.

Schennen, S., Wetterich, S., Schirrmeister, L., Schwamborn, G., & Tronicke, J. (2022). Seasonal impact on 3D GPR performance for surveying Yedoma ice complex deposits. *Frontiers in Earth Science, 10*, 741524. https://doi.org/10.3389/feart.2022.741524.

Schirrmeister, L., Froese, D., Tumskoy, V., Grosse, G., & Wetterich, S. (2013). Yedoma: Late Pleistocene ice-rich syngenetic permafrost of Beringia. In S. A. Elias, & C. J. Mock (Eds.), *Encyclopedia of quaternary science* (2nd ed., pp. 542–552).

Schroeder, D. M., & Steinbrügge, G. (2021). Alternatives to liquid water beneath the south polar ice cap of Mars. *Geophysical Research Letters, 48*(19). https://doi.org/10.1029/2021GL095912. e2021GL095912.

Schroeder, D. M., et al. (2020). Five decades of radioglaciology. *Annals of Glaciology, 61*(81), 1–13. https://doi.org/10.1017/aog.2020.11.

Selvans, M. M., Plaut, J. J., Aharonson, O., & Safaeinili, A. (2010). Internal structure of Planum Boreum, from Mars advanced radar for subsurface and ionospheric sounding data. *Journal of Geophysical Research, 115*, E09003. https://doi.org/10.1029/2009JE003537.

Seu, R., et al. (2007). SHARAD sounding radar on the Mars reconnaissance orbiter. *Journal of Geophysical Research, 112*, E05S05. https://doi.org/10.1029/2006JE002745.

Sinha, P., & Horgan, B. (2022). Sediments within the icy north polar deposits of Mars record recent impacts and volcanism. *Geophysical Research Letters, 49*. https://doi.org/10.1029/2022GL097758. e2022GL097758.

Smith, I. B., & Holt, J. W. (2015). Spiral trough diversity on the north pole of Mars, as seen by shallow radar (*SHARAD*). *Journal of Geophysical Research, 120*, 362–387. https://doi.org/10.1002/2014JE004720.

Smith, I. B., Lalich, D. E., Rezza, C., Horgan, B. H. N., Whitten, J. L., Nerozzi, S., & Holt, J. W. (2021). A solid interpretation of bright radar reflectors under the Mars south polar ice. *Geophysical Research Letters, 48*(15). https://doi.org/10.1029/2021GL093618.

Smith, I. B., et al. (2020). The holy grail: A road map for unlocking the climate record stored within Mars' polar layered deposits. *Planetary and Space Science, 184*, 104841. https://doi.org/10.1016/j.pss.2020.104841.

Soare, R. J., Séjourné, A., Pearce, G., Costard, F., & Osinski, G. R. (2011). The Tuktoyaktuk Coastlands of Northern Canada: A possible "Wet" periglacial analogue of Utopia Planitia, Mars. *Geological Society of America Special Paper, 483*. https://doi.org/10.1130/2011.2483(13).

Soare, R. J., Costard, F., Williams, J.-P., Gallagher, C., Hepburn, A. J., Stillman, D., … Godi, E. (2024). *Evidence, arguments, and cold-climate geomorphology that favour periglacial cycling at the Martian mid-to-high latitudes in the Late Amazonian Epoch*. In R. J. Soare, J.-P. Williams, C. Ahrens, F. E. G. Butcher, & M. R. El-Maarry (Eds.), *Ices in the solar system, a volatile-driven journey from the inner solar system to its far reaches*. Elsevier Books.

Solomatin, V. I., & Belova, N. G. (2012). Proof of the glacier origin of tabular massive ice. *Proceedings of the Tenth International Conference on Permafrost*, *10*(2), 427–431.

Sori, M. M., & Bramson, A. M. (2019). Water on Mars, with a grain of salt: Local heat anomalies are required for basal melting of ice at the south pole today. *Geophysical Research Letters*, *46*(3), 1222–1231. https://doi.org/10.1029/2018GL080985.

Stoll, N., Eichler, J., Horhold, M., Shigeyama, W., & Weikusat, I. (2021). A review of the microstructural location of impurities in polar ice and their impacts of deformation. *Front. Earth. Sci.*. https://doi.org/10.3389/feart.2020.615613.

Stuurman, C. M., Osinski, G. R., Holt, J. W., Levy, J. S., Brothers, T. C., Kerrigan, M., & Campbell, B. A. (2016). SHARAD detection and characterization of subsurface water ice deposits in Utopia Planitia, Mars. *Geophys. Res. Lett.*, *43*(18), 9484–9491. https://doi.org/10.1002/2016GL070138.

Sylvestre, T., Copland, L., Demuth, M. N., & Sharp, M. (2013). Spatial patterns of snow accumulation across Belcher Glacier, Devon Ice Cap, Nunavut, Canada. *J. Glaciol.*, *59*(217), 874–882. https://doi.org/10.3189/2013JoG12J227.

Taber, S. (1930). The mechanics of frost heaving. In U.S. Army Corp of Engineers, P. B. Black, & M. J. Hardenburg (Eds.), *Historical perspectives in frost heave research: The early works of S. Taber and G. Beskow. Special Report 91-23* (pp. 9–26). 1991, 159 p.

Vasil'chuk, Y. K., & Murton, J. B. (2016). Stable isotope geochemistry of massive ice. *Geography, Environment Sustainability*, *03*(09), 4–24. https://doi.org/10.15356/2071-9388_03v09_2016_01.

Weber, J. R., & Andrieux, P. (1970). Radar soundings on the penny ice cap, Baffin Island. *Journal of Glaciology*, *9*, 49–54.

Wetterich, S., Tumskoy, V., Rudaya, N., Andreev, A. A., Opel, T., Meyer, H., Schirrmeister, L., & Hüls, M. (2004). Ice complex formation in arctic East Siberia during the MIS3 interstadial. *Quatern. Sci. Rev.*, *84*, 39–55. https://doi.org/10.1016/j.quascirev.2013.11.009.

Whitten, J. L., & Campbell, B. A. (2018). Lateral continuity of layering in the Mars south polar layered deposits from SHARAD sounding data. *Journal of Geophysical Research*. https://doi.org/10.1029/2018JE005578.

Wilcoski, A. X., & Hayne, P. O. (2020). Surface roughness evolution and implications for the age of the north polar residual cap of Mars. *Journal of Geophysical Research*, *125*(12).

Williams, R. M., Ray, L. E., Lever, J. H., & Burzynski, A. M. (2014). Crevasse detection in ice sheets using ground penetrating radar and machine learning. *IEEE Journal of Selected Topics in Applied Earth Observations and Remote Sensing*, *7*(12), 4836–4848. https://doi.org/10.1109/JSTARS.2014.2332872.

Winter, A., Steinhage, D., Creyts, T. T., Kleiner, T., & Eisen, O. (2019). Age stratigraphy in the East Antarctic ice sheet inferred from radio-echo sounding horizons. *Earth System Science Data*, *11*, 1069–1081. https://doi.org/10.5194/essd-11-1069-2019.

Yan, Y., Bender, M. L., Brook, E. J., et al. (2019). Two-million-year-old snapshots of atmospheric gases from Antarctic ice. *Nature*, *574*, 663–666. https://doi.org/10.1038/s41586-019-1692-3.

Ceres—A volatile-rich dwarf planet in the asteroid belt

7

Margaret E. Landis[a], Julie Castillo-Rogez[b], and Caitlin J. Ahrens[c]

[a]*Laboratory for Atmospheric and Space Physics, University of Colorado, Boulder, CO, United States,* [b]*Jet Propulsion Laboratory, California Institute of Technology, Pasadena, CA, United States,* [c]*NASA Goddard Space Flight Center, Greenbelt, MD, United States*

Abstract

Well before the arrival of the Dawn mission in 2015, the dwarf planet Ceres, the largest object in the main asteroid belt, was known to be water-rich and suspected to host water-driven geophysical and geological processes. Dawn confirmed the significant role of water in shaping the chemical and physical evolution of Ceres and found evidence that it accreted organics and volatiles such as carbon dioxide and ammonia. Water-rock interaction led to chemical fractionation and the production of brines (salt-rich liquid) containing significant carbonate and bicarbonate, as well as ammonium and chlorides. Local enhancements in water ice and other salt deposits suggest that some amount of ongoing volatile exposure and loss is occurring in the present day from likely extensive subsurface layers of material. Although the Dawn spacecraft ran out of fuel in 2018 and remains silently in a ~50-year stable orbit, the analysis of the extensive dataset returned by the mission keeps expanding our understanding of this large, water-rich body and has fostered in situ and sample return mission concepts about investigating Ceres' past and present habitability potential that may be considered for flight in the next decade.

1 Pre-Dawn evidence for a volatile-rich Ceres

1.1 Discovery and early telescopic observations

Ceres was discovered by Giuseppe Piazzi at the Observatorio Astronomico di Palermo in January 1801 as part of a larger star cataloging campaign. While the orbit of Ceres was initially uncertain, it was eventually identified to be at an orbital distance between Mars and Jupiter. The discovery of a new planet was announced in 1802, as it lacked a tail and was in a location predicted by the now-defunct and -debunked Titus-Bode Law. However, arguments were made by scientists including Herschel (1832) that Ceres' small mass and position within a field of numerous other planetary bodies implied that Ceres was representative of a different type of object than planets Earth, Jupiter, and Uranus. Ceres was 'demoted' to the class of asteroids. It was later re-promoted to a dwarf planet along with Pluto and other bodies in the Kuiper Belt (e.g., Binzel, 2006).

Ceres' discovery and classification as an object in our Solar System (summarized by Serio et al., 2002) reflects our changing understanding of how small bodies fit into the overall picture of solar system history and evolution. A frequent target of telescopic campaigns even without its planet status, much was known before Dawn's arrival that suggested that Ceres was unusual among other carbonaceous asteroids. In particular, many studies reported the presence of a 3 μm spectral feature on Ceres, which had interpretations ranging from global water ice frost to water incorporated into the mineral structure of different species (e.g., Rivkin et al., 2011 and references therein). Notably, King et al. (1992) suggested a match for this spectral feature with ammoniated phyllosilicates, which the Dawn observations confirmed. Other interpretations, such as the presence of brucite, a magnesium hydroxide, and cronstedtite, an iron-rich form of serpentine by Milliken and Rivkin (2009), were discarded based on the broad spectral coverage of the Dawn infrared measurements (De Sanctis et al., 2015). On the other hand, Rivkin et al. (2006) and Milliken and Rivkin (2009) identified the presence of an absorption band at 4 μm interpreted as magnesium-rich carbonates, which was later on confirmed by the Dawn mission. Carbonates are evidence of water-rock alteration and these early papers paved the way for framing Ceres as a body of astrobiological significance.

1.2 Density and relaxed shape from Hubble Space Telescope (HST) data

The pre-Dawn hypothesis that Ceres contains a large fraction of water comes from the early realization that its density is low. Based on a variety of pre-Dawn observations (occultations and mutual interactions between large asteroids), Ceres' density was known to be about $2 \, g/cm^3$ (see McCord & Sotin, 2005 for a review of the state of knowledge of Ceres' physical properties before Dawn). This density corresponds to about equal volume fractions of anhydrous rock and ice, which makes Ceres the most water-rich object in the inner solar system after Earth. Since hydrated rocks may contain no more than 20% (molar) of water in their structure, Ceres should still contain "free water," i.e., water in the form of ice or liquid. Based on analogies with the evolution of icy moons, McCord and Sotin (2005) also inferred that Ceres should have differentiated a rocky mantle that is likely hydrated and an outer shell dominated by water.

Independent constraints on Ceres' water content and internal structure came from observations with the HST. Shape measurements by Thomas et al. (2005) indicated that Ceres' shape is relaxed to hydrostatic equilibrium and matches that expected for an oblate ellipsoid consistent with Ceres' angular velocity. Thomas et al. (2005) also showed that the shape is consistent with a rocky mantle and water-rich shell, but the uncertainties on the radii determination limited the extent of this interpretation. Castillo-Rogez and McCord (2010) showed that the shape was incompatible with the presence of a metallic core, providing an early hint that Ceres' thermal evolution remained relatively mild and that the internal partitioning of materials may not be complete.

1.3 Exosphere

In the late 1980s and 1990s, well before the Dawn mission, Ceres was observed to have an OH exosphere (A'Hearn & Feldman, 1992). Preliminary surface thermal models suggested it was possible for water ice to survive near the surface (Fanale & Salvail, 1989). Small bodies in the solar system producing some amount of volatile vapor and dust are not unheard of (see El-Maarry, 2024, pp. 261–282). However, the largest asteroid in the asteroid belt potentially showing signs of comet-like activity implied a potentially current water and volatile processing on or near the surface. This was in contrast to other S-type asteroids explored by spacecraft in that decade, like (433) Eros by the NEAR-Shoemaker mission.

An exosphere was not positively replicated again until 2014, when Küppers et al. (2014), using the Herschel Space Observatory, reported the detection of water vapor, rather than OH, around Ceres. Observations by Küppers et al. (2014) were performed from 2012 to 2013. Taking into account the previous positive detections in 1992, these multi-decade temporal spacing between detections of an exosphere suggest transient activity. The production rates were estimated to be upward of ~6 kg/s based on the modeling of the observed spectra (Küppers et al., 2014). However, if these rates persisted over the age of the solar system, it would cause 10s of meter scale collapse of the surface (Landis et al., 2017; Schorghofer et al., 2016).

Additional missions such as Hayabusa 2 (Sugita et al., 2019) and OSIRIS-REX (Kaplan et al., 2020) indicated that aqueous alteration on the C-type parent bodies may have been common. Ceres, the largest of the asteroids, is certainly a unique C type, though the opportunity to observe it up close would constrain what types of aqueous alteration could have been common on larger, internally differentiated C types. Improved thermal and vapor diffusion modeling (e.g., Hayne & Aharonson, 2015; Schorghofer, 2008) before Dawn's arrival continued to support the idea that if present, water ice could be trivial to trap below Ceres' surface. Therefore, the alteration of Ceres' surface by water in the present day, perhaps related to its water vapor exosphere, would be observable by Dawn.

The postulated sources of these exospheric detections ranged from the mundane, like a sufficiently large impact crater exposing water ice in the near-surface region and/or sputtering from micrometeorites or solar particle events (e.g., Villarreal et al., 2017), to the more exotic, like plumes of material from the interior like on Enceladus (e.g., Titus, 2015). The discovery and characterization of a group of objects with orbits like asteroids but with dust tails like comets, known as Main Belt Comets (e.g., El-Maarry, 2024, pp. 261–282; Hsieh & Jewitt, 2006), lend credence to the idea that the queen of the asteroid belt may be more volatile rich, and that volatile-rich asteroids have played a major role in the chemistry of the asteroid belt.

2 Dawn mission background and timeline

Dawn was the ninth project in the National Aeronautics and Space Administration's (NASA's) Discovery Program. The Dawn mission proposal was selected by NASA in 2001 and relied on solar electric propulsion to reach and orbit two targets in the main asteroid belt. A launch in 2007 would take advantage of an optimal configuration of Vesta and Ceres, which Dawn could achieve thanks to the flexibility offered by solar electric propulsion (Fig. 1). A history of the Dawn mission can be found in Russell and Raymond (2011).

Dawn started its approach to Ceres in December 2014 and was captured in orbit on March 6, 2015. As the most up-close observations of Ceres to date, how these observations were taken provides a key background to their interpretation as later reported in the chapter. We briefly summarize the various phases of the mission and refer the reader to Rayman (2020, 2022) for more detail. The Dawn science orbits at Ceres (Table 1) progressed from a high-altitude (~13,600 km) rotation characterization phase to the Survey phase (4400 km), then high altitude mapping orbit (1470 km), and then a low altitude mapping orbit (385 km) followed by a first extended mission at that same altitude. Near-global mapping with the Framing Camera was achieved in each of these orbits at a resolution going from 1.3 km/px to 30 m/px. The Dawn mission was extended to improve elemental measurements and coverage with VIR and in color filters. The last phase of the mission consisted of a highly elliptical orbit with a periapsis at about 35 km from Ceres' surface along a swath that encompassed Occator crater.

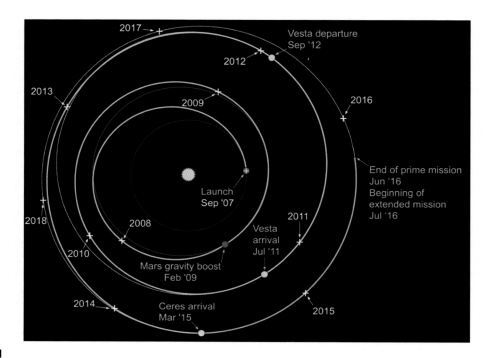

FIG. 1

Dawn's interplanetary trajectory following its launch in September 2007. After a Mars gravity assist in February 2009, Dawn reached Vesta in July 2011. The spacecraft left Vesta in September 2012 for a 2.5-year journey to Ceres. Dawn's prime mission at Ceres was followed by two extended missions until the spacecraft ran out of fuel in October 2018.

Credit: Planetary Society based on NASA/JPL-Caltech.

The Dawn spacecraft ran out of fuel in October 2018 and remains in orbit around Ceres. Due to planetary protection concerns associated with the presence of ice in the shallow subsurface at mid-latitudes and the recent exposure of brines at Occator crater, Dawn's orbit was designed so that the spacecraft does not impact Ceres' surface for at least 50 years (which includes a margin over the 20-year non-impact requirement set by NASA). This gives a future team a 50-year window to send a follow-up mission before Dawn potentially hits the surface and provides one more impact on its heavily cratered surface.

2.1 Instruments

Dawn's scientific measurements included panchromatic stereo and multispectral images; neutron, near ultraviolet, visible, infrared, and γ-ray spectra; and gravimetry (Fig. 2). To acquire these data, Dawn's instrument payload comprised a γ-ray and neutron detector (GRaND) (Prettyman et al., 2011), a visible and infrared mapping spectrometer (VIR) (De Sanctis et al., 2011), and a pair of identical cameras (framing camera #2, or FC2, the prime unit, and FC1, the backup) (Sierks et al., 2011). Gravimetry

Table 1 Main characteristics of the Dawn science phases at Ceres.

Science phase	Start date	End date	Duration (days)	Altitude (km)	Period (h)
Approach	December 26, 2015	April 24, 2015	119	n/a	n/a
Capture	March 6, 2015		n/a	60,600	n/a
RC3	April 24, 2015	May 9, 2015	15	13,600	15.2 day
Survey	June 5, 2015	June 30, 2015	27	4400	3.1 h
HAMO	August 17, 2015	October 23, 2015	57	1470	18.8 h
LAMO	December 16, 2015	June 30, 2016	197	385	5.4 h
XMO1	July 1, 2016	September 2, 2016	63	385	5.4
XMO2	October 17, 2016	November 4, 2016	18	1480	19
XMO3	December 5, 2016	February 23, 2017	80	7500×9300	185
XMO4	April 13, 2017	June 3, 2017	51	14,000×53,000	1350
XMO5	June 23, 2017	April 16, 2018	297	5400×38,000	730
XMO6	May 14, 2018	May 31, 2018	17	450×4730	37
XMO7	June 9, 2018	October 31, 2018	144	35×4000	27

was accomplished via the telecommunications subsystem and did not require dedicated flight hardware (Konopliv et al., 2011).

All of these instruments had the capability to detect water in various forms. The infrared channel of the VIR spectrometer was designed to cover a wavelength range from 1 to 5 μm, which encompasses absorption bands for water in hydroxyl form in silicate (~2.72 μm) and in ice (1.28, 1.65, ~2 μm, ~3 μm), as well as carbonate (~3.90–4.00 μm) and organic absorption bands (between 3.2 and 3.6 μm). The GRaND instrument could sense the presence of water up to ~1 m depth, using the principle of neutron moderation to perform nuclear spectroscopy. Global imaging of Ceres in the visible image and color revealed many expressions of the presence of water in the subsurface, both spectrally and geomorphologically. The topography data, derived from repeat imaging of the surface, was used to assess the mechanical properties of the crust from geomorphology and inferences from shape and gravity data. Photometry observations obtained in the visible provided insights into the surface properties and stratigraphic relationships. Lastly, the gravity observations, combined with shape data, were used to constrain the structure of the interior. Working in concert, these remote sensing techniques provided the comprehensive inventory of volatiles we have today.

2.2 Data coverage and types

Throughout its prime and extended missions, Dawn performed a near-global coverage of Ceres in the visible, color, infrared, gamma rays and neutron coverage; tracking of the spacecraft from the ground was also used for gravity science (Fig. 2). Examples of properties derived from these datasets are illustrated in Fig. 3. These include visible imaging (at pixel scales as good as 3.5 m/pix), color maps used to infer constraints on the surface lithology and age (e.g., Li et al., 2019; Schmedemann et al., 2016), hydrogen and iron abundances (average and variability on a regional scale), topography at a resolution of 100 m (Park et al., 2019), and gravity maps and anomalies at a harmonic degree of 18 on a global scale

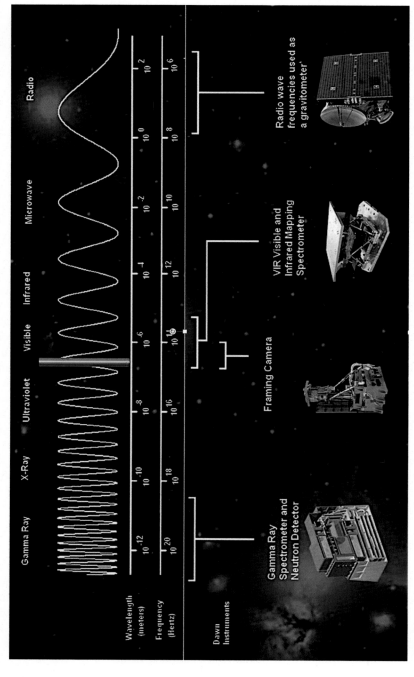

FIG. 2

Dawn mission instruments and the parts of the electromagnetic spectrum they sample. The Framing Camera and VIR sample the electromagnetic spectrum while elemental spectroscopy realizes galactic cosmic rays and subsurface neutron transport. Gravity science was performed using the spacecraft telecommunications subsystem. These various instruments brought complementary information on the properties of Vesta and Ceres: elemental abundances (GRaND), mineralogy (VIR), lithology and geology (from albedo, color, FC), and internal structure (gravity).

Based on a figure from Shantanu Naidu, source https://solarsystem.nasa.gov/missions/dawn/overview/.

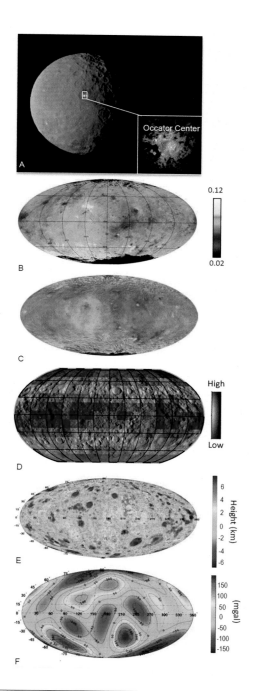

FIG. 3

Examples of data products returned by Dawn at Ceres. These products are highly complementary and their combined analysis is leading to inferences on the chemical and physical evolution of the dwarf planet, as summarized in this chapter. (A) Visible mapping, (B) 3.1 μm band depth, (C) composite color map (infrared, red, blue filters), (D) hydrogen content, (E) topography mapping, and (F) Bouguer anomalies.

and up to 45 on a local scale (covering Occator crater) (Park et al., 2016, 2020). These many datasets led to a very comprehensive geological mapping of the dwarf planet (e.g., a special issue by Williams et al., 2018) which provides critical context for compositional and morphometric observations. In the course of the mission, repeated observations of Juling crater were performed in order to track the variability in the VIR detection of water in the Northern wall of the crater. A global FC mosaic of Ceres with key locations labeled is given in Fig. 4.

3 State of knowledge after Dawn

3.1 Volatile composition

Dawn revealed the presence of a large fraction of water on Ceres, in various forms detailed in Section 3.3. The Dawn VIR spectrometer also identified minerals derived from more volatile compounds. Ammonium has been found in several forms: in the interlayers of phyllosilicates (De Sanctis et al., 2015) and in salts, in particular ammonium chlorides (Raponi et al., 2019) and potentially carbonate (De Sanctis et al., 2016). Ammonium is likely to have evolved from ammonia, some of which partitioned to form ammonium in the early ocean, depending on environmental conditions and in particular the pH (Castillo-Rogez et al., 2022b). The leading hypothesis on the origin of ammonia is that it was accreted in the form of ice (Castillo-Rogez et al., 2018; De Sanctis et al., 2015), indicating an origin where the temperatures were favorable to low-temperature alteration. It is also possible that ammonia was produced from the breakdown of organic compounds (e.g., amides, amino acids) (Pizzarello & Williams, 2012) at relatively low temperatures (~300°C) and pressures (100 MPa) that were likely achieved in the course of Ceres' history (Castillo-Rogez & McCord, 2010).

Ceres also features carbon in multiple forms (Fig. 5): carbonates associated with rocks (magnesite and dolomite, De Sanctis et al., 2019), carbonates evolved from brines sourced from deep liquids (likely in the bicarbonate form, Castillo-Rogez et al., 2018), and amorphous organic compounds (De Sanctis et al., 2017; Marchi et al., 2019; Prettyman et al., 2019). The presence of hydrates of carbon compounds has also been suggested based on the properties of the ice shell (Bland et al., 2016): its high viscosity—several orders of magnitude greater than ice—indicates the high abundance of a strong material, like silicate and salts. On the other hand, the low density of the order of 1.2–1.3 g/cm^3 suggests no more than 30% in volume of these stronger materials. Clathrates have a relatively low density (between 0.9 and up to 1.6 g/cm^3) and match the strength observations (Ermakov et al., 2017a; Fu et al., 2017). However, there is no direct evidence for the presence of this material, and alternative explanations have been suggested. In particular, Qi et al. (2018) demonstrated that rock particles located at ice grain boundaries can prevent grain boundary sliding and match the Dawn observations.

3.2 Exosphere

The Dawn mission was not equipped to detect an exosphere or outgassing, so only indirect information has been obtained in trying to confirm the Herschel Space Observatory (Küppers et al., 2014) water vapor estimates. Dawn, upon arrival at Ceres, may have detected a bow shock using the GRaND instrument (Russell et al., 2016), though confirmation observations of a bow shock were unable to be made. This may have coincided with a solar electric particle (SEP) event, though again without a space weather instrument on Dawn, this was difficult to directly confirm. Another tantalizing but

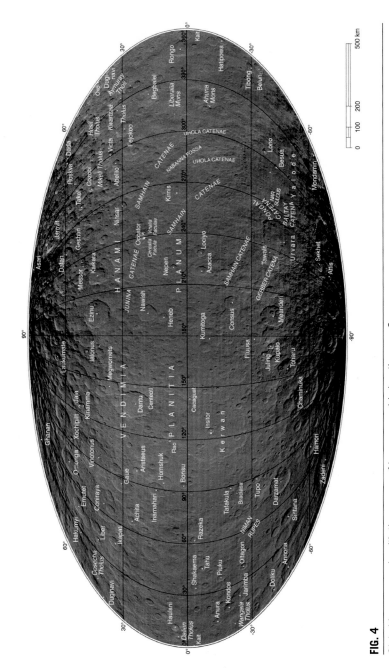

FIG. 4

Global image mosaic with the names of key geographic locations on Ceres.

Reproduced from Williams et al. (2018).

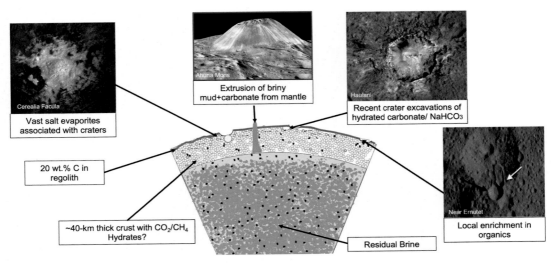

FIG. 5

Volatiles in Ceres have been found in multiple forms. The low ice shell density points to a large fraction of water and possibly clathrate hydrates. A residual layer of brines has been suggested at the interface between the crust and the rocky mantle. Salts evolved from carbon dioxide and ammonia are also found in many sites, including Occator faculae, Ahuna Mons, and Haulani crater (e.g., Palomba et al., 2019; Stein et al., 2019).

Images credit: NASA/JPL-Caltech/UCLA/MPS/DLR/IDA.

potentially inconclusive observation was documented by Nathues et al. (2015), in which off-nadir observations of Occator crater showed brightness variability and may have indicated the presence of a haze made from lofted regolith particles from active sublimation. Occator's location on Ceres roughly matches the "Piazzi" region on Ceres as argued by Küppers et al. (2014) to show an increase in the water band line depth, and potentially being their source region for the detected transient exosphere. However, other researchers argue in additional photometric analysis that the Occator haze may have been an over-interpretation of the data (e.g., Schröder et al., 2017). More direct observations (for example, emission in the ultraviolet and submillimeter spectroscopy) are required to confirm that sufficient vapor is being produced at Cerealia Tholus. Thermal modeling using updated Ceres orbital parameters from Dawn suggests that lofting the typical regolith size particles is difficult (Landis et al., 2017), but later observations using GRaND (Prettyman et al., 2021) do suggest that Occator crater is relatively hydrogen-rich compared to its surroundings, implying a higher water-equivalent hydrogen (WEH) content. Occator crater is undoubtedly a key clue to understanding Ceres' surface volatile cycle, but no conclusive direct observations suggest that water vapor outgassing was occurring there during the Dawn mission.

The most compelling rationale for an exosphere of some degree is the very common presence of water ice on or near the surface (see Section 3.3). Thermal modeling since Fanale and Salvail (1989) continued to find that conditions on Ceres were favorable for an amount of vapor loss up to a few tenths of a kg/s (e.g., Landis et al., 2017, 2019; Prettyman et al., 2017; Schorghofer et al., 2016) and that water ice spectral strength detections varied at a steep scarp at Juling crater (Formisano et al., 2018; Raponi et al., 2019). A significant fraction of this water vapor would be lost to space compared with being trapped at the poles (Schorghofer et al., 2017), indicating that most of the water vapor lost from Ceres could form an exosphere. The few tenths of a kilogram/second of water vapor production (at most) from sources like the persistent global ice table could produce is not close to the inferred rate of Küppers et al. (2014), though transient events may significantly enhance this (Landis et al., 2019; Villarreal et al., 2017). The timing of the Küppers et al. (2014) detections may fit well with SEP sources sputtering ice from the near-surface (Villarreal et al., 2017), though how much water vapor could be liberated using that method remains unclear. Small impacts are typical in the solar system, though Dawn did not detect any during the course of the mission [though typical small impacts on Earth year timescales may only be a few 10s of meters in diameter or less (Landis et al., 2019)]. Atmosphere-less Ceres is almost certainly losing its near-surface water to the vacuum of space in the present day, though through which mechanism and how much still has not been measured in situ.

Since Dawn's arrival at Ceres, additional telescopic observations have been made. Many telescopic observations have placed upper limits on the production of water and other species of vapor through non-detections. Ceres' total vapor production possibly lies in the range of $<10^{25}$–10^{28} molecules/s from the surface (e.g., McKay et al., 2017; Roth, 2018; Schorghofer et al., 2021, and references therein). One observation that was made by the Very Large Telescope during an SEP event close to Ceres' perihelion did not show a positive detection of Ceres' exosphere (Rousselot et al., 2019). Based on the slopes of currently observed water ice patches on Ceres (e.g., Combe et al., 2016, 2019; Landis et al., 2019), the maximum epoch of vapor production may be delayed from perihelion due to the steep, poleward-facing slopes that host the water ice patches confirmed by Dawn on Ceres' surface. This exposes the central paradox of linking geological and spectroscopic observations of Ceres' current surface to transient events like positive but still marginal water vapor exosphere detections: the water ice patches that survive the longest and that are easiest to observe lose an amount of vapor unlikely to match the significant quantity inferred by Küppers et al. (2014). However, if that inferred production rate still is the best interpretation of the Herschel Space Observatory spectral data is still under study at the time of writing.

3.3 Expressions of water at the surface and shallow subsurface

One major result from the Dawn mission is that water or hydrogen at the surface and near-surface of Ceres was ubiquitous. Ranging from nearly pure water ice exposed in craters and by landslides (Combe et al., 2016, 2019) to the water in phyllosilicates and hydrated salt species, Dawn confirmed that Ceres' water takes many forms.

3.3.1 Ice-rich sites

The first surface water ice detections on Ceres were made at Oxo crater (42.21°N, 359.60°E, ~10 km). The water ice detection was on the poleward-facing wall of the crater. At ~42°N latitude, GRaND data suggest that the stable ice table is within ~1 m of the surface, so the exposure of water ice (Fig. 6) in the crater walls was not surprising but rather a direct confirmation of inferred near-surface water ice from other remote sensing datasets.

The other large crater with exposed water ice on the walls is Juling crater. Juling crater (−35.90°N, 168.48°E, diameter ~20 km) displays ice on its north wall, which acts as a cold trap. The amount of

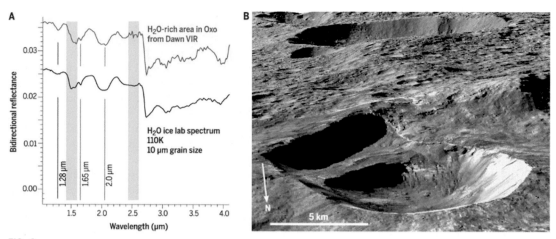

FIG. 6

Detection of water at Oxo crater based on absorption features at 1.28, 1.65, and 2 μm observed by the visible and infrared mapping spectrometer on the Dawn mission (A). Note that the laboratory spectrum is scaled vertically for legibility. Water ice is associated with the bright areas on the West world of the crater observed by the Dawn Framing Camera (B).

Credit: Combe et al. (2016).

ice present on the north wall was monitored over a 9-month baseline, which revealed changes in the strength of the water ice spectral line that were interpreted to be the occurrence of a water cycle between water ice found on Juling's floor and its northern wall (Fig. 7) (Raponi et al., 2018). However, how local vapor transport like this could occur given the predicted high loss rate to space of liberated vapor (e.g., Schorghofer et al., 2017) is still under study.

Six additional detections of surface-exposed water outside any permanently or persistently shadowed regions are reported by Combe et al. (2016, 2019). These smaller water ice detections are associated with smaller impact craters or landslides. The total water vapor production from these ice sites does not match the water vapor production rate inferred by Küppers et al. (2014) (see Section 3.2). Rather, the exposed water ice patches are hosted by poleward-facing slopes. Small landslides and impacts are ubiquitous on solid planetary surfaces, suggesting that the present configuration is likely reflective of water ice patches that are generally colder and more armored against vapor loss (e.g., Landis et al., 2019). It is well within the realm of plausibility that more water ice patches could have been present on the surface in the past, with a regolith sublimation lag deposit obscuring them from detection with the FC or VIR observing systems. Additionally, the floor deposits at Oxo crater appear to be the result of slumps of material, opening the possibility that H_2O ice could be re-exposed in the crater walls over geologic time.

3.3.2 Permanently shadowed regions

In addition to the mostly well-lit but cold surfaces on Ceres, there is also water ice in the permanently or persistently shadowed regions (PSRs) that Dawn was able to characterize at the north pole (Fig. 8). Due to the mission duration and timing, a complete survey of the PSRs at Ceres' south pole was not

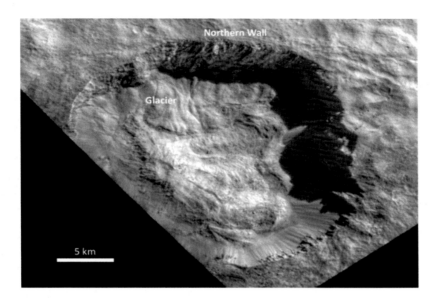

FIG. 7

20-km Juling Crater is located at 35.9°S and 168.48°E. The northern wall where ice has been identified by VIR and the large adjacent glacier, the likely vapor source, are indicated.

Image credit: NASA/JPL-Caltech/UCLA/MPS/DLR/IDA.

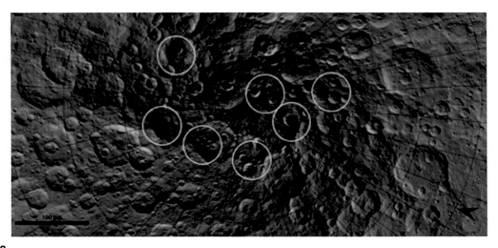

FIG. 8

North polar regions of Ceres (scale bar 100 km). *Yellow circles* denote the most concentrations (but not all) of permanently shadowed regions. For a complete inventory, the reader is referred to Ermakov et al. (2017b), Platz et al. (2016), and Schorghofer et al. (2016). Ceres' relatively low current obliquity of ~4° is likely to have been constant in time, so these regions may have only transiently been cold traps for volatiles released from the mid-latitudes of Ceres. Notably, due to Dawn's time of arrival, a similar characterization cannot be made for the south pole, as it was in shadow for the mission duration.

Background based on images from NASA/JPL-Caltech/UCLA/MPS/DLR/IDA.

possible. PSRs are found generally on low-obliquity solid surface objects, where the topographic relief is sufficient that some locations on the surface receive no or very little direct illumination. Some reflected infrared incident radiation can reach within PSRs, and on Ceres, the obliquity changes predicted are sufficient enough to change the precise area and margins that each PSR may have and, in some cases, may fully illuminate the former PSR. For example, the obliquity of Ceres likely varies over time (between 10° and 12° on ~100 kyr periods) (Ermakov et al., 2017b). Therefore, most of these PSRs on Ceres are most realistically thought of as "persistently" rather than permanently shadowed regions over 100 kyr or greater timescales.

The bright deposits, inferred to be volatiles, have spectral signatures consistent with water with some fraction of additional ammoniated material and do not sit at the inferred coldest part of the PSR on Ceres (Ermakov et al., 2017b; Platz et al., 2016; Schorghofer et al., 2016). This may indicate that sun can leak into the margins of these PSRs over these obliquity cycles, or that the temperatures of the floor change enough during the obliquity cycle to remove volatiles. What this indicates for the long-term preservation of volatile records of Ceres' poles is still under study.

Unlike the Moon, the source of volatiles for the PSRs on Ceres is most likely the loss of water from the global ice table. Some fraction of that loss can be trapped at the poles (Schorghofer et al., 2016), though the released vapor is overwhelmingly lost to space. The ammoniated species potentially being present in the VIR spectra of the PSR ice (Platz et al., 2016) is intriguing. Ammoniated species could potentially be a component of the volatiles lost in faculae-forming eruptive events, or from Ceres' regolith that has been mixed into the PSR. That the PSR ice on Ceres has albedos different enough in scattered light images (e.g., Platz et al., 2016) suggests that there is relatively little regolith contamination in the PSRs, but studies of the long-term nature of accumulated volatile deposits in PSRs remain future work at the time of writing.

3.4 Evidence for a water-rich crust from geological features

As noted above, there is geophysical evidence that Ceres' crust is dominated by water and hydrated minerals. The density of 1.2–1.3 g/cm^3 inferred by Ermakov et al. (2017a) indicates a volume fraction of water of up to 75%. Water may be in the form of ice or clathrate hydrates. The denser phase in the crustal material may be phyllosilicates, carbonates, and hydrated salts. The crust's average thickness is 40 km (Ermakov et al., 2017a).

Based on the GRaND data, Ceres' global ice table is likely stable to significant vapor loss at depths less than 1 m at latitudes greater than about 40° (Prettyman et al., 2017). Buried water ice stability zones should be roughly N/S hemispherically symmetrical, though the southern hemisphere does show slightly less hydration in the GRaND data (Prettyman et al., 2017). Geomorphological features bridge the gap between the surface-exposed water ice (e.g., Combe et al., 2019) and the GRaND results (up to ~1-m depth sensitivity) and interior models. These features have typical depth sensitivity due to the mechanical properties of icy regolith or for the necessity of vapor loss to cause eruptive or mass-wasting events, which has been used by previous authors to infer the hundreds of m- to km-scale depth properties of Ceres (e.g., Sizemore et al., 2019, and references therein).

Finally, it is important to note that from the GRaND neutron data, hydrogen is indicated by neutron moderation, which can be turned into a water-equivalent hydrogen value. With the latitudinal variation where more neutrons are moderated at higher (colder) latitudes, a global ice table is strongly implied (Prettyman et al., 2017). However, at the equator, there is still a small percentage of water-equivalent hydrogen present. This can be explained in several ways, notably from there being some H-bearing materials that are generally mixed into Ceres' regolith (Marchi et al., 2019). While water ice is an appealing volatile-rich mineral to focus on, the general composition of Ceres strongly implies that hydrated materials and volatiles both have played a major role in shaping the chemistry of the surface.

Several categories of geological landforms have been identified on Ceres that are in some way linked to subsurface ice. These include central pit craters; relaxed craters; domes; mounds; landslides and ejecta; pitted materials; scarps and depressions; and fractures (including grooves and channels; see Sizemore et al., 2019). All of these classes are widespread across the dwarf planet, each with its own rheological and compositional observations. These features tend to be sensitive to rheology and composition in the upper meters to km-scale in the subsurface. From geophysical arguments, it is possible to know the bulk composition and structure of the upper ~10 km of the crust (Bland et al., 2016; Fu et al., 2017). Therefore, these features can help start to bridge the gap in in-depth information on subsurface hydration between the surface VIR and upper ~1 m GraND observations and the deeper crustal structure.

These features observed from high-resolution imaging can be used to not only measure the current volatile loss (especially where landslides and impacts are most prevalent for this process) and probe the possible mechanisms involved, but also the geologic and compositional evolution of cryovolcanic features.

It is worth noting generally and in this chapter that "cryovolcanism" does not fully imply a genetic process similar to rocky volcanism on Earth, Mars, Venus, and Vesta. Silicate volcanism is the common type of volcanism for the inner planets of the solar system and of the asteroid Vesta, based on silicate chemistry and volcanic edifices (Ruesch et al., 2016). In contrast to silicate volcanism on terrestrial bodies, where SiO_2 dominates the composition of the melt, Cerean volcanism consists of a cold environment with an abundance of ices and brine. One such example appears to be Occator crater. For icier bodies, warm interiors are a necessity to produce brines that may be involved in the extrusion of geological constructs. Ceres is a volatile-rich dwarf planet without tidal dissipation, so the prospect of long-term cryovolcanism may not involve actual endogenic activity. Instead, the activity may be driven by passive processes, such as compressive stresses developing in a deep brine reservoir as a consequence of freezing. Another open question is what fraction of, if any, lithic material plays in Ceres' melts. Unlike lithic magmas dominated by SiO_2 and other rock-forming molecules that become positively buoyant when melted, a cryovolcanic melt dominated by liquid water may not be so when compared to a water ice-dominated crust. The salts undoubtedly play a role in the composition and physical properties of melts on Ceres. Worlds where "cryovolcanism" may occur may have fundamentally different physics than terrestrial geologists are accustomed to, in spite of producing similarly shaped surface features.

Alternative processes may mix interior materials into the near-surface or surface region of Ceres, besides cryovolcanism. Freeze–thaw processes may be similar to those that generate pingos, dome-shaped mounds with a core of ice overlain by soil (Gallagher, 2024, pp. 31–72) on the Earth, and could enhance ice lenses (e.g., Schmidt et al., 2020). Some crater morphologies may be explained by salt diapirism, rather than magmatic-style intrusion (Bland et al., 2019). The mechanics of producing positive topography features on icy bodies are key to understanding how these processes may work on icy moons, and therefore further study and understanding of how these familiar features form on bodies where SiO_2 and other lithic materials are not the main fluid components of the melt.

3.4.1 Relaxed and central pit craters

Impact crater relaxation occurs when the target material flows over geologic time, due to the strength and heating properties of the target. This can often occur on icy giant planet satellites, where a combination of water ice and tidal heating makes conditions for the viscous relaxation of craters more favorable. Pre-Dawn predictions of crater relaxation included scenarios where the water ice content was high enough to only preserve impact craters near the poles (Bland, 2013). The majority of cerean impact craters are not relaxed. Relaxed craters are defined as some crater morphology with a particular depth-to-diameter (d/D) ratio consistent with rheological simulations of ice-dominated crater relaxation (e.g., Bland et al., 2016). The five large craters on Ceres that fit this description are Kerwan, Kumitoga, Coniraya, Omongo, and Geshtin (Sizemore et al., 2019, and references therein) (e.g., Fig. 9). Interestingly, these relaxed craters are in close proximity to sharper-rimmed topography, indicating localized compositional heterogeneities on vertical and lateral scales. Because there is a low percentage of impact craters that are relaxed, this suggests that Ceres is less ice-rich on a global scale than anticipated in the most extreme model cases (Bland, 2013).

Craters on Ceres, particularly in the diameter range of ~70–150 km, exhibit central depressions or pits, rather than central peaks (Scully et al., 2018). On larger bodies like Callisto and Ganymede, central pits are observed in the crater diameter range of 30–60 km while central peaks are observed for smaller craters (Passey & Shoemaker, 1982; Schenk, 1993). On Mars, central pits occur sporadically across a variety of craters (Barlow, 2006). However, it is possible that while the final morphologies of central pit craters have similar appearances, their formation mechanisms may be significantly different on different planets.

FIG. 9

Geshtin crater (center of the image; 57°N, 258.81°E) as an example of a relaxed crater. Note the difference in crater rim morphology compared to the sharper-rimmed crater offset from the center of the crater. Scale bar 10 km.

Image credit: NASA/JPL-Caltech/UCLA/MPS/DLR/IDA.

Several models regarding the formation and development of these central pit craters have been proposed, each dependent on the availability of the volatiles and ice present. These include pits forming from rapid heating of subsurface volatiles (Wood et al., 1978); collapsing from mechanically weak ice (Melosh, 1982); uplifting from impacts into the brittle-ductile layer (Sizemore et al., 2019); or draining of a liquid water reservoir from a central uplift within the crater (Bray et al., 2012; Croft, 1983). The best examples of these central pit craters can be found in the Gaue crater (Fig. 9; 30.81°N, 86.16°E) and Occator crater, where the potential for a reservoir of liquid water is the most likely mechanism for their emplacement (Scully et al., 2018).

3.4.2 Large domes and small mounds

Large domes are classified based on their size and morphology, consisting of at least 32 observed domical to conical positive-relief (~1–6 km) mountains with diameters of ~30–120 km (Sizemore et al., 2019, and references therein). These features are assumed to be relatively ancient, mainly due to their weathered appearance and cratered flanks. The ones presently observable are not modified significantly by large impacts. Most of these large domes are not typically distinguishable from the surrounding terrain regarding texture and albedo (e.g., Sizemore et al., 2019, and references therein). Salt diapirism or briny cryovolcanism have both been proposed for the emplacement mechanisms of these large domes, especially at Cosecha Tholus, and similar features (Buczkowski et al., 2016). Generally, most domes on Ceres are small and degraded, and their formation mechanisms appear to have been active farther in the past.

The exception to this is Ahuna Mons (height 5 km; diameter 20 km; −10.48°N, 316.20°E) (Fig. 10), a nearly crater-free and high-albedo positive-relief structure (Platz et al., 2018; Ruesch et al., 2016).

FIG. 10

Example of a central pit crater (Gaue crater) from the global low altitude mapping orbit (LAMO) mosaic overlaid with the high altitude mapping orbit (HAMO) colorized topography. Scale bar 25 km.

Images making up the data products, credit: NASA/JPL-Caltech/UCLA/MPS/DLR/IDA.

Sori et al. (2017) had suggested an evolution from Ahuna Mons to large mound-type features over time, due to the domes' postulated rheological properties and the general lack of evidence on Ceres for extrusive volcanism. Other models also interpret Ahuna Mons to be a young cryovolcanic edifice (Ruesch et al., 2016), though due to its small size and high relief, the crater counts are limited in area and diameter range and highly confident absolute age determinations are not possible.

Regardless of the lack of knowledge of the material properties at Ceres, the origin hypotheses for these large domes (Fig. 11) in some way involve some upward motion of brines and ice (or some mixture of ice-brine-silicate slurries) (Castillo-Rogez et al., 2019). The proximity of these large domes to large impact structures (>150 km in diameter) may also be conducive to the cryovolcanic origins of these domes in that the impact could have penetrated deep layers of mobile reservoir materials (or produced melt pockets), and where fracture systems could serve as conduits for the movement of material (Bowling et al., 2019; Hesse & Castillo-Rogez, 2019).

Small mounds are classified apart from large domes due to their topographic relief of hundreds of meters in size, diameters of <10 km, and composed of morphologically diverse features (whereas the large dome features are more homogenous). The best examples of small mounds are observed in Yalode (−42.58°N, 292.48°E), Kerwan (−10.77°N, 123.99°E), and Occator craters (e.g., Fig. 12). These small mounds are hypothesized to be candidate cryovolcanic features, based on both their comical profiles and settings in large basin rims and central rings, surface fracturing, and exhibiting bright carbonate material (faculae; Quick et al., 2019; Stein et al., 2019). Other small mounds have been found in smooth crater material, which may be interpreted as impact melt slurries of soluble (and insoluble)

FIG. 11

Ahuna Mons (here viewed from the top) is a prime example of a geologic feature within the "large domes" classification. Its height ranges between 4 and 5 km. Scale bar 7.5 km.

Image credit: NASA/JPL-Caltech/UCLA/MPS/DLR/IDA.

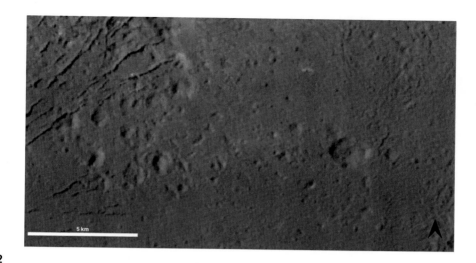

FIG. 12

Small mounds in the crater floor of Occator crater (17.926°N, 240.129°E). Note the proximity of these small mounds near an extensive fracture system to the northwest of the mound cluster. Scale bar 5 km.

Image credit: NASA/JPL-Caltech/UCLA/MPS/DLR/IDA.

salts, silicates, and ice (Schmidt et al., 2020; Scully et al., 2018). These smaller domes may be similar to pingos on Earth, though pingos involve a frost-heave process that requires liquid water, an unlikely factor in Ceres pingo-like dome formation.

3.4.3 Fluidized landslides

Three main types of fluidized landslide features have been identified on Ceres (Fig. 13), based on their setting, thickness, run-out length, and texture (Buczkowski et al., 2016; Schmidt et al., 2017). *Type-1* landslides are impact-triggered structures resulting from the deep brittle failure and ductile creep of ice-rich material, typically occurring at 70° latitude to poleward locations (Chilton et al., 2019; Schmidt et al., 2017). These landslides exhibit distinct lobate toes, and steep slopes and are vertically thick (hundreds of meters in scale) (e.g., Buczkowski et al., 2016; Chilton et al., 2019; Duarte et al., 2019). *Type-2* landslides are characterized by their longer run-out distances, wider toes, and thinner vertical profiles on shallower slopes. These typically occur in proximity to crater rims that exhibit alcoves or scars and are interpreted to have partial fluidization (Chilton et al., 2019). *Type-3* landslides may be fluidized ejecta due to their platy-textured, layered morphology. These occur typically equatorward of 60° in both hemispheres, which may be arising from latitudinal trends in ice concentration and temperature variations (Chilton et al., 2019).

3.4.4 Pits, depressions, and fractures

Pits are morphological features on the surface of Ceres that are linked to sublimation mechanisms (see example in Fig. 14). This was of particular interest as this may be a way to distinguish between impact-related geological features and gas-phase volatile loss features, even though both scenarios require some influence and availability of water or ice. Occator and Yalode craters exhibit some of these

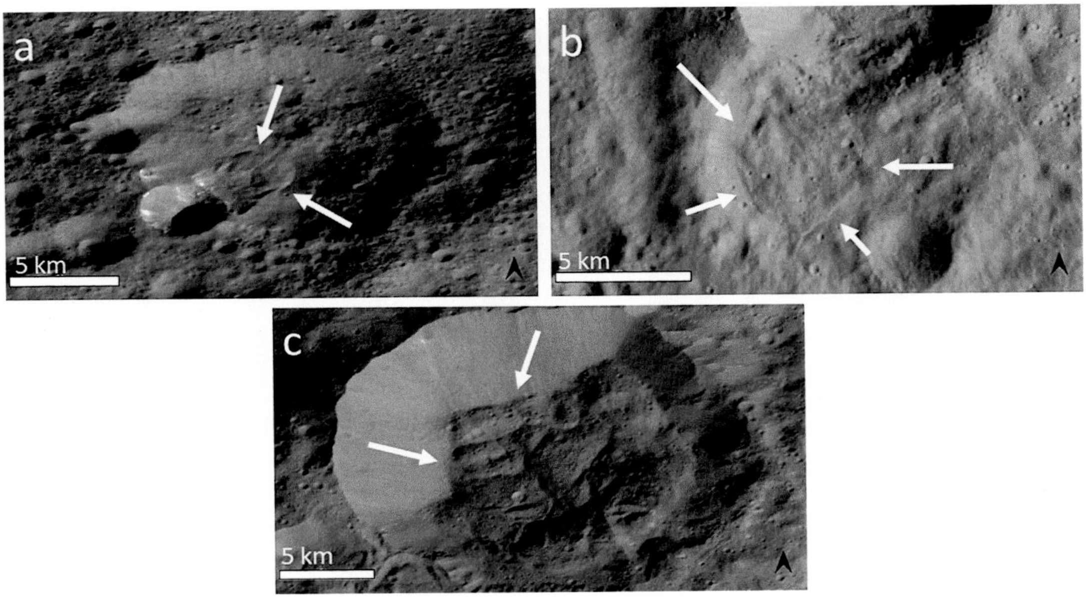

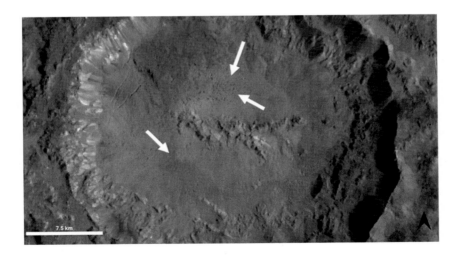

FIG. 13

Examples of landslide and ejecta features (indicated by *white arrows*) as discussed by Buczkowski et al. (2016), Chilton et al. (2019), and Duarte et al. (2019). (a) Type-1 landslide with a prominent tongue-like broad lobate feature; (b) Type-2 landslide; and (c) Type-3 landslide with lobate fluidized-ejecta features.

Image credit: NASA/JPL-Caltech/UCLA/MPS/DLR/IDA.

FIG. 14

Pitted terrains in Haulani crater (5.80°N, 10.77°E) are indicated by the *white arrows*. Scale bar 7.5 km.

Image credit: NASA/JPL-Caltech/UCLA/MPS/DLR/IDA.

interesting depressions and scarps, namely within the Occator ejecta and elongated depressions that may be fractured via sublimation of ice (Sizemore et al., 2019 and references therein). Pitted materials are usually within much younger (<700 Myr) impact craters, whereas more degraded pits are within older host craters. From temperature models (see Marchi et al., 2013; Schenk et al., 2020; Sizemore et al., 2017), these pitted materials are likely produced in locations where $T > 300$ K and are compositionally dominated by H_2O phases (as ices or clathrates).

Ceres' pitted materials are commonly associated with fractures and linear grooves in the host crater floor (Buczkowski et al., 2019; Schenk et al., 2020; Sizemore et al., 2017, 2019). These fractures and grooves can occur on crater ejecta and crater wall interiors (Fig. 15) but also independently at locations such as Nar Sulcus (e.g., Hughson et al., 2019). Such crater ejecta observed on Ceres and associated fracture and groove features suggest that there is some level of volatile involvement (Crown et al., 2018; Schenk et al., 2020; Sizemore et al., 2019), and generally the material strength required to maintain fractures and grooves can be determined from observations (Hughson et al., 2019). Channels and irregular depressions and grooves within the crater floors of Urvara (−45.66°N, 249.24°E) and Yalode, or within the Occator ejecta, may represent localized saturation of the impact melts by fluids (or pore saturating via sublimation) or excess ice (Crown et al., 2018).

3.4.5 Salt-rich sites

Bright deposits called *faculae* are found across Ceres' surface. With only albedo from approach FC images, it was difficult to determine if these faculae were water ice or salts, though water ice was known to be unstable to sublimation on Ceres' surface from a variety of thermal arguments. The faculae have since been confirmed to be salts with spectroscopic data, and they are likely tracers of past volatile activity. Stein et al. (2017) recorded more than 300 bright spots in association with all kinds of features, in particular with craters (Fig. 16). Several possible origins for these features were suggested:

FIG. 15

Fractures and grooves in the floor of Urvara crater. Scale bar 10 km.

Image credit: NASA/JPL-Caltech/UCLA/MPS/DLR/IDA.

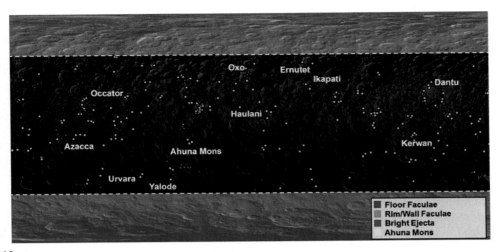

FIG. 16

Faculae observed on Ceres' surface are classified as a function of their geological landscape. The names refer to the largest craters that display faculae. In total, 300 white spots have been recorded by Stein et al. (2017).

concentrated salt residues following local melting by impact-produced heat, generally applying to large craters (>10s km); excavation of salt-rich slabs by impacts or landslides; upwelling of deep brines via fractures. Stein et al. (2017) attribute the absence of evaporites in some large impacts as evidence of lateral heterogeneities in subsurface composition.

Among the many sites associated with large faculae, Occator crater is the most outstanding (Fig. 17). The crater displays two main facula regions about 10 km in extent each. Cerealia Facula found in the center of the crater is a vast area dominated by sodium carbonate with a secondary component of ammonium chloride. A central dome (Cerealia Tholus) displays evidence of hydrohalite (hydrated sodium chloride). The Vinalia Faculae are a group of carbonate-rich sites in part associated with fractures extending in the eastern side of the crater. Sodium carbonate is associated with aluminum-rich phyllosilicates, ammonium chloride, and sodium chloride (Bramble & Hand, 2021; Raponi et al., 2018), in contrast to the regolith of Ceres, dominated by magnesium/calcium carbonate and other phyllosilicates (De Sanctis et al., 2015). De Sanctis et al. (2020) showed that Occator crater also displays the signature of hydrohalite at Cerealia Facula, the large central bright area at Occator crater (19.82°N, 239.33°E). These authors interpreted this discovery as evidence of ongoing brine exposure, as salt hydrates are not expected to be long-lived on the surface of airless bodies (e.g., Bu et al., 2018, 2019; McCord et al., 2001). However, if the activity or just mass wasting of dehydrated over-material can reveal younger, more hydrated material is not well known.

The emplacement, including the source and processes involved, of Occator crater's bright faculae have several remaining questions. One major hypothesis for its emplacement is through brine effusion in a brine-limited, impact-induced hydrothermal system (Scully et al., 2020). The impact would have produced fracturing which would then have enabled brines to reach the surface. There are three main faculae features within Occator crater: Cerealia, Pasola, and Vinalia. Cerealia and Pasola are the central

FIG. 17

Faculae and associated fracture system on the floor of Occator crater. The faculae are primarily composed of sodium carbonate and ammonium chloride (Raponi et al., 2019). Hydrohalite has been identified at the top of Cerealia Tholus (De Sanctis et al., 2020). The Cerealia and Pasola Faculae are located in a pit at the center of the crater while the Vinalia Faculae are in multiple discrete locations.

Image credit: NASA/JPL-Caltech/UCLA/MPS/DLR/IDA.

faculae in the center of Occator and postdate the central pit, and are primarily sourced from an impact-induced melt chamber (Scully et al., 2020). Vinalia Faculae are to the east of Cerealia and Pasola within the crater floor and are comparatively elongated and display a greater ballistic pattern than the other faculae. Vinalia may have been sourced from a laterally extensive deep reservoir of brines (Ruesch et al., 2019; Schenk et al., 2019, 2020).

Scully et al. (2020) have documented numerous localized bright material features surrounding Cerealia and Vinalia Faculae, and have found that faculae tend to occur within general regions within the host crater floor surrounding the central pit (Ruesch et al., 2019; Schenk et al., 2019, 2020; Scully et al., 2020). Compared to terrestrial impact-derived hydrothermal deposits, such deposits occur in crater-fill materials, within the margins of central uplifts, the ejecta, and the crater rim (Marzo et al., 2010; Osinski et al., 2013; Scully et al., 2020).

Other faculae on Ceres were divided into wall and floor faculae with varying amounts of carbonates and phyllosilicates, which have been argued to likely be part of a chemical and physical modification sequence (e.g., Palomba et al., 2019; Stein et al., 2019). The occurrence of many carbonate-rich sites

(Carrozzo et al., 2018) suggests this material is abundant, for example, in the form of lenses in the shallow subsurface. However, Stein et al. (2019) pointed out that shallow carbonate lenses were likely erased by gardening and lateral mixing over a timescale of a few 100 My. Hence, recent exposures of carbonate outcrops, which can be 10s km long, require a mechanism for carbonate to be emplaced in the shallow subsurface in geologically recent times. Two types of processes are being studied: the redistribution of material by impacts large enough to heat the material and decrease its viscosity (differential loading, Bland et al., 2019) and the percolation of brines via fractures or solid-state salts embedded in montes (Stein et al., 2019). Detailed modeling is in progress but the abundances of carbonates inferred at the enriched sites are similar (5–10 vol%) to the abundance found at Ahuna Mons, which suggest a shared brine source among these features. Notably, many of these materials do not appear hydrated in most cases, though some signs of hydration outside of Occator crater were observed by Raponi et al. (2018).

In the case of Occator crater, modeling of the impact predicts the formation of a briny melt chamber, if the target material were rich enough in water ice and salts, to be ~20 km in diameter and extend into the subsurface ~20 km in depth (Bowling et al., 2019; Raymond et al., 2020). The presence of fractures and widespread faculae leads to hypotheses on potential pathways to the surface for the faculae-forming brines that are likely opened by impact-induced fracturing within the crater (Fagents, 2003). Partial crystallization of the melt chamber that led to excess pressures could also initiate fracturing (Scully et al., 2019). The recent and potentially ongoing exposure of hydrohalite, rather than its dehydrated halite form expected after long-term exposure to the vacuum of space, on Cerealia Tholus suggests that the melt chamber below Occator crater is still at least partially molten or that Occator's halite is sourced from the deep brine, for example, by fractures opened by the crater-forming impact (Raymond et al., 2020). Regardless of the implications for continued melt chamber activity implied by the detection of hydrohalite, this observation indicates that faculae may begin their life in geologic time with more hydration than presently observed generally on the surface. The hydrated salt story may be inextricable on the local scale from the water ice/activity history of a region.

Outside of complex craters, it may be possible that liquid water solutions with or without dissolved salts and suspended silicates could enable ground ice (or water) and putative cryovolcanic systems (Schmidt et al., 2020). This would also enable flows within complex craters to armor themselves and stay mobile longer on Ceres' surface (Quick et al., 2019). While in silicate volcanism the ascension of melt occurs from the lower density of the melt plus pressure built up by thermal expansion, cryogenic systems instead have water ice contraction when melting, so the resultant liquid is not necessarily buoyant. The addition of brines and fluids (like ammonia) would be influenced by topography and thermal expansion due to freezing (Schmidt et al., 2020). Cryo-hydrologic activity and processes are crucial for understanding the context and timing of such events and the conditions in which these systems have occurred. Simulations have predicted that some of the faculae are hydrothermal deposits that were emplaced as ballistic flows originating from localized brine sources rather than one centralized source region (Scully et al., 2019, 2020), in which this process is defined as brine effusion, which encompasses both gases and liquids as "fluids" involved in the formation evolution.

These salts and other hydrogen-rich minerals (not necessarily water ice), and the older materials that make up noncomplex crater central faculae, such as those in crater walls, likely have been processed over significant periods of time through a series of impacts. However, their formation is directly linked to how materials melt, reform, and are available from a brine source within Ceres' interior. This process by which these materials are refreshed on the surface may be linked to repeated major impacts on the surface, which have also been invoked to explain local concentrations of hydrogen at Occator and perhaps some of the global asymmetries in hydrogen distribution generally in the upper meter of Ceres' surface (Prettyman et al., 2021).

4 Ceres' evolution and current interior structure

Several previous reviews have highlighted the role of water in shaping Ceres' evolution, starting with McCord and Sotin (2005). In this paper, the authors pointed out the similarities between the physical properties of Ceres and icy moons in the outer solar system. For example, Ceres is about the size of Saturn's moon Dione, and Pluto's moon Charon. Its density is slightly higher than those moons, but as noted in Section 1, it allows for about 25 wt% water ice. McCord and Sotin (2005) demonstrated that if Ceres formed early enough to have accreted ^{26}Al, a powerful but short-lived radioisotope, then its volatiles could melt on a global scale, leading to the separation of a water-rich shell from a rocky core. The latter was likely to be hydrated and thus have low density, as demonstrated later on by the Dawn mission. McCord and Sotin (2005) provided critical arguments for pursuing the Dawn mission, and a follow-on study by Castillo-Rogez and McCord (2010) expanded on the McCord and Sotin predictions in order to interpret the HST observations of Ceres' shape and refined density (Thomas et al., 2005). Key results included the possibility for clathrates and porosity in Ceres' rocky mantle, and the prospect for the preservation of cold temperatures brines until at present. The Dawn results confirmed several of the early predictions, in particular the differentiated interior and the preservation of liquid, at least locally. This section highlights the role of volatiles in shaping Ceres' internal evolution, by definition something that at the time of writing can only be inferred from remote sensing data and the current state.

4.1 Role of water in shaping Ceres' interior evolution and current state

A summary of Ceres' interior structure derived from the geophysical and topography observations from Ermakov et al. (2017a) and Fu et al. (2017) can be found in Fig. 18. Ceres' overall shape is dictated by the strength of materials deeper than other specific surface topography may be sensitive to, providing constraints on the bulk composition of Ceres over multi-km-scale depths.

Fu et al. (2017) modeled the topography relaxation of Ceres using finite-element modeling. This led to quantifying the mechanical structure of the upper 100 km of Ceres. Fu et al. (2017) confirmed the presence of a strong crust on a global scale overlaying a much weaker layer. The viscosity of that layer is bounded at 10^{21} Pa s but is otherwise poorly constrained. This layer was interpreted by Fu et al. (2017) as evidence of the presence of liquid on a global scale. This is consistent with other evidence of the deep sourcing of brine-derived material (e.g., Ruesch et al., 2019 for Ahuna Mons, Raymond et al., 2020 for Occator's faculae). However, in the absence of geophysical evidence, for example, electromagnetic sounding like in the case of Europa's ocean, the occurrence of a global-scale liquid layer inside Ceres remains debated.

Gravity data indicate porosity in the mantle, at least in an upper layer, filled with brines or salts. This possibility was predicted from the interpretation of global isostatic relaxation observations

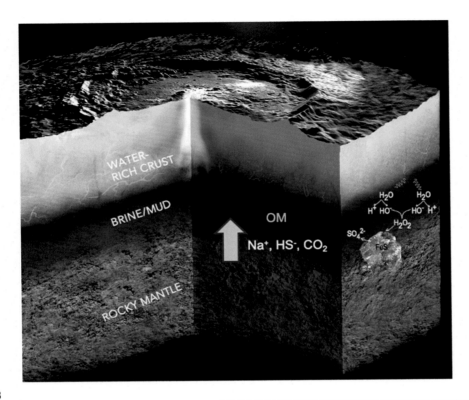

FIG. 18

The average interior structure of Ceres derived from the combined interpretation of the Dawn data. The gravity and topography inversion suggests Ceres' interior is layered in a ~40 km crust on top of a rocky mantle. The low density of the mantle suggests a high porosity fraction and brines rich in rock particles and organic matter (OM) may be present at the interface between the crust and mantle.

Credit: Raoul Ranoa for the background image.

(Fu et al., 2017). Melwani Daswani and Castillo-Rogez (2022) pointed out that the mean density derived from Ceres' mantle, between 2400 and 2800 kg/m³ (Park et al., 2016; Mao & McKinnon, 2020), is too low to be explained by solid rock. These authors tracked the evolution of fluids released from rock thermal metamorphism. The detailed modeling of thermal metamorphism in the mantles of water-rich bodies via tracking of the petrology and released fluids was first introduced by Melwani Daswani et al. (2021) for Europa. These authors used the PerpleX software, a thermodynamics (Gibbs) code extensively used in terrestrial studies. Melwani Daswani and Castillo-Rogez (2022) showed that various volatiles should be released from the rocky mantle as it warms up, in particular carbon dioxide (CO_2) and bisulfide (HS^-) (Fig. 19). Hence, the composition of metamorphic fluids is consistent with the production of sodium carbonate and can explain in particular the large abundance of that material in recent craters.

An alternative explanation for geophysical observations of Ceres relates the mantle's low density to a high content of organic compounds (Zolotov, 2020). A similar explanation has been offered for

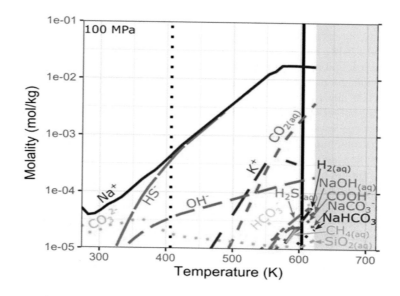

FIG. 19

Result of the simulation of metamorphic fluids released from Ceres' rocky mantle as a function of temperature. These fluids are loaded with sodium ion (Na$^+$), bisulfide (HS$^-$), carbonate ions (CO$_3^{2-}$), carbon dioxide (CO$_2$) and a variety of minor compounds. (Note: this model did not track the fate of ammonia that may be released from the breakdown of organic matter.)

From Melwani Daswani & Castillo-Rogez (2022)

interpreting the low density of Titan's rocky mantle (Néri et al., 2020). These models are based on the assumption that objects formed in the outer solar system, including potentially Ceres, could accrete a large abundance of cometary pebbles (Johansen & Lambrechts, 2017). The Rosetta mission revealed that comet 67P's dust could contain up to 45% in mass refractory organics (or macromolecular organics) (Bardyn et al., 2017). Assuming the findings at 67P are representative of comets in general, then pebble accretion could indeed lead to a large fraction of organic compounds in icy moons and dwarf planets.

Additional constraints on Ceres' interior are derived from gravity data returned during the Dawn extended mission (Park et al., 2020). Thanks to its low-perihelion elliptical orbit (Section 2), Dawn acquired local gravity data to a harmonic degree of ~45, or a spatial resolution better than 80 km. Although the coverage is limited to a regional band, it encompasses key landmarks, such as Occator crater. These observations allowed finer resolution on the interface between the icy shell and the deeper interior. In particular, Park et al. (2020) identified an increase in density toward the bottom of the crust, but that gradient has not been quantified yet.

4.2 Implications for large ice-rich body evolution

Several findings from the Dawn results and associated studies have implications for our understanding of icy bodies in the 1000 km size class. These are large enough that internal differentiation of an ice-rich shell from a dense, rocky interior is expected if they underwent global melting. On the other hand, the low gravity may promote the long-term suspension of fine particles that may accrete in the ice shell (Travis et al., 2018). On another water-rich body, Saturn's moon Enceladus, a high concentration of organic matter found in plume material sourced from a deep ocean (Postberg et al., 2018) has led to the suggestion that organic compounds could concentrate at interfaces as part of colloids. Travis et al. (2018) argue that this colloidal material, which could also include silicates and salts, could be incorporated into the shell of Ceres. It seems like additional research on colloidal solutions in the context of water worlds is necessary, especially to understand how much liquid water and organics can be processed at more astrobiologically relevant temperatures within large icy bodies.

Observations by the Dawn mission of water, carbon, and nitrogen in various forms have also shed light on material cycling in water-rich bodies. In Ceres, carbon was likely accreted as ices (CO_2, CH_4) and organics and yielded carbonates from interaction with rock. This process is expected to develop fast, a few million years at most from meteorite studies (e.g., Dyl et al., 2012).

If the partial pressure of dissolved gases was sufficient (Zolotov & Kargel, 2009), then these compounds could be used in the formation of clathrate hydrates that accreted at the top or bottom of the early ocean depending on their densities. Then carbonates and organics trapped in the rocky mantle and subject to temperatures in excess of ~650 K could break down in the longer term and release CO_2 and CH_4 to the ocean, potentially in sufficient amounts to alter its environment (redox). A similar evolution can be outlined for nitrogen, first accreted in the form of ices (especially ammonia hydrates) that first produced ammonium when released in water. This led to the formation of ammoniated phyllosilicates and salts. A late release of nitrogen from the breakdown of organics could enrich a residual ocean in ammonium. In these cycles, water drives chemistry and is a vector of material, for example, during thermal metamorphism. These observations led Castillo-Rogez et al. (2022b) to investigate the controls in aqueous systems whose chemistry is driven by the products of carbon dioxide and ammonia in solution.

5 Ceres origin and implications for the origin of the Main Belt

5.1 Telltale of origin

The composition of Ceres provides clues about its origin based on marker volatiles and its high content of carbon compounds. De Sanctis et al. (2015) introduced early on the possibility that Ceres' volatile material came from the outer solar system. Several pre-Dawn studies paved the way for the possible migration of Ceres from that region (e.g., Mousis et al., 2008). Other studies introduced the counterpoint that carbon dioxide and ammonia could be released from the heating of carbonates and organic compounds, in which case the observed mineralogy could still be explained for a formation in situ of the dwarf planet (McSween Jr et al., 2018). A caveat of the latter model is that the temperatures required for the breakdown of carbonates and organics (>600 K at 100s MPa) are reached several hundreds of Myr after Ceres formation (e.g., Melwani Daswani & Castillo-Rogez, 2022), by which time the bulk of the ocean would have been frozen and the conditions for ammonium-potassium exchange in phyllosilicates might not have been met (see Neveu et al., 2017). While the influx of carbon dioxide to a residual ocean could represent a source for the carbonates found in recent evaporite sites (i.e., locally, Melwani Daswani & Castillo-Rogez, 2022), the processes that would lead to the exposure of ammoniated phyllosilicates on a global scale are unknown. Global resurfacing with material from the upper mantle would require an intense heating event and some form of volcanism, or a major impact event for which there is little evidence outside of Kerwan crater. A more likely origin for the regolith has been suggested by Neveu and Desch (2015): these authors point out that in Ceres' early history when the crust was presumably thin, it could undergo foundering.

Other studies have emphasized the implications of the high content in organics inferred from the Dawn mission. In particular, Zolotov (2020) suggested that Ceres' ice shell density could be explained by a mixture of organics and hydrated rock with some porosity but no ice. Whether this model can explain the Dawn observations remains to be demonstrated, especially the latitudinal variation dominating the hydrogen content in the upper ~1 m of the regolith (Prettyman et al., 2017).

5.2 Ceres migration—State of understanding

Recent models of planetesimals' migration provide a framework for explaining the presence of water-rich material in the main belt of asteroids. The last decade has seen an emphasis on the origin of main belt asteroids from the destabilization of planetesimal belts in the regions between Jupiter and Neptune (e.g., Raymond & Izidoro, 2017).

Instead, the most recent Ceres-specific studies (e.g., de Sousa et al., 2022) support an origin between the orbits of the giant planets. The high concentration of carbon in Ceres, which has been attributed to a cometary origin (Zolotov, 2020), could instead be evidence of the pebble accretion process. de Sousa et al. (2022) tested a possible origin of Ceres in the Jupiter-Saturn region. They considered that 3600 Ceres-sized bodies could form in this region. The probability that one of these objects could be captured in the asteroid belt is between ~2.8×10^{-5} and ~1.2×10^{-3}. Accounting for the depletion of the asteroid belt during giant planet instability, these authors can explain the presence of a single Ceres in the outer belt with a probability of ~50%. A migration of Ceres from the Kuiper belt was also proposed (McKinnon, 2008), but the capture of such a massive object in the main belt was proven to be of low probability (10%, McKinnon, 2008).

This question of origin will likely not be resolved with the available data. However, it does benefit from being addressed as part of the broader context offered by small body populations. Besides the presence of many asteroids with hydrated minerals on their surfaces, the recent discovery of spectrally red objects that may be of Trans Neptunian Object origin (Hasegawa et al., 2021), ice-rich bodies (Vernazza et al., 2020) and the frequent occurrence of a ~3.1 μm absorption band attributed to ammoniated phyllosilicates (Kurokawa et al., 2022) suggest that a large fraction of volatile-rich asteroids could have migrated from between the orbits of Jupiter and Neptune. A summary of Main Belt Comets (MBCs) is included in El-Maarry (2024, pp. 261–282). Ceres itself may be the largest example of these additional icy small bodies in the Main Belt, and if not the case, any hypothesis for explaining these icy asteroids must also account for Ceres' current orbit and composition.

End-to-end simulations of Ceres' accretion, migration, and internal evolution based on state-of-the-art models are needed to better constrain Ceres' accretional environment.

A future mission to Ceres should aim to search for markers of origin, such as isotopic signatures that could indicate a colder formation of the dwarf planet's materials, e.g., $^{13}C/^{12}C$ and $^{15}N/^{14}N$. A major caveat, though, is that Ceres' relatively warm evolution, protracted water-rock interactions, and cycling of materials likely led to a modification of these ratios. Future research is needed to track this fractionation which may be due to the various processes that shaped Ceres' evolution.

6 Summary

The Dawn observations have led to important results on Ceres' interiors with implications for better understanding the class of mid-sized (~500–1000 km diameters) icy bodies. They have also left tantalizing hints regarding the possible habitability of Ceres through time and potentially also at present. Up-close observations of the largest and water-rich asteroid are a direct data point to understand the migration of icy objects in the solar system. While it may be possible that Ceres formed in situ, the presence of ammoniated species, significant amounts of water ice, and a plausible dynamical path to place Ceres in the outer Main Belt challenge the once common distinction of asteroids being rocky and comets being icy. However, major gaps remain in our knowledge of Ceres, in particular the extent of liquid in its interior, the extent of processing of its organic matter, and its origin.

Future mission concepts have been introduced in the literature to address these and other questions pertaining to Ceres' habitability potential. We direct the readers to a summary of these concepts in Li and Castillo-Rogez (2022) and Farnsworth et al. (2024, pp. 315–356). Here we highlight those concepts that have been subject to advanced studies in the context of informing the program of space agencies. The CALICO study is a concept currently developed under the M-class (competed) program of the European Space Agency, currently under assessment (https://www.cosmos.esa.int/web/call-for-missions-2021/update-on-the-f2-and-m7-mission-opportunity). If this mission is selected for development—to be announced by the end of 2023—then it would launch in ~2036 and land in Occator crater to search for habitability markers.

The Ceres Sample Return concept (Fig. 20) was selected by a study by NASA with the goal to inform the Planetary Decadal and Astrobiology Survey 2023–2032. The study was delivered in 2020 (Castillo-Rogez et al., 2022a) and was deemed scientifically interesting and technically viable by the decadal survey committee. The concept is featured in the New Frontiers 6 program with other exciting concepts, such as Centaur Orbiter and Lander, Venus in situ mission, Titan Orbiter, and Enceladus plume sampling mission. The New Frontiers program is for completed mission proposals at the ~$1B level (currently). Famous examples of New Frontiers class missions are New Horizons, OSIRIS-REx, and Juno. The next opportunity for this mission to be proposed is uncertain at this time. Assuming a Ceres sample return mission was selected by the end of the 2020s, it would launch around 2035 and return samples by the 2050s.

Barring sample return or landed spacecraft analysis, however, other observations may fill critical gaps in our understanding of how Ceres is currently losing volatiles. Improvements to the ground and space-based telescopes may allow for the observation of smaller amounts of water vapor around Ceres and other MBCs. So far, only the variability of the MBC tail extent with a heliocentric distance was diagnostic of the role of water vapor in driving cometary activity. JWST has the potential to provide direct detection of water vapor (Kelley et al., 2021). Continuing the monitoring of solar wind conditions and understanding how SEPs propagate into the asteroid belt will help resolve the role that sputtering plays in space weathering and releasing water from Ceres' near-surface. These observations could be made by JUICE on its way to Jupiter or as measurements of opportunity on not yet launched asteroid missions.

The extensive observations of the Dawn mission revealed a body whose physical, chemical, and geological evolution has been shaped by water. Like Saturn's moon Enceladus, Ceres shows evidence of aqueous geochemistry involving carbon ices, ammonia, and organic matter. A major difference between the two bodies is that Ceres' liquid is likely in a thin (a few kilometers) residual layer, or even

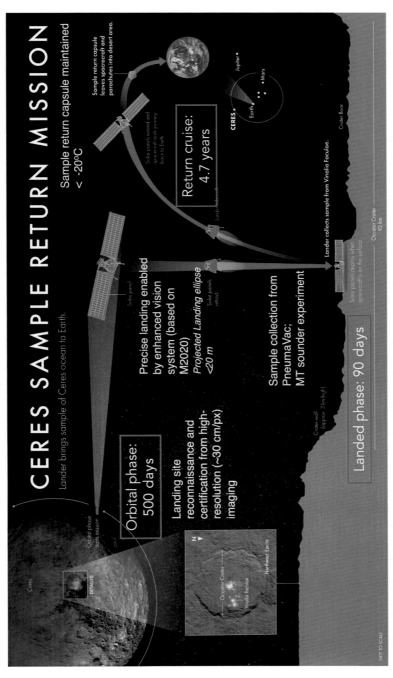

FIG. 20

Main phases in the Ceres sample return mission as described in Castillo-Rogez et al. (2022a). This concept includes an orbital phase that would search for surface changes since the Dawn observations. It also includes as part of its in situ phase an investigation about probing Ceres' deep interior with electromagnetic sounding to map the radial (down to 50 km) and lateral extent of brines below the region of Occator crater.

regional pockets, while the icy moon's ocean is tens of kilometers thick (e.g., Thomas et al., 2016). Indeed, although Ceres is twice as big as Enceladus, it does not benefit from a major heat source, like tidal heating. Hence, knowledge gained at Ceres can inform the evolution of other past or current ocean worlds rich in non-water volatiles, such as Pluto and the Uranian moons (Castillo-Rogez et al., 2023), as one extreme case where the chemistry may be close but with limited internal heating after the initial accretion and Al^{26}.

Acknowledgments

Part of this work was carried out at the Jet Propulsion Laboratory, California Institute of Technology, under a contract with the National Aeronautics and Space Administration (80NM0018D0004). The authors wish to thank the numerous scientists, engineers, and support and operations staff who made the Dawn mission observations possible. The authors also wish to thank B.E. Schmidt for helpful discussions on cryovolcanism.

References

A'Hearn, M. F., & Feldman, P. D. (1992). Water vaporization on Ceres. *Icarus*, *98*(1), 54–60.

Bardyn, A., Baklouti, D., Cottin, H., Fray, N., Briois, C., Paquette, J., ... Hilchenbach, M. (2017). Carbon-rich dust in comet 67P/Churyumov-Gerasimenko measured by COSIMA/Rosetta. *Monthly Notices of the Royal Astronomical Society*, *469*(Suppl_2), S712–S722.

Barlow, N. G. (2006). Impact craters in the northern hemisphere of Mars: Layered ejecta and central pit characteristics. *Meteoritics & Planetary Science*, *41*(10), 1425–1436.

Binzel, R. P. (2006). Definition of a planet: Prague 2006 IAU resolutions. *Minor Planet Bulletin*, *33*, 106–107.

Bland, M. T. (2013). Predicted crater morphologies on Ceres: Probing internal structure and evolution. *Icarus*, *226*(1), 510–521.

Bland, M. T., Buczkowski, D. L., Sizemore, H. G., Ermakov, A. I., King, S. D., Sori, M. M., ... Russell, C. T. (2019). Dome formation on Ceres by solid-state flow analogous to terrestrial salt tectonics. *Nature Geoscience*, *12*(10), 797–801.

Bland, M. T., Raymond, C. A., Schenk, P. M., Fu, R. R., Kneissl, T., Pasckert, J. H., ... Russell, C. T. (2016). Composition and structure of the shallow subsurface of Ceres revealed by crater morphology. *Nature Geoscience*, *9*(7), 538–542.

Bowling, T. J., Ciesla, F. J., Davison, T. M., Scully, J. E., Castillo-Rogez, J. C., Marchi, S., & Johnson, B. C. (2019). Post-impact thermal structure and cooling timescales of Occator crater on asteroid 1 Ceres. *Icarus*, *320*, 110–118.

Bramble, M. S., & Hand, K. P. (2022). Spectral evidence for irradiated sodium chloride on the surface of 1 Ceres. *Geophysical Research Letters*, *49*(3). e2021GL096973.

Bray, V. J., Schenk, P. M., Melosh, H. J., Morgan, J. V., & Collins, G. S. (2012). Ganymede crater dimensions— Implications for central peak and central pit formation and development. *Icarus*, *217*(1), 115–129.

Bu, C., Lopez, G. R., Dukes, C. A., McFadden, L. A., Li, J. Y., & Ruesch, O. (2019). Stability of hydrated carbonates on Ceres. *Icarus*, *320*, 136–149.

Bu, C., Rodriguez Lopez, G., Dukes, C. A., Ruesch, O., McFadden, L. A., & Li, J. Y. (2018). Search for sulfates on the surface of Ceres. *Meteoritics & Planetary Science*, *53*(9), 1946–1960.

Buczkowski, D. L., Schmidt, B. E., Williams, D. A., Mest, S. C., Scully, J. E. C., Ermakov, A. I., ... Russell, C. T. (2016). The geomorphology of Ceres. *Science*, *353*(6303), aaf4332.

Buczkowski, D. L., Scully, J. E., Quick, L., Castillo-Rogez, J., Schenk, P. M., Park, R. S., ... Russell, C. T. (2019). Tectonic analysis of fracturing associated with Occator crater. *Icarus, 320,* 49–59.

Carrozzo, F. G., De Sanctis, M. C., Raponi, A., Ammannito, E., Castillo-Rogez, J., Ehlmann, B. L., ... Russell, C. T. (2018). Nature, formation, and distribution of carbonates on Ceres. *Science advances, 4*(3), e1701645.

Castillo-Rogez, J., Brophy, J., Miller, K., Sori, M., Scully, J., Quick, L., ... Zacny, K. (2022a). Concepts for the future exploration of dwarf planet Ceres' habitability. *The Planetary Science Journal, 3*(2), 41.

Castillo-Rogez, J. C., Melwani Daswani, M., Glein, C. R., Vance, S. D., & Cochrane, C. J. (2022b). Contribution of non-water ices to salinity and electrical conductivity in ocean worlds. *Geophysical Research Letters, 49*(16). e2021GL097256.

Castillo-Rogez, J. C., Hesse, M. A., Formisano, M., Sizemore, H., Bland, M., Ermakov, A. I., & Fu, R. R. (2019). Conditions for the long-term preservation of a deep brine reservoir in Ceres. *Geophysical Research Letters, 46*(4), 1963–1972.

Castillo-Rogez, J. C., & McCord, T. B. (2010). Ceres' evolution and present state constrained by shape data. *Icarus, 205*(2), 443–459.

Castillo-Rogez, J., Neveu, M., McSween, H. Y., Fu, R. R., Toplis, M. J., & Prettyman, T. (2018). Insights into Ceres's evolution from surface composition. *Meteoritics & Planetary Science, 53*(9), 1820–1843.

Castillo-Rogez, J., Weiss, B., Beddingfield, C., Biersteker, J., Cartwright, R., Goode, A., ... Neveu, M. (2023). Compositions and interior structures of the large moons of Uranus and implications for future spacecraft observations. *Journal of Geophysical Research, Planets, 128*(1). e2022JE007432.

Chilton, H. T., Schmidt, B. E., Duarte, K., Ferrier, K. L., Hughson, K. H. G., Scully, J. E. C., ... Raymond, C. A. (2019). Landslides on Ceres: Inferences into ice content and layering in the upper crust. *Journal of Geophysical Research, Planets, 124*(6), 1512–1524.

Combe, J. P., McCord, T. B., Tosi, F., Ammannito, E., Carrozzo, F. G., De Sanctis, M. C., ... Russell, C. T. (2016). Detection of local H2O exposed at the surface of Ceres. *Science, 353*(6303), aaf3010.

Combe, J. P., Raponi, A., Tosi, F., De Sanctis, M. C., Carrozzo, F. G., Zambon, F., ... Russell, C. T. (2019). Exposed H2O-rich areas detected on Ceres with the dawn visible and infrared mapping spectrometer. *Icarus, 318,* 22–41.

Croft, S. K. (1983). A proposed origin for palimpsests and anomalous pit craters on Ganymede and Callisto. *Journal of Geophysical Research: Solid Earth, 88*(S01), B71–B89.

Crown, D. A., Sizemore, H. G., Yingst, R. A., Mest, S. C., Platz, T., Berman, D. C., ... Russell, C. T. (2018). Geologic mapping of the Urvara and Yalode quadrangles of Ceres. *Icarus, 316,* 167–190.

De Sanctis, M. C., Ammannito, E., McSween, H. Y., Raponi, A., Marchi, S., Capaccioni, F. A. B. R. I. Z. I. O., ... Russell, C. T. (2017). Localized aliphatic organic material on the surface of Ceres. *Science, 355*(6326), 719–722.

De Sanctis, M. C., Ammannito, E., Migliorini, A., Lazzaro, D., Capria, M. T., & McFadden, L. (2011). Mineralogical characterization of some V-type asteroids, in support of the NASA Dawn mission. *Monthly Notices of the Royal Astronomical Society, 412*(4), 2318–2332.

De Sanctis, M. C., Ammannito, E., Raponi, A., Frigeri, A. L., Ferrari, M. A., Carrozzo, F. G., Ciarniello, M., Formisano, M., Rousseau, B., Tosi, F., & Zambon, F. (2020). Fresh emplacement of hydrated sodium chloride on Ceres from ascending salty fluids. *Nature Astronomy, 4*(8), 786–793.

De Sanctis, M. C., Ammannito, E., Raponi, A., Marchi, S., McCord, T. B., McSween, H. Y., ... Russell, C. T. (2015). Ammoniated phyllosilicates with a likely outer solar system origin on (1) Ceres. *Nature, 528*(7581), 241–244.

De Sanctis, M. C., Raponi, A., Ammannito, E., Ciarniello, M., Toplis, M. J., McSween, H. Y., ... Russell, C. T. (2016). Bright carbonate deposits as evidence of aqueous alteration on (1) Ceres. *Nature, 536*(7614), 54–57.

De Sanctis, M. C., Vinogradoff, V., Raponi, A., Ammannito, E., Ciarniello, M., Carrozzo, F. G., ... Russell, C. T. (2019). Characteristics of organic matter on Ceres from VIR/Dawn high spatial resolution spectra. *Monthly Notices of the Royal Astronomical Society, 482*(2), 2407–2421.

de Sousa, R. R., Morbidelli, A., Gomes, R., Neto, E. V., Izidoro, A., & Alves, A. A. (2022). Dynamical origin of the dwarf planet Ceres. *Icarus, 379*, 114933.

Duarte, K. D., Schmidt, B. E., Chilton, H. T., Hughson, K. H. G., Sizemore, H. G., Ferrier, K. L., … Raymond, C. A. (2019). Landslides on Ceres: Diversity and geologic context. *Journal of Geophysical Research: Planets, 124*(12), 3329–3343.

Dyl, K. A., Bischoff, A., Ziegler, K., Young, E. D., Wimmer, K., & Bland, P. A. (2012). Early solar system hydrothermal activity in chondritic asteroids on 1–10-year timescales. *Proceedings of the National Academy of Sciences, 109*(45), 18306–18311.

El-Maarry, M. R. (2024). *Small icy bodies in the inner Solar System.* In R. J. Soare, J.-P. Williams, C. Ahrens, F. E. G. Butcher, & M. R. El-Maarry (Eds.), *Ices in the solar system, a volatile-driven journey from the inner solar system to its far reaches.* Elsevier Books.

Ermakov, A. I., Fu, R. R., Castillo-Rogez, J. C., Raymond, C. A., Park, R. S., Preusker, F., … Zuber, M. T. (2017a). Constraints on Ceres' internal structure and evolution from its shape and gravity measured by the Dawn spacecraft. *Journal of Geophysical Research, Planets, 122*(11), 2267–2293.

Ermakov, A. I., Mazarico, E., Schröder, S. E., Carsenty, U., Schorghofer, N., Preusker, F., … Zuber, M. T. (2017b). Ceres's obliquity history and its implications for the permanently shadowed regions. *Geophysical Research Letters, 44*(6), 2652–2661.

Fagents, S. A. (2003). Considerations for effusive cryovolcanism on Europa: The post-Galileo perspective. *Journal of Geophysical Research, Planets, 108*(E12).

Fanale, F. P., & Salvail, J. R. (1989). The water regime of asteroid (1) Ceres. *Icarus, 82*(1), 97–110.

Farnsworth, K. K., Dhingra, R. D., Ahrens, C. J., Nathan, E. M., & Magaña, L. O. (2024). *Titan, Enceladus, and other icy moons of Saturn.* In R. J. Soare, J.-P. Williams, C. Ahrens, F. E. G. Butcher, & M. R. El-Maarry (Eds.), *Ices in the solar system, a volatile-driven journey from the inner solar system to its far reaches.* Elsevier Books.

Formisano, M., Federico, C., De Sanctis, M. C., Frigeri, A., Magni, G., Raponi, A., & Tosi, F. (2018). Thermal stability of water ice in Ceres' craters: The case of Juling crater. *Journal of Geophysical Research: Planets, 123*(9), 2445–2463.

Fu, R. R., Ermakov, A. I., Marchi, S., Castillo-Rogez, J. C., Raymond, C. A., Hager, B. H., … Russell, C. T. (2017). The interior structure of Ceres as revealed by surface topography. *Earth and Planetary Science Letters, 476*, 153–164.

Gallagher, C. (2024). *Glaciation and glacigenic geomorphology on Earth in the Quaternary Period.* In R. J. Soare, J.-P. Williams, C. Ahrens, F. E. G. Butcher, & M. R. El-Maarry (Eds.), *Ices in the solar system, a volatile-driven journey from the inner solar system to its far reaches.* Elsevier Books.

Hasegawa, S., Marsset, M., DeMeo, F. E., Bus, S. J., Geem, J., Ishiguro, M., Im, M., Kuroda, D., & Vernazza, P. (2021). Discovery of two TNO-like bodies in the asteroid belt. *The Astrophysical Journal Letters, 916*(1), L6.

Hayne, P. O., & Aharonson, O. (2015). Thermal stability of ice on Ceres with rough topography. *Journal of Geophysical Research, Planets, 120*(9), 1567–1584.

Herschel, W. (1832, December). Observations on the two lately discovered celestial bodies. In *Vol. 1. Abstracts of the papers printed in the philosophical transactions of the Royal Society of London* (pp. 80–82). London: The Royal Society.

Hesse, M. A., & Castillo-Rogez, J. C. (2019). Thermal evolution of the impact-induced cryomagma chamber beneath Occator crater on Ceres. *Geophysical Research Letters, 46*(3), 1213–1221.

Hsieh, H. H., & Jewitt, D. (2006). A population of comets in the main asteroid belt. *Science, 312*(5773), 561–563.

Hughson, K. H. G., Russell, C. T., Schmidt, B. E., Travis, B., Preusker, F., Neesemann, A., et al. (2019). Normal faults on Ceres: Insights into the mechanical properties and thermal history of Nar Sulcus. *Geophysical Research Letters, 46*, 80–88. https://doi.org/10.1029/2018GL080258.

Johansen, A., & Lambrechts, M. (2017). Forming planets via pebble accretion. *Annual Review of Earth and Planetary Sciences, 45*, 359–387.

Kaplan, H. H., Lauretta, D. S., Simon, A. A., Hamilton, V. E., DellaGiustina, D. N., Golish, D. R., … Enos, H. L. (2020). Bright carbonate veins on asteroid (101955) Bennu: Implications for aqueous alteration history. *Science, 370*(6517), eabc3557.

Kelley, M. S., Bodewits, D., Hsieh, H. H., & Milam, S. N. (2021). First detection of volatiles from a main-belt comet. *JWST Proposal. Cycle, 1,* 2037.

King, T. V., Clark, R. N., Calvin, W. M., Sherman, D. M., & Brown, R. H. (1992). Evidence for ammonium-bearing minerals on Ceres. *Science, 255*(5051), 1551–1553.

Konopliv, A. S., Asmar, S., Bills, B. G., Mastrodemos, N., Park, R. S., Raymond, C. A., … Zuber, M. T. (2011). The Dawn gravity investigation at Vesta and Ceres. *Space Science Reviews, 163,* 461–486.

Küppers, M., O'Rourke, L., Bockelée-Morvan, D., Zakharov, V., Lee, S., von Allmen, P., … Moreno, R. (2014). Localized sources of water vapour on the dwarf planet (1) Ceres. *Nature, 505*(7484), 525–527.

Kurokawa, H., Shibuya, T., Sekine, Y., Ehlmann, B. L., Usui, F., Kikuchi, S., & Yoda, M. (2022). Distant formation and differentiation of outer main belt asteroids and carbonaceous chondrite parent bodies. *AGU Advances, 3*(1). e2021AV000568.

Landis, M. E., Byrne, S., Combe, J. P., Marchi, S., Castillo-Rogez, J., Sizemore, H. G., … Russell, C. T. (2019). Water vapor contribution to Ceres' exosphere from observed surface ice and postulated ice-exposing impacts. *Journal of Geophysical Research, Planets, 124*(1), 61–75.

Landis, M. E., Byrne, S., Schörghofer, N., Schmidt, B. E., Hayne, P. O., Castillo-Rogez, J., … Russell, C. T. (2017). Conditions for sublimating water ice to supply Ceres' exosphere. *Journal of Geophysical Research, Planets, 122*(10), 1984–1995.

Li, J. Y., & Castillo-Rogez, J. C. (2022). *Ceres: An ice-rich world in the inner solar system.* Jan 17.

Li, J. Y., Schröder, S. E., Mottola, S., Nathues, A., Castillo-Rogez, J. C., Schorghofer, N., … Russell, C. T. (2019). Spectrophotometric modeling and mapping of Ceres. *Icarus, 322,* 144–167.

Mao, X., & McKinnon, W. B. (2020). Spin evolution of Ceres and Vesta due to impacts. *Meteoritics & Planetary Science, 55*(11), 2493–2518.

Marchi, S., Bottke, W. F., Cohen, B. A., Wünnemann, K., Kring, D. A., McSween, H. Y., … Russell, C. T. (2013). High-velocity collisions from the lunar cataclysm recorded in asteroidal meteorites. *Nature Geoscience, 6*(4), 303–307.

Marchi, S., Raponi, A., Prettyman, T. H., De Sanctis, M. C., Castillo-Rogez, J., Raymond, C. A., … Yamashita, N. (2019). An aqueously altered carbon-rich Ceres. *Nature Astronomy, 3*(2), 140–145.

Marzo, G. A., Davila, A. F., Tornabene, L. L., Dohm, J. M., Fairén, A. G., Gross, C., … McKay, C. P. (2010). Evidence for Hesperian impact-induced hydrothermalism on Mars. *Icarus, 208*(2), 667–683.

McCord, T. B., Orlando, T. M., Teeter, G., Hansen, G. B., Sieger, M. T., Petrik, N. G., & Van Keulen, L. (2001). Thermal and radiation stability of the hydrated salt minerals epsomite, mirabilite, and natron under Europa environmental conditions. *Journal of Geophysical Research, Planets, 106*(E2), 3311–3319.

McCord, T. B., & Sotin, C. (2005). Ceres: Evolution and current state. *Journal of Geophysical Research, Planets, 110*(E5).

McKay, A. J., Bodewits, D., & Li, J. Y. (2017). Observational constraints on water sublimation from 24 Themis and 1 Ceres. *Icarus, 286,* 308–313.

McKinnon, W. B. (2008). On the possibility of large KBOs being injected into the outer asteroid belt. In *AAS/division for planetary sciences meeting abstracts# 40. 38-03.*

McSween, H. Y., Jr., Emery, J. P., Rivkin, A. S., Toplis, M. J., Castillo-Rogez, J. C., Prettyman, T. H., … Russell, C. T. (2018). Carbonaceous chondrites as analogs for the composition and alteration of Ceres. *Meteoritics & Planetary Science, 53*(9), 1793–1804.

Melosh, H. J. (1982). A schematic model of crater modification by gravity. *Journal of Geophysical Research: Solid Earth, 87*(B1), 371–380.

Melwani Daswani, M., & Castillo-Rogez, J. C. (2022). Porosity-filling metamorphic brines explain Ceres's low mantle density. *The Planetary Science Journal, 3*(1), 21.

Melwani Daswani, M., Vance, S. D., Mayne, M. J., & Glein, C. R. (2021). A metamorphic origin for Europa's ocean. *Geophysical Research Letters, 48*(18). e2021GL094143.

Milliken, R. E., & Rivkin, A. S. (2009). Brucite and carbonate assemblages from altered olivine-rich materials on Ceres. *Nature Geoscience, 2*(4), 258–261.

Mousis, O., Alibert, Y., Hestroffer, D., Marboeuf, U., Dumas, C., Carry, B., … Selsis, F. (2008). Origin of volatiles in the main belt. *Monthly Notices of the Royal Astronomical Society, 383*(3), 1269–1280.

Nathues, A., Hoffmann, M., Schaefer, M., Le Corre, L., Reddy, V., Platz, T., … Vincent, J. B. (2015). Sublimation in bright spots on (1) Ceres. *Nature, 528*(7581), 237–240.

Néri, A., Guyot, F., Reynard, B., & Sotin, C. (2020). A carbonaceous chondrite and cometary origin for icy moons of Jupiter and Saturn. *Earth and Planetary Science Letters, 530*, 115920.

Neveu, M., & Desch, S. J. (2015). Geochemistry, thermal evolution, and cryovolcanism on Ceres with a muddy ice mantle. *Geophysical Research Letters, 42*(23), 10–197.

Neveu, M., Desch, S. J., & Castillo-Rogez, J. C. (2017). Aqueous geochemistry in icy world interiors: Equilibrium fluid, rock, and gas compositions, and fate of antifreezes and radionuclides. *Geochimica et Cosmochimica Acta, 212*, 324–371.

Osinski, G. R., Tornabene, L. L., Banerjee, N. R., Cockell, C. S., Flemming, R., Izawa, M. R., … Southam, G. (2013). Impact-generated hydrothermal systems on earth and Mars. *Icarus, 224*(2), 347–363.

Palomba, E., Longobardo, A., De Sanctis, M. C., Stein, N. T., Ehlmann, B., Galiano, A. N. N. A., … Russell, C. T. (2019). Compositional differences among bright spots on the Ceres surface. *Icarus, 320*, 202–212.

Park, R. S., Konopliv, A. S., Bills, B. G., Rambaux, N., Castillo-Rogez, J. C., Raymond, C. A., … Nathues, A. (2016). Interior structure of dwarf planet Ceres from measured gravity and shape. *Nature, 537*, 515–517.

Park, R. S., Konopliv, A. S., Ermakov, A. I., Castillo-Rogez, J. C., Fu, R. R., Hughson, K. H. G., … Russell, C. T. (2020). Evidence of non-uniform crust of Ceres from Dawn's high-resolution gravity data. *Nature Astronomy, 4*(8), 748–755.

Park, R. S., Vaughan, A. T., Konopliv, A. S., Ermakov, A. I., Mastrodemos, N., Castillo-Rogez, J. C., … Zuber, M. T. (2019). High-resolution shape model of Ceres from stereophotoclinometry using Dawn imaging data. *Icarus, 319*, 812–827.

Passey, Q. R., & Shoemaker, E. M. (1982). Craters and basins on Ganymede and Callisto: Morphological indicators of crustal evolution. In D. Morrison (Ed.), *Satellites of Jupiter* (pp. 340–378). Tucson: University of Arizona Press.

Pizzarello, S., & Williams, L. B. (2012). Ammonia in the early solar system: An account from carbonaceous meteorites. *The Astrophysical Journal, 749*(2), 161.

Platz, T., Nathues, A., Schorghofer, N., Preusker, F., Mazarico, E., Schröder, S. E., … Russell, C. T. (2016). Surface water-ice deposits in the northern shadowed regions of Ceres. *Nature Astronomy, 1*(1), 1–6.

Platz, T., Nathues, A., Sizemore, H. G., Crown, D. A., Hoffmann, M., Schäfer, M., … Preusker, F. (2018). Geological mapping of the Ac-10 Rongo quadrangle of Ceres. *Icarus, 316*, 140–153.

Prettyman, T. H., Feldman, W. C., McSween, H. Y., Jr., Dingler, R. D., Enemark, D. C., Patrick, D. E., … Reedy, R. C. (2011). Dawn's gamma ray and neutron detector. *Space Science Reviews, 163*, 371–459.

Prettyman, T. H., Yamashita, N., Ammannito, E., Ehlmann, B. L., McSween, H. Y., Mittlefehldt, D. W., … Russell, C. T. (2019). Elemental composition and mineralogy of Vesta and Ceres: Distribution and origins of hydrogen-bearing species. *Icarus, 318*, 42–55.

Prettyman, T. H., Yamashita, N., Landis, M. E., Castillo-Rogez, J. C., Schörghofer, N., Pieters, C. M., … Russell, C. T. (2021). Replenishment of near-surface water ice by impacts into Ceres' volatile-rich crust: Observations by Dawn's gamma ray and neutron detector. *Geophysical Research Letters, 48*(15). e2021GL094223.

Prettyman, T. H., Yamashita, N., Toplis, M. J., McSween, H. Y., Schörghofer, N., Marchi, S., … Russell, C. T. (2017). Extensive water ice within Ceres' aqueously altered regolith: Evidence from nuclear spectroscopy. *Science, 355*(6320), 55–59.

Postberg, F., Khawaja, N., Abel, B., Choblet, G., Glein, C. R., Gudipati, M. S., … Waite, J. H. (2018). Macromolecular organic compounds from the depths of Enceladus. *Nature, 558*(7711), 564–568.

Qi, C., Stern, L. A., Pathare, A., Durham, W. B., & Goldsby, D. L. (2018). Inhibition of grain boundary sliding in fine-grained ice by intergranular particles: Implications for planetary ice masses. *Geophysical Research Letters, 45*(23), 12–757.

Quick, L. C., Buczkowski, D. L., Ruesch, O., Scully, J. E., Castillo-Rogez, J., Raymond, C. A., … Sykes, M. V. (2019). A possible brine reservoir beneath Occator crater: Thermal and compositional evolution and formation of the Cerealia dome and Vinalia faculae. *Icarus, 320,* 119–135.

Raponi, A., De Sanctis, M. C., Carrozzo, F. G., Ciarniello, M., Castillo-Rogez, J. C., Ammannito, E., … Russell, C. T. (2019). Mineralogy of Occator crater on Ceres and insight into its evolution from the properties of carbonates, phyllosilicates, and chlorides. *Icarus, 320,* 83–96.

Raponi, A., De Sanctis, M. C., Frigeri, A., Ammannito, E., Ciarniello, M., Formisano, M., … Russell, C. T. (2018). Variations in the amount of water ice on Ceres' surface suggest a seasonal water cycle. *Science Advances, 4*(3), eaao3757.

Rayman, M. D. (2020). Lessons from the Dawn mission to Ceres and Vesta. *Acta Astronautica, 176,* 233–237.

Rayman, M. D. (2022). Dawn at Ceres: The first exploration of the first dwarf planet discovered. *Acta Astronautica, 194,* 334–352.

Raymond, C. A., Ermakov, A. I., Castillo-Rogez, J. C., Marchi, S., Johnson, B. C., Hesse, M. A., … Russell, C. T. (2020). Impact-driven mobilization of deep crustal brines on dwarf planet Ceres. *Nature Astronomy, 4*(8), 741–747.

Raymond, S. N., & Izidoro, A. (2017). Origin of water in the inner solar system: Planetesimals scattered inward during Jupiter and Saturn's rapid gas accretion. *Icarus, 297,* 134–148.

Rivkin, A. S., Li, J. Y., Milliken, R. E., Lim, L. F., Lovell, A. J., Schmidt, B. E., McFadden, L. A., & Cohen, B. A. (2011). The surface composition of Ceres. *Space Science Reviews, 163,* 95–116.

Rivkin, A. S., Volquardsen, E. L., & Clark, B. E. (2006). The surface composition of Ceres: Discovery of carbonates and iron-rich clays. *Icarus, 185*(2), 563–567.

Roth, L. (2018). Constraints on water vapor and sulfur dioxide at Ceres: Exploiting the sensitivity of the hubble space telescope. *Icarus, 305,* 149–159.

Rousselot, P., Opitom, C., Jehin, E., Hutsemékers, D., Manfroid, J., Villarreal, M. N., … Mousis, O. (2019). Search for water outgassing of (1) Ceres near perihelion. *Astronomy & Astrophysics, 628,* A22.

Ruesch, O., Platz, T., Schenk, P., McFadden, L. A., Castillo-Rogez, J. C., Quick, L. C., … Russell, C. T. (2016). Cryovolcanism on ceres. *Science, 353*(6303), aaf4286.

Ruesch, O., Quick, L. C., Landis, M. E., Sori, M. M., Čadek, O., Brož, P., … Russell, C. T. (2019). Bright carbonate surfaces on Ceres as remnants of salt-rich water fountains. *Icarus, 320,* 39–48.

Russell, C., & Raymond, C. (2011). The Dawn mission to minor planets 4 Vesta and 1 Ceres. *Space Science Reviews, 163*(1–4).

Russell, C. T., Raymond, C. A., Ammannito, E., Buczkowski, D. L., De Sanctis, M. C., Hiesinger, H., … Yamashita, N. (2016). Dawn arrives at Ceres: Exploration of a small, volatile-rich world. *Science, 353*(6303), 1008–1010.

Schenk, P. M. (1993). Central pit and dome craters: Exposing the interiors of Ganymede and Callisto. *Journal of Geophysical Research, Planets, 98*(E4), 7475–7498.

Schenk, P., Scully, J., Buczkowski, D., Sizemore, H., Schmidt, B., Pieters, C., … Raymond, C. (2020). Impact heat driven volatile redistribution at Occator crater on Ceres as a comparative planetary process. *Nature Communications, 11*(1), 1–11.

Schenk, P., Sizemore, H., Schmidt, B., Castillo-Rogez, J., De Sanctis, M., Bowling, T., … Russell, C. (2019). The central pit and dome at Cerealia facula bright deposit and floor deposits in Occator crater, Ceres: Morphology, comparisons and formation. *Icarus, 320,* 159–187.

Schmedemann, N., Kneissl, T., Neesemann, A., Stephan, K., Jaumann, R., Krohn, K., … Russell, C. T. (2016). Timing of optical maturation of recently exposed material on Ceres. *Geophysical Research Letters, 43*(23), 11987.

Schmidt, B. E., Hughson, K. H., Chilton, H. T., Scully, J. E., Platz, T., Nathues, A., … Raymond, C. A. (2017). Geomorphological evidence for ground ice on dwarf planet Ceres. *Nature Geoscience, 10*(5), 338–343.

Schmidt, B. E., Sizemore, H. G., Hughson, K. H. G., Duarte, K. D., Romero, V. N., Scully, J. E. C., … Russell, C. T. (2020). Post-impact cryo-hydrologic formation of small mounds and hills in Ceres's Occator crater. *Nature Geoscience*, *13*(9), 605–610.

Schorghofer, N. (2008). The lifetime of ice on main belt asteroids. *The Astrophysical Journal*, *682*(1), 697.

Schorghofer, N., Byrne, S., Landis, M. E., Mazarico, E., Prettyman, T. H., Schmidt, B. E., … Russell, C. T. (2017). The putative Cerean exosphere. *The Astrophysical Journal*, *850*(1), 85.

Schorghofer, N., Mazarico, E., Platz, T., Preusker, F., Schröder, S. E., Raymond, C. A., & Russell, C. T. (2016). The permanently shadowed regions of dwarf planet Ceres. *Geophysical Research Letters*, *43*(13), 6783–6789.

Schorghofer, N., Williams, J. P., Martinez-Camacho, J., Paige, D. A., & Siegler, M. A. (2021). Carbon dioxide cold traps on the moon. *Geophysical Research Letters*, *48*(20). e2021GL095533.

Schröder, S. E., Mottola, S., Carsenty, U., Ciarniello, M., Jaumann, R., Li, J. Y., … Russell, C. T. (2017). Resolved spectrophotometric properties of the Ceres surface from Dawn Framing Camera images. *Icarus*, *288*, 201–225.

Scully, J. E., Bowling, T., Bu, C., Buczkowski, D. L., Longobardo, A., Nathues, A., … Jaumann, R. (2019). Synthesis of the special issue: The formation and evolution of Ceres' Occator crater. *Icarus*, *320*, 213–225.

Scully, J. E., Buczkowski, D. L., Neesemann, A., Williams, D. A., Mest, S. C., Raymond, C. A., … Ermakov, A. I. (2018). Ceres' Ezinu quadrangle: A heavily cratered region with evidence for localized subsurface water ice and the context of Occator crater. *Icarus*, *316*, 46–62.

Scully, J. E. C., Schenk, P. M., Castillo-Rogez, J. C., Buczkowski, D. L., Williams, D. A., Pasckert, J. H., … Russell, C. T. (2020). The varied sources of faculae-forming brines in Ceres' Occator crater emplaced via hydrothermal brine effusion. *Nature Communications*, *11*(1), 1–11.

Serio, G. F., Manara, A., Sicoli, P., & Bottke, W. F. (2002). *Giuseppe Piazzi and the discovery of Ceres*. University of Arizona Press.

Sierks, H., Keller, H. U., Jaumann, R., Michalik, H., Behnke, T., Bubenhagen, F., … Tschentscher, M. (2011). The Dawn framing camera. *Space Science Reviews*, *163*, 263–327.

Sizemore, H. G., Platz, T., Schorghofer, N., Prettyman, T. H., De Sanctis, M. C., Crown, D. A., … Raymond, C. A. (2017). Pitted terrains on (1) Ceres and implications for shallow subsurface volatile distribution. *Geophysical Research Letters*, *44*(13), 6570–6578.

Sizemore, H. G., Schmidt, B. E., Buczkowski, D. A., Sori, M. M., Castillo-Rogez, J. C., Berman, D. C., … Raymond, C. A. (2019). A global inventory of ice-related morphological features on dwarf planet Ceres: Implications for the evolution and current state of the cryosphere. *Journal of Geophysical Research, Planets*, *124*(7), 1650–1689.

Sori, M. M., Byrne, S., Bland, M. T., Bramson, A. M., Ermakov, A. I., Hamilton, C. W., … Russell, C. T. (2017). The vanishing cryovolcanoes of Ceres. *Geophysical Research Letters*, *44*(3), 1243–1250.

Stein, N. T., Ehlmann, B. L., Palomba, E., De Sanctis, M. C., Nathues, A., Hiesinger, H., … Russell, C. T. (2019). The formation and evolution of bright spots on Ceres. *Icarus*, *320*, 188–201.

Sugita, S., Honda, R., Morota, T., Kameda, S., Sawada, H., Tatsumi, E., … Tsuda, Y. (2019). The geomorphology, color, and thermal properties of Ryugu: Implications for parent-body processes. *Science*, *364*(6437), eaaw0422.

Thomas, P. C., Parker, J. W., McFadden, L. A., Russell, C. T., Stern, S. A., Sykes, M. V., & Young, E. F. (2005). Differentiation of the asteroid Ceres as revealed by its shape. *Nature*, *437*(7056), 224–226.

Thomas, P. C., Tajeddine, R., Tiscareno, M. S., Burns, J. A., Joseph, J., Loredo, T. J., … Porco, C. (2016). Enceladus's measured physical libration requires a global subsurface ocean. *Icarus*, *264*, 37–47.

Titus, T. N. (2015). Ceres: Predictions for near-surface water ice stability and implications for plume generating processes. *Geophysical Research Letters*, *42*(7), 2130–2136.

Travis, B. J., Bland, P. A., Feldman, W. C., & Sykes, M. V. (2018). Hydrothermal dynamics in a CM-based model of Ceres. *Meteoritics & Planetary Science*, *53*(9), 2008–2032.

Vernazza, P., Jorda, L., Ševeček, P., Brož, M., Viikinkoski, M., Hanuš, J., … Maestre, J. L. (2020). A basin-free spherical shape as an outcome of a giant impact on asteroid Hygiea. *Nature Astronomy*, *4*(2), 136–141.

Villarreal, M. N., Russell, C. T., Luhmann, J. G., Thompson, W. T., Prettyman, T. H., A'Hearn, M. F., … Raymond, C. A. (2017). The dependence of the Cerean exosphere on solar energetic particle events. *The Astrophysical Journal Letters, 838*(1), L8.

Williams, D. A., Buczkowski, D. L., Mest, S. C., Scully, J. E., Platz, T., & Kneissl, T. (2018). Introduction: The geologic mapping of Ceres. *Icarus, 316,* 1–13.

Wood, C. A., Head, J. W., & Cintala, M. J. (1978). Interior morphology of fresh Martian craters—The effects of target characteristics. In *Proceedings of the Ninth Lunar and Planetary Science Conference (Vol. 3), Houston, TX, March 13–17, 1978 (A79-39253 16-91). New York: Pergamon Press, Inc., 1978, Vol. 9* (pp. 3691–3709).

Zolotov, M. Y. (2020). The composition and structure of Ceres' interior. *Icarus, 335,* 113404.

Zolotov, M. Y., & Kargel, J. S. (2009). *On the chemical composition of Europa's icy shell, ocean, and underlying rocks. Vol. 431.* Tucson, AZ: University of Arizona Press.

Small icy bodies in the inner Solar System

Mohamed Ramy El-Maarry

*Space and Planetary Science Center and Department of Earth Sciences, Khalifa University, Abu Dhabi,
United Arab Emirates*

Abstract

Small icy bodies in the inner solar system, particularly the Asteroid "Main" Belt, offer a unique platform for testing various Solar System formation models by providing clues of the early physical and chemical conditions in the primary disk. Despite theoretically being more likely to be located in the outer Asteroid Belt, icy bodies are widely distributed across the inner Solar System and show a variety of compositions, orbital properties, and levels of activity. This is mainly because their small sizes and irregular shapes increase the likelihood of their drifting from their points of origin due to both gravitational and nongravitational effects. Collectively, these processes lead to the migration of small bodies from as far as the Trans-Neptunian region to the inner Solar System, including the Near-Earth environment, and may have been an essential pathway in the delivery of water and volatiles to Earth. The resulting long-term evolution of their orbital properties complicates the synthesis of knowledge that can be acquired from studying small icy bodies. However, ongoing and future space exploration missions, increased Earth-based observation density and accuracy, and continuously improving simulations are going to result in a better characterization of the distribution of ice in the inner solar system and more accurate models of the planetary system's formation.

1 Introduction

Small bodies offer important and unique, insights into the early physical and chemical conditions of our solar system (see Prolog I). However, given their small [typically several km to a few 100s of km in size (e.g., de León et al., 2018)], they have a high tendency to drift away with time from their initial formation zone, due to gravitational effects of larger nearby planets (Section 5), perturbations caused by uneven thermal emissions from their small irregular shapes, and collisions (Section 6). The resulting long-term evolution of their orbital properties complicates the synthesis of knowledge that can be acquired from studying them, as their exact point of origin becomes uncertain. This uncertainty is particularly impactful when studying *icy* small bodies because their primordial distribution and the stability of their ice content are directly controlled by the physical and thermal conditions that were prevalent in the early Solar System. On the other hand, continued scientific and technological advances in remote sensing methods, including Earth-based observations, as well as constant improvements in models of solar system formation allow for a better characterization of the distribution of ice in small

Ices in the Solar System. https://doi.org/10.1016/B978-0-323-99324-1.00012-2

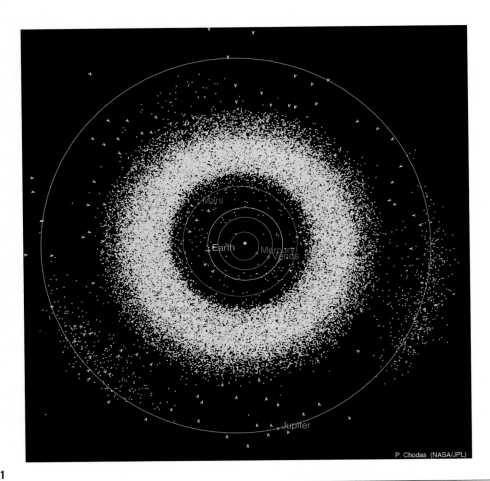

FIG. 1

Diagram of the inner Solar System to illustrate the scope of this chapter. The positions of all known asteroids and comets (on January 1, 2018) are also marked here as *yellow dots* and sunward-pointing *white wedges*, respectively. The diagram is available from NASA's Small-Body Database with credit to P. Chodas (https://ssd.jpl.nasa.gov/tools/sbdb_lookup.html#/).

bodies, and a better understanding of the migration pathways and the dynamical evolution of small bodies. As a result, fitting new observations into a more accurate framework of ice distribution in the solar system is continually improving.

This chapter focuses on small icy (i.e., ice-bearing, regardless of its content) bodies in the inner Solar System (Fig. 1), which for the purposes of this discussion is defined as the zone between, and encompassing, the orbits of Mercury (inward) and Jupiter (outward). In this zone, the majority of small bodies are concentrated in the Main "Asteroid" Belt (MB). Small bodies beyond that zone and extending beyond to the outer regions of the Solar System, including the Kuiper Belt and beyond, are discussed in Ahrens et al. (2024, pp. 357–376). Ceres, the largest body in the MB with a diameter of ~1000 km, is not considered a small body despite sharing many characteristics with small icy bodies in the MB; it is instead the focus of Landis et al. (2024, pp. 221–260). The main questions to be addressed in this chapter are:

- What are the "types" of icy small bodies in the inner solar system, and how are they distributed?
- What is the observational evidence for ice in small bodies, and how can this be improved?
- What are the origins and migrational pathways of small icy bodies?
- How do icy small bodies evolve with time?

2 Icy bodies: Distinct types or a broad continuum?

The two main types of small bodies in the inner Solar System are asteroids (predominantly situated in the MB) and comets which have a broader distribution owing to their wider range of orbital elements and origin from the outer Solar System (Ahrens et al., 2024, pp. 357–376). From a compositional perspective, comets are predominantly ice-rich, and while it has been proven difficult to quantify the actual ice content (or commonly dust-to-ice ratio in cometary literature) from remote sensing data, a comet can be described qualitatively as ice-"rich" based on its ability to sustain long-term, periodic ice-sublimation-driven activity over multiple orbits.

On the other hand, asteroids were considered (at least historically) to be typically "dry" with little to no ice content and no associated activity. However, it is now known that a large subset of asteroids is hydrated (Sections 3 and 4).

Asteroids are grouped into different taxonomic groups based on their spectral properties.

Volatile-bearing C-type asteroids (likely related to carbonaceous chondrites) are predominant in the central and outer parts of the MB (2.5–3.3 AU), while stony, anhydrous S-type asteroids (related to ordinary chondrites) are found mainly in the inner MB region. For the purposes of this discussion, it is enough to focus here on those two major groups (a more detailed discussion of the numerous taxonomic groups is available, for example, in Asphaug, 2009).

As a classical view, one can postulate that the asteroids formed "in situ" such that the inner asteroids that are mostly dry formed in the inner Solar System, while the ice-bearing asteroids formed in the outer MB or beyond. However, observational evidence argues against such a simple framework, particularly (a) the very low mass density of the MB with the current population accounting for only ~0.05 Earth mass (e.g., DeMeo & Cary, 2014; Raymond, 2023), when it is expected to have contained at least an Earth mass in its primordial state (e.g., Bitsch et al., 2015; Hayashi, 1981), (b) the high range of eccentricities and inclinations of small bodies (e.g., Morbidelli et al., 2015; Petit et al., 2002), and finally (c) the partial mixing of asteroid types. These characteristics collectively suggest that the MB has been subjected to extensive excitation and gravitational perturbations leading to significant orbital evolution.

In conjunction with the continuously improving understanding of the volatile/hydrated minerals-bearing nature of C-type asteroids through spectroscopic observations, there is a similarly growing number of observations of "active" asteroids (Section 4), a few of which also show repeated activity, that are referred to as "Main Belt Comets" (MBCs). MBCs are diluting what used to be a far clearer distinction between asteroids and comets. A dynamical classification that has been commonly used is the Tisserand parameter (Kosai, 1992; Kresak, 1982). This can be defined as

$$T_J = \frac{a_J}{a} + 2\left[\left(1-e^2\right)\frac{a}{a_J}\right]^{\frac{1}{2}}\cos(i)$$

where T_J is the Tisserand parameter with respect to Jupiter (J), and a, e, and i are the orbital parameters for the semimajor axis, eccentricity, and inclination, respectively. Under this dynamic classification, asteroids are bodies that have $T_J > 3$, whereas comets typically have values of $T_J < 3$. This chapter follows these distinctions and adopts the terminology first presented by Jewitt (2012).

As can be seen in Fig. 2, there are now asteroids that are known to be at least ice-bearing (if not ice-rich), and that display activity, sometimes repeated, like comets. So this chapter deals with two main types of small bodies: (1) icy asteroids, primarily situated in the MB, and (2) comets whose orbits are mainly within the inner solar system, primarily Jupiter Family Comets (JFCs), which incidentally have T_J between 2 and 3 (Fig. 2; Levison, 1996). Halley-Type Comets (HTCs) and Long-Period Comets (LPCs), which have a wider spatial distribution extending beyond the inner Solar System, as well as bodies in dynamical orbital transition between the outer and inner Solar System (e.g., centaurs) are addressed in more detail in Ahrens et al. (2024, pp. 357–376). Finally, while most of the discussions in this chapter, and indeed the book overall, focus on water-ice, comets are known to contain other types of volatiles, particularly CO_2, and CO ices (e.g., Bockelée-Morvan et al., 2004; Harrington et al., 2022).

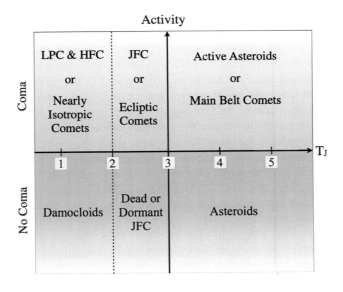

FIG. 2

Diagram showing the terminology used in this chapter for describing different types of small bodies based on inferred activity and the Tisserand parameter. The chapter mainly deals with icy or ice-bearing small bodies in the inner solar system, primarily the Main Belt, with a $T_J > 2$. Acronyms are defined in the main text.

The figure is taken from Jewitt (2012), and used with permission from original author.

Unless otherwise stated, the word "ice" in this chapter is going to refer to water-ice. However, other types of ices are also briefly discussed when needed, especially when assessing ice observations in JFCs.

3　So where is the ice?

The challenge in detecting or measuring ice in small bodies comes from the fact that those bodies are too small and lack a substantial and permanent atmosphere to allow for temperature and pressure conditions that would allow ice to be stable for a prolonged time if exposed to the surface. Indeed, thermal modeling work by Schorghhofer (2008) shows that most bodies in the MB are expected to have dry surfaces, and any ice in such bodies would be buried underneath an insulating low thermal conductivity layer of regolith. So, in order to map the potential distribution of ice in small bodies, there is a need to rely on either theoretical considerations of ice stability, both in terms of distance (both current and primordial) from the Sun and physical properties (e.g., density estimates, porosity, thermal inertia) of those potential ice-bearing bodies, or on direct observations under opportunistic conditions.

It is thought that most ice-bearing asteroids would have originated in the outer part of the MB, beyond the so-called "snow line", i.e., the minimum heliocentric distance that would allow for water-ice condensation. This line is typically assumed to be around 3 AU, with studies suggesting that it could be as close as 2.5 AU (e.g., Sonnett et al., 2011). However, the snow line may have shifted during the Solar System formation (e.g., Harsono et al., 2015), particularly during the protosolar nebular phase from as near as 0.7 AU to many tens of AU (e.g., Garaud & Lin, 2007; Min et al., 2011). On the other hand, Levison et al. (2009) suggest that inward, followed by outward, migration of the giant planets may have caused contamination of the MB by icy Trans-Neptunian scattered bodies (Section 4). In addition, nongravitational forces such as the Yarkovsky effect (Section 5), coupled with orbital resonances with Jupiter, may have pushed icy small bodies from their original locations in the MB, or beyond, into the inner Solar System, including the Near-Earth region.

4　How to know if a small body is "icy"?

There are three types of observations that have been used either to directly measure or interpret the presence of ice in small bodies in the MB, or the inner Solar System in general: (1) Observations of (recurrent) activity, (2) Earth-based and space telescope spectral observations (primarily infrared spectroscopy and sub-mm/radio observations), and (3) "in situ" measurements from space missions in relatively close proximity by appropriate remote sensing tools including infrared spectroscopy, radar, and nuclear spectroscopy. Below, a brief outline of each of these methods is provided along with a discussion of their most significant outcomes and limitations.

4.1　Observations of recurrent activity

Activity has commonly been used to differentiate between asteroids and comets and to indicate the presence/absence of sublimating ice. This is no longer the case as an ever-increasing number of observations show that asteroids, like comets, can be active, whether by ice sublimation or not. For example, asteroids may exhibit observable mass loss due to impacts, rotational instability (if rotation speeds increase to the point of centrifugal forces exceeding the cohesive forces or material strength of the body), radiation pressure, electrostatic ejection, and thermal fracturing (e.g., Jewitt, 2012; Jewitt & Hsieh, 2022).

Even in the case of comets, "activity" is a term that has been used mainly by astronomers to describe ice sublimation and the resulting interactions with levitated dust grains. However, high-resolution and long-term observations from spacecrafts that visited comets clearly show that comets display a "geologic-like" diversity of weathering and erosional processes that affect surface materials (e.g., Cheng et al., 2013; El-Maarry et al., 2017, 2019; Sunshine et al., 2016), thereby generalizing the usage of the word "activity" for comets and small bodies in general.

Consequently, the most robust method of linking activity in asteroids to ice sublimation has been observing unambiguous recurrent (over two or more orbits) mass loss when bodies are at or close to perihelion. Asteroids that display this behavior have been named "Main-Belt Comets" (MBCs, e.g., Hsieh & Jewitt, 2006; Jewitt, 2012; Orofino, 2022; Snodgrass et al., 2017). Fig. 3 displays observations for MBCs that have displayed recurrent activity.

It should be noted that MBCs' activity levels are very low compared to those of classical comets, and as such it may not be possible to directly detect water-ice signatures from them using Earth-based observations. However, recurrent activity particularly close to perihelion is a compelling piece of evidence that MBCs become active mainly through ice sublimation (e.g., Jewitt & Hsieh, 2022; Snodgrass

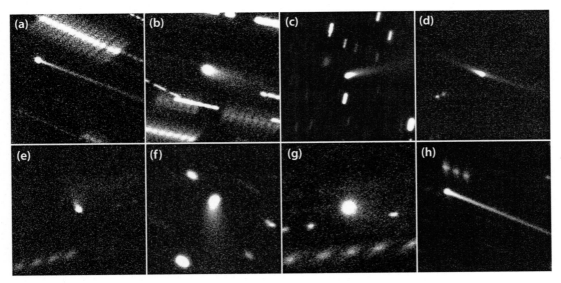

FIG. 3

Images of MBCs that have been confirmed to display recurrent activity. These include (a) 133P/Elst-Pizarro, (b) 238P/Read, (c) 259P/Garradd, (d) 288P/(300,163) 2006 VW$_{139}$, (e) 313P/Gibbs, (f) 324P/La Sagra, (g) 358P/PANSTARRS, and (h) 433P/(248,370) 2005 QN$_{173}$.

The figure is taken from Jewitt and Hsieh (2022), and used with permission from original author.

et al., 2017). The clear drawback of this technique is that more than one measurement near perihelion is needed, which could take up to a decade or more for outer Belt bodies with 5–6 year-orbits (Snodgrass et al., 2017).

4.2 Earth-based and space telescope spectral observations

Spectroscopy in various wavelengths offers a direct way of detecting volatiles in a solar system body. Spectroscopic techniques can be used to identify surface ice deposits, but they are particularly effective in studying active comets, which can contain ice grains in their comae during intense ice sublimation-related activity. Gas comae of comets are usually observed through fluorescence emission bands of various species in a wide range of wavelengths. Water itself has strong signatures in the Near-Infrared (NIR) at 2.7 μm. However, it is impossible to use this wavelength for detection from Earth because of the interference from atmospheric water. Instead, astronomers usually use the strong CN radical signature in the optical range as a proxy for water production since the ratio of water production to CN in the coma of typical JFCs is 350 (Snodgrass et al., 2017). Although this technique is indeed very useful in characterizing the production rates of JFCs and other comets, it may not be a suitable way for accurately characterizing water-ice in MBCs and other weakly active asteroids as the water/CN ratio in comets would not necessarily hold for MBCs.

A direct way of detecting and measuring water is by measuring its photo-dissociated fragments (such as the OH, H, and O radicals) in the UV range. Emission bands for such radicals are also observable in the optical range, but as was mentioned above, Earth-based observations of these radicals are challenging, while not impossible, and it would yield better results to observe them from space (see Section 4.3). It is also possible to detect water through the molecular emission lines in the IR and sub-mm/radio range. Such measurements allow direct detection of molecular water and its characterization instead of relying on assumptions made from the measurement of daughter radicals or closely associated species (like CN). Finally, in all these measurements, direct or inferred, what is measured is the water in a gaseous state in the coma resulting from the activity. However, it is possible to directly measure water-ice on the surface of a small body or even on solid grains in the coma through absorption features in the IR and NIR. Again, this technique would yield more accurate data in space (or in close proximity to the body), but is also possible from Earth. In addition, measurements in the 3 μm range allow not only for the detection of water-ice, but also hydrated minerals, which offer additional insights into the composition and evolutionary history of small bodies.

It is important to note that with respect to this category of observations, the body need not be "active" (Section 4.1). An asteroid can be water-rich without ever being subjected to processes/triggers or conditions that would cause ice sublimation and associated activity, such as impacts or rotational instabilities. For instance, measurements at 3 μm have been used to identify water-ice on asteroids (24) Themis, and (65) Cybele (Takir & Emery, 2012). This result has been recently challenged by O'Rourke et al. (2020) based on a combination of thermophysical modeling and lack of detection of water vapor in Herschel observations. Nevertheless, even in that case, the 3–3.1 μm could yet be attributed to volatiles (O'Rourke et al., 2020). Both Themis and Cybele are a part of asteroid families, which are a group of asteroids that share common orbital elements, and are probably of similar origin (Hirayama, 1918; see Novaković et al., 2022 for an extensive recent review). This is important for two reasons. First, if one member of an asteroid family is found to be hydrated or volatile-bearing, there is a strong probability that other members of its family have similar compositions, thereby expanding our repository of hydrated bodies in the MB. Secondly, most of the known MBCs also belong to asteroid families (e.g., Nesvorný et al., 2008; Novaković et al., 2014, 2022), which makes them highly relevant to future studies seeking to better characterize ice distribution in the MB (Section 6).

An important limitation of remote sensing observations, particularly in the N-IR region, is that they are sensitive to only the top 10s of microns of a surface. So rather than detecting primordial buried ice, the so-called 3 μm signature may instead be caused by exogenic water-ice. For instance, (16) Psyche was found to display signatures consistent with water (or OH) on its surface (Takir et al., 2017). However, Psyche is classified as a metal-rich (M-type) asteroid, which is thought to originate from the metallic core of a fragmented differentiated body. Because of that, Psyche's 3 μm signature has been reported to indicate that either Psyche is not a metal-rich asteroid, or that the source of its hydrated signature is caused by impacts from carbonaceous-rich materials over its lifetime (Takir et al., 2017). The latter is the more likely explanation given the numerous and diverse lines of evidence [e.g., high radar albedo (Shepard et al., 2017), high thermal inertia (Matter et al., 2013), and high density (Kuzmanoski, & Koračević, 2002) values] in favor of a metal-rich composition. Similarly, 3 μm signatures have been detected on supposedly anhydrous Near-Earth Asteroids (NEAs) 433 Eros and 1036 Ganymed, and have been also attributed possibly to exogenic sources, either through impacts from comets or carbonaceous impactors, or through solar wind interaction with silicate-rich surfaces (Rivkin et al., 2017). Finally, as noted above, measurements at or near 3 μm may not necessarily be caused by water-ice, but rather other (e.g., ammonium-bearing) volatiles or salts (e.g., O'Rourke et al., 2020; Poch et al., 2020). Therefore, it is favorable to have additional lines of evidence apart from spectroscopy (especially in the N-IR) to support the conclusion that a small body is icy, or that its ice content is primordial.

4.3 "In situ" measurements from space missions

Observing from space naturally offers better observing conditions and a higher signal-to-noise ratio, as well as an increase in overall accuracy, but is understandably more expensive, more difficult to maintain in the long term, and may not offer the high temporal frequency that Earth-based observations provide. Several flyby missions have been successful in directly detecting water molecules and other volatiles in the comae of JFCs. In particular, the presence of hypervolatiles such as CO_2 and CO suggests that other forms of ice are common in comets and indicates that JFCs are sourced originally from the outer regions of the Solar System (the Kuiper belt and beyond) as CO, for instance, is difficult to trap at temperatures above 25–50 K (e.g., Jewitt et al., 2007).

Water-ice in comets can either be crystalline or amorphous, as amorphous ice is thermodynamically stable at temperatures below ~110 K (Jewitt et al., 2007). Indeed, the presence of amorphous ice in comets has been suggested from IR spectra based on the absence of absorptions at 1.65 um, which are indicative of crystalline ice. Evidence of short-term exposed water-ice on cometary surfaces is available from the Rosetta mission at comet 67P/Churyumov-Gerasimenko, which observed over the mission's lifetime numerous features including irregular bright patches in smooth terrains (Fig. 4) (e.g., Barucci et al., 2016; El-Maarry et al., 2017; Fornasier et al., 2016b; Groussin et al., 2015), bright boulders (Pommerol et al., 2015; Fig. 5), and even bright freshly exposed cliff walls following collapses (Pajola et al., 2017), which were found to show spectrophotometric signatures consistent with crystalline water-ice (Fig. 6).

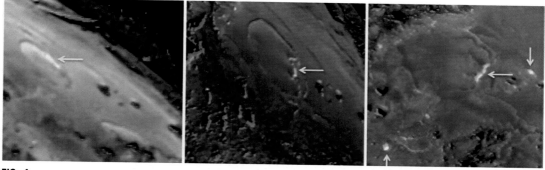

FIG. 4

Color composite images showing distinct circular patterns in the smooth deposits of comet 67P with bright exposures that have been interpreted to represent exposed water-ice (*yellow arrows*). The three images cover roughly the same area over three periods: June 18, 2015 (*left*), July 2, 2015 (*middle*), and July 11, 2015 (*right*). For scale, the distance between the two prominent boulders closest to the circular patterns in the left panel is approx. 100 m.

Figure adapted from Groussin et al. (2015).

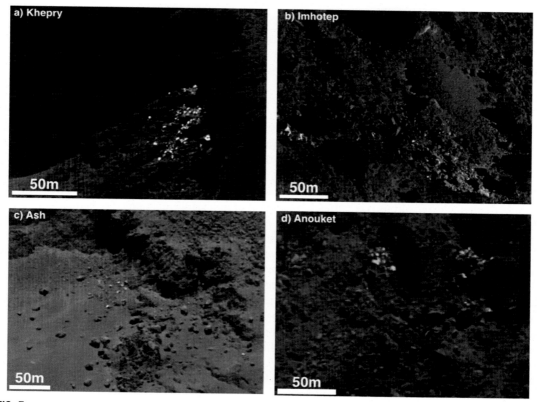

FIG. 5

Images from multiple regions on comet 67P displaying clusters of bright boulders. The boulders are situated at the bottom of cliffs and have most probably been exposed following mass wasting events. The names on the left corner signify the names of the regions (El-Maarry et al., 2015) on the comet where the clusters have been observed.

Figure source: Pommerol et al. (2015).

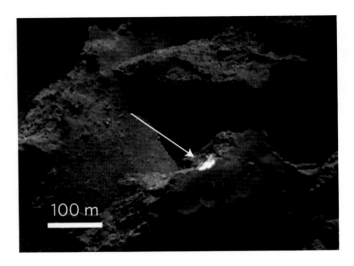

FIG. 6

Image of the so-called Aswan cliff on the northern hemisphere of comet 67P following a cliff collapse estimated to have taken place in July 2015. The event exposed fresh bright cliff walls, whose spectral signatures are consistent with exposures of crystalline water-ice. The image was taken in December 2015.

Figure adapted from Pajola et al. (2017).

While evidence for water-ice and other volatiles is unambiguous in JFCs and comets in general, to date no mission to an asteroid has made direct measurements or detections of water-ice, except for the dwarf planet Ceres (Landis et al., 2024, pp. 221–260). Many missions have flown by asteroids on their way to other solar system bodies but have not been successful in detecting water-ice either. There are a number of reasons for this lack of detection. Primarily, asteroids that have been visited by spacecraft are either genuinely anhydrous (i.e., s-types, Section 2) or at least were assumed to be predominantly anhydrous, so the missions were not designed to detect or investigate water-ice at these bodies. In addition, most asteroids (including even ice-rich small bodies) are expected to display dry surfaces (Section 3), unless they have icy interiors, and have through one mechanism or more become active. Nevertheless, space missions provide the possibility to make multi-instrument measurements that can provide information about ice content. For instance, while no direct evidence for water-ice has been found on Vesta, the largest asteroid and second largest body in the MB after Ceres, water-ice presence could be inferred from the high subsurface hydrogen content as measured by the DAWN mission's gamma rays and neutron spectrometer (Prettyman et al., 2012), hydrated signatures in the N-IR (De Sanctis et al., 2012), and geomorphological observations of possible gully networks (Scully et al., 2014).

Since active asteroids and MBCs are a relatively new discovery, a dedicated space mission is yet to explore one of these bodies with the appropriate payload. The Rosetta mission provided a unique aspect in being the only rendezvous mission to explore a comet over a long period of time, which has allowed it to analyze volatiles on 67P at an unprecedented spatial accuracy and temporal frequency. However, our knowledge of ice in the MB remains currently limited as most of the focus has been understandably on JFCs.

Closer to Earth, missions to NEAs have provided a wealth of information about bodies that are believed to have originated in the MB or beyond (see Section 5). Spectral analysis by the Hayabusa 2 and OSIRIS-REx missions to Ryugu and Bennu, respectively, show signatures consistent with an origin in the outer MB or beyond. At Ryugu, spectral analysis in the N-IR displays signatures consistent with OH-bearing minerals (Kitazato et al., 2019), and on Bennu, there are widespread signatures of hydrated minerals (e.g., Hamilton et al., 2019).

While no direct detections of water-ice have been made on NEAs, observations of hydrated minerals and overall spectral signatures consistent with C-type asteroids give a strong indication that NEAs may have experienced long-term volatile-driven processes before their long dynamical journey from the MB. This journey is discussed further in the next section.

5 "How did you get here, and where do you come from?"

Small bodies in the inner Solar System are an integral piece in the understanding of the Solar System formation. They are an essential parameter in simulations because their origins are key to constraining the physical, chemical, and dynamical conditions in the protoplanetary disk. However, their small size and mass make them more susceptible to both gravitational and nongravitational influences. So when considering the origins of icy bodies in the MB, it is important to be mindful of their probable complex dynamical history.

According to Wetherill (1992)'s model, the disruption and subsequent depletion of the MB occurred due to perturbations from the early planetary embryos (Moon- to Mars-sized) that formed in the inner Solar System. In this model, Jupiter and Saturn are assumed to form and remain in their current location. Alternatively, in the more recent Grand Tack model (Walsh et al., 2011), early migrations of Jupiter and Saturn (first inwardly, until Jupiter and Saturn get into a 3:2 resonance, and then outwardly, until the planetary disk is devoid of gas) are responsible for the MB's current distribution and depleted density (Fig. 7, see also Morbidelli et al., 2015). According to this model, C-type asteroids originated in the giant planets region before eventually drifting into their current locations (e.g., Kruijer et al., 2017; Walsh et al., 2011). Yet another provocative model was proposed by Raymond and Izidoro (2017a, 2017b). They propose that the primordial MB was "empty" to start with and that planetesimals were implanted into the MB region from the inner Solar System (in the case of the s-types) and the outer regions (in the case of the c-types) through a combined effect of gravitational scattering during the planets' formation, orbit destabilization, and aerodynamic gas drag (Raymond & Nesvorný, 2022). A detailed discussion of the numerous solar system formation models is beyond the scope and intent of this chapter and can instead be found in the recent reviews, for example, by Raymond and Nesvorný (2022) and Raymond et al. (2020).

The *current* orbital distribution of asteroids and small bodies in general in the MB is controlled by resonances (Nesvorný, 2018), primarily with Jupiter. Bodies that are in, or drift into, the orbital resonance of 2:1, 3:1, and 7:3 with Jupiter will undergo amplified variations in their eccentricity and eventually get ejected from the MB. Conversely, bodies in a 3:2 resonance are dynamically stable and, in fact, form a distinctive population (the Hilda asteroids). Even in the current configuration, the MB is continuously losing mass. Bodies drift into the abovementioned resonance orbits with Jupiter mainly through a nongravitational effect called the Yarkovsky (Yark) effect (Vokrouhlický et al., 2015; Yarkovsky, 1901). Yark is a dynamical force that arises due to lighting effects on a rotating body.

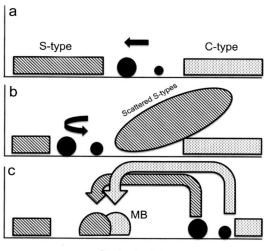

FIG. 7

A schematic diagram showing the effect of migration of Jupiter and Saturn (the two *large black dots*) on the MB. First, the planets migrate inward (a) till Saturn reaches its current mass, so they start to migrate outward instead (b). The *dashed and dotted areas* represent the orbital distributions (semimajor axis and eccentricity) of S-type and C-type asteroids, respectively. As Jupiter and Saturn continue their outward migration to their current location, an injection of scattered S-type and C-type asteroids into the MB takes place, leading to the current observable distribution (c).

Adapted from Morbidelli et al. (2015) and used by the permission of the University of Arizona Press.

As the hotter hemisphere (subjected to sunlight) emits more photons than the colder hemisphere, this generates a net force that pushes the body away from the Sun. This effect depends on numerous parameters but mainly affects small bodies, and is negligible for bodies larger than ~10 km (Asphaug, 2009). Over time, small bodies may drift into the resonant orbits and be expelled from the MB. Another relevant and equally important nongravitational effect is the Yarkovsky-O'Keefe-Radzievskii-Paddack (YORP) effect, which describes how the abovementioned thermal emission can also generate a rotational torque that can increase the rotation rate of small irregular bodies, possibly to a point where the body may fragment due to the increased centrifugal forces (e.g., Bottke et al., 2006; Rubincam, 2000; Vokrouhlický et al., 2015).

From the above, it should be apparent that a lack of knowledge of the full range of dynamical interactions and their durations complicates the attempts to understand the origin of icy or active bodies in general. For instance, it may be very difficult to constrain the origin of bodies very close to a Tisserand parameter of 3 (Section 2) as perturbations may have shifted their orbital characteristics over the boundary of 3 in either direction (e.g., Jewitt & Hsieh, 2022). This has led several studies to extend the limit between asteroids and comets into a more dynamic boundary of 3.05 or 3.08 (e.g., Jewitt et al., 2015; Tancredi, 2014). All the MBCs discovered so far have a range of T_j from 3.100 to 3.376 (Orofino, 2022), which should place them firmly in the "asteroid domain," and indicate that they formed initially in the outer MB. However, if the asteroid belt has been modified or affected substantially by the migration of the giant planets, the Tisserand parameter may no longer serve as a robust indicator of the dynamical origin of small bodies in the MB.

Similarly in the case of comets, JFCs are thought to have formed originally in the Scattered Disk (see Ahrens et al., 2024, pp. 357–376) and migrated through time and gravitational interactions into their current orbits. However, a study by Fernández and Sosa (2015) conducted on 58 JFCs on Near-Earth orbits suggests that several of those comets could have originated instead in the MB. Statistical analysis comprising all known asteroids and comets and their orbital characteristics indicates, in fact, that bodies with "cometary-like" orbits in the 2.8 to 3 T_j range are substantially dominated by MB contamination (Jewitt & Hsieh, 2022). A spectroscopic survey by Licandro et al. (2008) of over 40 bodies with cometary dynamical characteristics indicates that a few of these bodies show absorption features suggestive of S-type (silicate-bearing) asteroids.

To summarize, small bodies in the MB, including icy bodies, have a complex dynamical history, which complicates the efforts to investigate their origin and evolution. Indeed, the MB could be visualized as a section of a "highway" consisting of bodies that have (a) originated and stayed in the MB, (b) migrated from the Kuiper belt and beyond (e.g., JFCs), and (c) bodies that are drifting and migrating to the inner Solar System by way of disparate dynamical interactions (e.g., Morbidelli et al., 2002).

Given a long enough period of time, bodies may have migrated from as far away as the Kuiper Belt/ Scattered Disk to the near-Earth region of the Solar System. More than 100 JFCs are believed to have ended in the Near-Earth domain (Asphaug, 2009) such as 3552 Don Quixote (e.g., Mommert et al., 2014). Bottke et al. (2002) calculated on the basis of dynamical integration that an even higher number of bodies, up to 10% of Near-Earth Objects (NEOs), may be extinct JFCs. Finally, samples recently returned from the NEA Ryugu by the Hayabusa 2 mission were found to contain organics and hydrated minerals (Ito et al., 2022). In addition, a heavy abundance of certain Hydrogen and Nitrogen isotopes strongly implies that Ryugu was originally formed in the outer Solar System (Ito et al., 2022).

6 "What happens to you and your ice?"

The dynamical evolution of small bodies following the early stages of the Solar System formation including gas disk depletion and planetary migration is mainly dominated by collisions and thermal radiation-triggered (YARK and YORP) drift (e.g., Bottke et al., 2015; Brož et al., 2005). Collisions act with time to fragment the MB into smaller bodies. Consequently, collisional fragments are more susceptible to YARK and YORP. This further promotes their dynamical evolution and the potential for drifting into resonant orbits that push them out of the MB (e.g., Bottke et al., 2015).

The effect of collisions can be better understood by looking at asteroid family clustered fragments that are the result of collisional disruptions and can be identified through their similar orbital elements. For icy small bodies, collisions can have an even higher impact, as they may lead to exposure to previously buried ice and possibly trigger sublimation-driven activity. YORP can also have the same effect on small icy bodies. That is, if the body is small enough (~<5 km in size) for YORP to have a substantial effect, the body may spin up to the point of fragmenting and exposing previously covered ice, thereby activating the body. In fact, YORP-induced rotational instability has been suggested to be one of the main mechanisms by which MBCs become active (e.g., Jewitt & Hsieh, 2022).

As noted earlier, most MBCs and active asteroids in general are linked to collisional families, which further heightens the importance of studying both. A prominent example is 24 Themis. While not known to be active, Themis displays spectral properties consistent with c-types, with both hydrated and organic signatures (e.g., Campins et al., 2010; Florczak et al., 1999; Fornasier et al., 2016a; Rivkin & Emery, 2010). 24 Themis is part of the Themis dynamic family. The latter is thought to be the result of the fragmentation of a ~400 km-wide differentiated icy body (Castillo-Rogez & Schmidt, 2010; Marsset et al., 2016). The fragmentation is thought to have been triggered by an impact 2.5 ± 1.0 Gya (e.g., Durda et al., 2007; Novaković et al., 2022). At least three members of the Themis family are designated as active asteroids, and even possibly as MBCs (Novaković et al., 2022).

Families offer a unique insight into the collision history and dynamics in the MB because, in principle, it is possible from statistical analysis of the cluster as a whole and the magnitude of YARK-induced drift (if the fragments are small enough for it to be significant) to trace back the magnitude and time of the collision. However, this technique has been found to be most effective for very young families (e.g., Bottke et al., 2015; Novaković et al., 2022).

In conclusion, collisions have a significant effect on the long-term evolution of the MB. For icy small bodies, collisions can lead to fragmentation of large icy bodies triggering (a) possible activity due to ice exposure, and (b) if the fragments are small enough, high susceptibility to YARK and YORP effects, which can lead to these bodies further fragmenting (exposing ice in the process) and/or drifting out of the MB. This drifting may eventually lead to the migration of icy small bodies into the inner solar system, where they could have been involved in the delivery of water and volatiles to the terrestrial planets, including Earth (e.g., Meech & Raymond, 2020).

The presence of ice can not only drive further physical and chemical evolution in bodies beyond what is experienced by dry MB inhabitants, but may even lead to "geologic" surface changes. For small bodies, low gravity does not allow a permanent atmosphere to develop, thereby minimizing drivers for surface change. However, under the right external triggers, small icy bodies may undergo chemical evolution and surface modification. This is most evident in the case of comets, which is highly relevant to our discussion of JFCs as a subset of icy bodies in the inner Solar System.

The Rosetta mission to the JFC 67P/Churyumov-Gerasimenko from August 2014 to September 2016 remains the only cometary mission to monitor a comet for more than 2 years, including its perihelion passage (in August 2015), which has led to increased understating of cometary surface evolution on seasonal and longer-term scales. During the mission, many types of surface changes associated with activity were observed, including cliff collapses (Fig. 6), local erosion (Fig. 8), ejection, transport, and redeposition of materials, in addition to other transient surface changes (Fig. 9) (e.g., El-Maarry et al., 2017; Groussin et al., 2015; Hu et al., 2017; Pajola et al., 2017).

Comets may also undergo activity driven by volatiles other than water-ice (e.g., CO and CO_2 ice) at larger distances from the Sun because of the lower sublimation temperatures of these "hypervolatiles." In the longer term, this leads to chemical alterations of the primordial volatile inventory of these icy bodies.

In addition to long-term orbital evolution, a smaller yet significant dynamical evolution may occur in comets as they become active due to torques induced on the body by the activity, especially if the body is small enough and its shape and spin axis orientation allows for such orbital changes. Comet 67P underwent changes in its rotation period by around ~20 min during the lifetime of the Rosetta mission, which could have accounted for tectonic changes in its neck region (El-Maarry et al., 2017). Modeling

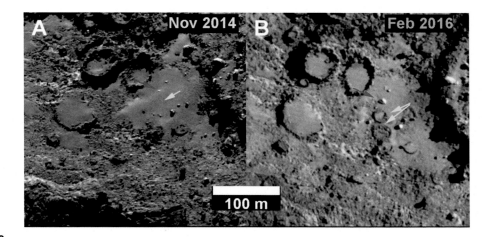

FIG. 8

Images of the so-called Imhotep region on comet 67P in 2014 (A) and 2016 (B) during the Rosetta mission showing substantial erosion during perihelion, which resulted in the exhumation of circular features (*yellow arrows*) and a large boulder (*red arrow*). The vertical extent of erosion in this particular location was estimated to be approx. 4 m.

Figure adapted from El-Maarry et al. (2017).

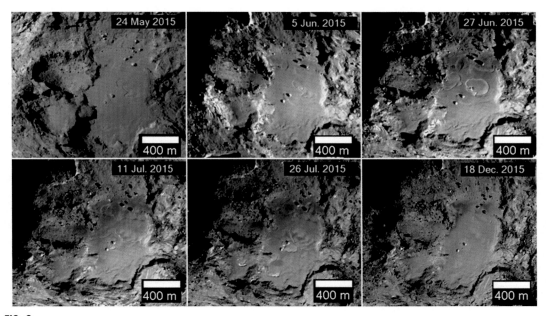

FIG. 9

A series of images of the Imhotep region on comet 67P showing transient changes in the smooth terrains in the form of evolving arcuate and circular patterns (*yellow arrows*), which were first reported in Groussin et al. (2015). Note the bright exposures in panel (E, also shown in Fig. 4), which have been interpreted to be crystalline ice.

Figure adapted from El-Maarry et al. (2017).

by Hirbayashi et al. (2016) has shown that activity-induced torques can be a fundamental factor in the splitting and reconfiguration of bilobed or multilobed small bodies, much like YORP's effect on small bodies in general.

In summary, icy bodies residing in the MB undergo long-term dynamical evolution just like dry members. However, they may also undergo extensive physical and chemical changes under conditions where the right triggers arise.

7 Future perspectives

The MB offers a unique platform for testing various solar System formation models. Yet in order to do so, there is a need to expand on space exploration as more observations are needed to constrain current models. Icy small bodies are at the forefront of this challenge because they could have been sourced from a radially wide range of locations in the primordial disk.

C-type asteroids are believed to have originated from the Jupiter and Saturn region and possibly beyond (e.g., Morbidelli et al., 2015; Raymond & Nesvorný, 2022; Walsh et al., 2011). They currently dominate the outer regions of the MB, between 2.5 and 3.2 au. Nearly all active asteroids and known MBCs fall within this range of heliocentric distances. Furthering the exploration of active asteroids and MBCs can provide more insights into this broad taxonomic group of asteroids. Studying icy bodies that are also part of dynamic families (e.g., Themis) may yield even more insights into the collisional history of the MB. Therefore, it is no wonder that there is a strong emphasis among the planetary science community to explore Themis and other members of its family in the near future (Landis et al., 2022). Other desirable future targets include sample-return missions to large MB bodies, "strategic" NEAs that can be linked with confidence to particular regions in the MB, or even the Kuiper Belt, and poorly explored/understood populations such as the Hilda asteroids, and Jupiter's irregular satellites (Bottke et al., 2020).

MBCs are an obvious target in that respect. Their recurrent activity is a direct indication of their icy nature and makes them easier to observe and to plan their observations. For instance, MBCs are found to have higher eccentricities on average than typical bodies in the MB, which would subject them to higher temperatures during perihelion and lower ones during aphelion. This would lead to higher levels of activity and, importantly, *higher chances of detection* at or close to their perihelion (Orofino, 2022).

There have certainly been positive developments in that regard: there were recently a couple of sample-return missions to NEAs: JAXA's Hayabusa 2 mission to Ryugu, and NASA's OSIRIS-REx mission to Bennu. In terms of future prospects, the Lucy mission (Levison et al., 2021) is on its way to explore multiple Jupiter Trojans. The Psyche mission (Elkins-Tanton et al., 2022), launching in 2023, will explore (16) Psyche in 2029 and hopefully shed more light on the origins of its $3\,\mu m$ signature. ESA's Comet Interceptor mission (Snodgrass & Jones, 2019), due to launch in 2029, plans to explore preferably a dynamically new comet. The United Arab Emirates is currently preparing a mission to explore seven MB asteroids in the 2030s (https://tinyurl.com/y42u6dw4), and the Chinese ZhengHe mission is expected to explore the active asteroid 311P/Pan STARRS in 2032.

These missions should provide in-depth in situ knowledge of a subset of small bodies. To fit these results into a much-needed bigger context, it is important to augment Earth- and Near-Earth-based capabilities, which is also making important steps forward. The Vera Rubin observatory is bound to increase the discoveries of small solar system objects (V. C. R. Collaboration et al., 2021), including dynamically new comets and MBCs. The long-awaited and now deployed James Webb Space Telescope with its IR observation range may allow a better understanding of the MB, detect ice and organics on the surface of asteroids (Norwood et al., 2016; Rivkin et al., 2016), and even possibly detect water in MBC comae (Kelley et al., 2016). The exploration of icy small bodies has only just begun gathering steam.

Acknowledgments

The author is grateful to William "Bill" Bottke for guidance on the literature that helped with this review and for discussions with Akos Kereszuri, Imre Toth, and the late Anny-Chantal Levasseur-Regourd. This chapter is dedicated to her. The author acknowledges support from the internal KU grant (8474000336-KU-SPSC) and is grateful for the anonymous review that has substantially improved the writing and content.

References

Ahrens, C. J., Lisse, C. M., Williams, J.-P., & Soare, R. J. (2024). *Geocryology of Pluto and the icy moons of Uranus and Neptune*. In R. J. Soare, J.-P. Williams, C. Ahrens, F. E. G. Butcher, & M. R. El-Maarry (Eds.), *Ices in the solar system, a volatile-driven journey from the inner solar system to its far reaches*. Elsevier Books.

Asphaug, E. (2009). Growth and evolution of Asteroids. *Annual Review of Earth and Planetary Sciences*, *37*, 413–448.

Barucci, A., et al. (2016). Detection of exposed H2O ice on the nucleus of comet 67P/Churyumov-Gerasimenko as observed by Rosetta OSIRIS and VIRTIS instruments. *Astronomy & Astrophysics*, *595*, A102.

Bitsch, B., Johansen, A., Lambrechts, M., & Morbidelli, A. (2015). The structure of protoplanetary discs around evolving young stars. *Astronomy and Astrophysics*, *575*, A28.

Bockelée-Morvan, D., Crovisier, J., Mumma, M. J., & Weaver, H. A. (2004). The composition of cometary volatiles. In M. C. Festou, H. U. Keller, & H. A. Weaver (Eds.), *745. Comets II* (pp. 391–423). University of Arizona Press, Tucson.

Bottke, W. F., Jr., Vokrouhlický, D., Rubincam, D. P., & Nesvorný, D. (2006). The Yarkovsky and Yorp effects: Implications for asteroid dynamics. *Annual Review of Earth and Planetary Sciences*, *34*, 157–191.

Bottke, W. F., Morbidelli, A., Jedicke, R., Petit, J.-M., Levison, H., et al. (2002). Debiased orbital and size distributions of the near-earth objects. *Icarus*, *156*, 399–433.

Bottke, W. F., et al. (2015). Collisional evolution of the asteroid belt. In P. Michel, F. E. DeMeo, & W. Bottke (Eds.), *Asteroids IV*. Univ. Arizona Press and LPI.

Brož, M., Vokrouhlický, D., Bottke, W., Nesvorný, D., Morbidelli, A., & Čapek, D. (2005). Non-gravitational forces acting on small bodies. *Proceedings of the International Astronomical Union*, *1*(S229), 351–365. https://doi.org/10.1017/S1743921305006848.

Campins, H., Hargrove, K., Pinilla-Alonso, N., Howell, E. S., Kelley, M. S., Licandro, J., et al. (2010). Water ice and organics on the surface of the asteroid 24 Themis. *Nature*, *464*(7293), 1320–1321.

Castillo-Rogez, J. C., & Schmidt, B. E. (2010). Geophysical evolution of the Themis family parent body. *Geophysical Research Letters*, *37*.

Cheng, A. F., Lisse, C. M., & A'Hearn, M. A. (2013). Surface geomorphology of Jupiter Family Comets: A geologic process perspective. *Icarus*, *222*, 808–817.

de León, J., Licandro, J., & Pinilla-Alonso, N. (2018). The diverse population of small bodies of the Solar system. In H. J. Deeg, & J. A. Belmonte (Eds.), *Handbook of exoplanets* (p. 3490). Springer International Publishing AG, part of Springer Nature.

De Sanctis, M. C., et al. (2012). Spectroscopic characterization of mineralogy and its diversity across Vesta. *Science, 336*, 697–700.

DeMeo, F. E., & Carry, B. (2014). Solar system evolution from compositional mapping of the asteroid belt. *Nature, 505*, 629–634.

Durda, D. D., Bottke, W. F., Jr., Nesvorný, D., Enke, B. L., Merline, W. J., Asphaug, E., & Richardson, D. C. (2007). Size–frequency distributions of fragments from SPH/N-body simulations of asteroid impacts: Comparison with observed asteroid families. *Icarus, 186*, 498–516.

Elkins-Tanton, L. T., et al. (2022). Distinguishing the origin of asteroid (16) psyche. *Space Science Reviews, 218*, 17.

El-Maarry, M. R., Groussin, O., Keller, H. U., Thomas, N., Vincent, J.-B., Mottola, S., Pajola, M., Otto, K., Herny, C., & Krasilnikov, S. (2019). Surface morphology of Comets and associated evolutionary processes: A review of Rosetta's observations of 67P/Churyumov-Gerasimenko. *Space Science Reviews, 215*(Issue 4). https://doi.org/10.1007/s11214-019-0602-1. 33 pp.

El-Maarry, M. R., et al. (2015). Regional surface morphology of comet 67P/Churyumov-Gerasimenko from Rosetta/OSIRIS images. *Astronomy & Astrophysics, 583*, A26.

El-Maarry, M. R., et al. (2017). Surface changes on comet 67P/Churyumov-Gerasimenko suggest a more active past. *Science, 355*(6332), 1392–1395. https://doi.org/10.1126/science.aak9384.

Fernández, & Sosa. (2015). Jupiter family comets in near- earth orbits: Are some of them interlopers from the asteroid belt? *Planetary and Space Science, 118*, 14–24.

Florczak, M., Lazzaro, D., Mothé-Diniz, T., Angeli, C. A., & Betzler, A. S. (1999). A spectroscopic study of the THEMIS family. *Astronomy & Astrophysics Supplement Series, 134*, 463–471.

Fornasier, S., Lantz, C., Perna, D., Campins, H., Barucci, M., & Nesvorný, D. (2016a). Spectral variability on primitive asteroids of the Themis and Beagle families: Space weathering effects or parent body heterogeneity? *Icarus, 269*, 1–14.

Fornasier, S., et al. (2016b). Rosetta's comet 67P/Churyumov-Gerasimenko sheds its dusty mantle to reveal its icy nature. *Science, 354*(6319), 1566–1570. https://doi.org/10.1126/science.aag2671.

Garaud, P., & Lin, D. N. C. (2007). The effect of internal dissipation and surface irradiation on the structure of disks and the location of the snow line around sun-like stars. *The Astrophysical Journal, 654*, 606–624.

Groussin, O., et al. (2015). Temporal morphological changes in the Imhotep region of comet 67P/Churyumov-Gerasimenko. *Astronomy & Astrophysics, 583*, A36. https://doi.org/10.1051/0004-6361/201527020.

Hamilton, V. E., et al. (2019). Evidence for widespread hydrated minerals on asteroid (101955) Bennu. *Nature Astronomy, 3*, 332–340.

Harrington, P. O., Womack, M., Fernandez, Y., & Bauer, J. (2022). A survey of CO, CO_2, and H_2O in Comets and centaurs. *The Planetary Science Journal, 3*, 247. https://doi.org/10.3847/PSJ/ac960d.

Harsono, D., Bruderer, S., & van Dishoeck, E. F. (2015). Volatile snowlines in embedded disks around low-mass protostars. *Astronomy and Astrophysics, 582*, A41.

Hayashi, C. (1981). Structure of the solar nebula, growth and decay of magnetic fields and effects of magnetic and turbulent viscosities on the nebula. *Progress of Theoretical Physics Supplement, 70*, 35–53.

Hirayama, K. (1918). Groups of asteroids probably of common origin. *Astronomical Journal, 31*, 185–188. https://doi.org/10.1086/104299.

Hsieh, H. H., & Jewitt, D. (2006). A population of comets in the main asteroid belt. *Science, 312*, 561–563.

Hirbayashi, M., et al. (2016). Fission and reconfiguration of bilobate comets as revealed by 67P/Churyumov-Gerasimenko. *Nature, 534*, 352–355.

Hu, X., et al. (2017). Seasonal erosion and restoration of the dust cover on comet 67P/Churyumov-Gerasimenko as observed by OSIRIS onboard Rosetta *Astronomy & Astrophysics, 604*, A114.

Ito, M., et al. (2022). A pristine record of outer Solar system materials from asteroid Ryugu's returned sample. *Nature Astronomy, 6*, 1163–1171.

Jewitt, D. (2012). The active Asteroids. *The Astronomical Journal, 143*, 66. https://doi.org/10.1088/0004-6256/143/3/66.

Jewitt, D., Chizmadia, L., Grimm, R., & Prialnik, D. (2007). Water in the small bodies of the Solar System. In B. Reipurth, D. Jewitt, & K. Keil (Eds.), *Protostars and planets V* University of Arizona Press.

Jewitt, D., & Hsieh, H. (2022). The asteroid-comet continuum. In K. Meech, & M. Combi (Eds.), *Comets III*. Tucson, AZ, USA: University of Arizona Press.

Jewitt, D., Hsieh, H., & Agarwal, J. (2015). *Asteroids IV* (pp. 221–241). Tucson, University of Arizona Press.

Kelley, M. S. P., Woodward, C. E., Bodewits, D., Farnham, T. L., Gudipati, M. S., Harker, D. E., Hines, D. C., Knight, M. M., Kolokolova, L., Li, A., de Pater, I., Protopapa, S., Russell, R. W., Sitko, M. L., & Wooden, D. H. (2016). Cometary science with the James Webb space telescope. *Publications of the Astronomical Society of the Pacific, 128*, 018009.

Kitazato, K., et al. (2019). The surface composition of asteroid 162173 Ryugu from Hayabusa2 near-infrared spectroscopy. *Science, 364*(6437), 272–275.

Kosai, H. (1992). Short-period comets and apollo–amor–Aten type asteroids in view of Tisserand invariant. *Celestial Mechanics and Dynamical Astronomy, 54*, 237–240.

Kresak, L. (1982). On the similarity of orbits of associated comets, asteroids and meteoroids. *Bulletin of the Astronomical Institutes of Czechoslovakia, 33*, 104–110.

Kruijer, T. S., Burkhardt, C., Budde, G., & Kleine, T. (2017). Age of Jupiter inferred from the distinct genetics and formation times of meteorites. *Proceedings of the National Academy of Science, 114*, 6712–6716.

Kuzmanoski, M., & Koračević, A. (2002). Motion of the asteroid (13206) 1997GC22 and the mass of (16) psyche. *Astronomy & Astrophysics, 395*, L17.

Landis, M., et al. (2022). The case for a Themis asteroid family spacecraft mission. *Planetary and Space Science, 212*, 105413.

Landis, M. E., Castillo-Rogez, J., & Ahrens, C. (2024). *Ceres—A volatile-rich dwarf planet in the asteroid belt.* In R. J. Soare, J.-P. Williams, C. Ahrens, F. E. G. Butcher, & M. R. El-Maarry (Eds.), *Ices in the solar system, a volatile-driven journey from the inner solar system to its far reaches.* Elsevier Books.

Levison, H. F. (1996). Comet taxonomy. In T. Rettig, & J. M. Hahn (Eds.), *Completing the inventory of the Solar System* (pp. 173–191). 107 of Astronomical Society of the Pacific Conference Series.

Levison, H. F., Bottke, W. F., Gounelle, M., Morbidelli, A., Nesvorný, D., & Tsiganis, K. (2009). Contamination of the asteroid belt by primordial trans-neptunian objects. *Nature, 460*, 364–366.

Levison, H. F., et al. (2021). Lucy Mission to the trojan Asteroids: Science goals. *Planetary Science Journal, 2*, 171. https://doi.org/10.3847/PSJ/abf840.

Licandro, J., et al. (2008). Spectral properties of asteroids in cometary orbits. *Astronomy & Astrophysics, 481*, 861–877.

Marsset, M., Vernazza, P., Birlan, M., DeMeo, F., Binzel, R. P., Dumas, C., et al. (2016). Compositional characterisation of the Themis family. *Astronomy and Astrophysics, 586*, A15.

Matter, A., Delbo, M., Carry, B., & Ligori, S. (2013). Evidence of a metal-rich surface for the asteroid (16) psyche from interferometric observations in the thermal infrared. *Icarus, 226*, 419–427.

Meech, K., & Raymond, S. N. (2020). Origin of Earth's water: Sources and constraints. In V. Meadows, G. Arney, D. D. Marais, & B. Schmidt (Eds.), *Planetary astrobiology*. Arizona University Press.

Min, M., Dullemond, C. P., Kama, M., & Dominik, C. (2011). The thermal structure and the location of the snow line in the Protosolar nebula: Axisymmetric models with full 3-D radiative transfer. *Icarus, 212*, 416–426.

Mommert, M., et al. (2014). The discovery of cometary activity in near-earth asteroid (3552) Don quixote. *The Astrophysical Journal, 781*, 25.

Morbidelli, A., Bottke, W. F., Jr., Froeschlé, C., & Michel, P. (2002). Origin and evolution of near-earth objects. In W. F. Bottke Jr.,, A. Cellino, P. Paolicchi, & R. P. Binzel (Eds.), *Asteroids III* (pp. 409–422). Tucson: University of Arizona Press.

Morbidelli, A., Walsh, K. J., O'Brien, D. P., Minton, D. A., & Bottke, W. F. (2015). The dynamical evolution of the asteroid belt. In P. Michel, et al. (Eds.), *Asteroids IV* (pp. 493–507). Tucson: Univ. of Arizona. https://doi.org/10.2458/azu_uapress_9780816532131-ch026.

Nesvorný, D. (2018). Dynamical evolution of the early Solar system. *Annual Review of Astronomy and Astrophysics*, *56*, 137–174. https://doi.org/10.1146/annurev-astro-081817-052028.

Nesvorný, D., Bottke, W. F., Vokrouhlický, D., Sykes, M., Lien, D. J., & Stansberry, J. (2008). Origin of the near-ecliptic circumsolar dust band. *Astrophysical Journal*, *679*(2), 143. https://doi.org/10.1086/588841.

Norwood, J., et al. (2016). Solar System observations with the *James Webb Space Telescope*. *Publications of the Astronomical Society of the Pacific*, *128*, 025004. https://doi.org/10.1088/1538-3873/128/960/025004.

Novaković, B., Hsieh, H. H., Cellino, A., Micheli, M., & Pedani, M. (2014). Discovery of a young asteroid cluster associated with P/2012 F5 (Gibbs). *Icarus*, *231*, 300–309. https://doi.org/10.1016/j.icarus.2013. 12.019.

Novaković, B., Vokrouhlický, D., Spoto, F., & Nesvorný, D. (2022). Asteroid families: Properties, recent advances, and future opportunities. *Celestial Mechanics and Dynamical Astronomy*, *134*, 34.

O'Rourke, L. O., et al. (2020). Low water outgassing from (24) Themis and (65) Cybele: 3.1 μm near-IR spectral implications. *Astrophysical Journal Letters*, *898*, L45.

Orofino, V. (2022). Main Belt Comets and other "Interlopers" in the Solar System. *Universe*, *8*, 518.

Pajola, M., et al. (2017). The pristine interior of comet 67P revealed by the combined Aswan outburst and cliff collapse. *Nature Astronomy*, *1*. https://doi.org/10.1038/s41550-017-0092.

Petit, J., Chambers, J., Franklin, F., & Nagasawa, M. (2002). Primordial excitation and depletion of the main belt. In W. F. Bottke Jr., et al. (Eds.), *Asteroids III* (pp. 711–738). Tucson: University of Arizona.

Poch, O., et al. (2020). Ammonium salts are a reservoir of nitrogen on a cometary nucleus and possibly on some asteroids. *Science*, *367*. https://doi.org/10.1126/science.aaw7462.

Pommerol, A., et al. (2015). OSIRIS observations of metre-size exposures of H_2O ice at the surface of 67P/Churyumov-Gerasimenko and interpretation using laboratory experiments. *Astronomy & Astrophysics*, *583*, A25.

Prettyman, T. H., et al. (2012). Elemental mapping by dawn reveals exogenic H in Vesta's regolith. *Science*, *338*, 242–246.

Raymond, S. N. (2023). The Solar System's ices and their origin. In R. J. Soare, J.-P. Williams, C. Ahrens, F. E. G. Butcher, & M. R. El-Maarry (Eds.), *Ices in the solar system, a volatile-driven journey from the inner solar system to its far reaches*. Elsevier Books.

Raymond, S. N., & Izidoro, A. (2017a). The empty primordial asteroid belt. *Science Advances*, *3*, e1701138.

Raymond, S. N., & Izidoro, A. (2017b). Origin of water in the inner Solar system: Planetesimals scattered inward during Jupiter and Saturn's rapid gas accretion. *Icarus*, *297*, 134–148.

Raymond, S. N., Izidoro, A., & Morbidelli, A. (2020). Solar system formation in the context of extrasolar planets. In V. Meadows, G. Arney, B. Schmidt, & D. J. Des Marais (Eds.), *Planetary astrobiology* The University of Arizona Press. 504 pp.

Raymond, S.N., Nesvorný, D. (2022). Origin and dynamical evolution of the asteroid belt. In S. Marchi, C. A. Raymond, & C. T. Russel (Eds.), *Vesta and ceres: Insights from the dawn mission for the origin of the solar system*. Cambridge University Press.

Rivkin, A. S., & Emery, J. P. (2010). Detection of ice and organics on an asteroidal surface. *Nature*, *464*, 1322.

Rivkin, A. S., Howell, E. S., Emery, J. P., & Sunshine, J. (2017). Evidence for OH or H_2O on the surface of 433 Eros and 1036 Ganymed. *Icarus*, *304*, 74–82.

Rivkin, A. S., et al. (2016). Asteroids and the James Webb Space Telescope. *Publications of the Astronomical Society of the Pacific*, *128*, 018003.

Rubincam, D. P. (2000). Radiative spin-up and spin-down of small asteroids. *Icarus*, *148*, 2–11.

Schorghhofer, N. (2008). The lifetime of ice on Main Belt Asteroids. *The Astrophysical Journal*, *682*, 697–705.

Scully, J. E. C., et al. (2014). Geomorphology and structural geology of saturnalia fossae and adjacent structures in the northern hemisphere of Vesta. *Icarus*, *244*, 23–40.

Shepard, M. K., et al. (2017). Radar observations and shape model of asteroid 16 psyche. *Icarus*, *281*, 388–403.

Snodgrass, C., & Jones, G. H. (2019). The European Space Agency's cometinterceptor lies in wait. *Nature Communications, 10*, 5418. https://doi.org/10.1038/s41467-019-13470-1.

Snodgrass, C., et al. (2017). The Main Belt Comets and ice in the Solar system. *Astronomy and Astrophysics Review, 25*, 5. https://doi.org/10.1007/s00159-017-0104-7.

Sonnett, S., Kleyna, J., Jedicke, R., & Masiero, J. (2011). Limits on the size and orbit distribution of main belt comets. *Icarus, 215*, 534–546.

Sunshine, J., Thomas, N., El-Maarry, M. R., & Farnham, T. L. (2016). Evidence for geologic processes on comets. *Journal of Geophysical Research, Planets, 121*. https://doi.org/10.1002/2016JE005119. http://onlinelibrary.wiley.com/doi/10.1002/2016JE005119/full.

Takir, D., & Emery, J. P. (2012). Outer main belt asteroids: Identification and distribution of four 3-μm spectral groups. *Icarus, 219*, 641–654. https://doi.org/10.1016/j.icarus.2012.02.022.

Takir, D., Reddy, V., Sanchez, J. A., Shepard, M. K., & Emery, J. P. (2017). Detection of water and/or hydroxyl on asteroid (16) psyche. *Astronomy Journal, 153*, 31.

Tancredi, G. (2014). A criterion to classify asteroids and comets based on the orbital parameters. *Icarus, 234*, 66–80.

Vera C. Rubin Observatory LSST Solar System Science Collaboration, et al. (2021). The scientific impact of the Vera C. Rubin Observatory's Legacy Survey of Space and Time (LSST) for Solar System science. Whitepaper #236 submitted to the Planetary Science and Astrobiology Decadal Survey 2023-2032. *Bulletin AAS*. https://doi.org/10.3847/25c2cfeb.d8909f28.

Vokrouhlický, D., Bootke, W. F., Chesley, S. R., Scheeres, D. J., & Statler, T. S. (2015). In I. V. Asteroids, P. Michel, et al. (Eds.), *The Yarkovsky and YORP effects* (pp. 509–531). Tucson: Univ. of Arizona. https://doi.org/10.2458/azu_uapress_9780816532131-ch027.

Walsh, K. J., Morbidelli, A., Raymond, S. N., O'Brien, D. P., & Mandell, A. M. (2011). A low mass for Mars from Jupiter's early gas-driven migration. *Nature, 475*, 206–209.

Wetherill, G. W. (1992). An alternative model for the formation of the asteroids. *Icarus, 100*, 307–325.

Yarkovsky, I. O. (1901). *The density of luminiferous ether and the resistance it offers to motion.* (in Russian). Bryansk, (published privately by the author).

Jupiter's ocean worlds: Dynamic ices and the search for life

Samuel M. Howell[a], Carver J. Bierson[b], Klára Kalousová[c], Erin Leonard[a], Gregor Steinbrügge[a], and Natalie Wolfenbarger[d]

[a]*NASA Jet Propulsion Laboratory, California Institute of Technology, Pasadena, CA, United States,* [b]*Arizona State University, Tempe, AZ, United States,* [c]*Charles University, Prague, Czechia,* [d]*Jackson School of Geosciences, University of Texas at Austin, Austin, TX, United States*

Abstract

From Europa's own style of plate tectonics to Ganymede's dichotomous terrain, to Callisto's ancient cratered blanket of dark material, the Myeariad icy processes of Jupiter's ocean worlds comprise unique geologic environments situated firmly at the forefront of planetary science and exploration. Since their discovery, the Galilean satellites have continually revolutionized astronomical thinking, from challenging the geocentric model of the universe to reveal the power of tidal interactions to drive change, to the revelation that global and potentially habitable subsurface oceans, and even life, may be commonplace in our solar system and beyond. In the coming decade, we will enter a new era of exploration for the Jovian ocean worlds with the arrival of two civilization-scale missions: NASA's Flagship-class Europa Clipper mission and ESA's L-Class JUICE mission. These missions will survey the surfaces, interiors, compositions, and space environments of Europa, Ganymede, and Callisto, providing new and exciting insight into the evolution and activity of icy formations tens to hundreds of kilometers thick over billions of years. Proposed concepts for future exploration seek to touch the icy surfaces, and even enter interior oceans. In this chapter, we review the exploration of the icy moons of Jupiter, informed by the Voyager and Galileo missions, as well as Earth- and space-based remote sensing. We highlight ongoing research into the formation and evolution of these complex hydrospheres, survey the unique icy geologic processes of these worlds, and discuss implications for the emergence of life within the ocean of Europa.

1 Introduction to the icy Galilean satellites

Perhaps nowhere else in our solar system does ice play as impactful a role in the evolution of worlds and the rise of habitability as at Jupiter. The outer ice shells of the ocean worlds Europa, Ganymede, and Callisto (Fig. 1), record billions of years of dynamic geologic and compositional change (Fig. 2), and in the case of Europa, recent or ongoing geologic activity capable of erasing the full surface over the past ~60 Myear (e.g., Bierhaus et al., 2009; Zahnle et al., 2003). These ice shells may dissipate tidal energy to maintain deep oceans, bring together the chemistry for life at Europa, and record changes in the compositional and orbital evolution of the Jupiter system through time. Named for one independent discoverer, the four Galilean moons of Jupiter have driven to the forefront of planetary science and exploration, motivated by their incredible activity and oceans, spanning a range of fiery and icy processes.

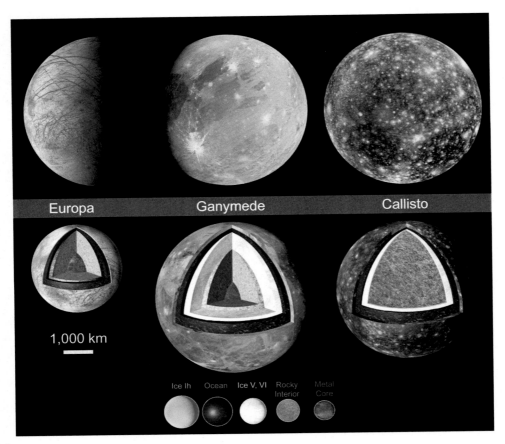

FIG. 1

The top row, not to scale, shows in color the icy surfaces of (left) Europa, acquired by *Galileo* and approximately true color (NASA/JPL-Caltech), (middle) Ganymede, acquired by *Juno* and approximately true color (NASA/JPL-Caltech/SwRI/MSSS), and (right) Callisto, acquired by *Voyager 2* and in enhanced color (NASA/JPL-Caltech). The bottom row, to scale, shows potential interior cutaways of these worlds, including outer ice shells, oceans, high-pressure ice shells, rocky mantles, and metallic cores.

The nearest Galilean satellite to Jupiter is Io, comparable in size and orbital distance from the parent body to Earth's moon, and the most volcanically active body in the solar system. Io is the only Galilean satellite that does not exhibit an outer water-ice shell thought to cover a subsurface liquid water ocean—though it may possess a magma ocean (Khurana et al., 2011)—and as such is not discussed in depth within this chapter.

Moving outward, we encounter Europa, measuring ~90% the diameter of Earth's moon. Europa's surface is geologically complex, and the paucity of impact craters implies an average surface age among the youngest in the solar system at ~60 Myear (e.g., Bierhaus et al., 2009; Zahnle et al., 2003). Separating the surface of Europa from a ~100 km deep interior ocean is an ice shell tens of kilometers

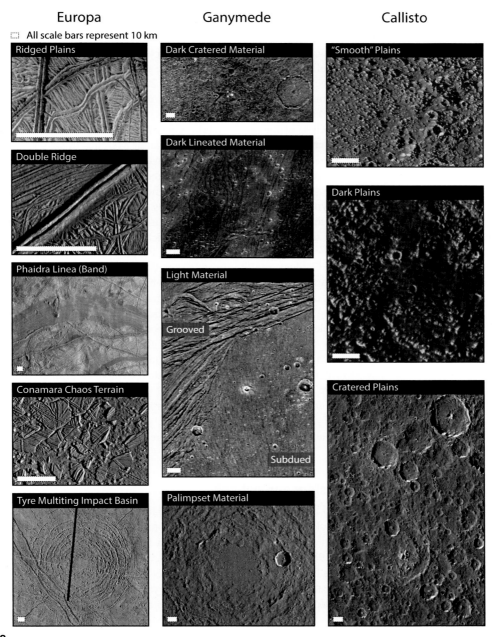

FIG. 2

Galileo spacecraft imagery of the water-ice shells of the Jovian ocean worlds, showing the archetypal geologic features of Europa (left column), Ganymede (middle column), and Callisto (right column) (NASA/JPL-Caltech). All scale bars represent 10 km. Note that the lighting angle, intensity, geometry, orientation, and resolution of the images are not consistent. Images of Ganymede are those identified by Patterson et al. (2010) as containing the type localities for Ganymede's landforms, and images of Callisto are those identified by Greeley et al. (2000a) as containing representative examples of landform types.

thick (e.g., Howell, 2021). Dynamic ice-shell activity, inferred from the surface geology, may help to convey oxidants produced radiolytically at the surface from Iogenic material and water ice to depth, reaching an ocean reduced by water-rock reactions (Hand et al., 2009, 2020; Vance & Melwani Daswani, 2020). The resulting redox gradient may create habitable environments that enable the emergence of past or extant life (Hand et al., 2009; Vance et al., 2016).

Further out still is the largest satellite of the solar system, Ganymede, which is the only satellite known to possess its own intrinsic magnetic field. Exceeding the planet Mercury in size, Ganymede is comparatively less than half of Mercury's mass. This low bulk density betrays a deep external layer of water and ice, which could include an outer ice shell tens of kilometers to more than 100 km thick, an ocean up to 750 km deep, and high-pressure ice phases that cap the seafloor and precipitate within the ocean (Vance et al., 2014). Ganymede's bimodal icy geology is divided into a heavily cratered, and therefore likely old, dark terrain that covers about 35% of the surface, and younger, bright terrain (Greeley et al., 2000a; Pappalardo et al., 2004). The origin of this dichotomy remains unknown but is likely related to the early history of the moon (Greeley et al., 2000a; Pappalardo et al., 2004).

Finally, we reach the furthest of the Galilean satellites from Jupiter, Callisto. Callisto is slightly smaller than the planet Mercury, with a dark surface evidenced by its considerably lower albedo than Europa, and slightly lower albedo than Ganymede, despite water ice being the primary constituent of its surface. The dark material may be a sublimation lag left behind by excited ice particles, or a remnant of a blanketing event early in the moon' history (Greeley et al., 2000b; Howard & Moore, 2008). The *Galileo* spacecraft measured a distinct induced magnetic field signal consistent with a global subsurface saltwater ocean (Khurana et al., 1998), which may also be explained by ionosphere induction (Hartkorn & Saur, 2017). It remains uncertain if Callisto is fully differentiated, or if at least some intermediate layer composed of silicate rock and ice exists (Hartkorn & Saur, 2017; Nagel et al., 2004).

In this chapter, we review the exploration of the icy moons of Jupiter, highlight ongoing research into the formation and evolution of these complex hydrospheres, survey the unique icy geologic processes of these worlds, and discuss implications for the emergence of life within the ocean of Europa.

1.1 Spacecraft exploration history and outlook

In 1977, the National Aeronautics and Space Administration (NASA) launched the dual *Voyager* spacecraft, beginning a tour of the outer solar system and beyond that continues today. The spacecraft encountered Jupiter and its satellites in 1979, teasing images of both fire and ice in the volcanic and frozen worlds they encountered. In 1989, NASA launched the *Galileo* spacecraft, orbiting Jupiter for the first time and visiting its satellites from 1995 to 2003.

As *Galileo* orbited Jupiter, it measured the time and spatially varying magnetic field produced by the gas giant, as well as interactions between the magnetic field and the satellites. This led to the fascinating discovery of an *intrinsic* magnetic field at Ganymede, the only satellite in the solar system that exhibits this behavior (Kivelson et al., 1996). Additionally, *Galileo* spacecraft measurements indicated that Europa, Ganymede, and Callisto exhibited *induced* magnetic fields resulting from their passage through Jupiter's dynamic magnetic environment (Khurana et al., 1998; Kivelson et al., 2000; Zimmer et al., 2000). These observations are only possible if these worlds possess global, electrically conductive layers, now thought to be global saltwater oceans hidden beneath their icy surfaces.

With one of the youngest surface ages in the solar system, Europa is the target of NASA's planned Europa Clipper flagship mission, with a planned launch in 2024 (Howell & Pappalardo, 2020). Europa, Ganymede, and Callisto are the targets of ESA's planned Jupiter Icy Moons Explorer (JUICE) L-class mission (Grasset et al., 2013), and Europa and Ganymede are near-term targets of NASA's New Frontiers class Juno mission.

Looking forward, researchers have begun thinking about how to touch the surfaces of these worlds, enabling in situ measurements that begin the search for biosignatures. For decades, missions have been developed internationally to land on Europa (Hand et al., 2017), Ganymede (Golubev et al., 2014), and Callisto (Li et al., 2021). Recently, studies have begun into architectures that would provide in situ access to Europa's ocean using "cryobots," robots capable of penetrating tens of kilometers of ice, to enable the direct detection of any potential extant or past life within the ocean (e.g., Howell et al., 2021).

1.2 Modern telescopic observation

Earth-based telescopic platforms also continue to play an important role in the study of these worlds. Although spacecraft can provide higher spatial resolution observations due to their relative proximity to these bodies, Earth-based measurements are capable of achieving higher spectral resolution measurements, unique observation geometries, and hemispherical-scale coverage. Though these spectral measurements are only sensitive to the uppermost layer of material on these bodies, they provide essential clues and context for processes and properties at depth.

Hubble Space Telescope (HST) observations have provided important constraints on the processes occurring at icy Galilean satellites. Detections of water, oxygen, and hydrogen suggest that these satellites have tenuous atmospheres generated through sputtering and sublimation of a water-ice-dominated surface (Alday et al., 2017; Mogan et al., 2021; Roth, 2021; Roth et al., 2016, 2021). At Europa, HST observations have resulted in the detection of water vapor plumes erupting from the surface (Paganini et al., 2020; Roth et al., 2014; Sparks et al., 2016) and the identification of a hypothesized chloride evaporate deposit (Trumbo et al., 2019). At Ganymede, HST observations of its auroral oscillations provide additional evidence for the presence of a deep ocean (Saur et al., 2015).

Atacama Large Millimeter/submillimeter Array (ALMA) observations have been used to generate thermal maps (inertia and emissivity) for the icy Galilean moons and identify anomalies that might be indicative of subsurface processes (Camarca & de Kleer, 2021; de Kleer et al., 2021; Trumbo et al., 2018). For example, an analysis using ALMA observations at Europa argued against the location of a hypothesized plume source, demonstrating the observed thermal anomaly could not be attributed to an excess heat flow and was instead the result of a local increase in thermal inertia (Trumbo et al., 2017).

Observations of the surface using the Very Large Telescope (VLT) suggest the distribution and composition of non-ice impurities varies across the icy Galilean satellites, although the major components include sulfuric acid, salts, and an unknown "darkening agent" at Ganymede and Callisto that could comprise hydrated silicates (Ligier et al., 2020, 2019; Ligier et al., 2016). Sulfuric acid hydrate is the dominant non-ice impurity at Europa, whereas the "darkening agent" is the dominant non-ice impurity at Ganymede and Callisto. Salts have been detected at both Europa and Ganymede with Keck Observatory and VLT. At Europa, possible chlorides and/or sulfate salts appear to be spatially correlated with resurfacing features (chaos) (Brown & Hand, 2013; Ligier et al., 2019; Trumbo et al., 2020).

At Ganymede, sulfate salts are also correlated with resurfacing features (sulci), whereas chloride salts are not (Ligier et al., 2019).

2 The unique geology of Jovian ice shells

The surface geology of the icy Jovian worlds provides the first window into their storied histories and current states. Broadly speaking, Myeariad landforms act to erase signs of age among Europa's nascent icy terrains, Ganymede's ancient formation-age dark terrains are split asunder by the much later tectonized light terrains, and Callisto's blanket of dark material reveals a stoic and unchanging world. In this section, we survey the unique icy geology of Jupiter's ocean worlds, including the hallmark icy landforms of Europa, Ganymede, and Callisto (Fig. 2).

It should be noted that the current state and practice of planetary geology is fundamentally different from Earth-based geology and that these differences are especially pronounced for the poorly characterized Jovian satellites. Planetary geology is almost purely performed through remote sensing data sets (e.g., images or other data collected from spacecraft), and there exists no ground truth in situ data for the icy satellites in the Jovian system. Additionally, when describing the geology of the *icy* Galilean moons Europa, Ganymede, and Callisto, it is important to note that the rocks of interest are predominantly water ice and not silicates. Geologic units and terrains described in this section are assumed to represent three-dimensional (3D) volumes of ice, each with a distinct relative brightness, texture, or unique structure. The actual depths of the units are unknown. We specifically note and emphasize that the morphologic differences between units could be due to a variety of factors, including compositional variation, tectonic deformation, or physical or chemical alteration by some other endogenic or exogenic process. This is in contrast to most geologic descriptions of Earth and other well-explored planetary bodies, where units are primarily based on lithological differences.

2.1 Europa's geology

Europa's surface consists of *regiones* (singular: *regio*), that have either a comparatively uniform relative high brightness or a low relative brightness that can appear mottled. Europa's high relative brightness terrains appear blue-white in corrected true color images taken by the Galileo spacecraft (Greeley et al., 2000a). This terrain is mostly composed of water ice along with a relatively small non-ice component (Dalton III et al., 2012). Europa's low relative brightness or mottled terrains appear orange-brown in corrected true color images taken by Galileo (Greeley et al., 2000a). This color variation is thought to result from its unique composition and possibly radiation processing of the surface materials. The composition of the low-brightness terrain is thought to be dominated by magnesium sulfate ($MgSO_4$) (T. B. McCord et al., 1999) or sodium chloride (NaCl) (Trumbo et al., 2019). The radiation environment imposed by Jupiter at Europa's orbit results in high-energy particles (~1–50 MeV) bombarding the surface, and therefore serving to process and alter the surface material (Cooper et al., 2001; Nordheim et al., 2017). Additionally, sulfur ions ejected from Io are hypothesized to impact Europa, primarily on the trailing hemisphere (Dalton III et al., 2013) which can result in surface color variations.

2.1.1 Archetypical European landforms

Ridged plains, covering ~60% of Europa's surface (Doggett et al., 1999; Greeley et al., 2000a), are thought to be one of the oldest units on Europa's surface because they fall in the stratigraphic background relative to essentially every other feature (Doggett et al., 1999; Figueredo & Greeley, 2004; Greeley et al., 2000a; Greenberg et al., 1998; Pappalardo et al., 1999; L. M. Prockter et al., 1999).

Ridge Linea are most commonly identified as double ridges (two ridges separated by a trough) or ridge complexes (more than two ridges separated by troughs), and could be formed through a variety of processes such as cryovolcanism (Fagents et al., 2000), tidal squeezing (Greenberg et al., 1998), linear diapirism (J. W. Head et al., 1999), shear heating (Kalousová et al., 2016; Nimmo & Gaidos, 2002), compression (Sullivan et al., 1998), wedging (Han & Melosh, 2010; Johnston & Montési, 2014; Melosh & Turtle, 2004), and compaction (Aydin, 2006); or even represent different stages of the same process (Greenberg et al., 1998).

Bands are linear to curvilinear landforms with a range of relative brightness and are primarily hypothesized to have formed by a process analogous to mid-ocean-ridge style rifting and spreading where a rift forms and the crustal blocks separate and spread laterally by new material rising at the spreading axis (L. M. Prockter et al., 2002).

Chaos terrain is characterized by blocks of crustal material, ranging from tens of kilometers to tens of meters in size, within a low albedo hummocky matrix. Lower albedo material on Europa typically has a higher concentration of non-ice material and is commonly inferred to have originated in the subsurface ocean (Dalton III et al., 2012; Shirley et al., 2010) and therefore chaos is often associated with surface-subsurface interaction and/or material exchange.

Craters on Europa are identified as quasi-circular depressions with a raised rim or complex annular structure. However, craters on Europa can also have little apparent topographic relief and/or an association with an annular structure consisting of discontinuous ridges and troughs. Craters typically include material within the excavated area, which may represent impact melt, or subsequent infilling due to post-impact cryomagmatic processes (Cooper et al., 2001; Moore et al., 1998; Schenk, 2002; Schenk & Turtle, 2009). A handful of the observed craters on Europa are associated with multi-ring structures and are hypothesized to be graben formed by the collapse of an impact crater where the impactor penetrated to the ocean or lower ice shell (Schenk, 2002).

2.1.2 Geologic history of Europa

Analysis of stratigraphic relationships among major landform types suggests that the surface of Europa has been shaped by a three-staged development: (1) the initial formation of ridged plains, followed by (2) the development of band-like structures, and finally (3) the creation of chaos features (Figueredo & Greeley, 2004; Greeley et al., 2000a; Leonard et al., 2018). The formation of the ridged plains is hypothesized to result from folding (J. W. Head et al., 1999; Leonard et al., 2020; Patel et al., 1999) or extensional faulting (Kattenhorn, 2002), but has not been widely investigated. The formation of band-like structures has been attributed and extensional process similar to mid-ocean-ridge spreading on Earth (L. M. Prockter et al., 2002; L. M. Prockter & Patterson, 2009). The spreading hypothesis is consistent with predicted band morphology and reconstructions of pre-band-formation structures that require significant ice-shell separation (S. M. Howell & Pappalardo, 2018; L. M. Prockter et al., 2002; L. M. Prockter & Patterson, 2009; Schenk & McKinnon, 1989). The formation of chaos has been intensely debated over time, with end-member hypotheses including melt-through, sill formation and collapse, and diapirism induced by thermal and/or chemical buoyancy (Collins & Nimmo, 2009; Greenberg et al., 1999; Michaut & Manga, 2014; Pappalardo et al., 1998; Schmidt et al., 2011).

2.2 Geology of Ganymede

Ganymede's surface is made up of two fundamental classes of units: dark materials, which cover 35% of the surface, and light materials covering the remainder of the surface (Collins et al., 2013). The separation between these two classes is based on sharp relative albedo contrasts at terrain boundaries (rather than absolute albedo). Dark materials are heavily cratered, though they are not thought to be representative of a primordial surface like on Callisto (Shoemaker et al., 1982). In the high-resolution images taken by the Galileo spacecraft, dark materials are made up of a loose, dark regolith thought to be derived from the sublimation of a more ice-rich crust, akin to Callisto (Moore et al., 1999). Light materials typically have a lower density of impact craters indicating that they are younger than dark materials. Most light materials are interpreted to be formed through extensional faulting (S. M. Howell & Pappalardo, 2018; Pappalardo et al., 1998; Pappalardo & Collins, 2005). There is debate within the community on whether light terrain is formed by: (i) cryovolcanic flooding of dark material with brighter ice, (ii) tectonic destruction of pre-existing surface features and exposure of brighter subsurface ice in fault scarps, or (iii) a combination of both of the previous processes (summarized in Pappalardo et al., 2004).

2.2.1 Archetypical Ganymede landforms

Dark, cratered material represents large areas of relatively low albedo material with moderate to high crater densities and makes up the majority of all dark material on Ganymede (Patterson et al., 2010). Heavily modified by impact processes, the low albedo, dark material that makes up this terrain is thought to form through sublimation lag deposits that then can undergo mass wasting.

Dark, lineated material is also a low albedo terrain, but contains sinuous grooves that appear shallow. This material has a similar appearance to grooved material and irregular material, but with a lower albedo, and is often found in close proximity to these terrains. The crater density of the lineated material terrain is only slightly greater than of the light materials, and Patterson et al. (2010) interpret this terrain to be a transitional terrain between cratered material and the light materials.

Light, grooved material is a relatively high albedo material that is dominated by repeating ridge-and-groove structures organized into lanes, or regions where the ridges and grooves all trend in the same direction and are roughly evenly spaced. Grooved material is hypothesized to form from dark or other light material units via extensional tectonism, potentially post-cryovolcanic resurfacing.

Light, subdued material has a relatively high albedo and is similar to grooved material but the grooves in this terrain are faint or undetectable (at image resolutions of 10^2–10^3 m/pixel) and the surface appears smooth. Often found in close proximity to grooved material and irregular material, the boundaries of subdued material are gradational with other light materials but sharp in contact with dark materials.

Craters on Ganymede often contain structures within that vary with increasing crater diameter from central peaks, to central pits, to central domes. The diameters at which craters transition between some interior structures do not appear to depend on the relative age of the craters. However, it has been suggested that dome morphology is more common in older craters (J. P. Kay et al., 2007; Patterson et al., 2010; Schenk et al., 2004). Typically recognized separately from crater materials, basin materials are made up of the deposits of the large crater basin, Gilgamesh (590 km diameter) (Patterson et al., 2010).

Palimpsest material consists of relatively high albedo, circular to elliptical structures. Palimpsest material is hypothesized to be the result of impacts into the crust of Ganymede during a time when there was a higher thermal gradient and/or a thinner brittle lithosphere (Shoemaker et al., 1982). Particularly, the centers of some palimpsests are characterized by smooth, circular to subcircular patches of high albedo material. Palimpsests lack rims but commonly contain scarps and internal, concentric ridges.

2.2.2 Geologic history of Ganymede

Cratered materials have the highest crater densities of the mapped units for Ganymede and are consistently superposed by all other units, leading to the inference that these are the oldest materials on Ganymede's surface (Patterson et al., 2010). Lineated material has a crater density similar to light material units supporting the interpretation that it is a transitional material that has undergone part of the resurfacing process that created light material units by the tectonic deformation of dark cratered material (Pappalardo et al., 2004). Light material crater densities are approximately half those of dark cratered material and consistently superpose dark materials and some of the oldest impact materials (Patterson et al., 2010). Almost all palimpsest materials are found within dark cratered material units and therefore thought to be generally older than light materials. However, there are a few palimpsests that superpose light material units, suggesting that palimpsest formation continued during the early stages of light material formation. Basin materials superpose light material and therefore are younger than the light materials.

2.3 Geology of Callisto

Images from Galileo and Voyager show that Callisto's surface is broadly blanketed in a dark material that is saturated with craters, implying the surface is ~4 Ga (Cassen et al., 1980; Greeley et al., 2000b; Passey & Shoemaker, 1982). The most prominent features on Callisto's surface are the large ringed features or multi-ring structures. Callisto's surface also contains numerous palimpsests and catenae, or crater chains (Passey & Shoemaker, 1982; Schenk, 1995; Schenk et al., 2004). Although most scarps, ridges, and fractures are associated with multi-ring structures, a few such features have no apparent associations and might represent tectonic events not related to impacts (Schenk, 1995; Schenk & McKinnon, 1987; Wagner & Neukum, 1994).

Spectra of the dark material that makes up the majority of Callisto's surface show absorption bands indicative of clays and carbon-rich materials (T. a McCord et al., 1997). Other absorptions in the spectra are inferred to be representative of tholin-like organic materials and CO_2 that may be trapped within the dark surface materials. The CO_2 absorption shows a correlation to the trailing side of the satellite, suggesting a link to bombardment by the co-rotating plasma; moreover, it is correlated with some bright impact craters (Hibbitts et al., 2000). CO_2 is a possible component of Callisto's icy subsurface, and its sublimation may contribute to the generation of the satellite's dark regolith.

2.3.1 Archetypical Callisto landforms

Cratered plains comprise most of Callisto's surface, and are thought to represent the ancient icy lithosphere consisting of ice, rock, and dust, all brecciated and 'gardened' by impact cratering over Callisto's surface history. Imaging by Galileo showed that surface degradation is an important process at small scales, manifested by landslides from crater rims and a generally older and degraded appearance compared to similar ancient terrain on Ganymede (Greeley et al., 2000b).

Light plains have a higher albedo than the cratered plains and form circular, elliptical, or irregular patches (Bender & Carroll, 1997). Thus, this material is typically interpreted to be ice-rich deposits excavated by impact processes (Greeley et al., 2000b). Once thought to be a smooth terrain based on Voyager data (Schenk, 1995), higher-resolution Galileo data revealed light plains as rugged at the sub-kilometer scale with an abundance of scarps and knobs that cause the apparently high albedo (Greeley et al., 2000b).

Dark smooth plains, perhaps the most enigmatic of Callisto's terrains, have only a handful of instances identified on Callisto's surface (Greeley et al., 2000b). The smooth dark plains material appears to mantle previously existing terrain, such as knobs and craters, and embays the surrounding terrain (Greeley et al., 2000b).

Craters. Callisto's surface is an exemplary display of a large variety of crater morphologies that can reveal the rheological properties of the surface and how these may have evolved through time. The smallest craters are simple bowl-shaped to flat-floored depressions that range in diameter from about 7 km down to ~50 m (the limit of Galileo image resolution). Central-peak craters range in size from about 5 to 40 km in diameter. The central-peak complexes are as large as 10 km across and 0.5 km high (Greeley et al., 2000b). Deposits akin to landslides are identified on the floors of several central-peak craters (Chuang & Greeley, 2000; Moore et al., 1999). Many, but interestingly not all, of the craters larger than ~25 km in diameter transition from central-peak to central pit morphologies (Schenk, 1993; Strom et al., 1981). Some craters greater than ~60 km show central domes (Schenk, 1995). As outlined by Schenk (1995), dome formation on icy satellites could have occurred after crater formation as diapiric intrusions of soft ice (Moore & Malin, 1988) or as part of the impact process by the rebound of the floor (Schenk, 1993).

2.3.2 Geologic history of Callisto

Callisto's surface history is typically defined by the formation of large multi-ring structures. From photogeologic mapping and estimated crater-age dating (Greeley et al., 2000b), the cratered plains and terrains associated with the Adlinda and Asgard multi-ring structures are found to be the oldest terrains on Callisto. Crater counting by Greeley et al. (2000b) also suggest that the bright plains formed immediately following the terrains associated with Valhalla and that the dark smooth plains formed soon afterward. The relatively close formation time for these terrains suggests that they all formed due to the Valhalla-forming impact. The impacts forming Lofn, Tindr, and Doh craters are among the latest events observed on Callisto.

Callisto's multi-ringed structures, palimpsests, and scattered central dome craters hint that Callisto's ice was warm and soft enough to flow in the ancient past, but only at substantial depth. There is little to no evidence of endogenically driven tectonic or cryovolcanic activity. Nonetheless, Callisto has crater-poor surfaces at high resolution—its small craters evidently have been erased more efficiently than on Ganymede—suggesting that the sublimation and mass wasting processes that subdued Callisto's topography might be active even at the present day (Greeley et al., 2000b).

3 Origin and evolution of the H$_2$O layers

3.1 Accretion of volatiles and early evolution

When comparing the four Galilean satellites, the varying icy and liquid water layers reveal a readily apparent gradient in bulk water inventory (Schubert et al., 1986). Io, closest to Jupiter, has no significant water-ice layer, while the outer satellites Ganymede and Callisto are approximately 30% water by mass. Europa is transitionary between these end-members, with H$_2$O contributing roughly 5%–10% of its total mass. There is not yet any clear consensus on why this transition exists, and the literature comprises healthy, ongoing debate.

At present, it is generally thought that the Galilean satellites accreted embedded within the proto-Jovian disk (Aydin, 2006; Canup & Ward, 2002). This disk would have been optically thick and heated internally by viscous dissipation (Canup & Ward, 2002). The dissipation would lead to a temperature gradient within the disk where the inner part of the disk, closer to Jupiter, is significantly warmer than the outermost part. This general configuration led to the hypothesis that there was an "ice line" within the proto-Jovian disk, preventing water from being accreted onto Io and Europa (Lunine & Stevenson, 1982).

It is possible that the inner satellites, Io and Europa, did accrete ice-rich material but it was subsequently removed from the surface. One such hypothesis is that, during accretion, the combination of the warm background disk and the energy of accretion itself was enough to keep the surface temperatures of the inner satellites high enough to sustain surface oceans. These surface oceans would produce a water atmosphere that would have been rapidly lost due to the low gravity of the satellites (Bierson & Nimmo, 2020). Later energy sources, such as tidal heating, are insufficient to produce enough mass loss to account for the observed present-day gradient in water inventory (Bierson & Steinbrügge, 2021). Therefore, if some process did remove an initial water inventory from Io and Europa, it must have been active during or shortly after accretion.

Parallel to work on the accretion of the Galilean satellites, there is significant literature examining the possibility that Europa's near-surface water inventory results from geochemical processes within the solid interior (Kargel et al., 2000; Melwani Daswani et al., 2021). Under this model, Europa accreted from hydrated silicates with no significant ocean or ice shell, and over time the warming of the interior from radiogenic decay caused the silicate minerals to dehydrate. The liberated water would then buoyantly rise to the surface. It remains unclear under such a model why Io does not exhibit a similar ocean to Europa.

3.2 Powering change within the ice

For every orbit around Jupiter that Ganymede completes, Europa completes two and Io completes four, comprising a behavior known as the Laplace resonance. This resonance leads to a coupled energy balance between the participating satellites, maintaining orbital eccentricity and motivating tidal interactions. In the early stages of evolution, accretional heating, the energy of differentiation, and the decay of short-lived radiogenic isotopes drive the heat budgets of these worlds, enabling various degrees of differentiation and structural change. The decay of long-lived radiogenic isotopes and internal frictional heating due to the tidal dissipation of gravitational potential energy become the decisive factors for the long-term thermal equilibrium of icy satellites. These sources and sinks of heat can be modulated for ocean worlds by the storage of energy in the phase of water: the latent heat of fusion required to melt ice, and released during the freezing of water, turns subsurface oceans and outer ice shells into heat budget "buffers."

3.2.1 Construction of the outer icy shells

The icy shells of the Galilean satellites are poorly characterized, and little is known for certain about their structure and state. The physical and thermal nature, including thickness and geodynamic state, of the Galilean ice shells are likely the product of a balance between heating (internally and from below) and radiative cooling from the surface. This is notably unlike ice sheets and ice shelves on Earth, where thick layers of ice are formed through the (meteoric) accumulation of snow. Cooling and subsequent freezing of the ocean is controlled by the transfer of heat to the overlying ice shell, with a heat flux influenced by the total thickness of the shell and the dominant forms of heat transfer across it (Billings & Kattenhorn, 2005; S. M. Howell, 2021; Sotin et al., 2009). Fluctuations in the internal heat flux could be driven by changes in orbital parameters (Hussmann & Spohn, 2004), ocean circulation (Soderlund et al., 2014), and the merging of convective cells (Peddinti & McNamara, 2019). The young surface age at Europa in particular is compatible with the recycling of the ice shell through catastrophic melting and freezing cycles (Green et al., 2021), which suggest that ice-shell growth could be occurring to this day.

Constraining the bulk composition of these ice shells represents an important step in understanding some key factors governing the dynamics and habitability of these worlds: (i) the distribution of water in the ice shell, (ii) the bulk thermophysical properties of the ice shell, and (iii) the composition of the sub-ice ocean.

Sea ice might be considered as an intuitive Earth analog for the growth of ice shells, given that it is also produced through the freezing of an ocean. In sea ice growth, a small fraction of oceanic material is retained as brine, trapped interstitially between essentially pure ice crystals. However, a critical difference between the formation of sea ice on Earth and the formation of ice shells of ocean worlds is the driving thermal gradient, which controls how quickly the ice forms. Even the highest modeled growth rates for ice shells are orders of magnitude lower than the seasonal sea ice growth cycles on Earth (Wolfenbarger et al., 2022a, 2022b).

Although sea ice formation may not represent a perfect analog for the growth of ice shells, more gradual freezing of the ocean does occur beneath ice shelves in Antarctica, where the temperature gradient is lower. In fact, a sample of seawater frozen to the base of the Ross Ice Shelf is thought to have formed in a temperature gradient environment which could approach conditions at the ice-ocean interface of the icy Galilean moons (Wolfenbarger et al., 2022a, 2022b). Importantly, the salinity profile derived from this sample demonstrated a similar asymptotic behavior as modeled for Europa's ice shell. The asymptotic salinity associated with this sample was determined to be approximately 7% of the ocean salinity, which was importantly a value similar to the so-called "critical porosity" of 5% often imposed as a percolation threshold in sea ice desalination models (Wolfenbarger et al., 2022a, 2022b).

Flushing of the ice shell through drainage of tidally generated meltwater could operate to further desalinate and even fractionate the ice shell (Wolfenbarger et al., 2022a, 2022b). The relative composition of sea ice is generally assumed to be representative of seawater, although processes like cryohydrate precipitation and differential diffusion of ions in brine networks could result in fractionation (Maus et al., 2011). The relatively early precipitation of sulfates that occur as seawater freezes has been implicated in the fractionation signals observed in certain sea ice cores, where sulfate salts are thought to be mobile in brine networks and redistributed by the flushing and refreezing of meltwater (Gjessing et al., 1993). The fractionation signatures observed in marine ice, which can maintain high permeability for significantly longer timescales than sea ice, appear to be consistent with a diffusion-related fractionation process. If brine networks can remain stable over geologic time in ice shells, the diffusion of ions along concentration gradients could be more significant than observed in ice on Earth (Wolfenbarger et al., 2022a, 2022b). Drainage and refreezing of meltwater could similarly redistribute mobile cryohydrates and may be an important factor to consider when using salts observed at the surface as a signature of processes occurring in the subsurface or the composition of the sub-ice ocean (Vance et al., 2019; Wolfenbarger et al., 2022b).

3.2.2 Radiogenic heating

The rocky masses of these worlds bring with them long-lived radiogenic isotopes that contribute to the deep, long-term heat budget, driven primarily by ^{238}U, ^{235}U, ^{232}Th, and ^{40}K, which have half-lives on the order of 10^9 year (Schubert et al., 1986; Spohn & Schubert, 2003). Given the unknown composition of the silicate interiors of Europa, Ganymede, and Callisto, CI chondrites are generally assumed to be a representative constituent, though many workers have explored a variety of potential source populations (e.g., Melwani Daswani et al., 2021). Assuming chondritic composition, the present-day radiogenic surface heat flux would be on the order of about 5–10 mW/m^2 for these icy worlds (Bland et al., 2009; Sotin et al., 2009). For the outer satellites Ganymede and Callisto, this likely constitutes the major heat source driving the thermal equilibrium of these moons and activity within their icy shells and ocean.

3.2.3 Tidal heating

For the inner satellites, Io and Europa, the amount of radiogenic heating is small compared to the heat dissipated by tidal forcing. Because planetary materials are not entirely rigid, they are periodically deformed by changes in their gravitational attraction to Jupiter. The driving tidal potential depends on the orbital parameters, such as the distance to Jupiter, the eccentricity of the orbit, and the mean motion that establishes the frequency of the deformation. Io is so close to Jupiter that tidal heating leads to internal melting and places Io in a prominent position among the most volcanically active worlds discovered. However, the reaction of a body to the tidal potential is also dependent on the interior structure and rheological properties.

For Europa, it is currently unclear whether the silicate shell is dissipative, and most of the tidal heating is commonly attributed to the ice shell. The total amount of tidal heat dissipated within Europa, therefore, depends on the thickness of the ice. Equilibrium models predict surface heat flows of ~10 mW/m^2 and ice layer thicknesses of around 20–30 km (e.g., Howell, 2021; Hussmann et al., 2022); significantly less than the total thickness of the H$_2$O-layer inferred from gravity science.

In contrast, at Ganymede, the significantly lower eccentricity (0.0015 on average, compared to 0.009 at Europa) and larger distance to Jupiter leads to a negligible amount of tidal dissipation. Hence, radiogenic heating dominates the heat budget of Ganymede, limiting the energy available to do useful work and drive geologic activity within the ice shell. The ice shell is consequently expected to be much thicker—on the order of 100 km—capping a mixture of liquid global subsurface ocean material and high-pressure ice phases that could extend for many hundreds of kilometers (Vance et al., 2014). It is possible, however, that past, transient periods of tidal heating have played an important role for these outer satellites (Nimmo & Manga, 2002; Bland et al., 2009).

3.2.4 Laplace resonance and thermal-orbital evolution

An open question remains as to whether the ice shells are currently in thermal equilibrium. The tidal energy gained by the satellites is compensated by a loss of orbital energy. However, a fascinating situation arises from Io, Europa, and Ganymede's orbits being coupled by the Laplace resonance. The tidal torques raised by Jupiter transfer orbital energy and angular momentum to Io, however, through the 4:2:1 orbital resonance, the energy is ultimately shared among the satellites. The satellites are consequently accelerated and their eccentricity is enhanced. Consequently, dissipation within the satellite's interiors has the opposite effect, dampening the eccentricity and reducing the orbital energy (e.g., Yoder & Peale, 1981). In this system, the amount of dissipated energy over time is not necessarily constant. For Io and Europa, models have been proposed in which the satellites undergo periodic phases of enhanced heating followed by phases with lower dissipation (Hussmann & Spohn, 2004). While Ganymede's present-day tidal dissipation might be negligible, it might have contributed a significant amount in the past in which a period of enhanced eccentricity caused runaway effect leading to a younger surface than its neighbor Callisto (Cameron et al., 2018, 2019, 2020; Malhotra, 1991; Showman & Malhotra, 1997).

4 Interior structures and dynamics of Jovian hydrospheres

4.1 Outer ice shells

Ocean world outer ice shells are generally regarded to comprise two fundamental layers (McKinnon, 1999; Pappalardo et al., 1998a, 1998b). As described below, these include a cold upper layer where geologic heat transfer occurs primarily through thermal conduction and a warm ductile layer that may experience solid-state convection. The brittle portion of the conductive layer is typically referred to as the "lithosphere" and the ductile portion of the conductive layer and the potential convective layer comprise the "asthenosphere," see Fig. 3 for an illustration of temperature profiles illustrating heat transfer within these layers and Fig. 4 for an illustration of processes within these layers on Europa.

The relative and total thicknesses of these layers on each body are poorly understood, and these thicknesses are critical to understanding the geologic and geophysical environment (Billings & Kattenhorn, 2005; Howell, 2021; Howell et al., 2023; Vance et al., 2018). Estimates of the thickness of these layers depend on assumptions about the global heat budget, composition, and planetary history.

4.1.1 Conductive icy layer: Tectonics and cryovolcanism

Near the surface, the ice is cold and the dominant mode of geologic heat transfer across this layer is thermal conduction. The upper portion of this conductive layer may deform through brittle and shear failure in response to geologic stresses over timescales of thousands to millions of years (e.g., Cameron et al., 2018, 2019; Howell & Pappalardo, 2018), through mass wasting processes (Mills et al., 2023), as well as through elastic bending and flexure (e.g., Nimmo et al., 2003). Brittle faulting and tectonism within the icy lithospheres of Europa and Ganymede can produce structures reminiscent of Earth's coupled system of plate tectonics (Fig. 4), though Howell and Pappalardo (2020) argued that the deformation of semi-rigid plates on ocean worlds (except potentially Neptune's moon Triton) cannot be self-sustaining, as is the case for Earth because the forces resisting self-driving plate tectonics are orders of magnitude stronger than those that drive the processes.

Extensional tectonism, where stresses within the mechanically strong upper icy lithosphere exceed the yield stresses and cause the shear failure of the upper portion of the ice, is the most common style observed on Ganymede and Europa (Pappalardo et al., 1998a, 1998b; Prockter et al., 2002). Extensional tectonics may be driven by ice-shell thickening through time (e.g., Nimmo, 2004), or the decoupling of the rotation of the ice shell from the rotation of the silicate interior (e.g., Nimmo & Manga, 2009). Europa's bands and

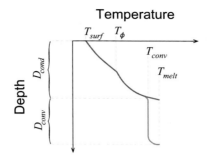

FIG. 3

Illustration showing outer ice-shell temperature profiles for (*red*) a thinner ice shell where geologic heat transfer occurs only through conduction and (*blue*) a thicker ice shell that exhibits isothermal convection at depth. Here, D_{cond} and D_{conv} are respectively the conductive and convective layer thicknesses, T_{surf} is the surface temperature, T_ϕ is the temperature to which the porous lithosphere extends, T_{conv} is the isothermal temperature of convection, and T_{melt} is the melting temperature.

After Howell (2021).

Ganymede's grooved terrains may form through extensional processes akin to mid-ocean-ridge spreading (Prockter et al., 2002). This process might cause intermittent rifting of the upper portion of the ice shell through tilt-block style faulting at heavily tectonized bands on both worlds (Howell & Pappalardo, 2018; Pappalardo et al., 1998a, 1998b), and at Europa's smooth bands plastic failure may allow the advection of frozen ocean material to the surface on timescales of $100 > \text{kyear}^{-1}$ Myear (Howell & Pappalardo, 2018).

Strike-slip tectonism, as is observed at transform plate boundaries like the San Andreas Fault on Earth, is also ubiquitous across the surfaces of Europa and Ganymede. Strike-slip motion can occur in response to extension, compression, or shearing, and is observed through the lateral offset of pre-existing features. On Europa, strike-slip motion associated with double ridge formation may result in frictional melting and drainage (e.g., Hammond, 2020; Kalousová et al., 2014). On Ganymede, the strike-slip motion may be a fingerprint of long-term changes in Ganymede's orbit (Cameron et al., 2020) and the reorientation of its icy shell with respect to the silicate interior (Cameron et al., 2018, 2019).

Convergent tectonism is the least-observed tectonism among the icy worlds of Jupiter. The plastic folding of the surface has been inferred for Europa's early ridged plains (Leonard et al. 2020) and regional-scale furrows (Prockter et al. 2000). The nature of any compressional past on Ganymede has been a subject of study for nearly half a century, and remains unclear to the present day (e.g., Bland & McKinnon, 2015; Lucchita, 1980). On Europa, however, the watershed discovery of a subduction-like process on Europa was proposed by Kattenhorn and Prockter (2014). They found a swath of surface area that, after plate reconstruction, indicated the loss of tabular zones ~100 km wide and ~1000 km long. This process was dubbed "subsumption," because the icy plate would be re-absorbed by warmer ice as it intrudes deeper into the shell as opposed to the hallmark densifying mineral phase changes of subduction on Earth. The nature of putative subsumption and proposed subduction on Europa has ignited continued debate over the plausibility of sustained, Earth-like plate tectonics (e.g., Howell & Pappalardo, 2019; Johnson et al., 2017; Witze, 2014).

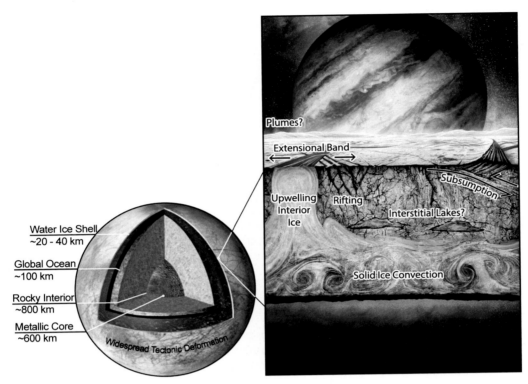

FIG. 4

Illustration of geological processes within the ice shell of Europa. Tectonic processes such as rifting, extensional band formation, and subsumption may create major faults that span lateral distances of more than 1000 km and extend to depths of 1–10 km. Below the brittle region of the ice shell, regions of coalesced melted water ice may form connected pore-networks or interstitial water bodies. At depth, the solid ice may convect due to the thermal expansion of ice near the warm ice-ocean interface providing thermal buoyancy, and the sinking of cold, dense ice near the brittle layer. Some of these processes may result in the interaction of liquid water with the shallow subsurface, and/or the eruption of water vapor plumes.

Modified from Howell and Pappalardo (2020).

Within the icy shells of Jupiter's ocean worlds, many workers have inferred the presence of shallow liquid water bodies on Europa that may be active today or may have been active in the recent past (Hammond, 2020; Lesage et al., 2020, 2021; Manga & Michaut, 2017; Schmidt et al., 2011), and potential cryovolcanism on Ganymede that may have occurred >1 Gyear before present day (Kay & Head, 1999; Patterson et al., 2010; Showman & Han 2004). Within the past decade, researchers have collected evidence for transient water vapor plumes on Europa through observational astronomy (Paganini et al., 2020; Roth et al., 2014; Sparks et al., 2016) and in magnetic field anomalies observed by the *Galileo* spacecraft (Jia et al., 2018) (Fig. 5).

Any past or present cryovolcanic activity would likely reflect shallow bodies of liquid water within the upper regions of the icy shell, where freezing stresses can build to eruption before they are viscously

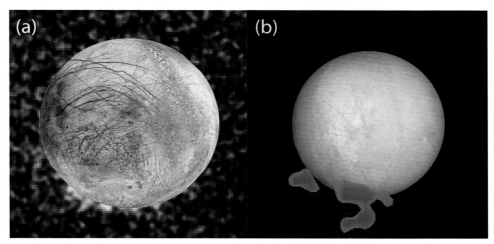

FIG. 5

Putative transient water vapor plume observations at Europa's south pole taken by the Hubbly Space Telescope in (a) January 2014, reported in Sparks et al. (2016) (NASA/ESA/W. Sparks (STScI)/USGS Astrogeology Science Center), and (b) in December 2012, reported in Roth et al. (2014) (NASA/ESA/L. Roth (SWRI)).

relaxed away (Lesage et al., 2022). Efforts continue to characterize the plausible compositional, thermal, and habitability implications of these bodies as they freeze (e.g., Naseem et al., 2023). However, no consensus yet exists on the plausibility of such reservoirs or their longevity, due to arguments that such a configuration is energetically unfavorable, as evidenced by the rapid freezing timescales of subsurface reservoirs (Buffo et al., 2020; Chivers et al., 2021), and the tendency of any melt in the shallow subsurface to quickly advect to the ocean below (Hesse et al., 2022; Kalousová et al., 2014, 2016).

4.1.2 Convective icy layer

Beneath the conductive layers of these outer ice shells may be a warm, nearly isothermal layer experiencing solid-state convection (Barr & McKinnon, 2007; Besserer et al., 2013; Grott et al., 2007; Pappalardo et al., 1998a, 1998b). Here, ice near the ice-ocean interface is warm, and at a lower density than ice near the base of the conductive layer, driving buoyant material from near the interface upwards and dense material from beneath the cool conductive layer downwards. Additionally, the ice may be heated from within by the frictional dissipation of tidal energy from orbital eccentricity (Ross & Schubert, 1987; Sotin et al., 2002; Vilella et al., 2020) and obliquity (Jankowski et al., 1989; Nimmo & Spencer, 2015).

The convective layer is likely free of vacuum-filled pores because the warm interior ice quickly relaxes over geologic timescales. At Europa, small amounts of melt may be produced and transported toward the underlying ocean (Kalousová et al., 2014; Tobie et al., 2003) or trapped within the convecting layer, providing up to a few percent water by volume trapped within the ice at depth (Vilella et al., 2020).

Freezing processes at the base of the convective layer in the thick icy shells of Jupiter's ocean worlds are likely too slow to allow for the incorporation of significant non-ice materials (Buffo et al., 2020), and therefore the convecting interior water ice may be relatively pure. However, it is possible for brines to form and persist at the ice-ocean interface, and where seawater has been injected directly into the ice shell (Buffo et al., 2021; Vance et al., 2019).

4.2 Ice-ocean interfaces

The ice-ocean interfaces of ocean worlds are poorly understood, in large part because of the difficulty in obtaining unique constraints on their structure and behavior. On Earth, this interface is crucial for participating in the global exchange of heat and salts (McPhee, 2008). On the ocean worlds, the ice-ocean interface may be one of the most crucial for understanding not only heat and material transport but also planetary habitability. Active geologic processes within planetary ice shells are critical for exchanging thermal and chemical energy between the surface and the ocean, potentially establishing or maintaining the chemical disequilibria required for life to emerge or persist (Hand et al., 2009, 2007; Vance et al., 2016).

Studies of the ice-ocean interfaces provide a wide range of predictions. Energy and mass balances adopted from Earth provide some constraints on the interface and compositional evolution of planetary ice shells (Buffo et al., 2021, 2020). These interfaces may host regions of crystalline mush up to several meters in thickness that may drive requirements for seismic or radar detection (Buffo et al., 2021). If the interfaces are indeed sharp, long-wavelength topography may arise from thermal convection within the ice shell, and short wavelength topography at the interface could include salty "brinicles" that protrude up to meters into the ocean (Vance et al., 2019). These brinicles may provide local habitats at the ice-ocean interface (Vance et al., 2019).

4.3 Interior oceans

The oceans of the solar system are central to the search for life in the universe (Howell et al., 2021; National Academies of Sciences, Engineering, and Medicine, 2022; Space Studies Board & National Research Council, 2012), spanning likely depths of a few tens to several hundreds of kilometers. They provide plentiful liquid water, the potential for sustained chemical disequilibrium at their icy and rocky interfaces, a warm habitat in a cold region of the solar system, and facilitate the transport of chemical constituents and thermal energy. Planetary oceans might be stratified, where they are poorly mixed and temperature and composition are strong functions of depth, or they may be vigorously convecting, quickly transporting material from the seafloor and ice-ocean interfaces and producing a homogeneous layer (Amit et al., 2020; Kvorka & Čadek, 2022; Soderlund, 2019; Soderlund et al., 2014; Hay & Matsuyama, 2019; Lobo et al., 2021; Thomson & Delaney, 2001). Broadly, salinities are generally considered to span approximately one order of magnitude above and below a few weight percent, approximately the salinity of Earth's oceans (Hand & Chyba, 2007).

The presence of oceans on the icy satellites and minor planets is inferred through measurements of magnetic induction, gravity, and orbit. No planetary ocean has been directly observed, and therefore the composition, state, and behavior of planetary oceans are poorly understood, however, understanding the physical state of the ocean is a key objective of ocean world exploration.

4.4 Seafloors

Due to Europa's relatively thin cryosphere and small mass, the ocean is likely in direct contact with the rocky interior, as on Earth. For more massive moons and moons with thicker cryospheres, like Ganymede and potentially Callisto, the pressure and temperature conditions at the base of the ocean are favorable to form high-pressure ice phases that are denser than the seawater (see the next section).

In cases where the ocean directly interacts with the rocky interior, water-rock reactions, such as serpentinization, may act to reduce the ocean and free volatile gasses from the rock into the water column (Hand et al., 2009). All known life relies on redox potentials as the primary currency for energy, enabling metabolism and reproduction. Serpentinization and periodic volcanism on Europa, for example, may have allowed reduction–oxidation (or redox) potentials to persist through time (Běhounková et al., 2021; Vance & Melwani Daswani, 2020).

Additionally, seafloor volcanic processes on bodies with active rocky interiors may produce structures similar to hydrothermal seafloor vents found on Earth. Hydrothermal processes act to concentrate minerals into localized hot thermal plumes that inject into the ocean. These vents thus act to concentrate mineral and thermal energy and are suggested as one candidate environment for the emergence of life on Earth and ocean worlds (Martin et al., 2008).

4.5 High-pressure ice layers

4.5.1 Ganymede's high-pressure ice layer

Ganymede's hydrosphere is estimated to be roughly 800–900 km thick (e.g., Anderson et al., 1996; Busarev et al., 2018; Vance et al., 2014) and the pressures between 1.5 and 1.7 GPa predicted at the rock/hydrosphere interface (Vance et al., 2014) are high enough to result in the crystallization of a layer of dense ice polymorphs. This high-pressure (HP) ice layer is estimated to be between 300 and 700 km thick (Vance et al., 2018), and its bulk seemingly separates the rocky mantle from the liquid water ocean—a setting which is often considered to impede Ganymede's habitability (Hand et al., 2020). The presence of this deep ice layer in Ganymede's interior also distinguishes it from Europa where the rock and ocean are adjoined.

The HP ice layer structure depends on Ganymede's hydrosphere thermal state and composition (e.g., Choblet et al., 2017; Kalousová et al., 2022). At the rock/hydrosphere interface, ice VI crystallizes to form a layer predicted to be several hundred kilometers thick (Vance et al., 2018), above which other HP ice phases might have crystallized. Fig. 6a shows some of the results from Vance et al. (2018)—the density profiles for pure H_2O oceans (solid lines) and oceans with 10 wt% of $MgSO_4$ (dashed lines) and two distinct thermal states: warm (for which the outer ice I shell is <100 km thick) and cold (for which the outer shell is thicker than 100 km). In the case of a relatively warm interior (red lines), the ocean is a few hundred kilometers thick, and only ice VI is expected to be present for both ocean compositions. On the other hand, in the cold case (blue lines), the ocean is only a few tens of kilometers thick, and ice V is predicted to crystallize above the ice VI layer for both, pure and salty hydrospheres.

Dynamic changes within Ganymede's HP ice layer are primarily driven by the availability of radiogenic heat produced within the rocky interior, and the possibility of thermal convection occurring in Ganymede's HP ice layer was first suggested nearly half a century ago (Poirier et al., 1981). Using a two-phase model capable of treating the melting process and the subsequent melt transport, Kalousová et al. (2018) investigated the dynamics of Ganymede's ice VI layer. Their results (Fig. 7) showed that most of the melt is accumulated in a partially molten layer at the top boundary (dark red in panel a), where it is extracted into the overlying ocean. Depending on the model parameters (mantle heat flux, HP ice layer thickness, HP ice viscosity), melt may also appear in the upwelling plumes as well as at the bottom boundary with the rocky mantle. In the latter case, volatiles present at the rocky interface could dissolve in water and be transferred into the ocean by the upwelling plumes. Such a scenario would allow material exchange between the rock and the ocean. Subsequent

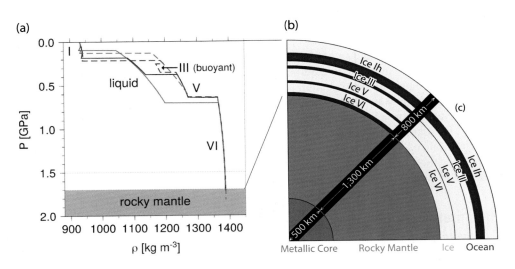

FIG. 6

(a) Density vs pressure for H_2O oceans (*solid lines*) and oceans with 10 wt% of $MgSO_4$ (*dashed lines*) and two thermal profiles (see text for details): warm (*red*) and cold (*blue*). Roman numerals indicate the particular ice phase. The top of the rocky mantle (*gray rectangle*) and the pressure at the rocky interface (1.5–1.7 GPa, *horizontal dashed lines*) are indicated. (b) Possible internal structure with a sandwiched hydrosphere. (c) Possible internal structure with a compact HP ice layer.

Panel (a) Data are from Vance (2017). Panels (b) and (c) reflect the results of Journaux et al. (2020).

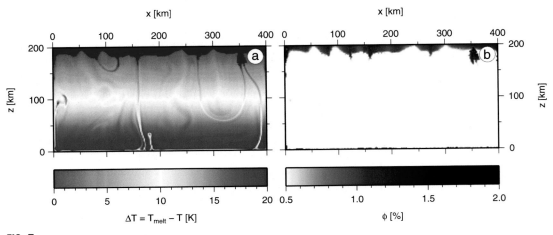

FIG. 7

(a) Temperature difference from the melting temperature ($\Delta T = T_{melt} - T$). *Dark red* color marks partially molten ice ($T = T_{melt}$). (b) Water content. The models shown are from Kalousová et al. (2018).

studies have shown that high-pressure melting may have been most important earlier in Ganymede's history when the mantle heat flux was larger and the HP ice layer thinner (Kalousová & Sotin, 2018; Kalousová et al., 2018).

4.5.2 Callisto's high-pressure ice layer

Callisto's interior is unlikely to be fully differentiated (Anderson et al., 2001), and the interpretation of Galileo's gravity data remains ambiguous because the satellite may not be in hydrostatic equilibrium (Gao & Stevenson, 2013). Furthermore, the presence of a deep ocean is debated, as the induced magnetic field (Zimmer et al., 2000) may also be explained by ionosphere induction (Hartkorn & Saur, 2017). Thus, the structure of Callisto's hydrosphere is poorly constrained. The compositional predictions of Vance et al. (2018) suggest that Callisto's hydrosphere is likely 300–450 km thick, with smaller values corresponding to the less-differentiated interior. In such a case, a layer of ice V ~100 km thick may be present, and potentially overlain by a thin ice III layer (depending on the thermal state and composition). Conversely, Journaux et al. (2020) proposed that there may not be high-pressure ice present in any phase.

Alternatively, if Callisto's interior is not hydrostatic, and Callisto is more differentiated than previously thought (Gao & Stevenson, 2013), the hydrosphere is likely thicker, with an ice VI layer crystallizing at the rock/hydrosphere interface, overlain by ice V and possibly also ice III (Vance et al., 2018). Given these uncertainties regarding Callisto's HP icy layers, no study has yet investigated their potential dynamics.

5 Ice and life

The hydrospheres of the Galilean satellites have likely persisted for billions of years, serving as warm, wet laboratories for chemical reactions to take place. On Europa, where no high-pressure ice phase is expected, the ocean has been in contact with the rocky seafloor throughout its lifetime. Geologic activity within Europa's rocky interior may drive volcanism to the present day (Běhounková et al., 2021; Vance & Melwani Daswani, 2020). Hydrothermal activity at the seafloor may act to concentrate mineral energy, as is the case for hydrothermal vents on Earth, during reducing serpentinization reactions (Vance & Melwani Daswani, 2020). At the same time, the bombardment of the surface of Europa by Iogenic sulfur and radiation acts to radiolytically forge oxidants (Hand et al., 2009, 2020). As Europa's recently or currently active geology acts to convey material from the surface to the ocean, a redox gradient may be sustained that could allow for the production of habitable environments within the ice and ocean (Greenberg, 2010; Hand et al., 2009, 2020). If the ocean is fully reduced, life may emerge at the ice-ocean interface where surface oxidants are first brought to depth; if the ocean is oxidant-limited, life may emerge where serpentinization reactions persist at the seafloor (Hand et al., 2009, 2020; Vance & Melwani Daswani, 2020).

Beyond the subsurface oceans, locations in the icy shells of these ocean worlds where water is stable represent potential habitats, potentially introduced through the geodynamic and geologic processes described within this chapter. Although the vast majority of the ice shell is below the pure ice pressure melting temperature, certain processes and impurities can facilitate the formation of liquid water in otherwise cryogenic environments. The presence of salts, acids, and organic material can depress the freezing point of a solution relative to pure water, allowing liquid water to be stable in equilibrium with ice in the form of brine (Marion & Kargel, 2007).

Chemical properties such as temperature, salinity, water activity, ionic strength, and chao/kosmotropicity represent parameters of focus in habitability investigations of aqueous systems on Earth and Mars (Fox-Powell et al., 2016). For brine in equilibrium with ice, a reduction in temperature corresponds to an increase in brine salinity, an increase in ionic strength, and a reduction in water activity. These changes in chemical properties translate to stressors which challenge the growth and reproduction of organisms that inhabit these environments (Fox-Powell et al., 2016). Extremophiles, such as psychrophiles and halophiles, have evolved strategies to survive and even thrive under these conditions (Deming & Young, 2017; Oren, 2011). Organisms inhabiting the icy shells of the Galilean moons would be subject to similar stressors and may rely on analogous survival strategies.

6 Conclusions and perspectives

The numerous icy layers and processes of Jupiter's ocean worlds are fascinating geologic environments at the forefront of planetary exploration. Since their discovery, the Galilean satellites have continually revolutionized astronomical thinking, from challenging the geocentric model of the universe to revealing the power of tidal interactions to drive change, to the revelation that global and potentially habitable subsurface oceans, and even life, may be commonplace in our solar system and beyond.

Europa, whose rate of resurfacing and level of icy tectonic activity may be unique within our solar system, may harbor the conditions required for the emergence of past or extant life. The icy interfaces and geologic transport processes of Europa are key to enabling its habitability.

Ganymede, the largest satellite of the solar system, ranks among the most enigmatic. Its bimodal surface reveals ancient activity at the time of formation, since disrupted only once, perhaps a billion or so years ago, and after billions of years of inactivity. Despite the apparently modest energy available to drive resurfacing, when compared to Europa, Ganymede likely maintains a liquid outer iron core that drives its present-day intrinsic magnetic field.

Callisto, despite similarities in its structure to Ganymede, has abstained from much of the activity of the other satellites. It has potentially failed to fully differentiate into metallic, rocky, and icy layers, and may continue to possess a gradient between its hydrosphere and rocky interior to the present day. Callisto's, ancient and blanketed with dark material, betrays no sign of interior activity since the time of its formation.

New observations of the thermal and mechanical states of these worlds, their composition and the properties of their interior oceans, and their current and past activity are likely to once again revolutionize how we understand our solar system. Potentially, exploring the icy processes and formations of Europa, Ganymede, and Callisto, and their interactions with Io and Jupiter, may even revolutionize our understanding of life in the universe.

Acknowledgments

Portions of this work were performed at the Jet Propulsion Laboratory, California Institute of Technology, under contract with the National Aeronautics and Space Administration. K.K. was supported by the Czech Science Foundation through project No. S.22-20388.

References

Alday, J., Roth, L., Ivchenko, N., Retherford, K. D., Becker, T. M., Molyneux, P., & Saur, J. (2017). New constraints on Ganymede's hydrogen corona: Analysis of Lyman-α emissions observed by HST/STIS between 1998 and 2014. *Planetary and Space Science, 148*, 35–44.

Amit, H., Choblet, G., Tobie, G., Terra-Nova, F., Čadek, O., & Bouffard, M. (2020). Cooling patterns in rotating thin spherical shells—Application to Titan's subsurface ocean. *Icarus, 338*, 113509.

Anderson, J. D., Jacobson, R. A., McElrath, T. P., Moore, W. B., Schubert, G., & Thomas, P. C. (2001). Shape, mean radius, gravity field, and interior structure of Callisto. *Icarus, 153*(1), 157–161. https://doi.org/10.1006/icar.2001.6664.

Anderson, J. D., Lau, E. L., Sjogren, W. L., Schubert, G., & Moore, W. B. (1996). Gravitational constraints on the internal structure of Ganymede. *Nature, 384*(6609), 541–543. https://doi.org/10.1038/384541a0.

Aydin, A. (2006). Failure modes of the lineaments on Jupiter's moon, Europa: Implications for the evolution of its icy crust. *Journal of Structural Geology, 28*(12), 2222–2236.

Barr, A. C., & McKinnon, W. B. (2007). Convection in ice I shells and mantles with self-consistent grain size. *Journal of Geophysical Research: Planets, 112*(E2).

Běhounková, M., Tobie, G., Choblet, G., Kervazo, M., Melwani Daswani, M., Dumoulin, C., & Vance, S. D. (2021). Tidally induced magmatic pulses on the oceanic floor of Jupiter's moon Europa. *Geophysical Research Letters, 48*, e2020GL090077.

Bender, K. C., & Carroll, R. D. (1997). *Geologic map of Callisto*. The Survey.

Besserer, J., Nimmo, F., Roberts, J. H., & Pappalardo, R. T. (2013). Convection-driven compaction as a possible origin of Enceladus's long wavelength topography. *Journal of Geophysical Research: Planets, 118*(5), 908–915.

Bierhaus, E. B., Zahnle, K., Chapman, C. R., Pappalardo, R. T., McKinnon, W. R., & Khurana, K. K. (2009). Europa's crater distributions and surface ages. *Europa*, 161–180.

Bierson, C. J., & Nimmo, F. (2020). Explaining the Galilean Satellites' density gradient by hydrodynamic escape. *The Astrophysical Journal Letters, 897*(2), L43.

Bierson, C. J., & Steinbrügge, G. (2021). Tidal heating did not dry out io and Europa. *The Planetary Science Journal, 2*(3), 89.

Bland, M. T., & McKinnon, W. B. (2015). Forming Ganymede's grooves at smaller strain: Toward a self-consistent local and global strain history for Ganymede. *Icarus, 245*, 247–262.

Bland, M. T., Showman, A. P., & Tobie, G. (2009). The orbital–thermal evolution and global expansion of Ganymede. *Icarus, 200*(1), 207–221.

Billings, S. E., & Kattenhorn, S. A. (2005). The great thickness debate: Ice shell thickness models for Europa and comparisons with estimates based on flexure at ridges. *Icarus, 177*(2), 397–412. https://doi.org/10.1016/j.icarus.2005.03.013.

Brown, M., & Hand, K. (2013). Salts and radiation products on the surface of Europa. *The Astronomical Journal, 145*(4), 110.

Buffo, J. J., Schmidt, B. E., Huber, C., & Meyer, C. R. (2021). Characterizing the ice-ocean interface of icy worlds: A theoretical approach. *Icarus, 360*, 114318.

Buffo, J. J., Schmidt, B. E., Huber, C., & Walker, C. C. (2020). Entrainment and dynamics of ocean-derived Impurities within Europa's ice shell. *Journal of Geophysical Research: Planets, 125*(10), e2020JE006394.

Busarev, V., Tatarnikov, A., & Burlak, M. (2018). Comparison and interpretation of spectral characteristics of the leading and trailing hemispheres of Europa and Callisto. *Solar System Research, 52*(4), 301–311.

Camarca, M., & de Kleer, K. (2021). *Fresh eyes on an old moon: New ALMA images of Callisto*. 53 (p. 308.06).

Cameron, M. E., Smith-Konter, B. R., Burkhard, L., Collins, G. C., Seifert, F., & Pappalardo, R. T. (2018). Morphological mapping of Ganymede: Investigating the role of strike-slip tectonics in the evolution of terrain types. *Icarus, 315*, 92–114.

Cameron, M. E., Smith-Konter, B. R., Collins, G. C., Patthoff, D. A., & Pappalardo, R. T. (2019). Tidal stress modeling of Ganymede: Strike-slip tectonism and Coulomb failure. *Icarus, 319*, 99–120.

Cameron, M. E., Smith-Konter, B. R., Collins, G. C., Patthoff, D. A., & Pappalardo, R. T. (2020). Ganymede, then and now: How past eccentricity may have altered tidally driven coulomb failure. *Journal of Geophysical Research: Planets, 125*(7), e2019JE005995.

Canup, R. M., & Ward, W. R. (2002). Formation of the Galilean satellites: Conditions of accretion. *The Astronomical Journal, 124*(6), 3404.

Cassen, P., Peale, S. J., & Reynolds, R. T. (1980). On the comparative evolution of Ganymede and Callisto. *Icarus, 41*(2), 232–239.

Chivers, C. J., Buffo, J. J., & Schmidt, B. E. (2021). Thermal and chemical evolution of small, shallow water bodies in Europa's ice shell. *Journal of Geophysical Research: Planets, 126*(5), e2020JE006692.

Choblet, G., Tobie, G., Sotin, C., Kalousová, K., & Grasset, O. (2017). Heat transport in the high-pressure ice mantle of large icy moons. *Icarus, 285*, 252–262. https://doi.org/10.1016/j.icarus.2016.12.002.

Chuang, F. C., & Greeley, R. (2000). Large mass movements on Callisto. *Journal of Geophysical Research: Planets, 105*(E8), 20227–20244.

Collins, G. C., & Nimmo, F. (2009). Chaotic terrain on Europa. In *Europa* University of Arizona Press.

Collins, G. C., Patterson, G. W., Head, J. W., Lucchitta, B. K., & Kay, J. P. (2013). *Global geologic map of Ganymede*. US Department of the Interior, US Geological Survey.

Cooper, J. F., Johnson, R. E., Mauk, B. H., Garrett, H. B., & Gehrels, N. (2001). Energetic ion and electron irradiation of the icy Galilean satellites. *Icarus, 149*(1), 133–159.

Dalton, J. B., III, Cassidy, T., Paranicas, C., Shirley, J. H., Prockter, L. M., & Kamp, L. W. (2013). Exogenic controls on sulfuric acid hydrate production at the surface of Europa. *Planetary and Space Science, 77*, 45–63.

Dalton, J. B., III, Shirley, J. H., & Kamp, L. W. (2012). Europa's icy bright plains and dark linea: Exogenic and endogenic contributions to composition and surface properties. *Journal of Geophysical Research: Planets, 117*(E3).

de Kleer, K., Butler, B., de Pater, I., Gurwell, M. A., Moullet, A., Trumbo, S., & Spencer, J. (2021). Ganymede's surface properties from millimeter and infrared thermal emission. *The Planetary Science Journal, 2*(1), 5.

Deming, J. W., & Young, J. N. (2017). The role of exopolysaccharides in microbial adaptation to cold habitats. In *Psychrophiles: From biodiversity to biotechnology* (pp. 259–284). Springer.

Doggett, T., Greeley, R., & Figueredo, P. (1999). Geologic stratigraphy and evolution of Europa's surface. In *Europa* (pp. 137–159).

Fagents, S. A., Greeley, R., Sullivan, R. J., Pappalardo, R. T., Prockter, L. M., Team, G. S., et al. (2000). Cryomagmatic mechanisms for the formation of Rhadamanthys Linea, triple band margins, and other low-albedo features on Europa. *Icarus, 144*(1), 54–88.

Figueredo, P. H., & Greeley, R. (2004). Resurfacing history of Europa from pole-to-pole geological mapping. *Icarus, 167*, 287–312. https://doi.org/10.1016/j.icarus.2003.09.016.

Fox-Powell, M. G., Hallsworth, J. E., Cousins, C. R., & Cockell, C. S. (2016). Ionic strength is a barrier to the habitability of Mars. *Astrobiology, 16*(6), 427–442.

Gao, P., & Stevenson, D. J. (2013). Nonhydrostatic effects and the determination of icy satellites' moment of inertia. *Icarus, 226*(2), 1185–1191. https://doi.org/10.1016/j.icarus.2013.07.034.

Gjessing, Y., Hanssen-bauer, I., Fujii, Y., Kameda, T., Kamiyama, K., & Kawamura, T. (1993). Chemical fractionation in sea ice and glacier ice. *Bulletin of Glacier Research, 11*, 1–8.

Golubev, Y. F., Grushevskii, A. V., Koryanov, V. V., & Tuchin, A. G. (2014). Gravity assist maneuvers of a spacecraft in Jupiter system. *Journal of Computer and Systems Sciences International, 53*(3), 445–463. https://doi.org/10.1134/S1064230714030083.

Grasset, O., Dougherty, M. K., Coustenis, A., Bunce, E. J., Erd, C., Titov, D., Blanc, M., Coates, A., Drossart, P., & Fletcher, L. N. (2013). JUpiter ICy moons Explorer (JUICE): An ESA mission to orbit Ganymede and to characterise the Jupiter system. *Planetary and Space Science, 78*, 1–21.

Greeley, R., Figueredo, P. H., Williams, D. A., Chuang, F. C., Klemaszewski, J. E., Kadel, S. D., Prockter, L. M., Pappalardo, R. T., Head, J. W., Collins, G. C., Spaun, N. A., Sullivan, R. J., Moore, J. M., Senske, D. A., Tufts, B. R., Johnson, T. V., Belton, M. J. S., & Tanaka, K. L. (2000a). Geologic mapping of Europa. *Journal of Geophysical Research: Planets (1991–2012)*, *105*(E9), 22559–22578. 10.1029/1999JE001173.

Greeley, R., Klemaszewski, J. E., & Wagner, R. (2000b). Galileo views of the geology of Callisto. *Planetary and Space Science*, *48*(9), 829–853. https://doi.org/10.1016/S0032-0633(00)00050-7.

Green, A., Montesi, L., & Cooper, C. (2021). The growth of Europa's icy shell: Convection and crystallization. *Journal of Geophysical Research: Planets*, *126*(4), e2020JE006677.

Greenberg, R. (2010). Transport rates of radiolytic substances into Europa's ocean: Implications for the potential origin and maintenance of life. *Astrobiology*, *10*(3), 275–283.

Greenberg, R., Geissler, P., Hoppa, G., Tufts, B. R., Durda, D. D., Pappalardo, R. T., Head, J. W., Greeley, R., Sullivan, R., & Carr, M. H. (1998). Tectonic processes on Europa: Tidal stresses, mechanical response, and visible features. *Icarus*, *135*, 64–78.

Greenberg, R., Hoppa, G. V., Tufts, B. R., Geissler, P., Riley, J., & Kadel, S. (1999). Chaos on Europa. *Icarus*, *141*(2), 263–286.

Grott, M., Sohl, F., & Hussmann, H. (2007). Degree-one convection and the origin of Enceladus' dichotomy. *Icarus*, *191*(1), 203–210.

Hammond, N. P. (2020). Estimating the magnitude of cyclic slip on strike-slip faults on Europa. *Journal of Geophysical Research: Planets*, *125*(7), no-no.

Han, L., & Melosh, H. (2010). Origin of Europa's ridges by incremental ice-wedging. *AGU Fall Meeting Abstracts*, *2010*, P33B–1577.

Hand, K. P., Carlson, R. W., & Chyba, C. F. (2007). Energy, chemical disequilibrium, and geological constraints on Europa. *Astrobiology*, *7*(6), 1006–1022.

Hand, K. P., & Chyba, C. F. (2007). Empirical constraints on the salinity of the European ocean and implications for a thin ice shell. *Icarus*, *189*(2), 424–438.

Hand, K. P., Chyba, C. F., Priscu, J. C., Carlson, R. W., & Nealson, K. H. (2009). Astrobiology and the potential for life on Europa. In R. T. Pappalardo, W. B. McKinnon, & K. Khurana (Eds.), *Europa* (p. 589). University of Arizona Press.

Hand, K. P., Murray, A., Garvin, J., Brinckerhoff, W., Christner, B., Edgett, K., Ehlmann, B., German, C., Hayes, A., & Hoehler, T. (2017). *Europa lander study 2016 report: Europa lander mission*. CA, USA: NASA Jet Propuls. Lab., La Cañada Flintridge. Tech. Rep. JPL D-97667.

Hand, K. P., Sotin, C., Hayes, A., & Coustenis, A. (2020). On the habitability and future exploration of ocean worlds. *Space Science Reviews*, *216*(5), 95. https://doi.org/10.1007/s11214-020-00713-7.

Hartkorn, O., & Saur, J. (2017). Induction signals from Callisto's ionosphere and their implications on a possible subsurface ocean. *Journal of Geophysical Research: Space Physics*, *122*(11), 11,677–11,697. https://doi.org/10.1002/2017JA024269.

Hay, H. C., & Matsuyama, I. (2019). Nonlinear tidal dissipation in the subsurface oceans of Enceladus and other icy satellites. *Icarus*, *319*, 68–85.

Head, J. W., Pappalardo, R. T., & Sullivan, R. (1999). Europa: Morphological characteristics of ridges and triple bands from Galileo data (E4 and E6) and assessment of a linear diapirism model. *Journal of Geophysical Research: Planets*, *104*, 24223–24236. https://doi.org/10.1029/1998JE001011.

Hesse, M. A., Jordan, J. S., Vance, S. D., & Oza, A. V. (2022). Downward oxidant transport through Europa's ice shell by density-driven brine percolation. *Geophysical Research Letters*, *49*(5), e2021GL095416.

Hibbitts, C., McCord, T., & Hansen, G. (2000). Distributions of CO2 and SO2 on the surface of Callisto. *Journal of Geophysical Research: Planets*, *105*(E9), 22541–22557.

Howard, A. D., & Moore, J. M. (2008). Sublimation-driven erosion on Callisto: A landform simulation model test. *Geophysical Research Letters*, *35*(3).

Howell, S. M. (2021). The likely thickness of Europa's icy shell. *The Planetary Science Journal*, 2(4), 129. https://doi.org/10.3847/PSJ/abfe10.

Howell, S. M., & Leonard, E. J. (2023). Ocean worlds: Interior processes and physical environments. In *Handbook of Space Resources* (pp. 873–906). Springer International Publishing.

Howell, S. M., & Pappalardo, R. T. (2018). Band formation and ocean-surface interaction on Europa and Ganymede. *Geophysical Research Letters*, 45(10). https://doi.org/10.1029/2018GL077594.

Howell, S. M., & Pappalardo, R. T. (2019). Can Earth-like plate tectonics occur in ocean world ice shells? *Icarus*, 322, 69–79.

Howell, S. M., & Pappalardo, R. T. (2020). NASA's Europa Clipper—A mission to a potentially habitable ocean world. *Nature Communications*, 11(1), 1311. https://doi.org/10.1038/s41467-020-15160-9.

Howell, S. M., Stone, W. C., Craft, K., German, C., Murray, A., Rhoden, A., & Arrigo, K. (2021). Ocean worlds exploration and the search for life. *Bulletin of the American Astronomical Society*, 53(4), 191.

Hussmann, H., Schubert, G., Steinbruegge, G., Sohl, F., & Kimura, J. (2022). Internal structure of Ganymede. In *Ganymede* Cambridge University Press.

Hussmann, H., & Spohn, T. (2004). Thermal-orbital evolution of Io and Europa. *Icarus*, 171(2), 391–410.

Jankowski, D. G., Chyba, C. F., & Nicholson, P. D. (1989). On the obliquity and tidal heating of Triton. *Icarus*, 80(1), 211–219.

Jia, X., Kivelson, M. G., Khurana, K. K., & Kurth, W. S. (2018). Evidence of a plume on Europa from Galileo magnetic and plasma wave signatures. *Nature Astronomy*, 2(6), 459–464.

Johnson, B. C., Sheppard, R. Y., Pascuzzo, A. C., Fisher, E. A., & Wiggins, S. E. (2017). Porosity and salt content determine if subduction can occur in Europa's ice shell. *Journal of Geophysical Research: Planets*, 122(12), 2765–2778.

Johnston, S. A., & Montési, L. G. (2014). Formation of ridges on Europa above crystallizing water bodies inside the ice shell. *Icarus*, 237, 190–201.

Journaux, B., Kalousová, K., Sotin, C., Tobie, G., Vance, S., Saur, J., Bollengier, O., Noack, L., Rückriemen-Bez, T., Van Hoolst, T., Soderlund, K. M., & Brown, J. M. (2020). Large ocean worlds with high-pressure ices. *Space Science Reviews*, 216(1), 7. https://doi.org/10.1007/s11214-019-0633-7.

Kalousová, K., Soderlund, K. M., Solomonidou, A., & Sotin, C. (2022). Structure and evolution of Ganymede's hydrosphere. In M. Volwerk, M. McGrath, X. Jia, & T. Spohn (Eds.), *Ganymede* Cambridge University Press.

Kalousová, K., & Sotin, C. (2018). Melting in high-pressure ice layers of large ocean worlds—Implications for volatiles transport. *Geophysical Research Letters*, 45(16), 8096–8103. https://doi.org/10.1029/2018GL078889.

Kalousová, K., Sotin, C., Choblet, G., Tobie, G., & Grasset, O. (2018). Two-phase convection in Ganymede's high-pressure ice layer—Implications for its geological evolution. *Icarus*, 299, 133–147. https://doi.org/10.1016/j.icarus.2017.07.018.

Kalousová, K., Souček, O., Tobie, G., Choblet, G., & Čadek, O. (2014). Ice melting and downward transport of meltwater by two-phase flow in Europa's ice shell. *Journal of Geophysical Research: Planets*, 119(3), 532–549. https://doi.org/10.1002/2013JE004563.

Kalousová, K., Souček, O., Tobie, G., Choblet, G., & Čadek, O. (2016). Water generation and transport below Europa's strike-slip faults. *Journal of Geophysical Research: Planets*, 121(12), 2444–2462. https://doi.org/10.1002/2016JE005188.

Kargel, J. S., Kaye, J. Z., Head, J. W., III, Marion, G. M., Sassen, R., Crowley, J. K., Ballesteros, O. P., Grant, S. A., & Hogenboom, D. L. (2000). Europa's crust and ocean: Origin, composition, and the prospects for life. *Icarus*, 148(1), 226–265.

Kattenhorn, S. A. (2002). Nonsynchronous rotation evidence and fracture history in the Bright Plains region, Europa. *Icarus*, 157(2), 490–506.

Kattenhorn, S. A., & Prockter, L. M. (2014). Evidence for subduction in the ice shell of Europa. *Nature Geoscience*, 7(10), 762–767.

Kay, J., & Head, J., III. (1999). Geologic mapping of the Ganymede G8 calderas region: Evidence for cryovolcanism. *Lunar and Planetary Science Conference, 1103.*

Kay, J. P., Collins, G. C., & Patterson, G. W. (2007). Comparison of crater classification schemes on Ganymede. *Lunar and Planetary Science Conference, 1338,* 2392.

Khurana, K. K., Jia, X., Kivelson, M. G., Nimmo, F., Schubert, G., & Russell, C. T. (2011). Evidence of a global magma ocean in Io's interior. *Science, 332*(6034), 1186–1189.

Khurana, K. K., Kivelson, M. G., Stevenson, D. J., Schubert, G., Russell, C. T., Walker, R. J., & Polanskey, C. (1998). Induced magnetic fields as evidence for subsurface oceans in Europa and Callisto. *Nature, 395*(6704), 777–780. https://doi.org/10.1038/27394.

Kivelson, M. G., Khurana, K. K., Russell, C. T., Volwerk, M., Walker, R. J., & Zimmer, C. (2000). Galileo magnetometer measurements: A stronger case for a subsurface ocean at Europa. *Science, 289*(5483), 1340–1343.

Kivelson, M. G., Khurana, K. K., Russell, C. T., Walker, R. J., Warnecke, J., Coroniti, F. V., Polanskey, C., Southwood, D. J., & Schubert, G. (1996). Discovery of Ganymede's magnetic field by the Galileo spacecraft. *Nature, 384*(6609), 537–541.

Kvorka, J., & Čadek, O. (2022). A numerical model of convective heat transfer in Titan's subsurface ocean. *Icarus, 376,* 114853.

Leonard, E. J., Pappalardo, R. T., & Yin, A. (2018). Analysis of very-high-resolution Galileo images and implications for resurfacing mechanisms on Europa. *Icarus, 312,* 100–120.

Leonard, E. J., Yin, A., & Pappalardo, R. T. (2020). Ridged plains on Europa reveal a compressive past. *Icarus, 343,* 113709. https://doi.org/10.1016/j.icarus.2020.113709.

Lesage, E., Massol, H., Howell, S. M., & Schmidt, F. (2022). Simulation of freezing cryomagma reservoirs in viscoelastic ice shells. *The Planetary Science Journal, 3*(7), 170.

Lesage, E., Massol, H., & Schmidt, F. (2020). Cryomagma ascent on Europa. *Icarus, 335,* 113369.

Lesage, E., Schmidt, F., Andrieu, F., & Massol, H. (2021). Constraints on effusive cryovolcanic eruptions on Europa using topography obtained from Galileo images. *Icarus, 361,* 114373.

Li, L., Jäggi, A., Wang, Y., Blanc, M., Zong, Q., Andre, N., Wang, C., Dandouras, I., Louarn, P., Hestroffer, D., Arnold, D., Guo, L., Wang, L., Blanc, M., Mousis, O., Reme, H., Vernazza, P., Zhang, L., & Desprats, W. (2021). *Gan De: A mission to search for the origins and workings of the Jupiter system. 43* (p. 253).

Ligier, N., Calvin, W., Carter, J., Poulet, F., Paranicas, C., & Snodgrass, C. (2020). *New insights into Callisto's surface composition with the ground-based near-infrared imaging spectrometer SINFONI of the VLT* (p. 1959).

Ligier, N., Paranicas, C., Carter, J., Poulet, F., Calvin, W., Nordheim, T., Snodgrass, C., & Ferellec, L. (2019). Surface composition and properties of Ganymede: Updates from ground-based observations with the near-infrared imaging spectrometer SINFONI/VLT/ESO. *Icarus, 333,* 496–515.

Ligier, N., Poulet, F., Carter, J., Brunetto, R., & Gourgeot, F. (2016). VLT/SINFONI observations of Europa: New insights into the surface composition. *The Astronomical Journal, 151*(6), 163.

Lobo, A. H., Thompson, A. F., Vance, S. D., & Tharimena, S. (2021). A pole-to-equator ocean overturning circulation on Enceladus. *Nature Geoscience, 14*(4), 185–189.

Lucchita, B. K. (1980). Grooved terrain on Ganymede. *Icarus, 44*(2), 481–501.

Lunine, J. I., & Stevenson, D. J. (1982). Formation of the Galilean satellites in a gaseous nebula. *Icarus, 52*(1), 14–39.

Malhotra, R. (1991). Tidal origin of the Laplace resonance and the resurfacing of Ganymede. *Icarus, 94*(2), 399–412.

Manga, M., & Michaut, C. (2017). Formation of lenticulae on Europa by saucer-shaped sills. *Icarus, 286,* 261–269. https://doi.org/10.1016/j.icarus.2016.10.009.

Marion, G. M., & Kargel, J. S. (2007). *Cold aqueous planetary geochemistry with FREZCHEM: From modeling to the search for life at the limits.* Springer Science & Business Media.

Martin, W., Baross, J., Kelley, D., & Russell, M. J. (2008). Hydrothermal vents and the origin of life. *Nature Reviews Microbiology*, 6(11), 805–814.

Maus, S., Müller, S., Büttner, J., Brütsch, S., Huthwelker, T., Schwikowski, M., Enzmann, F., & Vähätolo, A. (2011). Ion fractionation in young sea ice from Kongsfjorden. *Svalbard. Annals of Glaciology*, 52(57), 301–310.

McCord, T.a., Carlson, R., Smythe, W., Hansen, G., Clark, R., Hibbitts, C., Fanale, F., Granahan, J., Segura, M., Matson, D., et al. (1997). Organics and other molecules in the surfaces of Callisto and Ganymede. *Science*, 278(5336), 271–275.

McCord, T. B., Hansen, G. B., Matson, D. L., Johnson, T. V., Crowley, J. K., Fanale, F. P., Carlson, R. W., Smythe, W. D., Martin, P. D., & Hibbitts, C. A. (1999). Hydrated salt minerals on Europa's surface from the Galileo near-infrared mapping spectrometer (NIMS) investigation. *Journal of Geophysical Research: Planets*, 104(E5), 11827–11851.

McKinnon, W. B. (1999). Convective instability in Europa's floating ice shell. *Geophysical Research Letters*, 26(7), 951–954.

McPhee, M. (2008). *Air-ice-ocean interaction: Turbulent ocean boundary layer exchange processes*. Springer Science & Business Media.

Melosh, H., & Turtle, E. (2004). Ridges on Europa: Origin by incremental ice-wedging. *Lunar and Planetary Science Conference*, 2029.

Melwani Daswani, M., Vance, S. D., Mayne, M. J., & Glein, C. R. (2021). A metamorphic origin for Europa's ocean. *Geophysical Research Letters*, 48(18), e2021GL094143.

Michaut, C., & Manga, M. (2014). Domes, pits, and small chaos on Europa produced by water sills. *Journal of Geophysical Research: Planets*, 119(3), 550–573.

Mills, M. M., Pappalardo, R. T., Panning, M. P., Leonard, E. J., & Howell, S. M. (2023). Moonquake-triggered mass wasting processes on icy satellites. *Icarus*, 399, 115534.

Mogan, S. R. C., Tucker, O. J., Johnson, R. E., Vorburger, A., Galli, A., Marchand, B., Tafuni, A., Kumar, S., Sahin, I., & Sreenivasan, K. R. (2021). A tenuous, collisional atmosphere on Callisto. *Icarus*, 368, 114597.

Moore, J. M., Asphaug, E., Morrison, D., Spencer, J. R., Chapman, C. R., Bierhaus, B., Sullivan, R. J., Chuang, F. C., Klemaszewski, J. E., & Greeley, R. (1999). Mass movement and landform degradation on the icy Galilean satellites: Results of the Galileo nominal mission. *Icarus*, 140(2), 294–312.

Moore, J. M., Asphaug, E., Sullivan, R. J., Klemaszewski, J. E., Bender, K. C., Greeley, R., Geissler, P. E., McEwen, A. S., Turtle, E. P., & Phillips, C. B. (1998). Large impact features on Europa: Results of the Galileo nominal mission. *Icarus*, 135(1), 127–145.

Moore, J. M., & Malin, M. C. (1988). Dome craters on Ganymede. *Geophysical Research Letters*, 15(3), 225–228.

Nagel, K., Breuer, D., & Spohn, T. (2004). A model for the interior structure, evolution, and differentiation of Callisto. *Icarus*, 169(2), 402–412. https://doi.org/10.1016/j.icarus.2003.12.019.

National Academies of Sciences, Engineering, and Medicine. (2022). Origins, worlds, and life: A decadal strategy for planetary science and astrobiology 2023–2032.

Naseem, M., Neveu, M., Howell, S., Lesage, E., Daswani, M. M., & Vance, S.D. (2023). Salt distribution from freezing intrusions in ice shells on ocean worlds: Application to Europa. *The Planetary Science Journal*, 4(181) (22pp), Published by the American Astronomical Society.

Nimmo, F. (2004). Stresses generated in cooling viscoelastic ice shells: Application to Europa. *Journal of Geophysical Research: Planets*, 109(E12).

Nimmo, F., & Gaidos, E. (2002). Strike-slip motion and double ridge formation on Europa. *Journal of Geophysical Research: Planets*, 107, 5-1-5-8. https://doi.org/10.1029/2000JE001476.

Nimmo, F., & Manga, M. (2002). Causes, characteristics and consequences of convective diapirism on Europa. *Geophysical Research Letters*, 29(23), 24-1-24-4.

Nimmo, F., & Manga, M. (2009). Geodynamics of Europa's icy shell. In *Europa* (pp. 381–404). University of Arizona Press.

Nimmo, F., Giese, B., & Pappalardo, R. T. (2003). Estimates of Europa's ice shell thickness from elastically-supported topography. *Geophysical Research Letters*, *30*(5).

Nimmo, F., & Spencer, J. R. (2015). Powering Triton's recent geological activity by obliquity tides: Implications for Pluto geology. *Icarus*, *246*, 2–10.

Nordheim, T., Paranicas, C., & Hand, K. P. (2017). Europa's surface radiation environment and considerations for in-situ sampling and biosignature detection. *AGU Fall*, *52*. Meeting Abstracts http://adsabs.harvard.edu/abs/2017AGUFM.P52B..03N.

Oren, A. (2011). Thermodynamic limits to microbial life at high salt concentrations. *Environmental Microbiology*, *13*(8), 1908–1923.

Paganini, L., Villanueva, G. L., Roth, L., Mandell, A. M., Hurford, T. A., Retherford, K. D., & Mumma, M. J. (2020). A measurement of water vapour amid a largely quiescent environment on Europa. *Nature Astronomy*, *4*(3), 266–272.

Pappalardo, R. T., Belton, M. J. S., Breneman, H. H., Carr, M. H., Chapman, C. R., Collins, G. C., Denk, T., Fagents, S., Geissler, P. E., Giese, B., Greeley, R., Greenberg, R., Head, J. W., Helfenstein, P., Hoppa, G., Kadel, S. D., Klaasen, K. P., Klemaszewski, J. E., Magee, K., ... Williams, K. K. (1999). Does Europa have a subsurface ocean? Evaluation of the geological evidence. *Journal of Geophysical Research: Planets (1991–2012)*, *104*(E10), 24015–24055. 10.1029/1998JE000628.

Pappalardo, R. T., & Collins, G. C. (2005). Strained craters on Ganymede. *Journal of Structural Geology*, *27*(5), 827–838. https://doi.org/10.1016/j.jsg.2004.11.010.

Pappalardo, R. T., Collins, G. C., Head, J. W., Helfenstein, P., McCord, T. B., Moore, J. M., Prockter, L. M., Schenk, P. M., & Spencer, J. R. (2004). Geology of Ganymede. In F. D. Bagenal, T. E. Dowling, & W. B. McKinnon (Eds.), *Jupiter* (pp. 363–396). Cambridge University Press.

Pappalardo, R. T., Head, J. W., Collins, G. C., Kirk, R. L., Neukum, G., Oberst, J., Giese, B., Greeley, R., Chapman, C. R., Helfenstein, P., Moore, J. M., McEwen, A., Tufts, B. R., Senske, D. A., Breneman, H. H., & Klaasen, K. (1998a). Grooved Terrain on Ganymede: First Results from Galileo High-Resolution Imaging. *Icarus*, *135*, 276–302. https://doi.org/10.1006/icar.1998.5966.

Pappalardo, R. T., Head, J. W., Greeley, R., Sullivan, R. J., Pilcher, C., Schubert, G., Moore, W. B., Carr, M. H., Moore, J. M., Belton, M. J. S., & Goldsby, D. L. (1998b). Geological evidence for solid-state convection in Europa's ice shell. *Nature*, *391*(6665), 365–368. https://doi.org/10.1038/34862.

Passey, Q. R., & Shoemaker, E. M. (1982). Craters and basins on Ganymede and Callisto-morphological indicators of crustal evolution. *Satellites of Jupiter*, 379–434.

Patel, J. G., Pappalardo, R. T., Head, J. W., Collins, G. C., Hiesinger, H., & Sun, J. (1999). Topographic wavelengths of Ganymede groove lanes from Fourier analysis of Galileo images. *Journal of Geophysical Research: Planets*, *104*(E10), 24057–24074.

Patterson, G. W., Collins, G. C., Head, J. W., Pappalardo, R. T., Prockter, L. M., Lucchitta, B. K., & Kay, J. P. (2010). Global geological mapping of Ganymede. *Icarus*, *207*(2), 845–867. https://doi.org/10.1016/j.icarus.2009.11.035.

Peddinti, D. A., & McNamara, A. K. (2019). Dynamical investigation of a thickening ice-shell: Implications for the icy moon Europa. *Icarus*. https://doi.org/10.1016/j.icarus.2019.03.037.

Poirier, J. P., Sotin, C., & Peyronneau, J. (1981). Viscosity of high-pressure ice VI and evolution and dynamics of Ganymede. *Nature*, *292*(5820), 225–227. https://doi.org/10.1038/292225a0.

Prockter, L. M., Antman, A. M., Pappalardo, R. T., Head, J. W., & Collins, G. C. (1999). Europa: Stratigraphy and geological history of the anti-Jovian region from Galileo E14 solid-state imaging data. *Journal of Geophysical Research: Planets*, *104*(E7), 16531–16540.

Prockter, L. M., Head, J. W., Pappalardo, R. T., Sullivan, R. J., Clifton, A. E., Giese, B., Wagner, R., & Neukum, G. (2002). Morphology of Europan bands at high resolution: A mid-ocean ridge-type rift mechanism. *Journal of Geophysical Research: Planets*, *107*, 4-1-4-26. https://doi.org/10.1029/2000JE001458.

Prockter, L. M., Pappalardo, R. T., & Head, J. W., III. (2000). Strike-slip duplexing on Jupiter's icy moon Europa. *Journal of Geophysical Research: Planets*, *105*(E4), 9483–9488.

Prockter, L. M., & Patterson, G. W. (2009). Morphology and evolution of Europa's ridges and bands. In R. T. Pappalardo, W. B. McKinnon, & K. Khurana (Eds.), *Europa* (p. 589). University of Arizona Press.

Ross, M. N., & Schubert, G. (1987). Tidal heating in an internal ocean model of Europa. *Nature*, *325*(6100), 133–134.

Roth, L. (2021). A stable H2O atmosphere on Europa's trailing hemisphere from HST images. *Geophysical Research Letters*, *48*(20), e2021GL094289.

Roth, L., Ivchenko, N., Gladstone, G. R., Saur, J., Grodent, D., Bonfond, B., Molyneux, P. M., & Retherford, K. D. (2021). A sublimated water atmosphere on Ganymede detected from Hubble Space Telescope observations. *Nature Astronomy*, *5*(10), 1043–1051.

Roth, L., Saur, J., Retherford, K. D., Strobel, D. F., Feldman, P. D., McGrath, M. A., & Nimmo, F. (2014). Transient water vapor at Europa's south pole. *Science*, *343*(6167), 171–174.

Roth, L., Saur, J., Retherford, K. D., Strobel, D. F., Feldman, P. D., McGrath, M. A., Spencer, J. R., Blöcker, A., & Ivchenko, N. (2016). Europa's far ultraviolet oxygen aurora from a comprehensive set of HST observations. *Journal of Geophysical Research: Space Physics*, *121*(3), 2143–2170.

Saur, J., Duling, S., Roth, L., Jia, X., Strobel, D. F., Feldman, P. D., Christensen, U. R., Retherford, K. D., McGrath, M. A., Musacchio, F., et al. (2015). The search for a subsurface ocean in Ganymede with Hubble Space Telescope observations of its auroral ovals. *Journal of Geophysical Research*, *120*(3), 1715–1737.

Schenk, P. M. (1993). Central pit and dome craters: Exposing the interiors of Ganymede and Callisto. *Journal of Geophysical Research: Planets*, *98*(E4), 7475–7498.

Schenk, P. M. (1995). The geology of Callisto. *Journal of Geophysical Research: Planets*, *100*(E9), 19023–19040.

Schenk, P. M. (2002). Thickness constraints on the icy shells of the Galilean satellites from a comparison of crater shapes. *Nature*, *417*(6887), 419.

Schenk, P. M., Chapman, C. R., Zahnle, K., & Moore, J. M. (2004). Ages and interiors: The cratering record of the Galilean satellites. *Jupiter: The Planet, Satellites and Magnetosphere*, *2*, 427.

Schenk, P. M., & McKinnon, W. B. (1987). Ring geometry on Ganymede and Callisto. *Icarus*, *72*(1), 209–234.

Schenk, P. M., & McKinnon, W. B. (1989). Fault offsets and lateral crustal movement on Europa: Evidence for a mobile ice shell. *Icarus*, *79*(1), 75–100.

Schenk, P. M., & Turtle, E. P. (2009). Europa's impact craters: Probes of the icy shell. *Europa*, 181–198.

Schmidt, B. E., Blankenship, D. D., Patterson, G. W., & Schenk, P. M. (2011). Active formation of 'chaos terrain' over shallow subsurface water on Europa. *Nature*, *479*(7374), 502–505. https://doi.org/10.1038/nature10608.

Schubert, G., Spohn, T., & Reynolds, R. T. (1986). Thermal histories, compositions and internal structures of the moons of the solar system. In *Iau Colloq. 77: Some Background about Satellites* (pp. 224–292).

Shirley, J. H., Dalton, J. B., III, Prockter, L. M., & Kamp, L. W. (2010). Europa's ridged plains and smooth low albedo plains: Distinctive compositions and compositional gradients at the leading side–trailing side boundary. *Icarus*, *210*(1), 358–384.

Shoemaker, E. M., Lucchitta, B. K., Wilhelms, D. E., Plescia, J. B., & Squyres, S. W. (1982). The geology of Ganymede. *Satellites of Jupiter*, 435–520.

Showman, A. P., & Han, L. (2004). Numerical simulations of convection in Europa's ice shell: Implications for surface features. *Journal of Geophysical Research: Planets*, *109*(E1).

Showman, A. P., & Malhotra, R. (1997). Tidal evolution into the Laplace resonance and the resurfacing of Ganymede. *Icarus*, *127*(1), 93–111.

Soderlund, K. M. (2019). Ocean dynamics of outer solar system satellites. *Geophysical Research Letters*, *46*(15), 8700–8710.

Soderlund, K. M., Schmidt, B., Wicht, J., & Blankenship, D. (2014). Ocean-driven heating of Europa's icy shell at low latitudes. *Nature Geoscience*, *7*(1), 16–19.

Sotin, C., Head, J. W., III, & Tobie, G. (2002). Europa: Tidal heating of upwelling thermal plumes and the origin of lenticulae and chaos melting. *Geophysical Research Letters*, *29*(8). 74-1–74-4.

FIG. 2

Cassini-Huygens spacecraft undergoing vibration and thermal testing at NASA's Jet Propulsion Laboratory (JPL), prior to launch.

Image ID: IMG001942; Credit: NASA/JPL-Caltech; https://solarsystem.nasa.gov/resources/12942/
cassini-saturn-probe-undergoes-preflight-testing/.

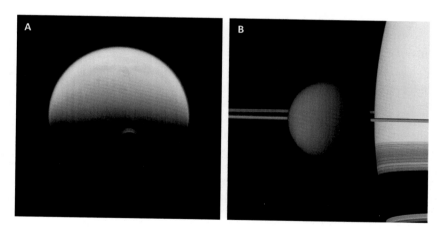

FIG. 3

Images of Titan from the Cassini ISS's narrow-angle camera. Titan's *orange color* is from a photochemical haze that prohibits the view of its surface in the visible spectrum. (a) A view of Titan's trailing hemisphere taken with the Cassini narrow-angle camera in 2013 from a distance of ~1.5 million km from Titan. Titan's south polar vortex is visible near the terminator. (b) Titan is seen in the foreground with Saturn on the right and Saturn's rings behind and center. This image was taken in 2011 at a distance of ~2.3 million km from Titan.

Panel (a): Image ID: PIA17177; Credit: NASA/JPL-Caltech/Space Science Institute; https://photojournal.jpl.nasa.gov/catalog/
PIA17177; panel (b): image ID: PIA14909; Credit: NASA/JPL-Caltech/Space Science Institute; https://photojournal.jpl.nasa.gov/
catalog/PIA14909.

0.35–1.7 km resolution) with several observation modes, most notably the Synthetic Aperture Radar (SAR) (Stofan et al., 2007). Together, these instruments could detect spatial variation and temporal changes in surface composition and material roughness, along with atmospheric composition.

Other instruments onboard Cassini that we refer to in this chapter include the Imagining Science Subsystem (ISS), Composite Infrared Spectrometer (CIRS), Ion and Neutral Mass Spectrometer (INMS), and Ultraviolet Imaging Spectrograph (UVIS). The ISS consisted of wide- and narrow-angle cameras, used for scientific data collection and optical navigation (Jaffe & Herrell, 1997). CIRS measured planetary radiation using three detectors (polarizing and Michelson interferometers) spanning the wavelengths 1 mm–7 μm (Jaffe & Herrell, 1997). The INMS was a quadrupole mass spectrometer designed to measure the chemical, elemental, and isotopic composition of neutral particles and low-energy ions throughout the Saturnian system (including Titan's atmosphere) (Jaffe & Herrell, 1997). Finally, UVIS measured ultraviolet emissions through two spectroscopic channels (extreme ultraviolet channel, and far ultraviolet channel), and had a measured brightness sensitivity level of 0.001 to several thousand Rayleighs (Jaffe & Herrell, 1997).

In 2004, Cassini-Huygens completed its orbital insertion at Saturn, and in 2005 the Huygens probe landed on the surface of Titan. This was the first time a robotic probe landed on a planetary body in the outer solar system and during its few hours of operations Huygens provided the first glimpse of Titan's surface (Fig. 4). This was the start of an incredible 13-year journey that made groundbreaking discoveries about the Saturnian system. When Cassini was nearly out of fuel, the decision was made to end Cassini's reign by plunging it into Saturn's atmosphere in 2017. This spectacle was called the "Cassini Grand Finale." NASA has plans to return to the Saturnian system, with Titan and Enceladus as their focus (see Section 10).

3 Titan

3.1 Overview

We begin our journey through the Saturnian system with Saturn's largest and most complex moon, Titan (Fig. 3). Titan was first discovered by Christiaan Huygens in 1655. Titan is the second-largest moon in our solar system (second to Jupiter's, Ganymede), and is larger than the planet Mercury, yet only ~40% of Mercury's mass. Titan orbits Saturn at a distance of over 1 million kilometers and, like Earth's moon, is tidally locked in a synchronous rotation with Saturn. It makes one full orbit around Saturn every ~16 Earth days and has the same length year as Saturn, ~29 Earth years. Titan orbits roughly along Saturn's equatorial plane, with an inclination of ~0.35° in relation to Saturn's equator, and a tilt of 27° to the Sun.

Titan has a subsurface global liquid water ocean capped by a water-ice crust (Griffith et al., 2003), similar to other icy moons (such as Enceladus and Jupiter's moon Europa; see Howell et al., 2024, pp. 283–314). Titan is also unique in the solar system as the only moon with a significant atmosphere (1.5 bar pressure; Fulchignoni et al., 2005), and the only planetary body other than Earth known to host liquid on its surface. Titan has a frigid surface temperature (~94 K; Fulchignoni et al., 2005), and as a result, water is in the solid phase on the surface and hydrocarbons such as methane and ethane are liquid. Titan has a hydrocarbon-based hydrologic cycle (Mitchell & Lora, 2016) and hosts stable polar hydrocarbon lakes and seas. Complex organic chemistry occurs in Titan's upper atmosphere (Section 3.5), leading to the presence of more complex molecules and ice materials than any other moon. Titan also has a limited surface temperature range (~5 K) because of Titan's slow rotation and hazy atmosphere that includes a relatively large methane component (~5% methane).

In this section, we begin by discussing Titan's interior structure (Section 3.2; water-ice exterior, liquid water interior, the potential for a high-pressure ice layer and clathrate layer), and then move into

FIG. 4

An image of Titan's surface returned by the Huygens probe in 2005. Rounded cobbles (likely composed of water ice) can be seen in the foreground. The rounded nature of the objects is evidence of erosion by a fluid.

Image ID: PIA07232; Credit: NASA/JPL/ESA/University of Arizona; https://photojournal.jpl.nasa.gov/catalog/PIA07232.

a discussion about the potential composition of Titan's surface from Cassini observations (Section 3.3). Next, we discuss the geologic states observed on Titan's water-ice crust (Section 3.3; e.g., labyrinth terrain, potential tectonic activity, and cryovolcanic flows), Titan's liquid hydrocarbon lakes (which occupy pits and depressions within the water-ice crust), and the potential for transient ice formation on Titan's surface (Section 3.4). We end with a discussion on ice in Titan's atmosphere and clouds (Section 3.5).

3.2 Titan's interior and subsurface ocean

Titan has a diameter of ~5,150 km, with a rocky, silicate core spanning ~4,000 km in diameter (see Fig. 5). Between its organic-rich, water-ice crust (Griffith et al., 2003) and core, lies a subsurface global ocean, identified through obliquity and gravitational tide measurements (Baland et al., 2014; Mitri et al., 2014; Nimmo & Pappalardo, 2016). The ice crustal thickness and interior ocean depth, however,

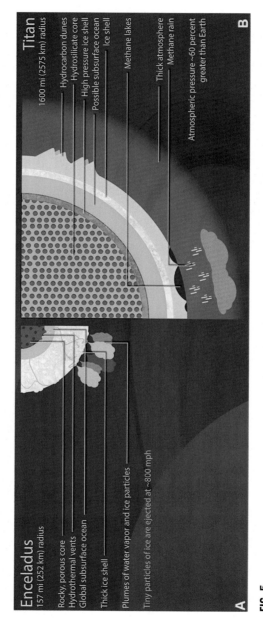

FIG. 5

Cross-section illustration of the interior of (a) Enceladus and (b) Titan. Note that the moons and features are not to scale.

Source: NASA: J Olmsted (STScI); modified https://solarsystem.nasa.gov/resources/2475/moons-active-worlds/.

are not fully constrained: 50–200 and 500–700 km, respectively (MacKenzie et al., 2021a; Nimmo & Pappalardo, 2016, and references therein), and the presence of circulation patterns in Titan's subsurface ocean is currently unknown (Lefevre et al., 2014).

Liquid and solid water are theoretically stable in Titan's interior ocean and are likely segregated into three layers: (1) low-pressure phase ice Ih, (2) liquid water, and (3) high-pressure polymorphs ice V and ice VI (Choukroun et al., 2010; Fortes, 2012; Hobbs, 2010). In the hydrosphere, just above the core, the pressure conditions may allow a dense, high-pressure ice layer to form (e.g., ice V and VI) (Journaux et al., 2020; Kalousová & Sotin, 2020). However, moment of inertia measurements (0.33–0.34) suggest that Titan is not fully differentiated, and therefore the presence of an interior ice layer (or lack thereof) remains unknown. The presence of a high-pressure ice layer may inhibit exchange between the ocean and the core, limiting the transfer of organics from the core into the interior ocean. However, convective mechanisms may assist in the movement of material through the ice layer, if present (Choblet et al., 2017; Sotin et al., 2021).

It is possible that a water-ice layer may decouple during the freezing process of the interior ocean, leaving behind a more ammonia-rich liquid (Jaumann et al., 2009a, 2009b; Sotin et al., 2009). The presence of salts (up to ~1 wt%; Tobie et al., 2005) and a few percent ammonia in Titan's subsurface ocean may help explain the ocean's high density (Mitri et al., 2014), although recent studies suggest the presence of liquid ammonia may not be a main component as salts may be sufficient to explain the high density and gravity field measurements. Tidal heating or the presence of a methane clathrate layer may explain the ocean's liquid state (Sotin et al., 2021).

Titan's outer water-ice shell may be capped below by a low-conductivity methane clathrate layer that is theoretically stable under Titan's surface and subsurface conditions (Kalousová & Sotin, 2020; Thomas et al., 2008). This clathrate layer acts as an insulator, with the thermal conductivity of the clathrate layer being more than an order of magnitude lower than that of water ice, which is important to support a liquid ocean and delay ocean crystallization (Carnahan et al., 2022). Over time, methane gas in methane clathrates can be replaced by ethane. Methane and ethane clathrates could potentially be destabilized if the lithosphere becomes weakened by faulting (e.g., impacts or tectonics), or upwelling of subsurface plumes (e.g., tidal heating; Sotin et al., 2005). The destabilization of the clathrate layer would lead to significant outgassing of methane and ethane, which could replenish methane gas in Titan's atmosphere (Kalousová & Sotin, 2020). Recent work suggests that Titan's core may be rich in organic material, leading to the hypothesis that Titan's atmospheric methane may be the result of outgassing from Titan's interior (Néri et al., 2020). Isotopic measurements of noble gases are necessary to explore the evolution and composition of Titan's ocean and outgassing from the interior (MacKenzie et al., 2021b).

3.3 Geologic features and ice structures Titan's surface

3.3.1 Surface composition and Cassini VIMS 5-μm bright features

Even though Titan's crust is predominately water ice, the exact composition of Titan's surface remains largely unknown. Cassini VIMS obtained spectral data of Titan's surface allowing us to identify some degree of compositional variation of Titan's surface. VIMS observations suggest two general categories of surface materials: organic-rich (organics originating from the atmosphere—see Section 3.5) and water-ice-rich (Barnes et al., 2008; Brossier et al., 2018; Griffith et al., 2003, 2019; Rodriguez et al., 2014; Soderblom et al., 2009; Solomonidou et al., 2018). Surface variations have been categorized into spectral classes based on VIMS spectral end-members (Jaumann et al., 2008a; Soderblom et al., 2007). At least two spectrally distinct types of dark material and two types of bright material have been identified. The dark material deemed "blue" (water-ice-rich) and "brown" (organic-rich) are based on false-color imaging.

The "5-µm bright unit" comprises features that are bright in the 5-µm VIMS window and are spectrally distinct from other bright materials on Titan's surface (Barnes et al., 2005; Jaumann et al., 2009b). VIMS 5-µm bright features can be observed as "bathtub rings" surrounding partially filled and dry lake beds (Barnes et al., 2009). These features are believed to be evaporites—a compositionally unique solid material that is left behind after a lake has evaporated (Barnes et al., 2009; MacKenzie et al., 2014). Although the composition of the material is currently unknown, many laboratory experiments and modeling efforts have aimed at understanding the chemistry of such evaporites (Cable et al., 2019, 2021b; Cordier et al., 2016; Czaplinski et al., 2019, 2020). One hypothesis is that the bright features are co-crystals, formed by interactions with a hydrocarbon solvent (liquid methane and/or ethane), and some trace species, such as benzene, butane, ethylene, and acetylene. Co-crystals have been confirmed to form in the laboratory, such as acetylene-butane (Cable et al., 2019) and acetylene-benzene (Czaplinski et al., 2020).

Another large 5-µm bright feature was discovered at 80°W and 20°S (Barnes et al., 2005). This feature is hypothesized to be carbon dioxide ice, sourced from either eroded underlying layers of carbon dioxide or more recent overlying deposits from cryovolcanism (Barnes et al., 2005).

3.3.2 Equatorial region

In Titan's equatorial region, ISS imaging revealed bright and dark regions and streaks generally trending west-to-east (Porco et al., 2005a). Many western boundaries of bright patches within larger dark regions are well defined and sometimes linear, while eastern boundaries of bright regions are more diffuse. One interpretation is that the dark material was deposited by west-to-east flow in the atmosphere or in surface liquids. A second interpretation is that the bright material is the result of erosion, and/or is being redeposited by west-to-east cryogenic flows. The linearity of the western boundaries of bright patches may also suggest tectonic modifications (Porco et al., 2005a).

The equatorial dark regions also consist of longitudinal dune fields (Lorenz et al., 2006), with bright features at the boundaries. These smaller, bright, features are called faculae (<500 km; Jaumann et al., 2009b). The low reflectivity spectral signatures (both in the near-infrared and microwave) of the surface in the equatorial region correspond to organic-rich material, similar to that of the dunes. The optimum particle size for grain mobility on the surface (i.e., saltation) is similar to terrestrial materials at ~250 µm (Greeley & Iversen, 1985). These may be produced by the breakdown of coarser materials from impact ejecta or fluvial sediments, or from the aggregation of smaller, finer-grained material, such as atmospheric haze particles (Jaumann et al., 2009b). The movement of the fluids on Titan's surface has also created fluvial valley networks (Poggiali et al., 2016) (Fig. 6a) and rounded cobbles (Fig. 4) observed by the Huygens probe in the equatorial region (Burr et al., 2013; Jaumann et al., 2008a; Langhans et al., 2012; Malaska et al., 2011; Radebaugh et al., 2016). The rounded cobbles are believed to be water ice that has been shaped by past fluid transport.

Cassini ISS and radar also identified circular or annular features, some accompanied by bright rings. These geologic features have been identified as impact craters and can be found within the Shangri-La region and other equatorial dark regions (e.g., Selk crater; Fig. 6b). Such craters with bright halos suggest the excavation of fresh subsurface icy material. Two of the largest are named

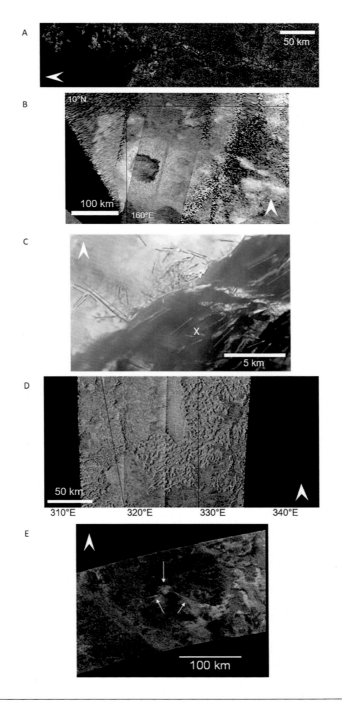

FIG. 6

See figure legend in next page.

(Continued)

FIG. 6

Geologic features on Titan's water-ice crust. *White arrows* point north in all images. (a) Dendritic fluvial patterns from liquid hydrocarbons, carved into the water-ice crust. (b) Selk impact crater in Titan's equatorial region in Cassini SAR data. Dunes can be seen at the bottom center of the image. (c) An example of tectonized terrain on Titan in IR. Inferred tectonic patterns are represented by *blue lines, with green arrows* representing the inferred surface drainage directions, and the drainage divide represented by the *red dotted line*. The Huygens landing site is indicated by the *white X*. (d) Sikun Labyrinth as imaged by Cassini SAR showing intersecting valleys and plateaus. The vertical seam in the image is a SAR beam artifact. (e) SAR observation of Ganesa Macula, a hypothesized cryovolcanic dome or shield. Arrows highlight Ganesa's central caldera and channels.

Panel (a): Radar image at Titan's north polar region adapted from Image ID: PIA16197; Credit: NASA/JPL-Caltech/ASI; https://photojournal.jpl.nasa.gov/catalog/PIA16197; panel (b): image via JMARS; panel (c): reprinted from Soderblom, L. A., Tomasko, M. G., Archinal, B. A., Becker, T. L., Bushroe, M. W., Cook, D. A., ..., Smith, P. H. (2007). Topography and geomorphology of the Huygens landing site on Titan. Planetary and Space Science, 55(13), 2015–2024. Copyright (2007) with permission from Elsevier; panel (d): Image via JMARS; panel (e): reprinted from Lopes, R. M. C., et al., Cryovolcanic features on Titan's surface as revealed by the Cassini Titan Radar Mapper. Icarus, 186(2), 395–412. Copyright (2007) with permission from Elsevier.

Menrva crater (450 km diameter; Elachi et al., 2006) and Sinlap crater (80 km diameter; Le Mouélic et al., 2008). Sinlap has a raised rim, flat floor, and notable impact ejecta blanket seen as radar dark streaks emanating from the crater. Sinlap's main ejecta blanket is enriched in water ice compared to its surroundings (Le Mouélic et al., 2008). Interestingly, Titan's relatively few craters seemingly lack topographic relaxation and central pits which have been observed at impact craters on other icy satellites (such as icy satellites of Jupiter) and rocky planets (Wood et al., 2010). The low number density of impact craters identified on Titan's surface suggests that Titan has a relatively young surface where resurfacing has taken place in geologically recent times (Hedgepeth et al., 2020; Neish & Lorenz, 2012).

3.3.3 Tectonics and labyrinth terrain

Tectonism on Titan is still not fully understood, but is expected because of Titan's radiogenic activity, eccentricity, and the presence of a subsurface ocean (Burkhard et al., 2022; Matteoni et al., 2020). There are features that are hypothesized to be tectonic in origin and appear partially degraded and embayed by surrounding plains (Fig. 6c). They include linear and ridge-like formations (e.g., chains of hills) and branching lineae (Lopes et al., 2007a). Some large lineae have north-south orientations and may have been created by compressional folding or extensional rifting (Lopes & Solomonidou, 2014; Mitri et al., 2010; Radebaugh et al., 2007), though additional imaging is needed to explore this hypothesis.

Labyrinth terrains (Fig. 6d) are comprised of circular-to-rectangular plateaus separated by steep-sided, ice-walled canyons that form maze-like patterns. The bulk composition of Titan's labyrinth terrains is estimated to be 14%–35% of the total surface organics (Malaska et al., 2020). Currently, 197 individual labyrinth terrain regions have been mapped, covering about 1.5% of Titan's total surface area, roughly the same as Titan's hydrocarbon lake coverage (Malaska et al., 2020). Four subtypes of labyrinth terrain have been defined including valleyed, polygonal, finely dissected, and Kronin (Malaska et al., 2020). There are subtle differences in labyrinthian textures constituting subtypes where valleyed is the most common type and Kronin is considered an outlier when compared to the other subtypes (Malaska et al., 2020). The origin of these large ice mazes is still unknown, though tectonism and/or dissolution may play an important role. Despite material and chemical differences between Titan's icy crust and Earth's rocky surface, many features familiar to Earth are found across Titan's surface.

3.3.4 Cryovolcanic flows

Cryovolcanism is similar to volcanism here on Earth, but instead of magma, cryovolcanoes eject aqueous solutions (e.g., liquid water or water-ammonia) or partially crystallized slurries from the planetary body's interior (Lopes et al., 2013). For Titan, cryovolcanic flows would likely be sourced from its subsurface ocean, with a composition of predominately liquid water with up to a few percent ammonia (Tobie et al., 2005) (see Section 3.2 for details on Titan's interior).

Surprisingly, active cryovolcanic activity has not been observed on Titan. Moreover, much evidence of past cryovolcanism is likely eroded or buried by organics, given that Titan has a relatively young surface (Jaumann et al., 2009b). Nonetheless, some of Titan's surface features (circular, bright, flows) have been hypothesized to be evidence of past cryovolcanic activity and are discussed below.

Ganesa Macula (Fig. 6e), Tortola Facula, and smaller features in Tui Regio have been hypothesized to be cryovolcanic in origin, based on morphology and brightness observations from Cassini VIMS (Barnes et al., 2006; Sotin et al., 2005). Hotei Arcus is exceptionally bright in the 5-μm wavelength and bears some resemblance to Tui Regio (Soderblom et al., 2009). Lobate features in Hotei Arcus may be solidified cryolava (Wall et al., 2009).

Another large circular feature named Ganesa Macula (180 km), is thought to be cryovolcanic in origin (Fig. 6e) (Lopes et al., 2007a). Both diffuse flows and caldera-like structures have been identified in the surrounding area. Fluvial channels have formed near Ganesa, but these channels have been observed without any association with cryovolcanic features (Lorenz et al., 2008a). Similar caldera structures and fluvial channels have been identified near Rohe Fluctus and Ara Fluctus. These flows are bright in Cassini VIMS and radar, which is consistent with the presence of fine-grained carbon dioxide ice (Le Corre et al., 2009). However, the composition and origin of these materials are largely unknown.

3.4 Titan's liquid hydrocarbons and potential for transient freezing

3.4.1 Titan's lakes and seas

The Cassini mission entered the Saturn system during Titan's southern hemisphere summer in 2005 when the south pole was illuminated by the Sun. The first images of the south pole showed a single lake—Ontario Lacus (Fig. 7). The northern hemisphere remained in darkness until Titan's spring equinox in August 2009. Over the next few years, Cassini radar revealed many filled lakes in the north polar region (Fig. 8), including hundreds of small lakes known as the north polar "lake district," as well as larger seas, such as Ligia Mare, Kraken Mare, and Punga Mare (e.g., Birch et al., 2016; Hayes et al., 2008; Lopes et al., 2007b; MacKenzie et al., 2019; Mitri et al., 2007; Stofan et al., 2007). The largest of these "seas" is Kraken Mare, measuring an area of ~400,000 km^2 and exceeding 100 m in depth (Poggiali et al., 2020). Meanwhile, Titan's second-largest sea, Ligeia Mare, has a surface area of ~130,000 km^2 and a depth of ~170 m (Mastrogiuseppe et al., 2014). Some of these seas are as large as our Great Lakes here on Earth. Cassini radar also revealed partially filled lakes and dry lake beds, with the vast majority of empty lake basins in the south polar region.

FIG. 7

Titan's south pole presented as a mosaic of false-color images from Cassini radar (polar stereographic projection). This image highlights Titan's surface (tan) and Titan's south polar lake, Ontario Lacus (*black/blue*). Notice the paucity of filled lakes compared to the north polar region shown in Fig. 8.

Credit: NASA/JPL-Caltech/USGS; reprinted from Rodriguez, S., et al., Science goals and new mission concepts for future exploration of Titan's atmosphere, geology, and habitability: Titan POlar scout/orbitEr and in situ lake lander and DrONe explorer (POSEIDON). Experimental Astronomy, 54, 911–973. Copyright (2022) modified https://creativecommons.org/licenses/by/4.0/.

Titan's surface conditions are much too cold for liquid water (1.5 bar, 89–94 K; Cottini et al., 2012; Fulchignoni et al., 2005; Jennings et al., 2019), and therefore, the liquid composition of the lakes and seas are various types of hydrocarbons, instead of water. Observational studies using Cassini VIMS identified ethane in Titan's liquids (Brown et al., 2008). It is now well accepted that Titan's lakes have a composition of predominantly methane and ethane with a percentage of dissolved atmospheric nitrogen (Battino et al., 1984; Brown et al., 2008; Malaska et al., 2017; Mastrogiuseppe et al., 2016, 2018a, 2018b; Poggiali et al., 2020). Cassini radar observational studies have determined that the composition of Titan's lakes and seas vary latitudinally, being more methane-rich in the north polar region and more ethane-rich in the south polar region (Mastrogiuseppe et al., 2014, 2016, 2018a, 2018b; Poggiali et al., 2020). The latitudinal distribution of Titan's filled-to-dry lake beds and

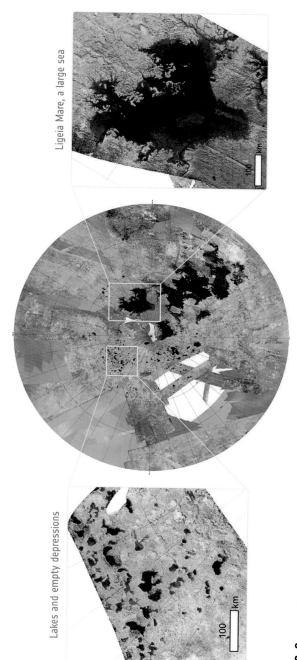

Ligeia Mare, a large sea

Lakes and empty depressions

FIG. 8

Titan's north pole is presented as a mosaic of false-color images from Cassini radar (*center*: polar stereographic projection). This image highlights Titan's surface (tan) and Titan's north polar hydrocarbon lakes and seas (*black/blue*). The *left panel* highlights part of the north polar lake district. The *right panel* highlights one of Titan's largest seas, Ligeia Mare. Notice the stark difference between the abundance of filled lakes in the north polar region compared to the singular filled lake in the south polar region (Fig. 7).

Credit: center: NASA/JPL-Caltech/ASI/USGS, image ID: PIA17655; left and right: NASA/ESA. Acknowledgment: T. Cornet, ESA: https://www.esa.int/ESA_Multimedia/Images/2015/06/Lakes_and_seas_on_Titan.

their compositional variations may be influenced by changes in Saturn's orbit over time (similar to the Milankovitch cycle on Earth) (Aharonson et al., 2009).

Interestingly, Titan's present-day surface temperature in the polar regions can get down to 89 K, a few degrees lower than the freezing point of both methane and ethane (91 and 90 K, respectively). Why are they not solid ice on the surface? Nitrogen gas readily dissolves into liquid methane, increasing in solubility with decreasing temperature (Battino et al., 1984; Hibbard & Evans Jr., 1968; Malaska et al., 2017; Steckloff et al., 2020). It is so soluble, in fact, that the dissolved nitrogen is ~15 mol% at 94 K and increases to ~30 mol% at 87 K (Malaska et al., 2017). The dissolved nitrogen suppresses the freezing point of methane, and therefore, methane will not be solid ice in the presence of nitrogen gas on Titan (Malaska et al., 2017; Steckloff et al., 2020). Ethane, on the other hand, has a very low affinity for nitrogen gas (<5 mol% below 89 K; Malaska et al., 2017) and freezes at ~90 K. However, laboratory experiments show that mixing liquid ethane with liquid methane, without the influence of nitrogen gas, will decrease the freezing point rapidly (Engle et al., 2021). Therefore, ethane ice could exist on Titan's surface (devoid of methane), while the presence of methane ice would be extremely unlikely.

3.4.2 Transient ice formation in Titan's seas?

Radar-bright features in two of Titan's seas, Ligeia Mare and Kraken Mare, were observed to appear and disappear over time (Fig. 9), earning the name "Magic Islands." Magic Islands were originally hypothesized to be caused by the presence of bubbles, waves, or floating/suspended solids such as transient ice formations (Hofgartner et al., 2014, 2016). Through laboratory and modeling endeavors, the most plausible explanation is the presence of waves (from the topographically funneled wind; Hayes et al., 2018), or nitrogen exsolution events in the form of bubbles (explosions of nitrogen bubbles induced by changes in liquid temperature or hydrocarbon composition; Cordier et al., 2017; Cordier & Liger-Belair, 2018; Farnsworth et al., 2019; Richardson et al., 2019). Since methane-ethane-nitrogen mixtures won't freeze under current Titan conditions, the periodic freezing of hydrocarbons is less likely (see Section 3.4.1). However, it is still possible that an unknown material might be freezing periodically in Titan's lakes.

3.4.3 Transient ice formation on Titan's surface?

Evidence of hydrocarbon rainfall was observed on multiple occasions by Cassini, including at the south pole (Turtle et al., 2009), equatorial regions (Barnes et al., 2013; Solomonidou et al., 2014; Turtle et al., 2011b), and recently at the north pole (Dhingra et al., 2019, 2020, 2021). Titan's active hydrocarbon-based hydrological cycle is similar to Earth's water cycle, which raises the question: does transient ice form on its surface similar to transient ice deposits on Earth?

After clouds had passed over Titan's equatorial region, darkening of the topography was observed in Cassini VIMS and ISS, evidence of hydrocarbon rain (Barnes et al., 2013; Dhingra et al., 2019; Turtle et al., 2009, 2011a, 2011b). After a few weeks to months, the darkened regions increased in albedo, well above the original albedo (Barnes et al., 2013). This brightening lasted a similar time frame before the surface reverted back to its original albedo (Barnes et al., 2013). The surface brightening was hypothesized to be caused by aeolian grain transport/sorting, chemical change of the surface, hail or snow, or freezing of the rainfall liquid on the surface due to evaporative cooling (Barnes et al., 2013). Because the albedo ultimately returned to normal, the freezing hypothesis is a plausible candidate, with the frozen liquid melting and drying over time (Barnes et al., 2013).

erupted from Enceladus' Tiger Stripes escapes into space (Dougherty et al., 2006; Porco et al., 2006), but this is enough to create and maintain Saturn's E ring. By studying the composition of Enceladus' plumes and Saturn's E ring, it is possible to learn about the interior of Enceladus.

Cassini was redirected to fly through and sample the expelled material using the INMS instrument onboard (Postberg et al., 2008, 2009, 2018). This was the first time the interior of an icy moon was indirectly analyzed in our solar system. The INMS revealed that the sample was dominated by water-ice and contained complex organic compounds with masses >200 amu, along with sodium and potassium salts (Postberg et al., 2008, 2009, 2018). Through these measurements it can be inferred that Enceladus' ocean is likely (1) in direct contact with the silicate interior of Enceladus, (2) rich in carbonate and/or bicarbonate, and (3) capped by a thin organic-rich layer at the ocean-surface interface (Postberg et al., 2008, 2009, 2018). Additionally, the positive identification of hydrogen (H_2) and the lack of nitrogen (N_2) in Enceladus' plumes (Waite et al., 2017), is evidence of hydrothermal activity and suggests that temperatures in this area are <500 K (from Cassini UVIS) where water interacts with a likely porous core.

Interestingly, the interior ocean may not be thermodynamically stable and may undergo a degree of periodic freezing (Fuller et al., 2016; Hemingway & Mittal, 2019; Lainey et al., 2012; Nimmo et al., 2018; Rudolph et al., 2022). Currently, there is uncertainty about the age and long-term stability of Enceladus' ocean (e.g., Roberts & Nimmo, 2008).

4.3 Enceladus' icy surface and geology

The SPT is the youngest region of Enceladus' surface, (<4 Ma; Neukum et al., 2006; Porco et al., 2006) and is nearly devoid of craters. Surrounding the Tiger Stripes is a ropy and hummocky terrain called the funicular plains (Fig. 11). Observations of the surface in this region, including warm thermal anomalies, and spectral measurements of carbon dioxide frost deposits, suggest rapid carbon dioxide outgassing from subsurface pockets (Matson et al., 2018). These carbon dioxide pockets may correlate to the areas beneath hundreds of thousands of large ice blocks (10–100 m across) mapped in the SPT that are concentrated in a 20 km radius of the Tiger stripes (Martens et al., 2015; Porco et al., 2006).

Outside the SPT, bright icy regolith appears to cover more heavily cratered terrains (>4.4 Ga old; Kirchoff & Schenk, 2009) and older tectonic features (e.g., Samarkand Sulcus; Crow-Willard & Pappalardo, 2015; Porco et al., 2006). The SPT lacks the bright regolith and appears darker, possibly due to the thermal processing of the ice grains and/or removal of bright icy regolith from steep slopes (Brown et al., 2006; Konstantinidis et al., 2015; Porco et al., 2006; Robidel et al., 2020). Another potential contributing factor to the relative SPT surface darkness is a variation in grain size, with larger, heavier, and perhaps salt-rich grains being deposited closer to the Tiger stripes (Kempf et al., 2010).

4.4 Comparing Enceladus to other moons

Enceladus offers a useful tool for comparative planetology among icy worlds. A longstanding question is why Enceladus is geologically active when its neighboring icy moon, Mimas, is heavily cratered and shows little activity (e.g., Czechowski & Witek, 2015; Neveu & Rhoden, 2017). Interestingly, both moons are similar in size and have a similar degree of tidal heating (e.g., Czechowski, 2014; Nathan et al., 2022; Neveu & Rhoden, 2017). Enceladus also offers a valuable point of comparison with Mab, a small moon of Uranus. Though Mab's surface was not resolved by Voyager 2, it orbits within Uranus' μ-ring, much like Enceladus orbits within Saturn's E ring. Both rings are broad, dominated by particles ~1 μm in size, and are most dense near the orbit of their embedded moon (Kempf et al., 2008; Showalter, 2020). It is unclear, however, how Mab influences Uranus' μ-ring and if it is similar to how Enceladus supplies Saturn's E ring (e.g., Beddingfield & Cartwright, 2022; Showalter, 2020).

5 Iapetus

5.1 Albedo variation on Iapetus

Iapetus (Fig. 12) was discovered by Giovanni Cassini in 1671, and it was noted that its brightness varied drastically during its orbit around Saturn. Voyagers 1 and 2 confirmed these observations by revealing that Iapetus had extreme variations in its surface albedo with the leading and equatorial hemispheres a dark, low albedo (0.03–0.05), and the trailing hemisphere and poles beaming nearly ten times brighter than water ice (albedo 0.5–0.6) (Buratti et al., 2005; Porco et al., 2005b) (Fig. 12a).

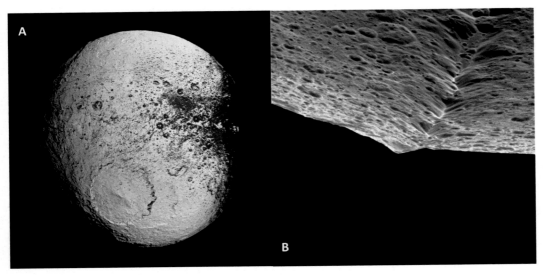

FIG. 12

Images of Iapetus by the Cassini orbiter taken in 2017. (a) This view is an orthographic projection of a false-color mosaic of 60 images. All images were taken by Cassini's narrow-angle camera at a distance of ~73,000 km. Notice the dark leading (*right side*) and bright trailing (*left side*) hemispheres of the Iapetus. (b) An unprocessed image of Iapetus' unique equatorial ridge, taken by Cassini's wide-angle camera at a distance of ~4,000 km.

Panel (a): Image ID: PIA08384; Credit: NASA/JPL/Space Science Institute; https://photojournal.jpl.nasa.gov/catalog/pia08384;

panel (b): Ciclops Image ID: 3771; Credit: NASA/JPL/Space Science Institute; https://ciclops.org/view.php%3Fid=3771.html.

It is hypothesized that Iapetus is predominantly water ice with the origins and composition of the dark region (called Cassini Regio; Spencer & Denk, 2010) still under debate with several proposed mechanisms (Jaumann et al., 2009a; Rivera-Valentin et al., 2011; Spencer & Denk, 2010). First, the dark material may be captured debris from a historically large impact in the Saturnian system, or accretion material from the formation of Titan. The next hypothesis is that interplanetary dust gradually altered the surface, or an ice layer that is being slowly removed and exposing the dark underlying substrate. Finally, the dark material may be emplaced by endogenic geological processes, such as cryovolcanism, or the exposure of darker, more viscous lag deposits of larger-grained ices (Jaumann et al., 2009a). It is possible that further darkening may have occurred because of chemical or irradiative processes. If the dark material on Iapetus is indeed irradiated ice, it would have originated as gases before being photolyzed by ultraviolet (UV) radiation and cosmic rays (Ahrens et al., 2022; Palmer & Brown, 2011). During this process, carbon dioxide molecules in the atmosphere are pulled apart and reorganized with H and O molecules. Thus, the low albedo, darker, regions contain C-bearing molecular species, including methane and possibly graphene-like minerals (Shi et al., 2015).

On Iapetus, carbon dioxide ice is generated through the photolysis of oxygen and carbon, and other compounds and ultimately condenses out at the poles (Ahrens et al., 2022), similar to Dione and Rhea. The obliquity of the Iapetus causes the carbon dioxide to switch hemispheres seasonally (Palmer & Brown, 2011). About 12% of the carbon dioxide polar cap is lost during a complete yearly cycle (1 Iapetus day = 79 Earth days). The presence of a seasonally varying exosphere is not unique to Iapetus and can be observed on other moons such as Dione and Rhea (e.g., Miles et al., 2022; Teolis et al., 2010; Teolis & Waite, 2016; references therein) (see Sections 6 and 8 for more details about these moons).

5.2 Iapetus' equatorial ridge

Iapetus is known as the "Walnut Moon" due to its enigmatic equatorial ridge (Fig. 12b). This ridge reaches a height of 20 km and extends over 75% of the moon's circumference (Singer et al., 2012). There are two major hypotheses behind the ridge formation: (1) an older formation from when Iapetus rotated much faster (e.g., Czechowski & Leliwa-Kopystynski, 2013; Kuchta et al., 2015) or (2) remnants of a collapsed ring (e.g., Damptz et al., 2018; Detelich et al., 2021; Ip, 2006; Stickle & Roberts, 2018). Regardless of how this ridge formed, it has been geomorphologically altered with a variety of slopes and peaks, some sharp and steep, while others are rounded (e.g., Giese et al., 2008; Lopez Garcia et al., 2014). Because not all the ridges are pristine, the icy material is not particularly stable. Landslides have been identified on steeper slopes, and they do not appear to be constructed of coherent material (Singer et al., 2012). These landslides provide an interesting example of degradation and erosional processes in an icy world. Landslides have also been found on other small bodies such as Pluto (see Ahrens et al., 2024, pp. 357–376) and Helene, another small moon of Saturn (Umurhan et al., 2016). By comparing these mass-wasting events, we can begin to understand the mobility of icy regolith, which may have broader implications for tectonic activity (Singer et al., 2012). For example, we can further understand how impact craters trigger movement and mass-wasting events, the behavior of non-volatile crustal material (in this case, water ice), and how tectonism may vary by lithospheric thickness.

6 Dione

Dione (Fig. 13) is a medium-sized Saturnian satellite, measuring 1,123 km in diameter, and was discovered by Giovanni Cassini in 1684. Voyager and Cassini images of Dione show a number of surface terrains, from cratered terrains and plains, tectonic features of linear and arcuate troughs, to wispy lineae streaks (Plescia, 1983). The higher-resolution images from Cassini (at 430 m/pixel) show many crosscutting relationships among these various tectonic features (Wagner et al., 2005).

A prominent surface feature on Dione is the "Wispy Terrain," bright ice cliffs across Dione's trailing hemisphere, likely formed by tectonic activity (Plescia, 1983). These bright features crosscut the cratered surface and smoother terrains (Plescia & Boyce, 1982). Their formation could be explained by an endogenic heat source. However, any present internal heat from tidal interactions and dissipation would be insufficient to explain these tectonic features (Dalle Ore et al., 2021). Constraining the time evolution of these features would help elucidate Dione's tidal evolution.

The tectonic-laden wispy terrains have exposed clean, pure water ice, and are associated with a higher abundance of crystalline water ice than the surrounding dark terrain (Fig. 13a; Newman et al., 2009). Mechanisms behind the crystalline ice (e.g., chemical) formations may be diverse, but are likely

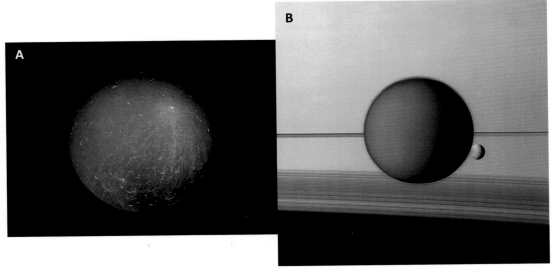

FIG. 13

Images of Dione taken by Cassini. (a) An image of Dione taken in 2012 with Cassini's narrow-angle camera at a distance of ~400,000 km. This image was taken using a polarized filter and a green spectral filter. (b) This image of Titan and Dione was taken in 2011. Here, we see Titan in the center, Dione to the right of Titan (passing behind Titan), and Saturn and its rings in the background.

Panel (a): Image ID: PIA17197; Credit: NASA/JPL-Caltech/Space Science Institute; https://photojournal.jpl.nasa.gov/catalog/
PIA17197; panel (b): image ID: PIA14910; Credit: NASA/JPL-Caltech/Space Science Institute; https://photojournal.jpl.nasa.gov/
catalog/PIA14910.

a result of temperature fluctuations, caused by processes such as impact cratering (where heat transfers from the impact directly onto the surface in a localized spot), cryovolcanic activity (driven by internal heat sources or heated lenses of viscous ice), thermal relaxation, or tidal interactions (McKinnon & Melosh, 1980; Moore et al., 2004).

Other possible explanations for the difference between the wispy and darker regions are ice crystallinity, ice composition, and grain size. Newman et al. (2009) found that the dark terrain has a lower crystallinity factor than the wispy and non-wispy terrains. They also found the wispy-to-dark-terrain spectral ratio fits that of crystalline ice 10:3, and the wispy terrain spectra of water ice have deeper spectral bands than the dark region. This suggests that the darker regions may have a smaller grain size than the other regions (Newman et al., 2009). Furthermore, wispy terrains are likely supported by tectonic episodes, which expose fresh ice with higher crystallinity or larger angular grain size. During this process, the surface was likely heated, which crystallized the existing amorphous ice and/or stimulated grain growth (Jenniskens & Blake, 1996). Indeed, regions of tectonic activity can be associated with both a higher level of crystallinity and larger grain sizes on Dione and like that on Enceladus (Jaumann et al., 2008b). Additionally, dark regions may be more amorphous and contaminated with a greater concentration of non-water ice material than the other regions (Newman et al., 2009). Spectroscopic observations from Cassini INMS have also detected an alternating polar exosphere, similar to Iapetus (Teolis & Waite, 2016).

It is possible that Dione's wispy terrain is a fossilized or cemented version of Enceladus' tiger stripes (Dalle Ore et al., 2021). Several studies suggest that Dione may be very similar to Enceladus when comparing geologic activity (Buratti et al., 2018; Dalle Ore et al., 2021). However, Cassini CIRS observations of Dione have not shown evidence of endogenic heat emission associated with the wispy terrains (Howett et al., 2014). Regardless, Dione could influence Enceladus' geologic processes; Enceladus and Dione are in an orbital resonance, and it has been observed that this small variability in resonance has led to variability in cryovolcanic activity on Enceladus (Ingersoll et al., 2020).

7 Hyperion

The small, 134.8 km mean radius, "sponge-like" moon, Hyperion (Fig. 14), was discovered in 1848 by two separate groups, William Lassell, and father-son-duo, William Cranch Bond and George Phillips Bond. Hyperion is non-spherical and is in orbital resonance with Titan. Hyperion is mostly composed of water ice, likely intermixed with other volatiles, such as methane or carbon dioxide. Clathrates may also be present and trace amounts of cyanide were detected in its craters (Cruikshank et al., 2007, 2010). Spectroscopy data from Cassini UVIS have identified two distinct surface units: (1) a spatially dominant high-albedo unit (most likely water ice) and (2) a dark material buried at the bottom of small craters. The dark material may be similar to the dark material found on Iapetus (see Section 5), or it may be related to observations of cyanide at the bottom of the craters (Jarvis et al., 2000).

Why does Hyperion have such a sponge-like appearance? The density of Hyperion is approximately half that of liquid water, caused by unique textural morphology. Its low density may be due to very porous water-ice composition (>40% void space), or other volatiles such as carbon dioxide may be present in the solid phase (Thomas et al., 2007). Because of Hyperion's low gravity, ice cannot crystallize and strengthen its surface. Interestingly, its rubble-pile-like morphology causes impactors to sink, creating depressions with soft rims instead of a typical impact structure (Thomas et al., 2007). Landslides

FIG. 14

A false-color image of Hyperion taken in 2005 by Cassini's narrow-angle camera from a distance of ~60,000 km. Spectral filters (*infrared, green, and ultraviolet*) were combined to enhance Hyperion's surface features.

Image ID: PIA07740; Credit: NASA/JPL/Space Science Institute; https://photojournal.jpl.nasa.gov/catalog/PIA07740.

within the impact craters have also been observed (Richardson et al., 2006). Moreover, impact ejecta is not seen on Hyperion. It is likely that some of the impacted material escapes into space.

8 Other Saturnian icy satellites

Other Saturnian satellites composed of water ice and rock are Rhea, Tethys, and Mimas (Fig. 15). Though we focus on just a few of the Saturnian icy moons in the earlier sections of this chapter, these additional icy moons have much to offer in understanding the surface geology of icy moons, because water ice is a versatile compound across varying surfaces.

Rhea (Fig. 15a), Saturn's second-largest moon, was discovered by Giovanni Cassini in 1672 and is an interesting mixture of chondritic and water-ice material. The internal structure of Rhea is poorly constrained (e.g., Anderson & Schubert, 2007; Castillo-Rogez, 2006; Czechowski, 2012; Iess et al., 2007). Interestingly, the leading side of Rhea has a higher albedo (opposite of Iapetus; see Section 5), possibly because of the accumulation of >1-μm ice particles during its orbit around Saturn (Dalle Ore

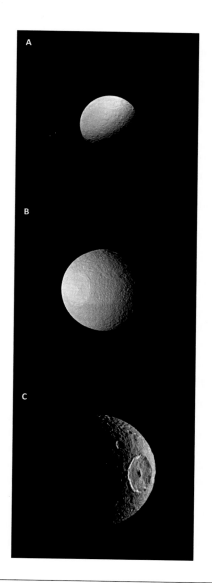

FIG. 15

Images of medium to large sized moons of Saturn taken by the Cassini orbiter. (a) An image of Rhea taken by the Cassini spacecraft wide-angle camera in 2010 at a distance of ~100,000 km. (b) An image of Tethys taken by the Cassini spacecraft narrow-angle camera in 2016 at a distance of ~400,000 km. This image highlights Odysseus Crater, seen on the left. (c) An image of Mimas taken in 2010 at a distance of ~100,000 km. This image highlights Herschel Crater, seen on the right.

Panel (a): ID: PIA 12702; Credit NASA/JPL/Space Science Institute; https://photojournal.jpl.nasa.gov/catalog/PIA12702; panel (b): ID: PIA 20518; Credit: NASA/JPL/Space Science Institute; https://photojournal.jpl.nasa.gov/catalog/PIA20518; panel (c): ID: PIA 12739; Credit: NASA/JPL/Space Science Institute; https://photojournal.jpl.nasa.gov/catalog/PIA12739.

et al., 2015; Schenk et al., 2011). Water ice is also concentrated in Rhea's fresh craters and on some steep tectonic scarps (e.g., Stephan et al., 2012). Rhea's exosphere is seasonally polar alternating, similar to Iapetus as observed from Cassini INMS (Teolis & Waite, 2016).

Tethys (Fig. 15b) was discovered by Giovanni Cassini, soon after Rhea, in 1684. Tethys has several water-ice geologic units, including plains (the youngest terrain), rifts (e.g., Ithaca Chasma), hilly cratered terrain (the oldest unit), and the Odysseus impact structure (Moore & Ahern, 1983; Stephan et al., 2016). Several hypotheses have been proposed for the creation of the rifts, including (1) stresses from the freezing of liquid water subsurface ocean and (2) tidal stresses from past high eccentricity (Giese et al., 2007; Hussmann et al., 2019; Moore & Ahern, 1983; Stephan et al., 2016).

Mimas (Fig. 15c) was discovered about 100 years later, in 1789, by William Herschel. Mimas is the smallest of Saturn's classical mid-sized satellites just outside the main ring system (within the inner part of the E ring) and has potential tidal interactions, though the existence of a subsurface ocean on Mimas is unknown (Rhoden et al., 2017; Rhoden & Walker, 2022; Schenk et al., 2011; Tajeddine et al., 2014). Voyager and Cassini images of Mimas show symmetrical v-shaped depressions (i.e., grooves), which may be an indication of global de-spinning (Schenk et al., 2011). Unlike Dione or Tethys, Mimas does not appear to have smooth plains (Kirchoff & Schenk, 2010), and it has a large crater, named Herschel Crater, measuring 130 km in diameter (e.g., Moore et al., 2004). The interior structure and geologic history of Mimas are particularly enigmatic in comparison to the other Saturnian moons. It has been suggested that Mimas formed later than other Saturnian moons, perhaps accreting from an earlier massive ring (e.g., Charnoz et al., 2011; Ida, 2019).

9 Prebiotic chemistry on icy moons

Amino acids, and other biologically relevant molecules, can be synthesized from organics in aqueous-phase reactions, such as hydrolysis reactions (Miller, 1955; see review by Cleaves, 2012; Raulin et al., 2012). Recent space flight missions have identified organic material on various icy bodies including comets, Pluto (see Ahrens et al., 2024, pp. 357–376), Charon, and most notably, Titan and Enceladus (e.g., Coates et al., 2007; Cruikshank et al., 2019, 2020; Lorenz et al., 2008b; McKay & Roth, 2021; Postberg et al., 2018). Liquid water is more difficult to obtain on the surface of icy moons, but transient liquid water environments are possible via cryovolcanic flows and impact melt pools (Poch et al., 2012; Neish et al., 2018). Impact events expel a tremendous amount of energy, heating, and potentially melting the water-ice crust (e.g., Crósta et al., 2021; Steinbrügge et al., 2020; Wakita et al., 2023). Impact craters on Titan could have impact melt lenses persisting for tens of thousands of years prior to refreezing (Artemieva & Lunine, 2003; Wakita et al., 2023). Organic matter can also migrate into subsurface water oceans from the surfaces of icy moons as a result of impactors that breach the ice crust (Crósta et al., 2021), or organics can leach from the core when interacting with the interior ocean (Sotin et al., 2021).

Various laboratories have simulated the prebiotic chemistry on icy moons, such as Titan, by synthesizing Titan-like organic aerosols (complex hydrocarbon chains), called tholins, in the laboratory (Cable et al., 2012; Yu et al., 2020). In various studies (e.g., Brassé et al., 2017; Cleaves et al., 2014; Neish et al., 2008, 2009, 2010; Poch et al., 2012; Ramírez et al., 2010), prebiotic molecules and oxygenated products were formed through hydrolysis reactions when tholins were mixed with water and water-ammonia solvents. Additionally, the addition of carbon dioxide gas in the production of tholins successfully incorporated oxygen into the organic product without hydrolysis reactions (Hörst et al., 2012; Sebree et al., 2018).

Thus, it is plausible that prebiotic chemistry (and those reaction pathways identified in these studies) could occur on icy moons within cryovolcanic flows and impact melt pools, prior to refreezing, and within the interior ocean of ocean worlds, if organic matter is indeed present.

10 Future missions to Titan and Enceladus

On Titan, impact craters, cryovolcanic flows, and the subsurface water ocean are the best candidate locations to identify prebiotic molecules and search for evidence of life as we know it (Poch et al., 2012; Neish et al., 2018). NASA's selected mission, *Dragonfly*, is a rotorcraft with a primary science goal to identify prebiotic molecules on Titan's surface (Barnes et al., 2021; Turtle et al., 2008) with a target location of Titan's Selk Crater (Fig. 6b) (Lorenz et al., 2021). *Dragonfly* is set to launch in the late-2020s and arrive at Titan in the mid-2030s (Barnes et al., 2021). This is the first mission that aims to sample the surface of an icy moon to search for the building blocks of life.

Enceladus is one of the primary candidates in the search for habitable worlds beyond Earth (e.g., Hendrix et al., 2019). The plumes offer a relatively accessible way to continue the assessment of Enceladus' habitability and search for signs of extant or extinct life (e.g., McKay et al., 2008, 2014; New et al., 2021). Indeed, the most recent 2023–2032 NASA's Decadal Survey for Planetary Science and Astrobiology details a Flagship-class mission concept for an Enceladus Orbilander that would sample Enceladus' plumes from orbit and capture icy material midfall from the plumes to the surface in search of biosignatures (MacKenzie et al., 2021b; National Academies of Sciences, Engineering, and Medicine, 2022). The Tiger stripes are the gateway to Enceladus' ocean and an exciting potential target for future missions to Enceladus (Affholder et al., 2021; Cable et al., 2021a; McKay et al., 2014; Parkinson et al., 2008).

11 Summary

Water ice is common in the Saturnian system and is incorporated into the structure and surfaces of many of Saturn's moons. Water ice is found on the surface of all the moons discussed in this chapter (Titan, Enceladus, Iapetus, Hyperion, Dione, Rhea, Tethys, and Mimas), while carbon dioxide, methane, and cyanide ice along with methane clathrates have also been identified or are hypothesized to exist on the surface or interior of a few of these icy moons. Titan's surface and atmosphere have a greater variety of ice compositions compared to other icy moons because of the complex chemistry in Titan's atmosphere. Titan and Enceladus are unique and deemed "ocean worlds" because they have interior liquid water oceans beneath solid water-ice crusts. Titan and Enceladus may also have a high-pressure ice layer between their core and the ocean. These moons are high-priority worlds in the search for habitable environments beyond Earth because they have subsurface water oceans and organic compounds (e.g., Hand et al., 2020; Hendrix et al., 2019).

All of Saturn's icy moons have surface structures and geology that are either cut into the water-ice crust or are directly influenced by the presence of ice. The ice present on icy moons may control surface geologic processes, impact internal ocean dynamics, and even influence the creation of a habitable environment.

Moreover, understanding the composition, phase, and grain size of the ice structures on icy moons is useful in determining the source of the surface material. By studying their geology and comparing different icy moons, we can have a better understanding of their origins, and discern their geologic pasts and futures. Furthermore, understanding the fault structures and tectonic activity on icy moons can give us insight into active endogenic/exogenic mechanisms, identify whether subsurface oceans exist, and improve interior modeling (e.g., Choblet et al., 2018; Simon et al., 2011; Teolis et al., 2010).

The Cassini-Huygens spacecraft has vastly improved our understanding of the Saturnian system but also left us with many intriguing unanswered questions. Future missions should aim to understand the origin of such ices, their geologic surface processes, and the potential habitability of such moons, particularly Titan and Enceladus. NASA's New Frontiers mission, *Dragonfly*, and NASA's flagship mission concept, an Enceladus Orbilander, are revolutionary missions that will aid in understanding the potential habitable environments on icy moons and pave the way for future missions directed to the outer solar system.

References

Affholder, A., Guyot, F., Sauterey, B., Ferrière, R., & Mazevet, S. (2021). Bayesian analysis of Enceladus's plume data to assess methanogenesis. *Nature Astronomy, 5*, 805–814. https://doi.org/10.1038/s41550-021-01372-6.

Aharonson, O., Hayes, A. G., Lunine, J. I., Lorenz, R. D., Allison, M. D., & Elachi, C. (2009). An asymmetric distribution of lakes on Titan as a possible consequence of orbital forcing. *Nature Geoscience, 2*(12), 851–854. https://doi.org/10.1038/ngeo698.

Ahrens, C. J., Lisse, C. M., Williams, J.-P., & Soare, R. J. (2024). *Geocryology of Pluto and the icy moons of Uranus and Neptune*. In R. J. Soare, J.-P. Williams, C. Ahrens, F. E. G. Butcher, & M. R. El-Maarry (Eds.), *Ices in the solar system, a volatile-driven journey from the inner solar system to its far reaches*. Elsevier Books.

Ahrens, C., Meraviglia, H., & Bennett, C. (2022). A geoscientific review on CO and CO2 ices in the outer solar system. *Geosciences, 12*(2), 51.

Anderson, C. M., Nna-Mvondo, D., Samuelson, R. E., Achterberg, R. K., Flasar, F. M., Jennings, D. E., & Raulin, F. (2017). Titan's high altitude south polar (HASP) stratospheric ice cloud as observed by Cassini CIRS. In *AAS/Division for Planetary Sciences Meeting Abstracts, vol. 49, 304.10*.

Anderson, C. M., & Samuelson, R. E. (2011). Titan's aerosol and stratospheric ice opacities between 18 and 500 μm: vertical and spectral characteristics from Cassini CIRS. *Icarus, 212*, 762–778.

Anderson, C. M., Samuelson, R. E., Achterberg, R. K., Barnes, J. W., & Flasar, F. M. (2014). Subsidence-induced methane clouds in Titan's winter polar stratosphere and upper troposphere. *Icarus, 243*, 129–138. https://doi.org/10.1016/j.icarus.2014.09.007.

Anderson, C. M., Samuelson, R. E., Bjoraker, G. L., & Achterberg, R. K. (2010). Particle size and abundance of HC3N ice in Titan's lower stratosphere at high northern latitudes. *Icarus, 207*(2), 914–922. https://doi.org/10.1016/j.icarus.2009.12.024.

Anderson, C. M., Samuelson, R. E., & Nna-Mvondo, D. (2018). Organic ices in Titan's stratosphere. *Space Science Reviews, 214*(8), 1–36.

Anderson, C. M., Samuelson, R. E., Yung, Y. L., & McLain, J. L. (2016). Solid-state photochemistry as a formation mechanism for Titan's stratospheric C4N2 ice clouds. *Geophysical Research Letters, 43*(7), 3088–3094. https://doi.org/10.1002/2016GL067795.

Anderson, J. D., & Schubert, G. (2007). Saturn's satellite Rhea is a homogeneous mix of rock and ice. *Geophysical Research Letters, 34*(2), L02202. https://doi.org/10.1029/2006GL028100.

Artemieva, N., & Lunine, J. (2003). Cratering on Titan: Impact melt, ejecta, and the fate of surface organics. *Icarus*, *164*(2), 471–480. https://doi.org/10.1016/S0019-1035(03)00148-9.

Baland, R. M., Tobie, G., Lefèvre, A., & Van Hoolst, T. (2014). Titan's internal structure inferred from its gravity field, shape, and rotation state. *Icarus*, *237*, 29–41. https://doi.org/10.1016/j.icarus.2014.04.007.

Barnes, J. W., Brown, R. H., Radebaugh, J., Buratti, B. J., Sotin, C., Le Mouelic, S., ... Nicholson, P. D. (2006). Cassini observations of flow-like features in western Tui Regio, Titan. *Geophysical Research Letters*, *33*(16), 1–5. https://doi.org/10.1029/2006GL026843.

Barnes, J. W., Brown, R. H., Soderblom, L., Buratti, B. J., Sotin, C., Rodriguez, S., ... Nicholson, P. (2007). Global-scale surface spectral variations on Titan seen from Cassini/VIMS. *Icarus*, *186*(1), 242–258.

Barnes, J. W., Brown, R. H., Soderblom, J. M., Soderblom, L. A., Jaumann, R., Jackson, B., ... Perry, J. (2009). Shoreline features of Titan's Ontario Lacus from Cassini/VIMS observations. *Icarus*, *201*(1), 217–225. https://doi.org/10.1016/j.icarus.2008.12.028.

Barnes, J. W., Brown, R. H., Soderblom, L., Sotin, C., Le Mouèlic, S., Rodriguez, S., ... Nicholson, P. (2008). Spectroscopy, morphometry, and photoclinometry of Titan's dunefields from Cassini/VIMS. *Icarus*, *195*(1), 400–414.

Barnes, J. W., Brown, R. H., Turtle, E. P., McEwen, A. S., Lorenz, R. D., Janssen, M., ... Sicardy, B. (2005). Planetary science: A 5-micron-bright spot on Titan: Evidence for surface diversity. *Science*, *310*(5745), 92–95. https://doi.org/10.1126/science.1117075.

Barnes, J. W., Buratti, B. J., Turtle, E. P., Bow, J., Dalba, P. A., Perry, J., ... Soderblom, L. A. (2013). Precipitation-induced surface brightenings seen on Titan by Cassini VIMS and ISS. *Planetary Science*, *2*(1), 1–22.

Barnes, J. W., Turtle, E. P., Trainer, M. G., Lorenz, R. D., MacKenzie, S. M., ... Stähler, S. C. (2021). Science goals and objectives for the Dragonfly Titan rotorcraft relocatable lander. *Planetary Science Journal*, *2*(130), 1–18. https://doi.org/10.3847/PSJ/abfdcf.

Battino, R., Rettich, T. R., & Tominaga, T. (1984). The solubility of nitrogen and air in liquids. *Journal of Physical and Chemical Reference Data*, *13*, 563. https://doi.org/10.1063/1.555713.

Beddingfield, C. B., & Cartwright, R. J. (2022). Miranda's thick regolith indicates a major mantling event from an unknown source. *The Planetary Science Journal*, *3*, 253. https://doi.org/10.3847/PSJ/ac9a4e.

Berton, M., Nathan, E., Karani, H., Girona, T., Huber, C., Williard, P. G., & Head, J. (2020). Experimental investigations on the effects of dissolved gases on the freezing dynamics of Ocean Worlds. *Journal of Geophysical Research: Planets*, *125*, 1–14. https://doi.org/10.1029/2020JE006528.

Birch, S. P. D., Hayes, A. G., Howard, A. D., Moore, J. M., & Radebaugh, J. (2016). Alluvial fan morphology, distribution and formation on Titan. *Icarus*, *270*, 238–247.

Brassé, C., Buch, A., Coll, P., & Raulin, F. (2017). Low-temperature alkaline pH hydrolysis of oxygen-free Titan Tholins: Carbonates' impact. *Astrobiology*, *17*, 8–26. https://doi.org/10.1089/ast.2016.1524.

Brossier, J. F., Rodriguez, S., Cornet, T., Lucas, A., Radebaugh, J., Maltagliati, L., ... Sotin, C. (2018). Geological evolution of Titan's equatorial regions: Possible nature and origin of the dune material. *Journal of Geophysical Research: Planets*, *123*(5), 1089–1112.

Brown, R. H., Clark, R. N., Buratti, B. J., Cruikshank, D. P., Barnes, J. W., Mastrapa, R. M. E., ... Sotin, C. (2006). Composition and physical properties of Enceladus' surface. *Science (New York, N.Y.)*, *311*(5766), 1425–1428. https://doi.org/10.1126/science.1121031.

Brown, R. H., Soderblom, L. A., Soderblom, J. M., Clark, R. N., Jaumann, R., ... Nicholson, P. D. (2008). The identification of liquid ethane in Titan's Ontario Lacus. *Nature*, *454*, 607–610. https://doi.org/10.1038/nature07100.

Buratti, B. J., Cruikshank, D. P., Brown, R. H., Clark, R. N., Bauer, J. M., Jaumann, R., ... Muradyan, A. (2005). Cassini visual and infrared mapping spectrometer observations of Iapetus: Detection of CO2. *The Astrophysical Journal Letters*, *622*(2), L149.

Buratti, B. J., Hansen, C. J., Hendrix, A. R., Esposito, L. W., Mosher, J. A., Brown, R. H., ... Nicholson, P. D. (2018). The search for activity on Dione and Tethys with Cassini VIMS and UVIS. *Geophysical Research Letters*, *45*(12), 5860–5866.

Burkhard, L. M., Smith-Konter, B. R., Fagents, S. A., Cameron, M. E., Collins, G. C., & Pappalardo, R. T. (2022). Strike-slip faulting on Titan: Modeling tidal stresses and shear failure conditions due to pore fluid interactions. *Icarus, 371*, 114700. https://doi.org/10.1016/j.icarus.2021.114700.

Burr, D. M., Taylor Perron, J., Lamb, M. P., Irwin, R. P., III, Collins, G. C., Howard, A. D., ... Black, B. A. (2013). Fluvial features on Titan: Insights from morphology and modeling. *Bulletin, 125*(3–4), 299–321.

Cable, M. L., Hörst, S. M., Hodyss, R., Beauchamp, P. M., Smith, M. A., & Willis, P. A. (2012). Titan Tholins: Simulating Titan organic chemistry in the Cassini-Huygens Era. *Chemical Reviews, 112*, 1882–1909. https://doi.org/10.1021/cr200221x.

Cable, M. L., Porco, C., Glein, C. R., German, C. R., MacKenzie, S. M., Neveu, M., ... Núñez, J. (2021a). The science case for a return to Enceladus. *The Planetary Science Journal, 2*(132). https://doi.org/10.3847/psj/abfb7a.

Cable, M. L., Runčevski, T., Maynard-Casely, H. E., Vu, T. H., & Hodyss, R. (2021b). Titan in a test tube: Organic co-crystals and implications for Titan mineralogy. *Accounts of Chemical Research, 54*(15), 3050–3059. https://doi.org/10.1021/acs.accounts.1c00250.

Cable, M. L., Vu, T. H., Malaska, M. J., Maynard-Casely, H. E., Choukroun, M., & Hodyss, R. (2019). A co-crystal between Acetylene and Butane: A potentially ubiquitous molecular mineral on Titan. *ACS Earth Space Chemistry, 3*(12), 2808–2815. https://doi.org/10.1021/acsearthspacechem.9b00275.

Carnahan, E., Vance, S. D., Hesse, M. A., Journaux, B., & Sotin, C. (2022). Dynamics of mixed clathrate-ice shells on Ocean Worlds. *Geophysical Research Letters, 49*(8). e2021GL097602.

Castillo-Rogez, J. C. (2006). Internal structure of Rhea. *Journal of Geophysical Research: Planets, 111*(11), 1–13. https://doi.org/10.1029/2004JE002379.

Charnoz, S., Crida, A., Castillo-Rogez, J. C., Lainey, V., Dones, L., Karatekin, Ö., ... Salmon, J. (2011). Accretion of Saturn's mid-sized moons during the viscous spreading of young massive rings: Solving the paradox of silicate-poor rings versus silicate-rich moons. *Icarus, 216*, 535–550. https://doi.org/10.1016/j.icarus.2011.09.017.

Choblet, G., Tobie, G., Kervazo, M., Castillo, J. C., Sotin, C., & Lainey, V. (2018, December). Mimas, Enceladus, Tethys, Dione: Contrasting geological paths for Saturn's inner icy moons. In *AGU Fall meeting abstracts (Vol. 2018, P33A-01)*.

Choblet, G., Tobie, G., Sotin, C., Kalousova, K., & Grasset, O. (2017). Heat transport in the high-pressure ice mantle of large icy moons. *Icarus, 285*, 252–262.

Choukroun, M., Grasset, O., Tobie, G., & Sotin, C. (2010). Stability of methane clathrate hydrates under pressure: Influence on outgassing processes of methane on Titan. *Icarus, 205*(2), 581–593.

Cleaves, H. J. (2012). Prebiotic chemistry: What we know, what we don't. *Evolution: Education and Outreach, 5*, 342–360. https://doi.org/10.1007/s12052-012-0443-9.

Cleaves, H. J., Neish, C., Callahan, M. P., Parker, E., Fernández, F. M., & Dworkin, J. P. (2014). Amino acids generated from hydrated Titan tholins: Comparison with Miller-Urey electric discharge products. *Icarus, 237*, 182–189.

Coates, A. J., Crary, F. J., Lewis, G. R., Young, D. T., Waite, J. H., Jr., & Sittler, E. C., Jr. (2007). Discovery of heavy negative ions in Titan's ionosphere. *Geophysical Research Letters, 34*(22). https://doi.org/10.1029/2007GL030978.

Coates, A. J., Wellbrock, A., Lewis, G. R., Jones, G. H., Young, D. T., Crary, F. J., & Waite, J. H., Jr. (2009). Heavy negative ions in Titan's ionosphere: Altitude and latitude dependence. *Planetary and Space Science, 57*(14–15), 1866–1871. https://doi.org/10.1016/j.pss.2009.05.009.

Cordier, D., Cornet, T., Barnes, J. W., MacKenzie, S. M., Le Bahers, T., Nna-Mvondo, D., Rannou, P., & Ferreira, A. G. (2016). Structure of Titan's evaporites. *Icarus, 270*, 41–56. https://doi.org/10.1016/j.icarus.2015.12.034.

Cordier, D., García-Sánchez, Justo-García, D. N., & Liger-Belair, G. (2017). Bubble streams in Titan's seas as a product of liquid N2+ CH4+ C2H6 cryogenic mixture. *Nature Astronomy, 1*, 0102. https://doi.org/10.1038/s41550-017-0102.

Cordier, D., & Liger-Belair, G. (2018). Bubbles in Titan's seas: Nucleation, growth and RADAR signature. *The Astrophysical Journal*, *859*, 1. https://doi.org/10.3847/1538-4357/aabc10.

Cottini, V., Nixon, C. A., Jennings, D. E., de Kok, R., Teanby, N. A., Irwin, P. G. J., & Flasar, F. M. (2012). Spatial and temporal variations in Titans surface temperatures from Cassini CIRS observations. *Planetary and Space Science*, *60*(1), 62–71.

Crósta, A. P., Silber, E. A., Lopes, R. M. C., Johnson, B. C., Bjonnes, E., Malaska, M. J., … Soderblom, J. M. (2021). Modeling the formation of Menrva impact crater on Titan: Implications for habitability. *Icarus*, *370*(August). https://doi.org/10.1016/j.icarus.2021.114679.

Crow-Willard, E. N., & Pappalardo, R. T. (2015). Structural mapping of Enceladus and implications for formation of tectonized regions. *Journal of Geophysical Research: Planets*, *120*. https://doi.org/10.1002/2015JE004818.

Cruikshank, D. P., Dalton, J. B., Dalle Ore, C. M., Bauer, J., Stephan, K., Filacchione, G., … Mennella, V. (2007). Surface composition of Hyperion. *Nature*, *448*(7149), 54–56.

Cruikshank, D. P., Materese, C. K., Pendleton, Y. J., Boston, P. J., Grundy, W. M., Schmitt, B., … Bray, V. J. (2019). Prebiotic chemistry of Pluto. *Astrobiology*, *19*(7), 831–848. https://doi.org/10.1089/ast.2018.1927.

Cruikshank, D. P., Meyer, A. W., Brown, R. H., Clark, R. N., Jaumann, R., Stephan, K., … Matson, D. L. (2010). Carbon dioxide on the satellites of Saturn: Results from the Cassini VIMS investigation and revisions to the VIMS wavelength scale. *Icarus*, *206*(2), 561–572.

Cruikshank, D. P., Owen, T. C., Ore, C. D., Geballe, T. R., Roush, T. L., de Bergh, C., … Emery, J. P. (2005). A spectroscopic study of the surfaces of Saturn's large satellites: H2O ice, tholins, and minor constituents. *Icarus*, *175*(1), 268–283. https://doi.org/10.1016/j.icarus.2004.09.003.

Cruikshank, D. P., Pendleton, Y. J., & Grundy, W. M. (2020). Organic components of small bodies in the solar system: Some results of the new horizons mission. *Life*, *10*, 126. https://doi.org/10.3390/life10080126.

Czaplinski, E. C., Gilbertson, W. A., Farnsworth, K. K., & Chevrier, V. F. (2019). Experimental study of ethylene evaporites under Titan conditions. *ACS Earth and Space Chemistry*, *3*(10), 2353–2362. https://doi.org/10.1021/acsearthspacechem.9b00204.

Czaplinski, E. C., Yu, X., Dzurilla, K., & Chevrier, V. (2020). Experimental investigation of the acetylene-benzene cocrystal on Titan. *The Planetary Science Journal*, *1*, 76. https://doi.org/10.3847/PSJ/abbf57.

Czechowski, L. (2012, September). Differentiation and melting of Rhea. In *European planetary science congress 2012* (pp. EPSC2012-783).

Czechowski, L. (2014). Some remarks on the early evolution of Enceladus. *Planetary and Space Science*, *104*, 185–199. https://doi.org/10.1016/j.pss.2014.09.010.

Czechowski, L., & Leliwa-Kopystynski, J. (2013). Remarks on the Iapetus' bulge and ridge. *Earth, Planets and Space*, *65*(8), 929–934. https://doi.org/10.5047/eps.2012.12.008.

Czechowski, L., & Witek, P. (2015). Comparison of early evolutions of Mimas and Enceladus. *Acta Geophysica*, *63*(3), 900–921. https://doi.org/10.1515/acgeo-2015-0024.

Dalle Ore, C. M., Cruikshank, D. P., Mastrapa, R. M., Lewis, E., & White, O. L. (2015). Impact craters: An ice study on Rhea. *Icarus*, *261*, 80–90.

Dalle Ore, C. M., Long, C. J., Nichols-Fleming, F., Scipioni, F., Valentín, E. G. R., Oquendo, A. J. L., & Cruikshank, D. P. (2021). Dione's Wispy terrain: A cryovolcanic story? *The Planetary Science Journal*, *2*(2), 83.

Damptz, A. L., Dombard, A. J., & Kirchoff, M. R. (2018). Testing models for the formation of the equatorial ridge on Iapetus via crater counting. *Icarus*, *302*, 134–144. https://doi.org/10.1016/j.icarus.2017.10.049.

De Kok, R., Irwin, P. G. J., Teanby, N. A., Nixon, C. A., Jennings, D. E., Fletcher, L., … Taylor, F. W. (2007). Characteristics of Titan's stratospheric aerosols and condensate clouds from Cassini CIRS far-infrared spectra. *Icarus*, *191*(1), 223–235. https://doi.org/10.1016/j.icarus.2007.04.003.

De Kok, R. J., Teanby, N. A., Maltagliati, L., Irwin, P. G., & Vinatier, S. (2014). HCN ice in Titan's high-altitude southern polar cloud. *Nature*, *514*(7520), 65–67.

Detelich, C. E., Byrne, P. K., Dombard, A. J., & Schenk, P. M. (2021). The morphology and age of the Iapetus equatorial ridge supports an exogenic origin. *Icarus, 367,* 114559. https://doi.org/10.1016/j.icarus.2021.114559.

Dhingra, R. D., Barnes, J. W., Hedman, M. M., & Radebaugh, J. (2019). Using elliptical Fourier descriptor analysis (EFDA) to quantify Titan lake morphology. *The Astronomical Journal, 158*(6), 230.

Dhingra, R. D., Barnes, J. W., Heslar, M. F., Brown, R. H., Buratti, B. J., Sotin, C., … Jaumann, R. (2020). Spatio-temporal variation of bright ephemeral features on Titan's north pole. *The Planetary Science Journal, 1*(2), 31.

Dhingra, R. D., Jennings, D. E., Barnes, J. W., & Cottini, V. (2021). Lower surface temperature at bright ephemeral feature site on Titan's north pole. *Geophysical Research Letters, 48*(7). e2020GL091708.

Dougherty, M. K., Khurana, K. K., Neubauer, F. M., Russell, C. T., Saur, J., Leisner, J. S., & Burton, M. E. (2006). Identification of a dynamic atmosphere at Enceladus with the Cassini magnetometer. *Science, 311*(5766), 1406–1409. https://doi.org/10.1126/science.1120985.

Dougherty, M. K., & Spilker, L. J. (2018). Review of Saturn's ice moons following the Cassini mission. *Reports on Progress in Physics, 81,* 065901. https://doi.org/10.1088/1361-6633/aabdfb.

Elachi, C., Allison, M. D., Borgarelli, L., Encrenaz, P., Im, E., Janssen, M. A., … Zebker, H. A. (2004). Radar: The Cassini Titan radar mapper. *The Cassini-Huygens Mission,* 71–110.

Elachi, C., Wall, S., Janssen, M., Stofan, E., Lopes, R., Kirk, R., … West, R. (2006). Titan radar mapper observations from Cassini's T3 fly-by. *Nature, 441*(7094), 709–713. https://doi.org/10.1038/nature04786.

Engle, A. E., Hanley, J., Dustrud, S., Thompson, G., Lindberg, G. E., Grundy, W. M., & Tegler, S. C. (2021). Phase diagram for the methane-ethane system and its implications for Titan's lakes. *The Planetary Science Journal, 2*(118), 1–10. https://doi.org/10.3847/PSJ/abf7d0.

Farnsworth, K. K., Chevrier, V. F., Steckloff, J. K., Laxton, D., Singh, S., Soto, A., & Soderblom, J. M. (2019). Nitrogen exsolution and bubble formation in Titan's lakes. *Geophysical Research Letters, 46*(23), 13658–13667. https://doi.org/10.1029/2019GL084792.

Filacchione, G., D'Aversa, E., Capaccioni, F., Clark, R. N., Cruikshank, D. P., Ciarniello, M., … Dalle Ore, C. M. (2016). Saturn's icy satellites investigated by Cassini-VIMS. IV. Daytime temperature maps. *Icarus, 271,* 292–313. https://doi.org/10.1016/J.ICARUS.2016.02.019.

Fortes, A. D. (2012). Titan's internal structure and the evolutionary consequences. *Planetary and Space Science, 60*(1), 10–17.

Fulchignoni, M., Ferri, F., Angrilli, F., Ball, A. J., Bar-Nun, A., Barucci, M. A., … Zarnecki, J. C. (2005). In situ measurements of the physical characteristics of Titan's environment. *Nature, 438*(7069), 785–791.

Fuller, J., Luan, J., & Quataert, E. (2016). Resonance locking as the source of rapid tidal migration in the Jupiter and Saturn moon systems. *Monthly Notices of the Royal Astronomical Society, 458*(4), 3867–3879. https://doi.org/10.1093/mnras/stw609.

Giese, B., Denk, T., Neukum, G., Roatsch, T., Helfenstein, P., Thomas, P. C., … Porco, C. C. (2008). The topography of Iapetus' leading side. *Icarus, 193,* 359–371. https://doi.org/10.1016/j.icarus.2007.06.005.

Giese, B., Wagner, R., Neukum, G., Helfenstein, P., & Thomas, P. C. (2007). Tethys: Lithospheric thickness and heat flux from flexurally supported topography at Ithaca Chasma. *Geophysical Research Letters, 34,* 1–5. https://doi.org/10.1029/2007GL031467.

Glein, C. R., & Waite, J. H. (2020). The carbonate geochemistry of Enceladus' Ocean. *Geophysical Research Letters, 47*(3). https://doi.org/10.1029/2019GL085885.

Goguen, J. D., Buratti, B. J., Brown, R. H., Clark, R. N., Nicholson, P. D., Hedman, M. M., … Blackburn, D. G. (2013). The temperature and width of an active fissure on Enceladus measured with Cassini VIMS during the 14 April 2012 South Pole flyover. *Icarus, 226,* 1128–1137. https://doi.org/10.1016/j.icarus.2013.07.012.

Greeley, R., & Iversen, J. D. (1985). *Wind as a geological process: On Earth, Mars, Venus and Titan (No. 4).* CUP Archive.

Griffith, C. A., Penteado, P., Rannou, P., Brown, R., Boudon, V., Baines, K. H., … Jaumann, R. (2006). Evidence for a polar ethane cloud on Titan. *Science, 313*(5793), 1620–1622.

Griffith, C. A., Penteado, P. F., Turner, J. D., Neish, C. D., Mitri, G., Montiel, N. J., … Lopes, R. (2019). A corridor of exposed ice-rich bedrock across Titan's tropical region. *Nature Astronomy, 3*(7), 642–648.

Griffith, C. A., et al. (2003). Evidence for the exposure of water ice on Titan's surface. *Science, 300*(5619), 628–630. https://doi.org/10.1126/science.1081897.

Hand, K. P., Sotin, C., Hayes, A., & Coustenis, A. (2020). On the habitability and future exploration of Ocean Worlds. *Space Science Reviews, 216*, 95. https://doi.org/10.1007/s11214-020-00713-7.

Hayes, A., Aharonson, O., Callahan, P., Elachi, C., Gim, Y., Kirk, R., … Wall, S. (2008). Hydrocarbon lakes on Titan: Distribution and interaction with a porous regolith. *Geophysical Research Letters, 35*(9).

Hayes, A. G., Soderblom, J. M., Barnes, J. W., Hofgartner, J. D., Poggiali, V., & Mastrogiuseppe, M. (2018). Wind, waves, and magic islands at Titan's largest sea: Kraken Mare. In *Lunar and planetary science conference XLIX, abstract #2065.*

Hedgepeth, J. E., Neish, C. D., Turtle, E. P., Stiles, B. W., Krik, R., & Lorenz, R. D. (2020). Titan's impact crater population after Cassini. *Icarus, 344*, 113664. https://doi.org/10.1016/j.icarus.2020.113664.

Hedman, M. M., Gosmeyer, C. M., Nicholson, P. D., Sotin, C., Brown, R. H., Clark, R. N., … Showalter, M. R. (2013). An observed correlation between plume activity and tidal stresses on Enceladus. *Nature, 500*(7461), 182–184. https://doi.org/10.1038/nature12371.

Helfenstein, P., & Porco, C. C. (2015). Enceladus' Geysers: Relation to geological features. *Astronomical Journal, 150*(3), 96. https://doi.org/10.1088/0004-6256/150/3/96.

Hemingway, D. J., & Mittal, T. (2019). Enceladus's ice shell structure as a window on internal heat production. *Icarus, 332*, 111–131. https://doi.org/10.1016/j.icarus.2019.03.011.

Hendrix, A. R., Hurford, T. A., Barge, L. M., Bland, M. T., Bowman, J. S., Brinckerhoff, W., … Vance, S. D. (2019). The NASA roadmap to Ocean worlds. *Astrobiology, 19*(1), 1–27. https://doi.org/10.1089/ast.2018.1955.

Hibbard, R. R., & Evans, A., Jr. (1968). *On the solubilities and rates of solution of gases in liquid methane.* Cleveland, OH: National Aeronautics and Space Administration Technical Note (US). Aug. 22 p. NASA TN Dd-4701 https://archive.org/details/nasa_techdoc_19680020605.

Hobbs, P. V. (2010). *Ice physics.* Oxford University Press, ISBN:9780199587711.

Hofgartner, J. D., Hayes, A. G., Lunine, J. I., Zebker, H., Lorenz, R. D., … Soderblom, J. M. (2016). Titan's "Magic Islands": Transient features in a hydrocarbon sea. *Icarus, 271*, 338–349. https://doi.org/10.1016/j.icarus.2016.02.022.

Hofgartner, J. D., Hayes, A. G., Lunine, J. I., Zebker, H., Stiles, B. W., … Wood, C. (2014). Transient features in a Titan sea. *Nature Geoscience, 7*(7), 493–496. https://doi.org/10.1038/ngeo2190.

Hörst, S. M. (2017). Titan's atmosphere and climate. *Journal of Geophysical Research: Planets, 122*, 432–482.

Hörst, S. M., Yelle, R. V., Buch, A., Carrasco, N., Cernogora, G., … Vuitton, V. (2012). Formation of amino acids and nucleotide bases in a Titan atmosphere simulation experiment. *Astrobiology, 12*(9), 809–817.

Howell, S. M., Bierson, C. J., Kalousová, K., Leonard, E., Steinbrügge, G., & Wolfenbarger, N. (2024). *Jupiter's ocean worlds: Dynamic ices and the search for life.* In R. J. Soare, J.-P. Williams, C. Ahrens, F. E. G. Butcher, & M. R. El-Maarry (Eds.), *Ices in the solar system, a volatile-driven journey from the inner solar system to its far reaches.* Elsevier Books.

Howett, C. J. A., Spencer, J. R., Hurford, T., Verbiscer, A., & Segura, M. (2014). Thermophysical property variations across Dione and Rhea. *Icarus, 241*, 239–247.

Howett, C. J. A., Spencer, J. R., Pearl, J., & Segura, M. (2010). Thermal inertia and bolometric Bond albedo values for Mimas, Enceladus, Tethys, Dione, Rhea and Iapetus as derived from Cassini/CIRS measurements. *Icarus, 206*(2), 573–593. https://doi.org/10.1016/j.icarus.2009.07.016.

Hurford, T. A., Helfenstein, P., Hoppa, G. V., Greenberg, R., & Bills, B. G. (2007). Eruptions arising from tidally controlled periodic openings of rifts on Enceladus. *Nature, 447*(7142), 292–294. https://doi.org/10.1038/nature05821.

Hussmann, H., Rodríguez, A., Callegari, N., & Shoji, D. (2019). Early resonances of Tethys and Dione: Implications for Ithaca Chasma. *Icarus, 319*, 407–416. https://doi.org/10.1016/J.ICARUS.2018.09.025.

Ida, S. (2019). The origin of Saturn's rings and moons. *Science, 364*(6445), 1028–1030. https://doi.org/10.1126/science.aaw3098.

Iess, L., Rappaport, N. J., Tortora, P., Lunine, J., Armstrong, J. W., Asmar, S. W., … Zingoni, F. (2007). Gravity field and interior of Rhea from Cassini data analysis. *Icarus, 190*, 585–593. https://doi.org/10.1016/j.icarus.2007.03.027.

Ingersoll, A. P., Ewald, S. P., & Trumbo, S. K. (2020). Time variability of the Enceladus plumes: Orbital periods, decadal periods, and aperiodic change. *Icarus, 344*, 113345.

Ingersoll, A. P., & Nakajima, M. (2016). Controlled boiling on Enceladus. 2. Model of the liquid-filled cracks. *Icarus, 272*, 319–326. https://doi.org/10.1016/j.icarus.2015.12.040.

Ip, W. H. (2006). On a ring origin of the equatorial ridge of Iapetus. *Geophysical Research Letters, 33*. https://doi.org/10.1029/2005GL025386.

Jaffe, L. D., & Herrell, L. M. (1997). Cassini/Huygens science instruments, spacecraft, and mission. *Journal of Spacecraft and Rockets, 34*(4), 509–521. https://doi.org/10.2514/2.3241.

Jarvis, K. S., Vilas, F., Larson, S. M., & Gaffey, M. J. (2000). Are hyperion and phoebe linked to Iapetus? *Icarus, 146*(1), 125–132.

Jaumann, R., Brown, R. H., Stephan, K., Barnes, J. W., Soderblom, L. A., Sotin, C., … Lorenz, R. D. (2008a). Fluvial erosion and post-erosional processes on Titan. *Icarus, 197*(2), 526–538.

Jaumann, R., Clark, R. N., Nimmo, F., Hendrix, A. R., Buratti, B. J., Denk, T., … Srama, R. (2009a). Icy satellites: Geological evolution and surface processes. In *Saturn from Cassini-Huygens* (pp. 637–681). Dordrecht: Springer.

Jaumann, R., Kirk, R. L., Lorenz, R. D., Lopes, R., Stofan, E., Turtle, E. P., … Tomasko, M. G. (2009b). Geology and surface processes on Titan. In *Titan from Cassini-Huygens* (pp. 75–140). Dordrecht: Springer.

Jaumann, R., Stephan, K., Brown, R. H., Buratti, B. J., Clark, R. N., McCord, T. B., … Porco, C. C. (2006). High-resolution CASSINI-VIMS mosaics of Titan and the icy Saturnian satellites. *Planetary and Space Science, 54*(12), 1146–1155.

Jaumann, R., Stephan, K., Hansen, G. B., Clark, R. N., Buratti, B. J., Brown, R. H., … Wagner, R. (2008b). Distribution of icy particles across Enceladus' surface as derived from Cassini-VIMS measurements. *Icarus, 193*(2), 407–419.

Jennings, D. E., Achterberg, R. K., Cottini, V., Anderson, C. M., Flasar, F. M., Nixon, C. A., … Calcutt, S. (2015). Evolution of the far-infrared cloud at Titan's south pole. *Astrophysical Journal Letters, 804*(2), 1–5. https://doi.org/10.1088/2041-8205/804/2/L34.

Jennings, D. E., Anderson, C. M., Samuelson, R. E., Flasar, F. M., Nixon, C. A., Bjoraker, G. L., … Calcutt, S. B. (2012a). First observation in the south of titan's far-infrared 220 cm^{-1} cloud. *Astrophysical Journal Letters, 761*(1), 1–4. https://doi.org/10.1088/2041-8205/761/1/L15.

Jennings, D. E., Anderson, C. M., Samuelson, R. E., Flasar, F. M., Nixon, C. A., Kunde, V. G., … Calcutt, S. B. (2012b). Seasonal disappearance of far-infrared haze in Titan's stratosphere. *Astrophysical Journal Letters, 754*(1), 1–4. https://doi.org/10.1088/2041-8205/754/1/L3.

Jennings, D. E., Tokano, T., Cottini, V., Nixon, C. A., Achterberg, R. K., … Segura, M. E. (2019). Titan surface temperatures during the Cassini mission. *The Astrophysical Journal Letters, 877*, L8.

Jenniskens, P., & Blake, D. F. (1996). Crystallization of amorphous water ice in the solar system. *The Astrophysical Journal, 473*(2), 1104.

Johnston, S. A., & Montési, L. G. J. (2017). The impact of a pressurized regional sea or global ocean on stresses on Enceladus. *Journal of Geophysical Research: Planets, 122*, 1258–1275. https://doi.org/10.1002/2016JE005217.

Journaux, B., Kalousová, K., Sotin, C., Tobie, G., Vance, S., Saur, J., … Brown, J. M. (2020). Large ocean worlds with high-pressure ices. *Space Science Reviews, 216*(1), 1–36.

Kalousová, K., & Sotin, C. (2020). The insulating effect of methane clathrate crust on Titan's thermal evolution. *Geophysical Research Letters, 47*(13). e2020GL087481.

Kempf, S., Beckmann, U., Moragas-Klostermeyer, G., Postberg, F., Srama, R., Economou, T., … Grün, E. (2008). The E ring in the vicinity of Enceladus. I. Spatial distribution and properties of the ring particles. *Icarus*, *193*(2), 420–437. https://doi.org/10.1016/j.icarus.2007.06.027.

Kempf, S., Beckmann, U., & Schmidt, J. (2010). How the Enceladus dust plume feeds Saturn's E ring. *Icarus*, *206*(2), 446–457. https://doi.org/10.1016/j.icarus.2009.09.016.

Kirchoff, M. R., & Schenk, P. (2009). Crater modification and geologic activity in Enceladus' heavily cratered plains: Evidence from the impact crater distribution. *Icarus*, *202*(2), 656–668. https://doi.org/10.1016/j.icarus.2009.03.034.

Kirchoff, M. R., & Schenk, P. (2010). Impact cratering records of the mid-sized, icy saturnian satellites. *Icarus*, *206*(2), 485–497.

Konstantinidis, K., Flores Martinez, C. L., Dachwald, B., Ohndorf, A., Dykta, P., Bowitz, P., … Förstner, R. (2015). A lander mission to probe subglacial water on Saturn's moon Enceladus for life. *Acta Astronautica*, *106*, 63–89. https://doi.org/10.1016/j.actaastro.2014.09.012.

Kuchta, M., Tobie, G., Miljković, K., Běhounková, M., Souček, O., Choblet, G., & Čadek, O. (2015). Despinning and shape evolution of Saturn's moon Iapetus triggered by a giant impact. *Icarus*, *252*, 454–465. https://doi.org/10.1016/j.icarus.2015.02.010.

Lainey, V., Karatekin, Ö., Desmars, J., Charnoz, S., Arlot, J. E., Emelyanov, N., … Zahn, J. P. (2012). Strong tidal dissipation in saturn and constraints on Enceladus' thermal state from astrometry. *Astrophysical Journal*, *752*(1). https://doi.org/10.1088/0004-637X/752/1/14.

Langhans, M. H., Jaumann, R., Stephan, K., Brown, R. H., Buratti, B. J., Clark, R. N., … Nelson, R. (2012). Titan's fluvial valleys: Morphology, distribution, and spectral properties. *Planetary and Space Science*, *60*(1), 34–51.

Larson, E. J. L., Toon, O. B., & Friedson, A. J. (2014). Simulating Titan's aerosols in a three dimensional general circulation model. *Icarus*, *243*, 400–419. https://doi.org/10.1016/j.icarus.2014.09.003.

Le Corre, L., Le Mouélic, S., Sotin, C., Combe, J. P., Rodriguez, S., Barnes, J. W., … Nicholson, P. D. (2009). Analysis of a cryolava flow-like feature on Titan. *Planetary and Space Science*, *57*(7), 870–879. https://doi.org/10.1016/j.pss.2009.03.005.

Le Mouélic, S., Paillou, P., Janssen, M. A., Barnes, J. W., Rodriguez, S., Sotin, C., … Wall, S. (2008). Mapping and interpretation of Sinlap crater on Titan using Cassini VIMS and RADAR data. *Journal of Geophysical Research: Planets*, *113*(E4).

Lefevre, A., Tobie, G., Choblet, G., & Čadek, O. (2014). Structure and dynamics of Titan's outer icy shell constrained from Cassini data. *Icarus*, *237*, 16–28.

Lopes, R. M., Kirk, R. L., Mitchell, K. L., LeGall, A., Barnes, J. W., Hayes, A., … Malaska, M. J. (2013). Cryovolcanism on Titan: New results from Cassini RADAR and VIMS. *Journal of Geophysical Research: Planets*, *118*(3), 416–435. https://doi.org/10.1002/jgre.20062.

Lopes, R. M. C., Mitchell, K. L., Stofan, E. R., Lunine, J. I., Lorenz, R., Paganelli, F., … Zebker, H. A. (2007a). Cryovolcanic features on Titan's surface as revealed by the Cassini Titan Radar Mapper. *Icarus*, *186*(2), 395–412. https://doi.org/10.1016/j.icarus.2006.09.006.

Lopes, R. M. C., Mitchell, K. L., Wall, S. D., Mitri, G., Janssen, M., Ostro, S., … Paganelli, F. (2007b). The lakes and seas of Titan. *Eos, Transactions American Geophysical Union*, *88*(51), 569–570. https://doi.org/10.1029/2007EO510001.

Lopes, R. M., & Solomonidou, A. (2014, November). Planetary geological processes. In *Vol. 1632 (1)*. *AIP conference proceedings* (pp. 27–57). American Institute of Physics.

Lopez Garcia, E. J., Rivera-Valentin, E. G., Schenk, P. M., Hammond, N. P., & Barr, A. C. (2014). Topographic constraints on the origin of the equatorial ridge on Iapetus. *Icarus*, *237*, 419–421. https://doi.org/10.1016/j.icarus.2014.04.025.

Lorenz, R. D., Lopes, R. M., Paganelli, F., Lunine, J. I., Kirk, R. L., Mitchell, K. L., … Wood, C. A. (2008a). Fluvial channels on Titan: Initial Cassini RADAR observations. *Planetary and Space Science*, *56*(8), 1132–1144. https://doi.org/10.1016/j.pss.2008.02.009.

Lorenz, R. D., MacKenzie, S. M., Neish, C. D., Le Gall, A., Turtle, E. P., … Karkoschka, E. (2021). Selection and characteristics of the dragonfly landing site near Selk Crater, Titan. *The Planetary Science Journal*, *2*, 24. https://doi.org/10.3847/PSJ/abd08f.

Lorenz, R. D., Mitchell, K. L., Kirk, R. L., Hayes, A. G., Aharonson, O., Zebker, H. A., … Stofan, E. R. (2008b). Titan's inventory of organic surface materials. *Geophysical Research Letters*, *35*(2).

Lorenz, R. D., Wall, S., Radebaugh, J., Boubin, G., Reffet, E., Janssen, M., … West, R. (2006). The sand seas of titan: Cassini RADAR observations of longitudinal dunes. *Science*, *312*(5774), 724–727. https://doi.org/10.1126/science.1123257.

MacKenzie, S. M., Barnes, J. W., Hofgartner, J. D., Birch, S. P. D., Hedman, M. M., … Sotin, C. (2019). The case for seasonal surface changes at Titan's lake district. *Nature Astronomy*, *3*, 506–510. https://doi.org/10.1038/s41550-018-0687-6.

MacKenzie, S. M., Barnes, J. W., Sotin, C., Soderblom, J. M., Le Mouélic, S., Rodriguez, S., … McCord, T. B. (2014). Evidence of Titan's climate history from evaporite distribution. *Icarus*, *243*, 191–207.

MacKenzie, S. M., Birch, S. P., Hörst, S., Sotin, C., Barth, E., Lora, J. M., … Coates, A. (2021a). Titan: Earth-like on the outside, ocean world on the inside. *The Planetary Science Journal*, *2*(3), 112.

MacKenzie, S. M., Neveu, M., Davila, A. F., Lunine, J. I., Craft, K. L., … Spilker, L. J. (2021b). The Enceladus orbilander mission concept: Balancing return and resources in the search for life. *Planetary Science Journal*, *2*(77), 1–18. https://doi.org/10.3847/PSJ/abe4da.

Malaska, M. J., Hodyss, R., Lunine, J. I., Hayes, A. G., Hofgartner, J. D., Hollyday, G., & Lorenz, R. D. (2017). Laboratory measurements of nitrogen dissolution in Titan lake fluids. *Icarus*, *289*, 94–105. https://doi.org/10.1016/j.icarus.2017.01.033.

Malaska, M., Radebaugh, J., Le Gall, A., Mitchell, K., Lopes, R., & Wall, S. (2011, March). High-volume meandering channels in Titan's south polar region. In *42nd annual lunar and planetary science conference (No. 1608, p. 1562)*.

Malaska, M. J., Radebaugh, J., Lopes, R. M., Mitchell, K. L., Verlander, T., Schoenfeld, A. M., … Cassini RADAR Team. (2020). Labyrinth terrain on Titan. *Icarus*, *344*, 113764.

Manga, M., & Wang, C. Y. (2007). Pressurized oceans and the eruption of liquid water on Europa and Enceladus. *Geophysical Research Letters*, *34*(7), L07202. https://doi.org/10.1029/2007GL029297.

Martens, H. R., Ingersoll, A. P., Ewald, S. P., Helfenstein, P., & Giese, B. (2015). Spatial distribution of ice blocks on Enceladus and implications for their origin and emplacement. *Icarus*, *245*, 162–176. https://doi.org/10.1016/J.ICARUS.2014.09.035.

Mastrogiuseppe, M., Hayes, A. G., Poggiali, V., Lunine, J. I., Lorenz, R. D., … Birch, S. P. D. (2018a). Bathymetry and composition of Titan's Ontario Lacus derived from Monte Carlo-based waveform inversion of Cassini RADAR altimetry data. *Icarus*, *300*, 203–209. https://doi.org/10.1016/j.icarus.2017.09.009.

Mastrogiuseppe, M., Hayes, A., Poggiali, V., Seu, R., Lunine, J. I., & Hofgartner, J. D. (2016). Radar sounding using the Cassini Altimeter: Waveform modeling and Monte Carlo approach for data inversion of observations of Titan's seas. *IEEE*, *54*, 5646–5656. https://doi.org/10.1109/TGRS.2016.2563426.

Mastrogiuseppe, M., Poggiali, V., Hayes, A., Lorenz, R., Lunine, J., … Zebker, H. (2014). The bathymetry of a Titan sea. *Geophysical Research Letters*, *41*(5), 1432–1437. https://doi.org/10.1002/2013GL058618.

Mastrogiuseppe, M., Poggiali, V., Hayes, A. G., Lunine, J. I., Seu, R., Di Achille, G., & Lorenz, R. D. (2018b). Cassini radar observation of Punga Mare and environs: Bathymetry and composition. *Earth and Planet Letters*, *496*, 89–95. https://doi.org/10.1016/j.epsl.2018.05.033.

Matson, D. L., Davies, A. G., Johnson, T. V., Combe, J.-P., McCord, T. B., Radebaugh, J., & Singh, S. (2018). Enceladus' near-surface CO2 gas pockets and surface frost deposits. *Icarus*, *302*, 18–26. https://doi.org/10.1016/J.ICARUS.2017.10.025.

Matteoni, P., Mitri, G., Poggiali, V., & Mastrogiuseppe, M. (2020). Geomorphological analysis of the southwestern margin of Xanadu, Titan: Insights on tectonics. *Journal of Geophysical Research: Planets*, *125*(12). e2020JE006407.

McKay, C. P., Anbar, A. D., Porco, C. C., & Tsou, P. (2014). Follow the Plume: The habitability of Enceladus. *Astrobiology*, *14*(4), 352–355. https://doi.org/10.1089/ast.2014.1158.

McKay, C. P., Porco, C. C., Altheide, T., Davis, W. L., & Kral, T. A. (2008). The possible origin and persistence of life on Enceladus and detection of biomarkers in the Plume. *Astrobiology*, *8*(5). https://doi.org/10.1089/ast.2008.0265.

McKay, A. J., & Roth, N. X. (2021). Organic matter in cometary environments. *Life (Basel)*, *11*(1), 37. https://doi.org/10.3390/life11010037.

McKinnon, W. B., & Melosh, H. J. (1980). Evolution of planetary lithospheres: Evidence from multiringed structures on Ganymede and Callisto. *Icarus*, *44*(2), 454–471.

Miles, G., Howett, C. J., Spencer, J., & Schenk, P. (2022). Sub-field of view surface thermal modeling of Cassini CIRS observations of Rhea during south polar winter. *Icarus*, *377*, 114910.

Miller, S. L. (1955). Production of some organic compounds under possible primitive earth conditions. *Journal of the American Chemical Society*, *77*(9), 2351–2361.

Mitchell, J. L., & Lora, J. M. (2016). The climate of Titan. *Annual Review of Earth and Planetary Sciences*, *44*, 353–380.

Mitri, G., Bland, M. T., Showman, A. P., Radebaugh, J., Stiles, B., Lopes, R. M., … Pappalardo, R. T. (2010). Mountains on Titan: Modeling and observations. *Journal of Geophysical Research: Planets*, *115*(E10). https://doi.org/10.1029/2010JE003592.

Mitri, G., Meriggiola, R., Hayes, A., Lefèvre, A., Tobie, G., Genova, A., … Zebker, H. (2014). Shape, topography, gravity anomalies and tidal deformation of Titan. *Icarus*, *236*, 169–177. https://doi.org/10.1016/j.icarus.2014.03.018.

Mitri, G., Showman, A. P., Lunine, J. I., & Lorenz, R. D. (2007). Hydrocarbon lakes on Titan. *Icarus*, *186*(2), 385–394.

Moore, J. M., & Ahern, J. L. (1983). The geology of Tethys. *Journal of Geophysical Research: Solid Earth*, *88*(S02), A577–A584.

Moore, J. M., Schenk, P. M., Bruesch, L. S., Asphaug, E., & McKinnon, W. B. (2004). Large impact features on middle-sized icy satellites. *Icarus*, *171*(2), 421–443.

Nathan, E., Huber, C., & Head, J. (2022). A stress-based framework for understanding the evolution of icy worlds. In *Lunar and planetary science conference 53, abstract #1075*.

National Academies of Sciences, Engineering, and Medicine. (2022). *Origins, worlds, and life: A decadal strategy for planetary science and astrobiology 2023-2032*. Washington, DC: National Academies Press. https://doi.org/10.17226/26522.

Neish, C. D., & Lorenz, R. D. (2012). Titan's global crater population: A new assessment. *Planetary and Space Science*, *60*(1), 26–33.

Neish, C. D., Lorenz, R. D., Turtle, E. P., Barnes, J. W., Trainer, M. G., … Malaska, M. J. (2018). Strategies for detecting biological molecules on Titan. *Astrobiology*, *18*(5), 571–585.

Neish, C. D., Somogyi, Á., Imanaka, H., Lunine, J. I., & Smith, M. A. (2008). Rate Measurements of the hydrolysis of complex organic macromolecules in cold aqueous solutions: Implications for prebiotic chemistry on the Early Earth and Titan. *Astrobiology*, *8*, 273–287.

Neish, C. D., Somogyi, Á., Lunine, J. I., & Smith, M. A. (2009). Low temperature hydrolysis of laboratory tholins in ammonia-water solutions: Implications for prebiotic chemistry on Titan. *Icarus*, *201*, 412–421.

Neish, C. D., Somogyi, A., & Smith, M. A. (2010). Titan's primordial soup: Formation of amino acids via low-temperature hydrolysis of tholins. *Astrobiology*, *10*(3), 337–347. https://doi.org/10.1089/ast.2009.0402.

Néri, A., Guyot, F., Reynard, B., & Sotin, C. (2020). A carbonaceous chondrite and cometary origin for icy moons of Jupiter and Saturn. *Earth and Planetary Science Letters*, *530*, 115920.

Neukum, G., Wagner, R., Wolf, U., & Denk, T. (2006). The cratering record and cratering chronologies of the saturnian satellites and the origin of impactors: Results from Cassini ISS data. *European Planetary Science Congress, 610.*

Neveu, M., & Rhoden, A. R. (2017). The origin and evolution of a differentiated Mimas. *Icarus, 296,* 183–196. https://doi.org/10.1016/j.icarus.2017.06.011.

New, J. S., Kazemi, B., Spathis, V., Price, M. C., Mathies, R. A., & Butterworth, A. L. (2021). Quantitative evaluation of the feasibility of sampling the ice plumes at Enceladus for biomarkers of extraterrestrial life. *Proceedings of the National Academy of Sciences of the United States of America, 118*(37), 1–6. https://doi.org/10.1073/pnas.2106197118.

Newman, S. F., Buratti, B. J., Brown, R. H., Jaumann, R., Bauer, J., & Momary, T. (2009). Water ice crystallinity and grain sizes on Dione. *Icarus, 203*(2), 553–559.

Niemann, H. B., Atreya, S. K., Demick, J. E., Gautier, D., Haberman, J. A., ... Raulin, O. F. (2010). Composition of Titan's lower atmosphere and simple surface volatiles as measured by the Cassini-Huygens probe gas chromatograph mass spectrometer experiment. *Journal of Geophysical Research: Planets, 115*(E12), E12006. https://doi.org/10.1029/2010JE003659.

Nimmo, F., Barr, A. C., Běhounková, M., & McKinnon, W. B. (2018). The thermal and orbital evolution of Enceladus: Observational constraints and models. In R. Doston (Ed.), *Enceladus and the icy Moons of Saturn* (pp. 79–94). Tucson: University of Arizona Press. https://doi.org/10.2458/azu.

Nimmo, F., & Pappalardo, R. T. (2016). Ocean worlds in the outer solar system. *Journal of Geophysical Research: Planets, 121*(8), 1378–1399.

Palmer, E. E., & Brown, R. H. (2011). Production and detection of carbon dioxide on Iapetus. *Icarus, 212*(2), 807–818.

Parkinson, C. D., Liang, M.-C., Yung, Y. L., Kirschivnk, J. L., Parkinson, C. D., Liang, M.-C., ... Kirschivnk, J. L. (2008). Habitability of Enceladus: Planetary conditions for life. *Origins of Life and Evolution of Biospheres, 38,* 355–369. https://doi.org/10.1007/s11084-008-9135-4.

Plescia, J. B. (1983). The geology of Dione. *Icarus, 56*(2), 255–277.

Plescia, J. B., & Boyce, J. M. (1982). Crater densities and geological histories of Rhea, Dione, Mimas and Tethys. *Nature, 295*(5847), 285–290.

Poch, O., Coll, P., Buch, A., Ramírez, S. I., & Raulin, F. (2012). Production yields of organics of astrobiological interest from H2ONH3 hydrolysis of Titans tholins. *Planetary and Space Science, 61*(1), 114–123.

Poggiali, V., Hayes, A. G., Mastrogiuseppe, M., Le Gall, A., Lalich, D., Gómez-Leal, I., & Lunine, J. I. (2020). The bathymetry of moray sinus at Titan's Kraken Mare. *Journal of Geophysical Research: Planets, 125.* https://doi.org/10.1029/2020JE006558. e2020JE006558.

Poggiali, V., Mastrogiuseppe, M., Hayes, A. G., Seu, R., Birch, S. P. D., Lorenz, R., ... Hofgartner, J. D. (2016). Liquid-filled canyons on Titan. *Geophysical Research Letters, 43*(15), 7887–7894.

Porco, C. C., Baker, E., Barbara, J., Beurle, K., Brahic, A., Burns, J. A., ... West, R. (2005a). Imaging of Titan from the Cassini spacecraft. *Nature, 434*(7030), 159–168.

Porco, C. C., Baker, E., Barbara, J., Beurle, K., Brahic, A., Burns, J. A., ... West, R. (2005b). Cassini imaging science: Initial results on Phoebe and Iapetus. *Science, 307*(5713), 1237–1242. https://doi.org/10.1126/science.1107981.

Porco, C. C., DiNino, D., & Nimmo, F. (2014). How the Geysers, Tidal stress, and thermal emission across the south polar terrain of Enceladus are related. *The Astronomical Journal, 148*(3), 45. https://doi.org/10.1088/0004-6256/148/3/45.

Porco, C. C., Helfenstein, P., Thomas, P. C., Ingersoll, A. P., Wisdom, J., West, R. D., ... Squyres, S. W. (2006). Cassini observes the active south pole of Enceladus. *Science, 311*(5766), 1393–1401. https://doi.org/10.1126/science.1123013.

Portyankina, G., Esposito, L. W., Aye, K. M., Hansen, C. J., & Ali, A. (2022). Modeling the complete set of Cassini's UVIS occultation observations of Enceladus' plume. *Icarus, 383,* 114918. https://doi.org/10.1016/j.icarus.2022.114918.

Postberg, F., Kempf, S., Hillier, J. K., Srama, R., Green, S. F., McBride, N., & Grün, E. (2008). The E-ring in the vicinity of Enceladus. II. Probing the moon's interior—The composition of E-ring particles. *Icarus, 193*(2), 438–454. https://doi.org/10.1016/j.icarus.2007.09.001.

Postberg, F., Kempf, S., Schmidt, J., Brilliantov, N., Beinsen, A., Abel, B., … Srama, R. (2009). Sodium salts in E-ring ice grains from an ocean below the surface of Enceladus. *Nature, 459*(7250), 1098–1101. https://doi.org/10.1038/nature08046.

Postberg, F., Khawaja, N., Abel, B., Choblet, G., Glein, C. R., Gudipati, M. S., … Waite, J. H. (2018). Macromolecular organic compounds from the depths of Enceladus. *Nature, 558*(7711), 564–568. https://doi.org/10.1038/s41586-018-0246-4.

Radebaugh, J., Lorenz, R. D., Kirk, R. L., Lunine, J. I., Stofan, E. R., Lopes, R. M., … Cassini Radar Team. (2007). Mountains on Titan observed by Cassini RADAR. *Icarus, 192*(1), 77–91.

Radebaugh, J., Lorenz, R. D., Lunine, J. I., Wall, S. D., Boubin, G., … The Cassini Radar Team. (2008). Dunes on Titan observed by Cassini Radar. *Icarus, 194*, 690–703.

Radebaugh, J., Ventra, D., Lorenz, R. D., Farr, T., Kirk, R., Hayes, A., Malaska, M. J., Birch, S., Liu, Z. Y.-C., Lunine, J., Barnes, J., Le Gall, A., Lopes, R., Stofan, E., Wall, S., & Paillou, P. (2016). Alluvial and fluvial fans on Saturn's moon Titan reveal processes, materials and regional geology. In D. Ventra, & L. E. Clarke (Eds.), *Geology and geomorphology of alluvial and fluvial fans: Terrestrial and planetary perspectives* Geological Society, London, Special Publications, 440. https://doi.org/10.1144/SP440.6.

Ramírez, S. I., Coll, P., Buch, A., Brassé, C., Poch, O., & Raulin, F. (2010). The fate of aerosols on the surface of Titan. *Faraday Discussions, 147*, 419–427.

Rannou, P., Lebonnois, S., Hourdin, F., & Luz, D. (2005). Titan atmosphere database. *Advances in Space Research, 36*(11), 2194–2198.

Raulin, F., Brassé, C., Poch, O., & Coll, P. (2012). Prebiotic-like chemistry on Titan. *Chemical Society Reviews, 41*(16), 5380–5393.

Rhoden, A. R., Henning, W., Hurford, T. A., Patthoff, D. A., & Tajeddine, R. (2017). The implications of tides on the Mimas ocean hypothesis. *Journal of Geophysical Research: Planets, 122*, 400–410. https://doi.org/10.1002/2016JE005097.

Rhoden, A. R., & Walker, M. E. (2022). The case for an ocean-bearing Mimas from tidal heating analysis. *Icarus, 376*. https://doi.org/10.1016/j.icarus.2021.114872.

Richardson, I. A., Hartwig, J. W., & Leachman, J. W. (2019). Experimental effervescence and freezing point depression measurements of nitrogen in liquid methane-ethane mixtures. *International Journal of Thermal Sciences, 137*, 534–538. https://doi.org/10.1016/j.ijthermalsci.2018.12.024.

Richardson, J. E., Veverka, J., & Thomas, P. C. (2006, September). Large impact features on Phoebe and Hyperion: Early analysis results. In *AAS/Division for Planetary Sciences Meeting Abstracts# 38, pp. 69-04*.

Rivera-Valentin, E. G., Blackburn, D. G., & Ulrich, R. (2011). Revisiting the thermal inertia of Iapetus: Clues to the thickness of the dark material. *Icarus, 216*(1), 347–358. https://doi.org/10.1016/j.icarus.2011.09.006.

Roberts, J. H., & Nimmo, F. (2008). Tidal heating and the long-term stability of a subsurface ocean on Enceladus. *Icarus, 194*, 675–689. https://doi.org/10.1016/j.icarus.2007.11.010.

Robidel, R., Le Mouélic, S., Tobie, G., Massé, M., Seignovert, B., Sotin, C., & Rodriguez, S. (2020). Photometrically-corrected global infrared mosaics of Enceladus: New implications for its spectral diversity and geological activity. *Icarus, 349*, 113848. https://doi.org/10.1016/j.icarus.2020.113848.

Rodriguez, S., Garcia, A., Lucas, A., Appéré, T., Le Gall, A., Reffet, E., … Turtle, E. P. (2014). Global mapping and characterization of Titan's dune fields with Cassini: Correlation between RADAR and VIMS observations. *Icarus, 230*, 168–179.

Rudolph, M. L., Manga, M., Walker, M., & Rhoden, A. R. (2022). Cooling crusts create concomitant cryovolcanic cracks. *Geophysical Research Letters, 49*(5), 1–11. https://doi.org/10.1029/2021gl094421.

Samuelson, R. E., Smith, M. D., Achterberg, R. K., & Pearl, J. C. (2007). Cassini CIRS update on stratospheric ices at Titan's winter pole. *Icarus, 189*(1), 63–71. https://doi.org/10.1016/j.icarus.2007.02.005.

Schenk, P. M., Hamilton, D. P., Johnson, R. E., McKinnon, W. B., Paranicas, C., Schmidt, J., & Showalter, M. R. (2011). Plasma, plumes and rings: Saturn system dynamics as recorded in global color patterns on its midsize icy satellites. *Icarus, 211*(1), 740–757. https://doi.org/10.1016/j.icarus.2010.08.016.

Sebree, J. A., Roach, M. C., Shipley, E. R., He, C., & Horst, S. M. (2018). Detection of prebiotic molecules in plasma and photochemical aerosol analogs using GC/MS/MS techniques. *Astrophysical Journal, 865*, 133. https://doi.org/10.3847/1538-4357/aadba1.

Shi, J., Grieves, G. A., & Orlando, T. M. (2015). Vacuum ultraviolet photon-stimulated oxidation of buried ice: Graphite grain interfaces. *The Astrophysical Journal, 804*(1), 24.

Showalter, M. R. (2020). The rings and small moons of Uranus and Neptune: Rings and Moons of Uranus and Neptune. *Philosophical Transactions of the Royal Society A: Mathematical, Physical and Engineering Sciences, 378*(2187), 1–12. https://doi.org/10.1098/rsta.2019.0482.

Simon, S., Saur, J., Neubauer, F. M., Wennmacher, A., & Dougherty, M. K. (2011). Magnetic signatures of a tenuous atmosphere at Dione. *Geophysical Research Letters, 38*(15).

Singer, K. N., McKinnon, W. B., Schenk, P. M., & Moore, J. M. (2012). Massive ice avalanches on Iapetus mobilized by friction reduction during flash heating. *Nature Geoscience, 5*(8), 574–578.

Soderblom, L. A., Brown, R. H., Soderblom, J. M., Barnes, J. W., Kirk, R. L., Sotin, C., ... Nicholson, P. D. (2009). The geology of Hotei Regio, Titan: Correlation of Cassini VIMS and RADAR. *Icarus, 204*(2), 610–618.

Soderblom, L. A., Tomasko, M. G., Archinal, B. A., Becker, T. L., Bushroe, M. W., Cook, D. A., ... Smith, P. H. (2007). Topography and geomorphology of the Huygens landing site on Titan. *Planetary and Space Science, 55*(13), 2015–2024. https://doi.org/10.1016/j.pss.2007.04.015.

Solomonidou, A., Coustenis, A., Lopes, R. M., Malaska, M. J., Rodriguez, S., Drossart, P., ... Schoenfeld, A. (2018). The spectral nature of Titan's major geomorphological units: Constraints on surface composition. *Journal of Geophysical Research: Planets, 123*(2), 489–507.

Solomonidou, A., Hirtzig, M., Coustenis, A., Bratsolis, E., Le Mouélic, S., Rodriguez, S., ... Moussas, X. (2014). Surface albedo spectral properties of geologically interesting areas on Titan. *Journal of Geophysical Research: Planets, 119*(8), 1729–1747.

Sotin, C., Jaumann, R., Buratti, B. J., Brown, R. H., Clark, R. N., Soderblom, L. A., ... Scholz, C. K. (2005). Release of volatiles from a possible cryovolcano from near-infrared imaging of Titan. *Nature, 435*(7043), 786–789. https://doi.org/10.1038/nature03596.

Sotin, C., Kalousová, K., & Tobie, G. (2021). Titan's interior structure and dynamics after the Cassini-Huygens mission. *Annual Review of Earth and Planetary Sciences, 49*(1), 579–607. https://doi.org/10.1146/annurev-earth-072920-052847.

Sotin, C., Mitri, G., Rappaport, N., Schubert, G., & Stevenson, D. (2009). Titan's interior structure. In R. H. Brown, J. P. Lebreton, & J. H. Waite (Eds.), *Titan from Cassini-Huygens*. Dordrecht: Springer. https://doi.org/10.1007/978-1-4020-9215-2_4.

Spahn, F., Schmidt, J., Albers, N., Hörning, M., Makuch, M., Seiß, M., ... Grün, E. (2006). Cassini dust measurement at Enceladus and implications for the origin of the E ring. *Science, 311*(5766), 1416–1418. https://doi.org/10.1126/science.1121375.

Spencer, J. R., & Denk, T. (2010). Formation of Iapetus' extreme albedo dichotomy by exogenically triggered thermal ice migration. *Science, 327*(5964), 432–435.

Spitale, J. N., Hurford, T. A., Rhoden, A. R., Berkson, E. E., & Platts, S. S. (2015). Curtain eruptions from Enceladus' south-polar terrain. *Nature, 521*(7550), 57–60. https://doi.org/10.1038/nature14368.

Steckloff, J. K., Soderblom, J. M., Farnsworth, K. K., Chevrier, V. F., Hanley, J., Soto, A., ... Engle, A. (2020). The evaporation-induced stratification of Titan's lakes. *Planetary Science Journal, 1*, 26. https://doi.org/10.3847/PSJ/ab974e.

Steinbrügge, G., Voigt, J. R. C., Wolfenbarger, N. S., Hamilton, C. W., Soderlund, K. M., Young, D. A., ... Schroeder, D. M. (2020). Brine migration and impact-induced cryovolcanism on Europa. *Geophysical Research Letters, 47*(21), 1–10. https://doi.org/10.1029/2020GL090797.

Stephan, K., Jaumann, R., Wagner, R., Clark, R. N., Cruikshank, D. P., Giese, B., ... Matson, D. L. (2012). The Saturnian satellite Rhea as seen by Cassini VIMS. *Planetary and Space Science, 61*, 142–160. https://doi.org/10.1016/j.pss.2011.07.019.

Stephan, K., Wagner, R., Jaumann, R., Clark, R. N., Cruikshank, D. P., Brown, R. H., ... Nicholson, P. D. (2016). Cassini's geological and compositional view of Tethys. *Icarus, 274*, 1–22. https://doi.org/10.1016/j.icarus.2016.03.002.

Stickle, A. M., & Roberts, J. H. (2018). Modeling an exogenic origin for the equatorial ridge on Iapetus. *Icarus, 307*, 197–206. https://doi.org/10.1016/j.icarus.2018.01.017.

Stofan, E. R., Elachi, C., Lunine, J. I., Lorenz, R. D., Stiles, B., Mitchell, K. L., ... West, R. (2007). The lakes of Titan. *Nature, 445*(7123), 61–64.

Tajeddine, R., Rambaux, N., Lainey, V., Charnoz, S., Richard, A., Rivoldini, A., & Noyelles, B. (2014). Constraints on Mimas' interior from Cassini ISS libration measurements. *Science, 346*(6207), 322–324. https://doi.org/10.1126/science.1255299.

Tajeddine, R., Soderlund, K. M., Thomas, P. C., Helfenstein, P., Hedman, M. M., Burns, J. A., & Schenk, P. M. (2017). True polar wander of Enceladus from topographic data. *Icarus, 295*, 46–60. https://doi.org/10.1016/j.icarus.2017.04.019.

Teanby, N. A., Irwin, P. G., Nixon, C. A., De Kok, R., Vinatier, S., Coustenis, A., ... Flasar, F. M. (2012). Active upper-atmosphere chemistry and dynamics from polar circulation reversal on Titan. *Nature, 491*(7426), 732–735.

Teolis, B. D., Jones, G. H., Miles, P. F., Tokar, B. A., Magee, J. H., ... Baragiola, R. A. (2010). Cassini finds an oxygen-carbon dioxide atmosphere at Saturn's icy moon Rhea. *Science, 330*, 1813. https://doi.org/10.1126/science.1198366.

Teolis, B. D., & Waite, J. H. (2016). Dione and Rhea seasonal exospheres revealed by Cassini CAPS and INMS. *Icarus, 272*, 277–289.

Thomas, P. C., Armstrong, J. W., Asmar, S. W., Burns, J. A., Denk, T., Giese, B., ... Veverka, J. (2007). Hyperion's sponge-like appearance. *Nature, 448*(7149), 50–53.

Thomas, C., Picaud, S., Mousis, O., & Ballenegger, V. (2008). A theoretical investigation into the trapping of noble gases by clathrates on Titan. *Planetary and Space Science, 56*(12), 1607–1617.

Thomas, P. C., Tajeddine, R., Tiscareno, M. S., Burns, J. A., Joseph, J., Loredo, T. J., ... Porco, C. C. (2016). Enceladus's measured physical libration requires a global subsurface ocean. *Icarus, 264*, 37–47. https://doi.org/10.1016/j.icarus.2015.08.037.

Tobie, G., Grasset, O., Lunine, J. I., Mocquet, A., & Sotin, C. (2005). Titan's internal structure inferred from a coupled thermal-orbital model. *Icarus, 175*(2), 496–502.

Turtle, E. P., Del Genio, A. D., Barbara, J. M., Perry, J. E., Schaller, E. L., McEwen, A. S., ... Ray, T. L. (2011a). Seasonal changes in Titan's meteorology. *Geophysical Research Letters, 38*(3).

Turtle, E. P., Perry, J. E., Hayes, A. G., Lorenz, R. D., Barnes, J. W., ... Stofan, E. R. (2011b). Rapid and extensive surface changes near Titan's equator: Evidence of April showers. *Science, 331*(6023), 1414–1417. https://doi.org/10.1126/science.1201063.

Turtle, E. P., Perry, J. E., McEwen, A. S., DelGenio, A. D., Barbara, J., West, R. A., ... Porco, C. C. (2009). Cassini imaging of Titan's high-latitude lakes, clouds, and south-polar surface changes. *Geophysical Research Letters, 36*(2).

Turtle, E. P., et al. (2008). Dragonfly: In situ exploration of Titan's organic chemistry and habitability. In *Lunar and Planetary Science XLIX, abstract #1641*.

Umurhan, O. M., White, O. L., Moore, J. M., Howard, A. D., & Schenk, P. (2016, December). Modeling surface processes occurring on moons of the outer solar system. In *AGU Fall meeting abstracts (Vol. 2016, pp. EP43D-08)*.

Vinatier, S., Schmitt, B., Bézard, B., Rannou, P., Dauphin, C., de Kok, R., ... Flasar, F. M. (2018). Study of Titan's fall southern stratospheric polar cloud composition with Cassini/CIRS: Detection of benzene ice. *Icarus, 310*, 89–104. https://doi.org/10.1016/j.icarus.2017.12.040.

Wagner, R. J., Neukum, G., Denk, T., Giese, B., Roatsch, T., & Cassini ISS Team. (2005, August). The geology of Saturn's satellite Dione observed by Cassini's ISS camera. In *AAS/Division for planetary sciences meeting abstracts# 37, pp. 36-02*.

Waite, J. H., Combi, M. R., Ip, W. H., Cravens, T. E., McNutt, R. L., Kasprzak, W., ... Tseng, W. L. (2006). Cassini ion and neutral mass spectrometer: Enceladus plume composition and structure. *Science, 311*(5766), 1419–1422. https://doi.org/10.1126/science.1121290.

Waite, J. H., Glein, C. R., Perryman, R. S., Teolis, B. D., ... Bolton, S. J. (2017). Cassini finds molecular hydrogen in the Enceladus plume: Evidence for hydrothermal processes. *Science, 356*(6334), 155–159.

Wakita, S., Johnson, B. C., Soderblom, J. M., Shah, J., Neish, C. D., & Steckloff, J. K. (2023). Modeling the formation of selk impact crater on Titan: Implications for dragonfly. *The Planetary Science Journal, 4*, 51. https://doi.org/10.3847/PSJ/acbe40.

Wall, S. D., Lopes, R. M., Stofan, E. R., Wood, C. A., Radebaugh, J. L., Hörst, S. M., ... Mitchell, K. L. (2009). Cassini RADAR images at Hotei Arcus and western Xanadu, Titan: Evidence for geologically recent cryovolcanic activity. *Geophysical Research Letters, 36*(4).

West, R. A., Del Genio, A. D., Barbara, J. M., Toledo, D., Lavvas, P., ... Perry, J. (2016). Cassini imaging science subsystem observations of Titan's south polar cloud. *Icarus, 270*, 399–408. https://doi.org/10.1016/j.icarus.2014.11.038.

Wood, C. A., Lorenz, R., Kirk, R., Lopes, R., Mitchell, K., Stofan, E., & Cassini RADAR Team. (2010). Impact craters on Titan. *Icarus, 206*(1), 334–344.

Yu, X., Hörst, S. M., He, C., McGuiggan, P., Kristiansen, K., & Zhang, X. (2020). Surface energy of the Titan aerosol analog "Tholin". *The Astrophysical Journal, 905*, 88. https://doi.org/10.3847/1538-4357/abc55d.

Yung, Y. L., Allen, M., & Pinto, J. P. (1984). Photochemistry of the atmosphere of Titan: Comparison between model and observations. *The Astrophysical Journal Supplement Series, 55*, 465–506.

Geocryology of Pluto and the icy moons of Uranus and Neptune

Caitlin J. Ahrens[a], Carey M. Lisse[b], Jean-Pierre Williams[c], and Richard J. Soare[d]

[a]*NASA Goddard Space Flight Center, Greenbelt, MD, United States,* [b]*Applied Physics Laboratory, Johns Hopkins University, Laurel, MD, United States,* [c]*Department of Earth, Planetary, and Space Sciences, University of California, Los Angeles, CA, United States,* [d]*Department of Geography, Dawson College, Montreal, QC, Canada*

Abstract

Before the flyby of the Pluto-Charon dwarf planetary system in 2015 by the New Horizons spacecraft, expectations were that the two bodies would be extremely cold, exhibit ancient and densely cratered surfaces, and would be geologically inactive. Surprisingly, the spacecraft delivered images and data that showed Pluto to have: (1) a surface that is sparsely cratered and, as such, relatively young in some regions, (2) landscape features and forms suggestive of relatively recent glaciation, albeit not based on H_2O ice, (3) geological traits that point to cryovolcanism and, despite Pluto's diminutive size, the possibility of a subsurface ocean. Similarly, Charon possesses an extensive canyon/fault system that rivals some of the largest in the solar system and smooth plains that are geologically complex. This points to significant geologic upheaval that included tectonic, and possibly cryovolcanic, resurfacing.

Pluto, the first discovered and best-characterized Kuiper Belt Object (*KBO*), had been considered an archetype for hypotheses concerning the geology and physico-chemical processes associated with these distal bodies. The evocative findings of the New Horizon mission at the Pluto-Charon system and beyond have begun to shift the baseline understanding of *KBOs* and Trans-Neptunian Objects. In this chapter, some of the keynote observations and findings of the New Horizon mission are explored and compared with the principal icy satellites of Uranus and Neptune, as well as with *KBO* Arrokoth.

1 Introduction and background

Before the New Horizons mission, only a handful of spacecraft had ventured into the outer solar system beyond the Jovian (~5.2 AU) and Saturnian (~9.5 AU) systems. Pioneer 10 marked the first spacecraft to venture beyond the asteroid belt and provide the first close encounter with Jupiter (Hall, 1974; Opp, 1974), followed by Pioneer 11, which was launched a year later and flew by both Jupiter and Saturn (Hall, 1975). These missions paved the way for a pair of spacecraft, Voyagers 1 and 2, launched in 1977 (Kohlhase & Penzo, 1977) that collectively conducted a grand tour of the outer solar system inclusive of the Jovian and Saturnian systems. Voyager 2 extended its exploration to the Uranian and Neptunian systems, the only spacecraft to visit these two planetary systems. These spacecraft continued to provide data on the solar wind and cosmic rays as they migrated into the outer reaches of the solar system. The New Horizon spacecraft, launched in 2006, transited the Pluto-Charon system (~34 AU) in 2015, and crossed paths with Kuiper Belt Object (*KBO*) Arrokoth (~45 AU) in 2019 (Fig. 1).

Ices in the Solar System. https://doi.org/10.1016/B978-0-323-99324-1.00016-X

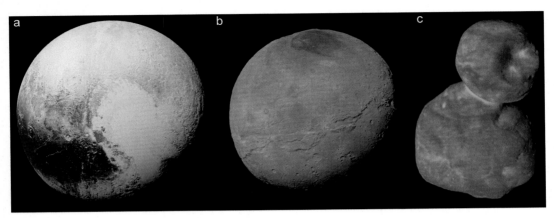

FIG. 1

Color images from the New Horizons mission processed to approximate "true color," colors as would be perceived by the human eye, of: (a) Pluto (mean radius ~1188 km) with the bright nitrogen-methane rich icy expanse, Sputnik Planitia, clearly visible, (b) Charon (mean radius ~606 km) showing striking tectonic features and the dark, reddish north-polar region known as Mordor Macula, and (c) Kuiper Belt Object Arrokoth (dimensions: 35.95 × 19.90 × 9.75 km), a primordial contact binary.

Image Credits: NASA/JHUAPL/SWRI.

The Kuiper Belt is a disk-shaped region comprised of an almost infinitely large set of small objects/bodies, many of which could be icy. It extends outwardly from the orbit of Neptune to ~55 AU and is thought to be homologous with the edge of the original proto-planetary disk (*PPD*). The Kuiper Belt, chemically distinct from the ice giant region (Brown, 2012; Levison & Stern, 2001), did not contain enough spatial mass density to condense a full-sized planet.

Pluto, categorized as a "dwarf" or minor planet, is the largest known Trans-Neptunian Object (*TNO*) in the Kuiper Belt. *TNOs* comprise objects that orbit the sun at a (average) distance greater than that of Neptune. Other large dwarf planets with similar densities and elliptical orbits display a variety of unusual shapes and spins or compositions. For example, ground (Earth)-based spectral observations have identified water ice (H_2O) on Haumea (~43 AU) and a methane (CH_4)-rich dwarf planet, Eris (~68 AU) (Dumas et al., 2011; Tegler et al., 2010). Haumea's elongated shape and rapid spin suggest that it originated from a much larger icy body that was impacted violently; Eris' surface shows extreme temperature fluctuations due to its highly eccentric orbit around the sun (Schaller & Brown, 2007).

Outer solar system satellites such as Ariel (Uranus) and Triton (Neptune) accreted a menagerie of ices, along with carbonaceous material, and rocky interstellar and solar nebular matter. *KBOs*, except for the largest bodies such as Pluto, Eris, and Sedna, probably lost their hypervolatiles (e.g., N_2, CH_4, and CO; Lisse et al., 2021; Schaller & Brown, 2007; Steckloff et al., 2021) within tens of Myrs after the *PPD* disk cleared due to solar heating; moreover, amorphous water ices may have crystallized in the interiors and this would have triggered the release of volatiles trapped in the amorphous ice (Brown, 2012; Lisse et al., 2022).

Outer solar system satellites, *KBOs*, centaurs (small bodies with either a perihelion or a semi-major axis between Jupiter and Neptune), and comets (icy small bodies with highly elliptical orbits rooted in the Kuiper Belt or the Oort Cloud) may have all had similar *PPD* icy feedstock origins in the early solar system; however, their evolutionary paths have been diversified due to differences in their thermal heating (a function of body size and distance from the Sun), collisions (mostly stochastic), and orbital scattering (a function of distance from the Sun and Neptune).

Pluto, Triton, and the icy moons under discussion here run a gamut of sizes with radii ranging from a few hundred to 1200 km (Fig. 2). Pluto and Triton have atmospheres, as they are large enough to gravitationally retain heavy volatiles at the local temperatures ($T \sim 35$ K); smaller Charon and the Uranian/Neptunian moons do not have such atmospheres. The densities and radii of Pluto and Triton, compared to the moons of Jupiter,

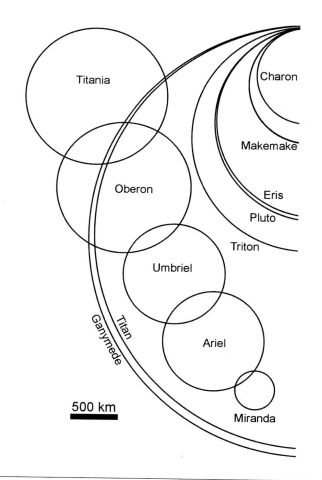

FIG. 2

Size comparison of Pluto/Charon, the icy moons of Uranus/Neptune, Titan/Ganymede, and KBO's Eris/Makemake.

Figure adapted from Croft and Soderblom (1991).

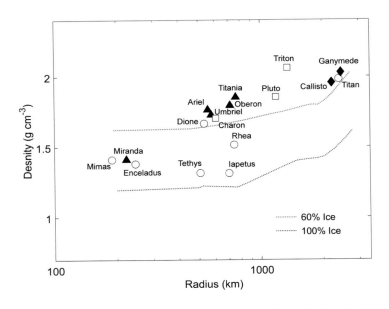

FIG. 3

Densities and radii of outer solar system satellites and Pluto with modeled ice curves for 60% and 100%-by-mass water ice to silicates. *Dark diamond symbols* denote Jovian moons, *open circle symbols* denote Saturnian moons, *dark triangles* for the Uranian satellites, and *open squares* for the Neptune and Pluto systems.

Adapted from Brown et al. (1991).

Saturn, and Uranus (Fig. 3), show that Pluto and Triton fall above the 60% silicate-ice mixture line, similar to the larger Galilean satellites and Titan. Pluto is ~70% chondritic (rock) and ~30% ice by mass (and about 44% rock by volume and 56% ice by volume; McKinnon et al., 2017). Pluto's size and temperature limit the gravitational escape of volatiles such as N_2, CH_4, and CO. This allows the ices to be retained and exist in equilibrium with a tenuous atmosphere (Schaller & Brown, 2007; Tan, 2022; Young et al., 2018). These constituents can be mobilized by climatic atmospheric cycles associated with Pluto's highly elliptical orbit (i.e., 248-Earth-year orbital period and a 2.8-Myr obliquity variability; Earle et al., 2018; Johnson et al., 2021).

Keynote questions raised in this chapter include: (1) What are the most common ice species on Pluto and what is their distribution? (2) Is cryovolcanism an active process and is it a plausible mechanism for explaining Pluto's relatively young surface at locations such as Sputnik Planitia? (3) Does Pluto harbor a subsurface ocean? (4) Are there significant commonalities between Pluto, the icy Uranian/Neptunian moons, and *KBOs* such as Arrokoth?

2 Pre-New Horizons understanding

Multi-year searches for astronomical objects beyond Neptune were very popular in the early 1900s (see Lunine et al., 2021 and references therein). Many of these searches were undertaken by Percival

Lowell at the Lowell Observatory in Flagstaff, Arizona. However, it was not until 1930 that Clyde Tombaugh officially identified Pluto as a possible planet using blink comparator techniques of plates (Lunine et al., 2021). The difficulty in exploring this remote object frustrated astronomers for decades. For example, Tombaugh spent the next 18 years looking for objects similar to Pluto; it was not until the early 1990s, with the advent of the large, dark sky, high altitude Hawaiian telescopes, that other *TNOs* were discovered.

By 1973, Pluto's obliquity had been calculated (Andersson & Fix, 1973). Later that same decade, Pluto's moon Charon was discovered (Christy & Harrington, 1978) and methane was detected on its surface (Cruikshank et al., 1976). The first definitive detection of Pluto's atmosphere occurred in 1988 by Earth-based observations of a stellar occultation and was estimated to have a surface pressure of ~1–10 μbar (Elliot et al., 1989). Additional discoveries continued in the 1990s and 2000s with improvements to ground-based observations and the advent of the Hubble Space Telescope (HST). This included the detection of nitrogen and carbon monoxide ices (Owen et al., 1993). While variations in the brightness of the surface had been mapped using light curves from ground-based photometric observations (e.g., Buie & Tholen, 1989; Buie et al., 1992), Stern et al. (1997) used the Hubble Space Telescope (HST) to collect the first direct images of the Pluto's surface. They identified distinct albedo regions pointing to the possibility of a geologically varied surface and/or heterogeneity in the ice types and behavior of the planet's surface ice.

The New Horizons spacecraft was launched on January 19, 2006, as part of a "Pluto-Kuiper Belt Express" mission (Stern, 2008). The spacecraft carried seven instruments (see Fig. 4 for a description of the payload) and successfully executed a flyby of the Pluto system in 2015. The spacecraft was then directed toward a KBO, Arrokoth, as a secondary mission. The latter was encountered in 2019 at 43.4 AU (Stern et al., 2019).

3 Ices in the outer solar system

3.1 Pluto's ices

The most important ice-forming molecules on Pluto are H_2O, CO, CO_2 (carbon dioxide), N_2, NH_3 (ammonia), and CH_4. These materials dominate the ice compositional landscape in the outer solar system and were derived initially from the dominant interstellar H, H_2, He, CO, and N_2 gas species via hydrogen addition reactions in gas- and grain-mediated chemistry (Ehrenfreund & Charnley, 2000). This would have been followed by sublimative loss of extremely volatile H, H_2, and He species. In solid form, they can exist either as quickly assembled amorphous species or as re-crystallized solids at higher temperature. For example, CO_2 undergoes an amorphous to crystalline phase change at ~24 K (Escribano et al., 2013; Umurhan et al., 2021), while H_2O undergoes an amorphous to crystalline phase change at ~100 K. The presence of crystalline water ice is a strong indication of recent or ongoing surface renewal. Since crystalline ice should be completely removed from surfaces within ~1 Myr via solar wind and cosmic ray sputtering (Cook et al., 2007), a "crystalline renewal" process from exogenic or endogenic mechanisms is required. The conversion of amorphous ice to crystalline at the cold temperatures of the satellites and *KBOs* is most likely from the thermal annealing caused by the deposition of micrometeorites. This has been modeled to occur on millions of years timescales (Desch et al., 2009). Alternatively, cryovolcanism can provide a mechanism for forming crystalline ice.

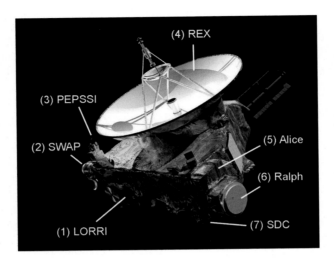

FIG. 4

The New Horizons spacecraft and its seven instruments. (1) The LOng-Range Reconnaissance Imager (*LORRI*) is a narrow-angle panchromatic camera with a bandpass of ~350 to 850 nm (Cheng et al., 2008). (2) Solar Wind Around Pluto (*SWAP*) is designed to measure the solar wind and pickup ions around Pluto (McComas et al., 2008). (3) The Pluto Energetic Particle Spectrometer Investigation (*PEPSSI*) is designed to measure pickup ions from Pluto's atmosphere and characterize the energetic particle environment along with ionosphere/solar wind interaction (McNutt et al., 2008). (4) The Radio Science Experiment (*REX*) is designed to measure the atmospheric state near the surface of Pluto by conducting radio occultation experiments (Tyler et al., 2008). (5) The Alice instrument is an ultraviolet imaging spectrometer designed to study Pluto's atmospheric composition and structure (Stern et al., 2008). (6) The Ralph instrument suite consists of a visible/near-infrared multispectral imager and a short wavelength infrared (SWIR) spectral imager used to map geology, composition, and surface temperatures (Reuter et al., 2008). (7) The Student Dust Counter (*SDC*) is an impact dust detector capable of detecting particles with masses $>10^{-12}$ g (Horányi et al., 2008).

Image credit: NASA/JHUAPL/SWRI.

Pluto's "bedrock" is water ice as observed in isolated outcrops. The surface is dominated by a spatially complicated distribution of CO, CH_4, and N_2, and nonvolatile complex organic materials (tholins) that coat the heavily crater equatorial regions (Grundy et al., 2016). Sublimation rates are strongly temperature-dependent with surface temperatures low enough on Pluto and Triton to allow CO to condense. However, CO ice mobilizes at low temperatures (~44 K), and there is a seasonal transience with CH_4 and N_2 via atmosphere vapor pressure equilibrium (Binzel et al., 2017; Lellouch et al., 2011). Ices at these surfaces are not pure and can be found in some interstitial mixtures or trapped within defects or voids in the crystalline structures (Mastrapa et al., 2013). CO does not produce pure deposits on Pluto; however, in order for CO to condense and form deposits requires a crystallization solution with N_2 or CH_4 (Ahrens et al., 2022 and references therein). CO and N_2 are miscible due to their similar molecular characteristics, such as volatility and polarity, but could separate through a solid-state distillation process (Tegler et al., 2019). Pluto's dark terrain (Cthulhu Macula, in particular) is a large reservoir of processed, irradiated organic matter and is thought to be comprised of tholins. The latter may also be responsible for the dark appearance of the north polar region of Charon (Grundy et al., 2016; Mandt et al., 2021).

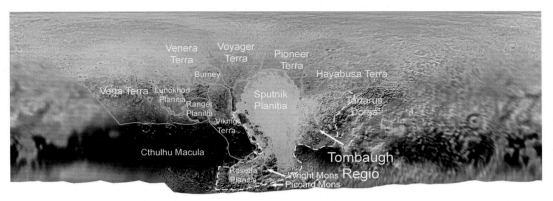

FIG. 5

Simple cylindrical global map of Pluto showing locations of major terrains and features (*IAU/USGS*). The boundaries on the map portray the approximate extent of the named features.

Basemap credit: NASA/JHUAPL/SWRI.

Based on New Horizons data, distinct geological units and terrains have been identified (Fig. 5). Pluto's strongest CO ice spectral signatures are found in central and southern Sputnik Planitia. Sputnik Planitia comprises a topographic depression or basin (~1492 km in diameter), quite possibly formed by a large impact (Schenk et al., 2015) within which CO, N_2, and CH_4 ice deposits have been identified (McKinnon et al., 2016; Stern et al., 2015). The accumulation of icy fill could be the result of the basin being ~5–10 km below the surrounding elevation datum (McKinnon et al., 2016). This relatively low elevation comprises a cold trap that stabilizes CO, N_2, and CH_4 ice in the solid phase; its relatively low latitude also contributes to the stability of these compounds (Earle et al., 2017). The basin's surface has no discernible superposed impact craters (Robbins et al., 2017); this suggests ongoing and constant resurfacing due to convection, glacial flow, and sublimation resulting from nitrogen's sensitivity to seasonal and atmospheric changes (Earle et al., 2017; Singer et al., 2019; White et al., 2017) (Fig. 6). Solid-state convection of predominately low-viscosity N_2 ice is inferred from the cellular morphology (polygonal and ovoid features ~10–40 km across) of the surface (McKinnon et al., 2016; Stern et al., 2015; White et al., 2017).

Typically, CO and CH_4 ices are observed on Pluto within the latitude range 40°–60°, crater floors, eastern Hayabusa Terra, and southwestern Venera Terra (Schmitt et al., 2017). The Venera, Burney, and Hayabusa regions are rich in CH_4 with mixtures of N_2 and CO. The Tartarus Dorsa region consists of rough, eroded uplands and CH_4-N_2 bladed terrain deposits; these terrains are comprised of ridges with complex bladed textures, possibly formed by the degradation and sublimation of a previously continuous deposit (Fig. 6).

Pluto's large moon Charon, unlike Pluto, is dominated by water ice at the surface (Buie et al., 1987). Charon's albedo is muted relative to Pluto. It shows less variation and a significantly lower value. This can be attributed to the lack of extensive mobile surface volatiles such as N_2, CO, and CH_4 (Stern et al., 2018).

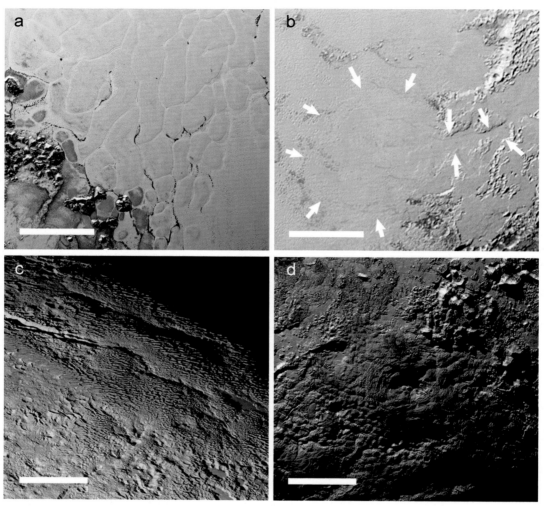

FIG. 6

(a) Close-up image of Sputnik Planitia. Note the lack of impact craters and the overall cellular texture of the icy surface, possibly due to convection that may be active to the present day. (b) Nitrogen-rich ice appears to have accumulated and flowed downslope into Sputnik Planitia through ~3–8 km-wide valleys (*arrows*) with the approximate extent of the flow front within Sputnik Planitia visible (*arrows*). (c) Bladed or "snakeskin" terrain consisting of ridged highlands covering the Tartarus Dorsa region. (d) Wright Mons, south of Sputnik Planitia, is a ~150 km wide, ~3–4 km-high feature with a > 5 km-deep summit depression with a concentric fabric and a distinctive hummocky texture on its flanks. Scale bars are 100 km in all frames.

Image credit: NASA/JHUAPL/SWRI.

3.2 Ices elsewhere

The Voyager 2 probe lacked spectral mapping capabilities; however, the composition of the ices at Uranus and Neptune has been studied by Earth-based telescopes. For example, crystalline water ice has been confirmed at the Uranus system, and CO_2, CO, N_2, and CH_4 ices have been identified at Triton (Buratti et al., 2011; Grundy & Young, 2004; Grundy et al., 2010), while Charon exhibits water ice but no indication of CH_4 ice, unlike Pluto and Triton (Cruikshank et al., 2015). Near-infrared spectra of Triton revealed that H_2O and N_2 are the dominant surface ices, with CH_4, CO_2, and CO also being present (Cruikshank et al., 1993). Spectral variations in longitude suggest that N_2 and CO ices are prevalent on Triton's Neptune-facing hemisphere, while CH_4 shows a different longitudinal variation (Buratti et al., 2011; Grundy et al., 2010). Seasonal evolution of the surface volatiles is also suggested by an evolution in absorption band amplitudes and visible and UV photometry (Buratti et al., 1994; Young & Stern, 2001) as well as by changes in atmospheric pressure inferred from stellar occultations (Elliot et al., 2000).

The satellites of Uranus and Neptune formed under conditions that favor CH_4 and CO, with N_2 or NH_3 as hydrates (Cartwright et al., 2020; Lunine & Stevenson, 1985). Partial melting produced CH_4, N_2 liquids, and even an H_2O-NH_3 peritectic melt. This can cause differences in fluid mobility, density contrasts, migration, and segregated pockets of material or cryomagma reservoirs (Ahrens, 2020; Jankowski & Squyres, 1988). Even if these liquid pockets freeze and solidify after accretional heat is lost, tidal and radiogenic heating may help them remobilize. Certain ices can also act as "antifreezes" in dissolved volatiles. This prevents water molecules from organizing themselves into a crystalline solid phase (Neveu et al., 2015). Usually, these volatiles are ammonia (NH_3) and methanol (CH_3OH).

The surfaces of the icy Uranian satellites consist of a mixture of H_2O bedrock and a darker, lower-albedo compound that is rich in carbon (Cartwright et al., 2015). Several mechanisms, such as sublimation, UV photolysis, charged particle sputtering, and micrometeorite bombardment, will remove CO_2 from surfaces on timescales shorter than the lifetime of the solar system (Grundy et al., 2006). Therefore, CO_2 at these surfaces must be deposited, exposed, or replenished in some way on relatively short timescales. The origin of this CO_2 ice could be derived from several environmental factors such as the accretion of primordial materials from impacts or exposure by tectonism, delivery from interstellar dust, sublimation of CO_2 from cold traps, or irradiative processes (Ahrens et al., 2022).

There have been no observations of CO_2 on Miranda, and the possible presence of CO_2 on Oberon is debatable (Cartwright et al., 2015; Grundy et al., 2006). CO_2 ice is not expected to deposit on Miranda due to its very low atmospheric escape velocity. Oberon spends part of its orbit outside Uranus' magnetic field, making radiolytic production comparably weak (Sori et al., 2017). However, CO_2 has been detected on Umbriel, Ariel, and Titania (Grundy et al., 2003, 2006). Umbriel has a sizable bright polar deposit of CO_2 (Sori et al., 2017). This may be due to locally concentrated cold-trapped CO_2.

4 Subsurface oceans

4.1 Radiogenic history of Pluto and a possible subsurface ocean

Questions abound on the state of Pluto's current and past heat flux and whether its surface is undergoing endogenic-driven change to the present day. Surface observations pointing to the former include the possibility of convecting nitrogen at Sputnik Planitia, forming extensive cellular patterns (Fig. 6).

Heat flux also determines the temperature gradient. In turn, this controls the rheologic and mechanical properties of surface and near-surface, manifesting different landforms. For example, brittle deformation is evident on Pluto as an extensional graben that formed radially from the Sputnik Planitia basin across the cryo-lithosphere (Nimmo & McKinnon, 2021). Impact craters on Enceladus and Ganymede often exhibit viscous relaxation, where craters produce stress gradients in the subsurface that result in ductile flow (Bland et al., 2012; Dombard & McKinnon, 2006).

Pluto's surface is dominated by N_2, NH_4, C ices, and H_2O ice along the southwestern and southern edges of Sputnik Planitia. Here, water ice mountains can reach elevations of kilometers (Grundy et al., 2016; Schmitt et al., 2017). The interior of Pluto is inferred from its density, which implies that it is composed of ~70% rock and 30% ice by mass. The absence of compressional tectonic structures indicates that it is fully differentiated (Nimmo & McKinnon, 2021).

Whether a subsurface ocean exists on Pluto is an open question. Part of the answer lies in identifying Pluto's initial interior conditions. Modeling of the coupled thermal and spin evolution of Pluto shows that a present-day subsurface ocean is possible under certain conditions that influence the balance between heat transfer and radiogenic heating. If the viscosity of the ice shell is high enough to suppress solid-state convection, thus reducing heat transfer from the interior, coupled with adequate heat production from radiogenic elements, a present-day ocean is possible. The presence of ammonia would sustain colder ocean temperatures and increase the likelihood of a subsurface ocean (Robuchon & Nimmo, 2011). Clathrates (molecules of one compound physically trapped within the solid crystal structure of another) can also provide insulation to maintain a liquid ocean due to their low thermal conductivity (Kamata et al., 2019).

Several observations support the possibility of a subsurface ocean (Nimmo & McKinnon, 2021). The predominance of extensional tectonic features and possible cryovolcanism point to a thickening ice shell causing a radial expansion and pressurizing fluids at depth enabling them to ascend and breach the surface (Dalle Ore et al., 2019; Martin & Binzel, 2021). Additionally, the absence of a detectible "fossilized bulge" suggests an ice lithosphere that was either too thin or too weak as Pluto spun down to retain a bulge. The reorientation of Sputnik Planitia, to move it close to the anti-Charon point, can also be explained by the presence of a subsurface ocean. On the other hand, this is not a necessary requirement of re-orientation. (Keane et al., 2016; Nimmo & McKinnon, 2021).

Thermal evolution modeling by Bierson et al. (2020) evaluated the initial conditions of Pluto's interior by comparing "cold-start" and "hot-start" scenarios. In the "cold-start" timeline, the silicate interior of Pluto warms due to radioactive decay and solid ice layer melts at the base. This leads to a reduction in volume and compression, followed by a late stage of extension as the interior cools and the ice shell base refreezes. Images of Pluto's oldest terrains do not show unambiguous evidence for compression having occurred. Extensional faulting appears to be widespread and the absence of compressional features favors the "hot-start" scenario (Bierson et al., 2020; Hammond et al., 2016).

For the "hot-start" model, Pluto forms with an initial subsurface ocean that leads to an early and rapid phase of extension, followed by a more modest, prolonged phase of extension. Evidence for relatively recent extension is abundant (McGovern et al., 2019; Moore et al., 2016) and a ridge-trough system has been identified as a possible ancient extensional feature (Schenk et al., 2018). Charon may have also possessed a subsurface ocean early in its history. The moon has two large chasms extending around much of the observed hemisphere that may have formed from a freezing primordial subsurface ocean. The southern hemisphere also has fewer craters, suggesting that a large-scale resurfacing event also occurred that may be cryovolcanic in origin (Moore et al., 2016).

4.2 Triton's ocean

Triton has a relatively young surface with few craters (Schenk & Zahnle, 2007; Smith et al., 1989), shows evidence of cryovolcanism and erupting plumes (Conrath et al., 1989; Croft et al., 1995), and hosts a nitrogen atmosphere in vapor pressure equilibrium with its surface (Broadfoot et al., 1989; Tyler et al., 1989). This makes it similar to Pluto in many ways (Fig. 7) including its comparable size and possibly shared origin as a Kuiper Belt dwarf-planet; the latter is deduced from a highly inclined, retrograde orbit around Neptune that points to Pluto being a captured KBO (McKinnon & Kirk, 2007).

Tidal heating following capture probably resulted in the differentiation of the body, and tidal dissipation of heat due to the eccentricity of the orbit immediately after capture could be retained over geologic timescales (Gaeman et al., 2012). Triton's orbit has long since been circularized. However, obliquity-driven tides due to its high inclination could have and may still provide tidal heating to maintain an ocean (Nimmo & Spencer, 2015).

Thermal models of Triton estimate the ice-shell crust to be ~150 km overlying an ocean of similar thickness (Nimmo & Spencer, 2015) and predict that subsurface oceans are possible around other mid-sized icy satellites and large TNOs (Hussmann et al., 2006). Were there to be an ocean on Triton, it could comprise remnant cometary materials such as chlorides, sodium, and ammonium (Castillo-Rogez et al., 2018). The cantaloupe terrain, a feature uniquely present on Triton, might have been formed by thermal diapirs that transported fluids from the subsurface ocean to the surface (Ruiz et al., 2007). Possible cryogenic features further suggest that communication between the surface and a subsurface liquid reservoir

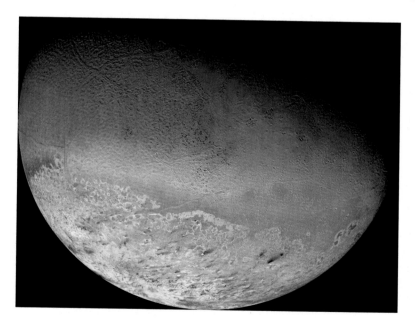

FIG. 7

Global color mosaic of Neptune's moon Triton (mean radius ~1353.4 km) taken by Voyager 2. The enigmatic cantaloupe terrain is visible in the top half of the image.

Image credit: NASA/JPL/USGS.

has occurred; given the young surface age of Triton, this could have persisted up to the present or the recent geologic past (Hansen et al., 2021).

5 Cryovolcanism

5.1 Cryovolcanism on Pluto

Landscape features on Pluto possibly formed by cryovolcanism are located to the south and south-west of Sputnik Planitia. The features are comprised of broad quasicircular mounds with large central depressions, observably deeper than the height of the edifices, and undulating/hummocky flanks (Fig. 6d) (Moore et al., 2016; Schenk et al., 2018; Singer et al., 2022). Wright Mons (−21.36°N, 173.24°E) measures 155 km in diameter, rising to 4.7 km above its base, with a central depression ~45 km across and ~3.5 km deep (White et al., 2021). Piccard Mons (−35.26°N, 176.79°E), to the south of Wright Mons, is ~280 km across. It rises ~5.0 km above its base, with a central depression up to ~120 km in diameter and at least ~11 km deep. In the surrounding area several cavi or shallow, irregularly shaped depressions are observed. They are thought to be ponded cryovolcanic materials and degraded forms of cryovolcanic structures (Ahrens & Chevrier, 2021; Singer et al., 2022).

Cryovolcanism involves magmatic processes occurring in volatile ices and fluids at very low temperatures and moderate pressures; this contrasts with the high temperature and pressure environments of silicate-based rocks and melts on terrestrial bodies. There are two main types of cryovolcanism—explosive and effusive. Explosive cryovolcanism is favored with hydrated cores (Neveu et al., 2015). N_2, CH_4, and H_2 are conducive to exsolution (bubbles) that may drive explosive reactions. Explosive volcanism can include geysers, but any sort of rapid, active pressurized bursting of material would be considered explosive. Effusive volcanism, by comparison, erupts at more modest rates (i.e., non-explosive), is more subdued in movement, and can include caldera formation, basin in-filling, and flows.

There seem to be no clear-cut markers of explosive volcanism on Pluto, i.e., ballistic fall deposit patterns, either radial or directional, or steep cones (Singer et al., 2022). Although there are no clear-cut markers of effusive volcanism either, i.e., flow fronts, streamlines, levees, or fractures/vent locations observed so far, some observations are supportive of this hypothesis (Singer et al., 2022). For example, lower elevation plains to the west of Wright Mons have an undulating/hummocky appearance similar to Wright Mons but are superposed by intersecting fractures, many of which appear to be shallow (Singer et al., 2022); these more modified terrains may represent an earlier episode of the process that created Wright Mons and the other large rises (Singer et al., 2022). The most prominent flow-like morphologies on Pluto are associated with glacier-like distribution of material, principally N_2, at the Sputnik Planitia basin (19.51°N, 178.69°E) (Fig. 6b). CH_4, and CO ices have also been detected at this location (Gladstone et al., 2016; Grundy et al., 2016).

Explaining the geochemical possibility of cryovolcanism on Pluto requires consideration of cryovolcanic liquids versus silicate melts—water-based cryovolcanic liquids are denser than the solid ice that they originally formed from, in comparison to silicate melts. Silicate melts naturally rise due to buoyancy, and cryovolcanic melts should sink (Neveu et al., 2015). For cryovolcanism, pressure gradients must overcome the melt's negative buoyancy and fracture the crust to erupt onto the surface (Fagents, 2003). Crustal behavior influences the resulting morphology of surface features. Undifferentiated crusts inhibit exsolution or cryovolcanism by acting more like a pressure seal (Neveu et al., 2015). Ice relaxation at fractures and cracks or freezing upon ascent also hinder cryovolcanism. However, clathrate layers can contribute to cryovolcanism by thermally insulating liquid layers, making mobility of ascending material easier (Kamata et al., 2019).

Antifreezes such as NH_3 and even salts can be dissolved in the water ice and this prevents the water molecules from organizing themselves into a crystalline solid phase (Neveu et al., 2015). As such, these materials can exist in the liquid phase well below the freezing points of pure water (Cruikshank et al., 2019; Robuchon & Nimmo, 2011). NH_3 content also influences the density and thermal evolution of cryomagmas (Ahrens, 2020; Martin & Binzel, 2021). The presence of NH_3 (observed at Pluto's surface) reduces the density difference between the liquid and solid phases of a mixture, allowing buoyant cryovolcanic liquids to bubble or plume to the surface with less overpressure during freezing (Martin & Binzel, 2021). In tandem with the over-pressurization of Pluto's subsurface ocean—were there to be one—through partial freezing, all of this might be able to generate a large volume of cryomagma at depth and cause it to ascend to the surface (Cruikshank et al., 2019; Dalle Ore et al., 2019; Fagents, 2003; Martin & Binzel, 2021).

5.2 Triton and the Uranian moons

Active cryovolcanism in the form of plume activity has been identified on the Jovian and Saturnian icy satellites Europa and Enceladus (Hansen et al., 2006; Roth et al., 2014). Triton also possesses landscape features that may be cryogenic in origin such as calderas, vents, lava flows and lakes, fissures, and sinuous rilles (Croft, 1990; Kargel, 1998; Schenk et al., 2021). Vent-like structures point to cryovolcanic mechanisms such as: (i) thermal stress driving wall failure, (ii) vaporization of volatiles from wall failure, and (iii) thermal melting and/or erosion of the vent walls. However, the understanding of volatile material for these thermophysical processes largely is unknown. Additionally, Voyager 2 observed several active geyser-like eruptions with columns of dark material reaching altitudes of ~8 km. Whether these are driven by internal heat sources is unclear; however, their location near the subsolar latitude during the Voyager 2 flyby of Triton led to speculation that the solar heating of subsurface nitrogen ice (i.e., solid-state greenhouse) engenders explosive venting to the surface (Kirk et al., 1990; Soderblom et al., 1990).

It seems likely that both Triton and Pluto experienced volcanic activity driven by some form of internal heat, Triton through tidal heating by Neptune and Pluto through radiogenic heating and/or trapped heat from the Charon-forming impact event (Singer et al., 2022). However, the emplacement processes and material compositions involved are different across both icy bodies. The much lower-relief volcanic plains on Triton suggests low-viscosity ices were involved (Kargel, 1998).

In the Uranian system, the morphologic characteristics of possible flow features suggest solid-state extrusion (Jankowski & Squyres, 1988). The extrusion of these flows is most likely from extensional stresses (Stephan et al., 2013). The sources of the flows are typically from linear fracture systems on the graben floors. Miranda and Ariel have relatively young surfaces, estimated to be <1 Gyr (Plescia, 1988; Zahnle et al., 2003); they were possibly formed by resurfacing processes that originated at the grabens (Kargel, 1998). Miranda also seems to have similar flow features.

The flow deposits at these moons are 0.1–1.0 km thick requiring highly viscous flows, in contrast to the Jovian and Saturnian moons with comparatively thin deposits formed by lower viscosity flows (Kargel, 1998). If thermal effects dominate the bulk viscosity of the systems, then one would expect the flows in the much colder Uranian and Neptunian systems to be much more viscous (Castillo-Rogez et al., 2023). However, the moons in these systems may also possess large amounts of unequilibrated comet-like volatiles, including ammonia-water-soluble constituents (and possibly methanol) that would influence melts and viscous flows.

6 Pluto and Arrokoth

Although there is a paucity of data on *KBOs*, the available data shows (cf., Barucci et al., 2021) that Pluto's composition, color, and albedo fall within the range of known measurements for the largest *KBOs* such as Haumea, Makemake, Orcus, Quaoar, and Eris (Barucci et al., 2021; Buratti et al., 2017). On the other hand, a comparison between Pluto, a relatively large, differentiated body about two-thirds the size of the Earth's Moon (~1200 km radius), and Arrokoth, an elongated contact-binary ~36 km long composed of two planetesimals (Fig. 1), alludes to a great diversity of objects to be encountered within the Kuiper Belt. Arrokoth's orbit has a semi-major axis of 44.2 AU and is among the most primitive objects in the solar system surveyed to date (Singer et al., 2019).

Arrokoth consists of two distinctly primitive lobes, connected by a neck junction. The alignment of the lobes suggests that they were originally a co-orbiting binary that merged at a low speed (<2 m/s) as they experienced a loss of angular momentum (McKinnon et al., 2020; Spencer et al., 2020). Pluto could have been formed out of the accretion, merger, and destructive dissolution of millions of Arrokoth-like objects.

Arrokoth shows no evidence of hosting any of the bulk hypervolatile cryogenic $CO/CH_4/N_2$ ices that cover Pluto's surface; instead, it is dominated by methanol, tholins, and likely water (Grundy et al., 2020). This finding has been used by Lisse et al. (2021, 2022) to propose that these hypervolatile ices are ephemeral in small *KBOs*, possibly on the order of a few 10^7 years. This suggests that Pluto could have been formed from Arrokoth-like bodies in a short time, or that the hypervolatiles expressed on its surface are from a thoroughly differentiated interior consistent with its spherical shape and subsurface ocean (Kamata et al., 2019). A formation time of ~4.5 Ga for Pluto would be consistent with the ~4 Ga age estimates of the Sputnik Planitia basin (Greenstreet et al., 2015; Singer et al., 2019) and the surface geomorphology of Charon (Spencer et al., 2021).

7 Summary and conclusions

Surprisingly enigmatic might be the best turn of phrase to describe the observations and data returned by the New Horizons spacecraft in its flyby of the Pluto-Charon system. Assumptions concerning the long-term core temperatures and associated dynamic activities, i.e., cryovolcanism, mountain building, geysers, etc., of small, outer solar system bodies need to be revisited if not re-postulated. Were subsurface oceans present on Pluto and Triton, as they are hypothesized on Enceladus, Europa, and Ceres, what would this say about our current paradigms of small-body evolution and differentiation? Is relatively recent if not current resurfacing, i.e., endogenically driven, on Pluto, Triton, Ceres, etc., more commonplace among small bodies than has been thought hitherto, even when interior outputs are extremely low, i.e., a few mW/m^2 of heat flow (Nimmo & McKinnon, 2021)? Despite its large distance from the sun, methane, nitrogen, and carbon monoxide on Pluto are mobilized by seasonal atmospheric changes/cycles associated with the dwarf planet's highly elliptical orbit and obliquity variances. Other *TNOs*, unobserved or insufficiently resolved, might exhibit similar seasonal changes/cycles.

All in all, Pluto comprises a cross-disciplinary template of processes, landforms, and landscapes for comparing if not contrasting small bodies in the outer solar system with one another. No less importantly, it is an evocative case study of wholly unanticipated but significantly impactful findings in planetary science. Further monitoring of Pluto, the icy moons of Uranus/Neptune, and *KBOs/TNOs* by ground-based telescopes and the newly commissioned James Webb Space Telescope (*JWST*) will evolve our understanding of these small bodies. Follow-up and dedicated missions to Pluto would be extremely useful. One such mission, drawn up but not yet mandated, is a lander named "Pluto Hop, Skip, and Jump" (Nock et al., 2021). Its aim is to deliver a low-mass, low-cost spacecraft on a high-speed interplanetary trajectory.

Dedicated missions to Uranus and Neptune along with their icy moons could be equally informative, for these moons seemingly are as disparate in their possible origins as they are in the processes driving their development. Toward this and other ends, *NASA* has designated the "Uranus Orbiter and Probe" as a high-priority flagship mission that is scheduled to be launched in the 2030s (National Academy of Sciences, Engineering, and Medicine, 2022).

References

Ahrens, C. J. (2020). Modeling cryogenic mud volcanism on Pluto. *Journal of Volcanology and Geothermal Research*, *406*, 107070.

Ahrens, C. J., & Chevrier, V. F. (2021). Investigation of the morphology and interpretation of Hekla Cavus, Pluto. *Icarus*, *356*, 114108.

Ahrens, C., Meraviglia, H., & Bennett, C. (2022). A geoscientific review on CO and CO2 ices in the outer solar system. *Geosciences*, *12*(2), 51.

Andersson, L. E., & Fix, J. D. (1973). Pluto: New photometry and a determination of the axis of rotation. *Icarus*, *20*(3), 279–283.

Barucci, M. A., Dalle Ore, C. M., & Fornasier, S. (2021). The Transneptunian objects as the context for Pluto: An astronomical perspective. *The Pluto System After New Horizons*, *21*.

Bierson, C. J., Nimmo, F., & Stern, S. A. (2020). Evidence for a hot start and early ocean formation on Pluto. *Nature Geoscience*, *13*(7), 468–472.

Binzel, R. P., et al. (2017). Climate zones on Pluto and Charon. *Icarus*, *287*, 30–36.

Bland, M. T., Singer, K. N., McKinnon, W. B., & Schenk, P. M. (2012). Enceladus' extreme heat flux as revealed by its relaxed craters. *Geophysical Research Letters*, *39*(17).

Broadfoot, A. L., Atreya, S. K., Bertaux, J. L., Blamont, J. E., Dessler, A. J., Donahue, T. M., et al. (1989). Ultraviolet spectrometer observations of Neptune and Triton. *Science*, *246*(4936), 1459–1466.

Brown, M. E. (2012). The compositions of Kuiper Belt objects. *Annual Review of Earth and Planetary Sciences*, *40*, 467–494.

Brown, R. H., Johnson, T., Synnott, S., Anderson, J., Jacobson, R., Dermott, S., & Thomas, P. (1991). Physical properties of the Uranian satellites. In J. T. Bergstralh, E. D. Miner, & M. S. Matthews (Eds.), *Uranus* University of Arizona Press.

Buie, M. W., Cruikshank, D. P., Lebofsky, L. A., & Tedesco, E. F. (1987). Water frost on Charon. *Nature*, *329*, 522–523.

Buie, M. W., & Tholen, D. J. (1989). The surface albedo distribution of Pluto. *Icarus*, *79*(1), 23–37.

Buie, M. W., Tholen, D. J., & Horne, K. (1992). Albedo maps of Pluto and Charon: Initial mutual event results. *Icarus*, *97*(2), 211–227.

Buratti, B. J., Goguen, J. D., Gibson, J., & Mosher, J. (1994). Historical photometric evidence for volatile migration on Triton. *Icarus*, *110*(2), 303–314. https://doi.org/10.1006/icar.1994.1124.

Buratti, B. J., et al. (2011). Photometry of Triton 1992–2004: Surface volatile transport and discovery of a remarkable opposition surge. *Icarus*, *212*(2), 835–846.

Buratti, B. J., et al. (2017). Global albedos of Pluto and Charon from LORRI new horizons observations. *Icarus*, *287*, 207–217.

Cartwright, R. J., Emery, J. P., Rivkin, A. S., Trilling, D. E., & Pinilla-Alonso, N. (2015). Distribution of CO2 ice on the large moons of Uranus and evidence for compositional stratification of their near-surfaces. *Icarus*, *257*, 428–456.

Cartwright, R. J., et al. (2020). Evidence for ammonia-bearing species on the Uranian satellite Ariel supports recent geologic activity. *The Astrophysical Journal Letters*, *898*(1), L22.

Castillo-Rogez, J., Neveu, M., McSween, H. Y., Fu, R. R., Toplis, M. J., & Prettyman, T. (2018). Insights into Ceres's evolution from surface composition. *Meteoritics & Planetary Science*, *53*(9), 1820–1843.

Castillo-Rogez, J., et al. (2023). Compositions and interior structures of the large moons of Uranus and implications for future spacecraft observations. *Journal of Geophysical Research: Planets*, *128*(1), e2022JE007432.

Cheng, A. F., Weaver, H. A., Conard, S. J., et al. (2008). Long-range reconnaissance imager on new horizons. *Space Science Reviews*, *140*, 189–215. https://doi.org/10.1007/s11214-007-9271-6.

Christy, J. W., & Harrington, R. S. (1978). The satellite of Pluto. *Astronomy Journal*, *83*, 1005–1008.

Conrath, B., Flasar, F. M., Hanel, R., Kunde, V., Maguire, W., Pearl, J., et al. (1989). Infrared observations of the Neptunian system. *Science*, *246*(4936), 1454–1459.

Cook, J. C., Desch, S. J., Roush, T. L., Trujillo, C. A., & Geballe, T. R. (2007). Near-infrared spectroscopy of Charon: Possible evidence for cryovolcanism on Kuiper belt objects. *The Astrophysical Journal*, *663*(2), 1406.

Croft, S. K. (1990). Physical cryovolcanism on Triton. *Lunar and Planetary Science Conference*, *21*.

Croft, S. K., Kargel, J. S., Kirk, R. I., Moore, J. M., Schenk, P. M., & Strom, R. G. (1995). The geology of Triton. In D. P. Cruikshank (Ed.), *Neptune and Triton* (pp. 879–947). Univ. Arizona Press.

Croft, S., & Soderblom, L. (1991). Geology of the Uranian satellites. In J. T. Bergstralh, E. D. Miner, & M. S. Matthews (Eds.), *Uranus* University of Arizona Press.

Cruikshank, D. P., Pilcher, C. B., & Morrison, D. (1976). Pluto: Evidence for methane frost. *Science*, *194*(4267), 835–837.

Cruikshank, D. P., et al. (1993). Ices on the surface of Triton. *Science*, *261*(5122), 742–745.

Cruikshank, D. P., et al. (2015). The surface compositions of Pluto and Charon. *Icarus*, *246*, 82–92.

Cruikshank, D. P., et al. (2019). Recent cryovolcanism in Virgil fossae on Pluto. *Icarus*, *330*, 155–168.

Dalle Ore, C. M., et al. (2019). Detection of ammonia on Pluto's surface in a region of geologically recent tectonism. *Science Advances*, *5*(5), eaav5731.

Desch, S. J., Cook, J. C., Doggett, T. C., & Porter, S. B. (2009). Thermal evolution of Kuiper belt objects, with implications for cryovolcanism. *Icarus*, *202*(2), 694–714.

Dombard, A. J., & McKinnon, W. B. (2006). Elastoviscoplastic relaxation of impact crater topography with application to Ganymede and Callisto. *Journal of Geophysical Research: Planets*, *111*(E1).

Dumas, C., Carry, B., Hestroffer, D., & Merlin, F. (2011). High-contrast observations of (136108) Haumea-A crystalline water-ice multiple system. *Astronomy & Astrophysics*, *528*, A105.

Earle, A. M., et al. (2017). Long-term surface temperature modeling of Pluto. *Icarus*, *287*, 37–46.

Earle, A. M., et al. (2018). Albedo matters: Understanding runaway albedo variations on Pluto. *Icarus*, *303*, 1–9.

Ehrenfreund, P., & Charnley, S. B. (2000). Organic molecules in the interstellar medium, comets, and meteorites: A voyage from dark clouds to the early earth. *Annual Review of Astronomy and Astrophysics*, *38*(1), 427–483.

Elliot, J. L., Dunham, E. W., Bosh, A. S., et al. (1989). Pluto's atmosphere. *Icarus*, *77*, 148–170.

Elliot, J. L., et al. (2000). The prediction and observation of the 1997 July 18 stellar occultation by Triton: More evidence for distortion and increasing pressure in Triton's atmosphere. *Icarus*, *148*(2), 347–369. https://doi.org/10.1006/icar.2000.6508.

Escribano, R. M., Caro, G. M. M., Cruz-Diaz, G. A., Rodríguez-Lazcano, Y., & Maté, B. (2013). Crystallization of CO2 ice and the absence of amorphous CO2 ice in space. *Proceedings of the National Academy of Sciences, 110*(32), 12899–12904.

Fagents, S. A. (2003). Considerations for effusive cryovolcanism on Europa: The post-Galileo perspective. *Journal of Geophysical Research: Planets, 108*(E12).

Gaeman, J., Hier-Majumder, S., & Roberts, J. H. (2012). Sustainability of a subsurface ocean within Triton's interior. *Icarus, 220*(2), 339–347.

Gladstone, G. R., et al. (2016). The atmosphere of Pluto as observed by new horizons. *Science, 351*(6279), aad8866.

Greenstreet, S., Gladman, B., & McKinnon, W. B. (2015). Impact and cratering rates onto Pluto. *Icarus, 258,* 267–288.

Grundy, W. M., & Young, L. A. (2004). Near-infrared spectral monitoring of Triton with IRTF/SpeX I: Establishing a baseline for rotational variability. *Icarus, 172*(2), 455–465. https://doi.org/10.1016/j.icarus.2004.07.013.

Grundy, W. M., Young, L. A., Spencer, J. R., Johnson, R. E., Young, E. F., & Buie, M. W. (2006). Distributions of H2O and CO2 ices on Ariel, Umbriel, Titania, and Oberon from IRTF/SpeX observations. *Icarus, 184*(2), 543–555.

Grundy, W. M., Young, L. A., Stansberry, J. A., Buie, M. W., Olkin, C. B., & Young, E. F. (2010). Near-infrared spectral monitoring of Triton with IRTF/SpeX II: Spatial distribution and evolution of ices. *Icarus, 205*(2), 594–604.

Grundy, W. M., Young, L. A., & Young, E. F. (2003). Discovery of CO2 ice and leading–trailing spectral asymmetry on the uranian satellite Ariel. *Icarus, 162*(1), 222–229. https://doi.org/10.1016/S0019-1035(02)00075-1.

Grundy, W. M., et al. (2016). Surface compositions across Pluto and Charon. *Science, 351*(6279), aad9189.

Grundy, W. M., et al. (2020). Color, composition, and thermal environment of Kuiper Belt object (486958) Arrokoth. *Science, 367*(6481), eaay3705.

Hall, C. F. (1974). Pioneer 10. *Science, 183,* 301–302. https://doi.org/10.1126/science.183.4122.301.

Hall, C. F. (1975). Pioneer 10 and Pioneer 11. *Science, 188,* 445–446. https://doi.org/10.1126/science.188.4187.445.

Hammond, N. P., Barr, A. C., & Parmentier, E. M. (2016). Recent tectonic activity on Pluto driven by phase changes in the ice shell. *Geophysical Research Letters, 43*(13), 6775–6782.

Hansen, C. J., Castillo-Rogez, J., Grundy, W., Hofgartner, J. D., Martin, E. S., Mitchell, K., et al. (2021). Triton: Fascinating moon, likely ocean world, compelling destination! *The Planetary Science Journal, 2*(4), 137.

Hansen, C. J., et al. (2006). Enceladus' water vapor plume. *Science, 311*(5766), 1422–1425.

Horányi, M., Hoxie, V., James, D., et al. (2008). The student dust counter on the new horizons mission. *Space Science Reviews, 140,* 387–402. https://doi.org/10.1007/s11214-007-9250-y.

Hussmann, H., Sohl, F., & Spohn, T. (2006). Subsurface oceans and deep interiors of medium-sized outer planet satellites and large trans-neptunian objects. *Icarus, 185*(1), 258–273.

Jankowski, D. G., & Squyres, S. W. (1988). Solid-state ice volcanism on the satellites of Uranus. *Science, 241*(4871), 1322–1325.

Johnson, P. E., et al. (2021). Modeling Pluto's minimum pressure: Implications for haze production. *Icarus, 356,* 114070.

Kamata, S., Nimmo, F., Sekine, Y., Kuramoto, K., Noguchi, N., Kimura, J., & Tani, A. (2019). Pluto's ocean is capped and insulated by gas hydrates. *Nature Geoscience, 12*(6), 407–410.

Kargel, J. S. (1998). Physical chemistry of ices in the outer solar system. In *Solar system ices* (pp. 3–32). Dordrecht: Springer.

Keane, J. T., Matsuyama, I., Kamata, S., & Steckloff, J. K. (2016). Reorientation and faulting of Pluto due to volatile loading within sputnik Planitia. *Nature, 540*(7631), 90–93.

Kirk, R. L., Brown, R. H., & Soderblom, L. A. (1990). Subsurface energy storage and transport for solar-powered geysers on Triton. *Science, 250*(4979), 424–429.

Kohlhase, C. E., & Penzo, P. A. (1977). Voyager mission description. *Space Science Reviews, 21,* 77–101. https://doi.org/10.1007/BF00200846.

Lellouch, E., Stansberry, J., Emery, J., Grundy, W., & Cruikshank, D. P. (2011). Thermal properties of Pluto's and Charon's surfaces from Spitzer observations. *Icarus, 214*(2), 701–716.

Levison, H. F., & Stern, S. A. (2001). On the size dependence of the inclination distribution of the main Kuiper Belt. *The Astronomical Journal, 121*(3), 1730.

Lisse, C. M., et al. (2021). On the origin & thermal stability of Arrokoth's and Pluto's ices. *Icarus, 356*, 114072.

Lisse, C. M., et al. (2022). 29P/Schwassmann–Wachmann 1: A Rosetta stone for amorphous water ice and CO↔CO2 conversion in centaurs and comets? *The Planetary Science Journal, 3*(11), 251.

Lunine, J. I., Stern, S. A., Young, L. A., Neufeld, M. J., & Binzel, R. P. (2021). Early Pluto science, the imperative for exploration, and new horizons. In *The Pluto System After New Horizons* (p. 9).

Lunine, J. I., & Stevenson, D. J. (1985). Thermodynamics of clathrate hydrate at low and high pressures with application to the outer solar system. *Astrophysical Journal Supplement Series, 58*(3), 493–531.

Mandt, K. E., Luspay-Kuti, A., Cheng, A., Jessup, K. L., & Gao, P. (2021). Photochemistry and haze formation. In *The Pluto System After New Horizons* (p. 279).

Martin, C. R., & Binzel, R. P. (2021). Ammonia-water freezing as a mechanism for recent cryovolcanism on Pluto. *Icarus, 356*, 113763.

Mastrapa, R. M., Grundy, W. M., & Gudipati, M. S. (2013). Amorphous and crystalline H 2 O-ice. In *The science of solar system ices* (pp. 371–408). New York, NY: Springer.

McComas, D., Allegrini, F., Bagenal, F., et al. (2008). The solar wind around Pluto (SWAP) instrument aboard *new horizons*. *Space Science Reviews, 140*, 261–313. https://doi.org/10.1007/s11214-007-9205-3.

McGovern, P. J., White, O. L., & Schenk, P. M. (2019). Tectonism across Pluto: Mapping and interpretations. In *Proc. Pluto System after New Horizons*. abstr. 2133.

McKinnon, W. B., & Kirk, R. L. (2007). Triton. In L.-A. McFadden, P. R. Weissman, & T. V. Johnson (Eds.), *Encyclopedia of the solar system* (pp. 483–502). Academic Press.

McKinnon, W. B., et al. (2016). Convection in a volatile nitrogen-ice-rich layer drives Pluto's geological vigour. *Nature, 534*(7605), 82–85.

McKinnon, W. B., et al. (2017). Origin of the Pluto–Charon system: Constraints from the New Horizons flyby. *Icarus, 287*, 2–11.

McKinnon, W. B., et al. (2020). The solar nebula origin of (486958) Arrokoth, a primordial contact binary in the Kuiper Belt. *Science, 367*(6481), eaay6620.

McNutt, R. L., Livi, S. A., Gurnee, R. S., et al. (2008). The Pluto energetic particle spectrometer science investigation (PEPSSI) on the new horizons Mission. *Space Science Reviews, 140*, 315–385. https://doi.org/10.1007/s11214-008-9436-y.

Moore, J. M., et al. (2016). The geology of Pluto and Charon through the eyes of new horizons. *Science, 351*(6279), 1284–1293.

National Academies of Sciences, Engineering, and Medicine. (2022). *Origins, worlds, and life: A decadal strategy for planetary science and astrobiology 2023–2032*. Washington, DC: The National Academies Press. https://doi.org/10.17226/26522.

Neveu, M., Desch, S. J., Shock, E. L., & Glein, C. R. (2015). Prerequisites for explosive cryovolcanism on dwarf planet-class Kuiper belt objects. *Icarus, 246*, 48–64.

Nimmo, F., & McKinnon, W. B. (2021). Geodynamics of Pluto. In *The Pluto System After New Horizons* (pp. 89–103).

Nimmo, F., & Spencer, J. R. (2015). Powering Triton's recent geological activity by obliquity tides: Implications for Pluto geology. *Icarus, 246*, 2–10.

Nock, K., Jacob, J., Hofgartner, W., & J., Warnecke, M. (2021). *Pluto hop, skip and jump Mission. NASA Innovative Advanced Concepts (NIAC) Symposium*. Global Aerospace Company, Oklahoma State University, NASA Innovative Advanced Concepts, *JPL*.

Opp, A. G. (1974). Pioneer 10 Mission: Summary of scientific results from the encounter with Jupiter. *Science, 183*, 302–303. https://doi.org/10.1126/science.183.4122.302.

Owen, T. C., et al. (1993). Surface ices and the atmospheric composition of Pluto. *Science, 261*(5122), 745–748.

Plescia, J. B. (1988). Cratering history of Miranda: Implications for geologic processes. *Icarus, 73*(3), 442–461.

Reuter, D. C., Stern, S. A., Scherrer, J., et al. (2008). Ralph: A visible/infrared imager for the new horizons Pluto/Kuiper Belt Mission. *Space Science Reviews, 140*, 129–154. https://doi.org/10.1007/s11214-008-9375-7.

Robbins, S. J., et al. (2017). Craters of the Pluto-Charon system. *Icarus, 287*, 187–206.

Robuchon, G., & Nimmo, F. (2011). Thermal evolution of Pluto and implications for surface tectonics and a subsurface ocean. *Icarus, 216*(2), 426–439.

Roth, L., Saur, J., Retherford, K. D., Strobel, D. F., Feldman, P. D., McGrath, M. A., & Nimmo, F. (2014). Transient water vapor at Europa's south pole. *Science, 343*(6167), 171–174.

Ruiz, J., Montoya, L., López, V., & Amils, R. (2007). Thermal Diapirism and the habitability of the icy Shell of Europa. *Origins of Life and Evolution of Biospheres, 37*(3), 287–295.

Schaller, E. L., & Brown, M. E. (2007). Volatile loss and retention on Kuiper belt objects. *The Astrophysical Journal, 659*(1), L61.

Schenk, P. M., & Zahnle, K. (2007). On the negligible surface age of Triton. *Icarus, 192*(1), 135–149.

Schenk, P. M., et al. (2015). A large impact origin for sputnik Planum and surrounding terrains, Pluto? *AAS/Division for Planetary Sciences Meeting Abstracts# 47, 47*, 200–206.

Schenk, P. M., et al. (2018). Basins, fractures and volcanoes: Global cartography and topography of Pluto from new horizons. *Icarus, 314*, 400–433.

Schenk, P. M., et al. (2021). Triton: Topography and geology of a Probable Ocean world with comparison to Pluto and Charon. *Remote Sensing, 13*(17), 3476.

Schmitt, B., et al. (2017). Physical state and distribution of materials at the surface of Pluto from new horizons LEISA imaging spectrometer. *Icarus, 287*, 229–260.

Singer, K. N., et al. (2019). Impact craters on Pluto and Charon indicate a deficit of small Kuiper Belt objects. *Science, 363*(6430), 955–959.

Singer, K. N., et al. (2022). Large-scale cryovolcanic resurfacing on Pluto. *Nature Communications, 13*(1), 1542.

Smith, B. A., et al. (1989). Voyager 2 at Neptune: Imaging science results. *Science, 246*(4936), 1422–1449.

Soderblom, L., et al. (1990). Triton's geyser-like plumes: Discovery and basic characterization. *Science, 250*(4979), 410–415.

Sori, M. M., Bapst, J., Bramson, A. M., Byrne, S., & Landis, M. E. (2017). A Wunda-full world? Carbon dioxide ice deposits on Umbriel and other Uranian moons. *Icarus, 290*, 1–13.

Spencer, J., Beyer, R. A., Robbins, S. J., Singer, K. N., & Nimmo, F. (2021). The geology and geophysics of Charon. In *The Pluto System After New Horizons* (pp. 395–412).

Spencer, J. R., et al. (2020). The geology and geophysics of Kuiper Belt object (486958) Arrokoth. *Science, 367*(6481), eaay3999.

Steckloff, J. K., Lisse, C. M., Safrit, T. K., Bosh, A. S., Lyra, W., & Sarid, G. (2021). The sublimative evolution of (486958) Arrokoth. *Icarus, 356*, 113998.

Stephan, K., Jaumann, R., & Wagner, R. (2013). Geology of icy bodies. In *The science of solar system ices* (pp. 279–367). New York, NY: Springer.

Stern, S. A. (2008). The new horizons Pluto Kuiper Belt Mission: An overview with historical context. *Space Science Reviews, 140*, 3–21. https://doi.org/10.1007/s11214-007-9295-y.

Stern, S. A., Buie, M. W., & Trafton, L. M. (1997). HST high-resolution images and maps of Pluto. *The Astronomical Journal, 113*, 827.

Stern, S. A., Grundy, W. M., McKinnon, W. B., Weaver, H. A., & Young, L. A. (2018). The Pluto system after new horizons. *Annual Review of Astronomy and Astrophysics, 56*, 357–392.

Stern, S. A., Slater, D. C., Scherrer, J., et al. (2008). ALICE: The ultraviolet imaging spectrograph aboard the new horizons Pluto–Kuiper Belt Mission. *Space Science Reviews, 140*, 155–187. https://doi.org/10.1007/s11214-008-9407-3.

Stern, S. A., et al. (2015). The Pluto system: Initial results from its exploration by new horizons. *Science, 350*(6258), aad1815.

Stern, S. A., et al. (2019). Initial results from the new horizons exploration of 2014 MU69, a small Kuiper Belt object. *Science*, *364*(6441), eaaw9771.

Tan, S. P. (2022). Low-pressure and low-temperature phase equilibria applied to Pluto's lower atmosphere. *Monthly Notices of the Royal Astronomical Society*, *515*(2), 1690–1698.

Tegler, S. C., et al. (2010). Methane and nitrogen abundances on Pluto and Eris. *The Astrophysical Journal*, *725*(1), 1296.

Tegler, S. C., et al. (2019). A new two-molecule combination band as a diagnostic of carbon monoxide diluted in nitrogen ice on Triton. *The Astronomical Journal*, *158*(1), 17.

Tyler, G. L., Linscott, I. R., Bird, M. K., et al. (2008). The new horizons radio science experiment (REX). *Space Science Reviews*, *140*, 217–259. https://doi.org/10.1007/s11214-007-9302-3.

Tyler, G. L., et al. (1989). Voyager radio science observations of Neptune and Triton. *Science*, *246*(4936), 1466–1473.

Umurhan, O. M., Ahrens, C. J., & Chevrier, V. F. (2021). Rheological and thermophysical properties and some processes involving common volatile materials found on Pluto's surface. In *The Pluto System After New Horizons* (pp. 195–255).

White, O. L., Moore, J. M., McKinnon, W. B., Spencer, J. R., Howard, A. D., Schenk, P. M., et al. (2017). Geological mapping of sputnik planitia on pluto. *Icarus*, *287*, 261–286.

White, O. L., et al. (2021). The geology of Pluto. *The Pluto System After New Horizons*, *55*.

Young, L. A., & Stern, S. A. (2001). Ultraviolet observations of Triton in 1999 with the space telescope imaging spectrograph: 2150-3180 Å spectroscopy and disk-integrated photometry. *The Astronomical Journal*, *122*(1), 449. https://doi.org/10.1086/322062.

Young, L. A., et al. (2018). Structure and composition of Pluto's atmosphere from the new horizons solar ultraviolet occultation. *Icarus*, *300*, 174–199.

Zahnle, K., Schenk, P., Levison, H., & Dones, L. (2003). Cratering rates in the outer solar system. *Icarus*, *163*(2), 263–289.

Index

Note: Page numbers followed by *f* indicate figures and *t* indicate tables.

Printed in the United States
by Baker & Taylor Publisher Services